COMPACT
WORLD
ATLAS

DK | Penguin
Random
House

FOR THE SEVENTH EDITION

SENIOR CARTOGRAPHIC EDITOR Simon Mumford

JACKET DESIGNER Dhirendra Singh   JACKET DESIGN DEVELOPMENT Sophia MTT

PRODUCER (PRE PRODUCTION) Jacqueline Street   SENIOR PRODUCER Angela Graef

PUBLISHER
Andrew Macintyre

PUBLISHING DIRECTOR
Jonathan Metcalf

ASSOCIATE PUBLISHING DIRECTOR
Liz Wheeler

ART DIRECTOR
Karen Self

First American edition 2001
Published in the United States by DK Publishing,
345 Hudson Street, New York, New York 10014

Reprinted with revisions 2002. Second edition 2003. Reprinted with revisions 2004.
Third edition 2005. Fourth edition 2009. Fifth edition 2012. Sixth edition 2015. Seventh edition 2018.

Copyright © 2001, 2002, 2003, 2004, 2005, 2009, 2012, 2015, 2018
Dorling Kindersley Limited
DK, a Division of Penguin Random House LLC

18 19 20 21 22 10 9 8 7 6 5 4 3 2 1
001–308089–Jun/2018

A catalog record for this book is available from the Library of Congress.
ISBN 978-1-4654-6886-4

DK books are available at special discounts when purchased in bulk for sales promotions, premiums,
fund-raising, or educational use. For details, contact: DK Publishing Special Markets,
345 Hudson Street, New York, New York 10014 or SpecialSales@dk.com

Printed and bound in Hong Kong

A WORLD OF IDEAS:
SEE ALL THERE IS TO KNOW

www.dk.com

Key to map symbols

Physical features

Elevation

- 19,686ft/6000m
- 13,124ft/4000m
- 9843ft/3000m
- 6562ft/2000m
- 3281ft/1,000m
- 1640ft/500m
- 820ft/250m
- 0
- Below sea level

△ Mountain

▽ Depression

◸ Volcano

)(Pass/tunnel

Sandy desert

Drainage features

Major perennial river

Minor perennial river

Seasonal river

Canal

Waterfall

Perennial lake

Seasonal lake

Wetland

Ice features

  Permanent ice cap/ice shelf

  Winter limit of pack ice

  Summer limit of pack ice

Borders

Full international border

Disputed de facto border

Territorial claim border

✕ ✕ ✕ Cease-fire line

Undefined boundary

Internal administrative boundary

Communications

Major road

Minor road

Railroad

✈ International airport

Settlements

◉ Above 500,000

◎ 100,000 to 500,000

○ 50,000 to 100,000

∘ Below 50,000

● National capital

● Internal administrative capital

Miscellaneous features

+ Site of interest

⊓⊔⊓⊔ Ancient wall

Graticule features

Line of latitude/longitude/Equator

Tropic/Polar circle

25° Degrees of latitude/longitude

Names

Physical features

Andes
Sahara | Landscape features
Ardennes

Land's End | Headland

Mont Blanc 4,807m | Elevation/volcano/pass

Blue Nile | River/canal/waterfall

Ross Ice Shelf | Ice feature

PACIFIC OCEAN
Sulu Sea | Sea features
Palk Strait

Chile Rise | Undersea feature

Regions

FRANCE | Country

BERMUDA (to UK) | Dependent territory

KANSAS | Administrative region

Dordogne | Cultural region

Settlements

PARIS | Capital city

SAN JUAN | Dependent territory capital city

Chicago
Kettering | Other settlements
Burke

Inset map symbols

Urban area

City

Park

▪ Place of interest

▫ Suburb/district

COMPACT
WORLD
ATLAS

Contents

The World's Regions

North &
Central America

South America

Africa

Europe

Australasia & Oceania

North & West Asia

South & East Asia

Index – Gazetteer

The Political World

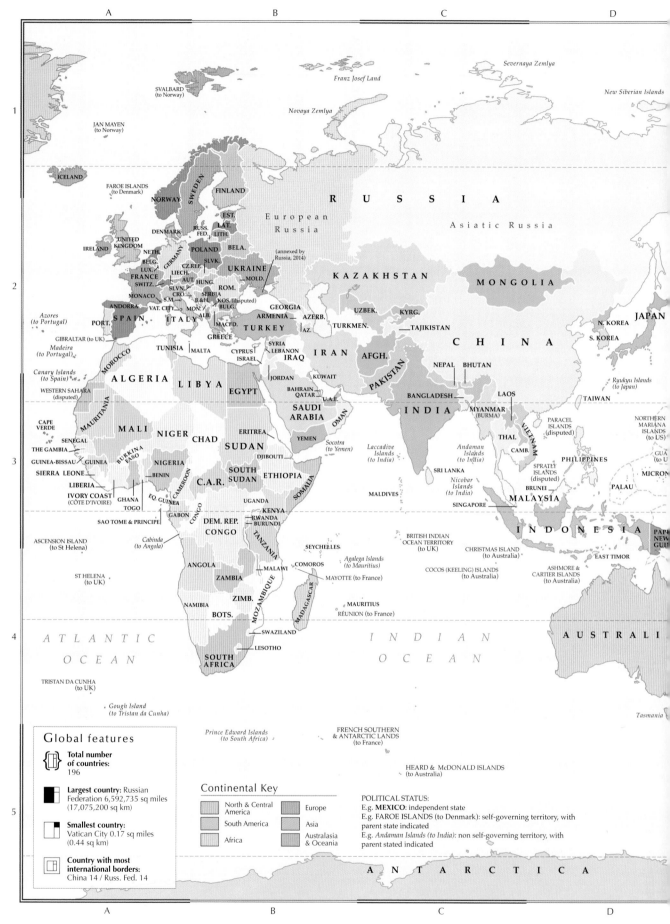

A · B · C · D

1

Severnaya Zemlya

New Siberian Islands

SVALBARD
(to Norway)

Franz Josef Land

JAN MAYEN
(to Norway)

Novaya Zemlya

ICELAND

FAROE ISLANDS
(to Denmark)

R   U   S   S   I   A

NORWAY SWEDEN FINLAND

European
Russia

Asiatic Russia

DENMARK RUSS.
FED. EST.
LAT.
LITH.

IRELAND UNITED
KINGDOM

NETH.
BELG.
LUX. GERMANY POLAND BELA.

KAZAKHSTAN

MONGOLIA

2

CZ.REP.
LIECH.
SWITZ.
FRANCE AUT. SLVK.
HUNG. UKRAINE MOLD.

(annexed by
Russia, 2014)

SLVN.
MONACO CRO.
S.M. SERBIA
B.&H. ROM.

N. KOREA JAPAN

Azores
(to Portugal)

ANDORRA VAT. CITY MON. KOS. (disputed)
BULG. GEORGIA
ARMENIA AZERB.

UZBEK. KYRG.

S. KOREA

PORT. SPAIN ITALY ALB.
MACED. TURKEY AZ. TURKMEN. TAJIKISTAN

C   H   I   N   A

GIBRALTAR (to UK) GREECE
Madeira
(to Portugal)

TUNISIA MALTA

SYRIA
CYPRUS LEBANON
ISRAEL IRAQ IRAN AFGH.

NEPAL BHUTAN

Ryukyu Islands
(to Japan)

Canary Islands
(to Spain)

MOROCCO

JORDAN KUWAIT

PAKISTAN

NORTHERN
MARIANA
ISLANDS
(to US)

WESTERN SAHARA
(disputed)

ALGERIA LIBYA EGYPT

BAHRAIN
QATAR U.A.E.

BANGLADESH LAOS

TAIWAN

3

CAPE
VERDE

MAURITANIA MALI NIGER CHAD

SAUDI
ARABIA OMAN

INDIA MYANMAR
(BURMA)

THAI. CAMB.

PARACEL
ISLANDS
(disputed)

PHILIPPINES

GUA
(to U

SENEGAL ERITREA YEMEN Socotra
(to Yemen) Laccadive
Islands
(to India) Andaman
Islands
(to India)

SPRATLY
ISLANDS
(disputed)

MICRON

THE GAMBIA

GUINEA-BISSAU GUINEA BURKINA
FASO SUDAN DJIBOUTI SRI LANKA Nicobar
Islands
(to India) BRUNEI

PALAU

SIERRA LEONE NIGERIA SOUTH
SUDAN ETHIOPIA MALDIVES SINGAPORE MALAYSIA

LIBERIA BENIN CAMEROON C.A.R. SOMALIA

IVORY COAST
(CÔTE D'IVOIRE) GHANA EQ. GUINEA UGANDA KENYA

I   N   D   O   N   E   S   I   A

TOGO GABON CONGO RWANDA
BURUNDI

SAO TOME & PRINCIPE

ASCENSION ISLAND
(to St Helena)

DEM. REP.
CONGO TANZANIA

Cabinda
(to Angola)

SEYCHELLES

BRITISH INDIAN
OCEAN TERRITORY
(to UK)

CHRISTMAS ISLAND
(to Australia)

PAP
NEW
GUI

Agalega Islands
(to Mauritius)

COCOS (KEELING) ISLANDS
(to Australia)

ASHMORE &
CARTIER ISLANDS
(to Australia)

EAST TIMOR

ANGOLA MALAWI COMOROS

MAYOTTE (to France)

ST HELENA
(to UK)

ZAMBIA MAURITIUS

RÉUNION (to France)

NAMIBIA ZIMB. MADAGASCAR

A   U   S   T   R   A   L   I

MOZAMBIQUE

BOTS.

4

A   T   L   A   N   T   I   C

OCEAN

SWAZILAND

LESOTHO

I   N   D   I   A   N

OCEAN

SOUTH
AFRICA

TRISTAN DA CUNHA
(to UK)

Tasmania

Gough Island
(to Tristan da Cunha)

Prince Edward Islands
(to South Africa)

FRENCH SOUTHERN
& ANTARCTIC LANDS
(to France)

Global features

**Total number
of countries:**
196

HEARD & McDONALD ISLANDS
(to Australia)

Largest country: Russian
Federation 6,592,735 sq miles
(17,075,200 sq km)

Continental Key

POLITICAL STATUS:

Smallest country:
Vatican City 0.17 sq miles
(0.44 sq km)

North & Central America	Europe
South America	Asia
Africa	Australasia & Oceania

E.g. **MEXICO**: independent state

E.g. FAROE ISLANDS (to Denmark): self-governing territory, with
parent state indicated

E.g. *Andaman Islands (to India)*: non self-governing territory, with
parent stated indicated

5

**Country with most
international borders:**
China 14 / Russ. Fed. 14

A   N   T   A   R   C   T   I   C   A

A · B · C · D

E   F   G   H

A R C T I C

O C E A N

Queen Elizabeth Islands

GREENLAND
(to Denmark)

Baffin Island

1

Arctic Circle

Alaska
(to US)

C   A   N   A   D   A

Aleutian Islands (to US)

*l Islands
Russia)*

ST PIERRE
& MIQUELON
(to France)

2

P A C I F I C

O C E A N

UNITED STATES
OF AMERICA

A T L A N T I C

O C E A N

BERMUDA
(to UK)

MIDWAY ISLANDS
(to US)

Guadalupe
(to Mexico)

PUERTO RICO (to US)

DOM. REP.

BRITISH VIRGIN ISLANDS (to UK)

VIRGIN ISLANDS (to US)

TURKS & CAICOS ISLANDS
(to UK)

ANGUILLA (to UK)

Tropic of Cancer

CAYMAN ISLANDS
(to UK)

THE
BAHAMAS

ST KITTS & NEVIS

Hawaii
(to US)

HONDURAS

CUBA

ANTIGUA & BARBUDA

WAKE ISLAND
(to US)

*Revillagigedo
Islands
(to Mexico)*

BELIZE

MONTSERRAT (to UK)

JAMAICA

GUADELOUPE (to France)

NAVASSA I.
(to US)

HAITI

DOMINICA

JOHNSTON ATOLL (to US)

GUATEMALA

MARTINIQUE (to France)

EL SALVADOR

CURAÇAO
(Neth.)

ST LUCIA

MARSHALL
ISLANDS

NICARAGUA

BARBADOS

ARUBA
(Neth.)

ST VINCENT & THE GRENADINES

WALLIS & FUTUNA
(to France)

KINGMAN REEF (to US)

COSTA RICA

GRENADA

*CLIPPERTON ISLAND
(to French Polynesia)*

PANAMA

TRINIDAD & TOBAGO

PALMYRA ATOLL (to US)

VENEZUELA

3

BAKER &
HOWLAND
ISLANDS
(to US)

JARVIS ISLAND
(to US)

COLOMBIA

FRENCH GUIANA
(to France)

*Galápagos Islands
(to Ecuador)*

Equator

NAURU

K   I   R   I   B   A   T   I

GUYANA

SURINAME

ECUADOR

OLOMON
LANDS

TUVALU

TOKELAU
(to NZ)

B   R   A   Z   I   L

SAMOA

P   E   R   U

VANUATU

AMERICAN
SAMOA
(to US)

BOLIVIA

EW
DONIA
France)

FIJI

TONGA

COOK
ISLANDS
(to NZ)

FRENCH POLYNESIA
(to France)

PARAGUAY

Tropic of Capricorn

*San Felix Island
(to Chile)*

NIUE (to NZ)

4

SEA ISLANDS
tralia)

Easter Island
(to Chile)

*Sala y Gomez
(to Chile)*

*San Ambrosia
Island
(to Chile)*

CHILE

NORFOLK ISLAND
(to Australia)

PITCAIRN,
HENDERSON,
DUCIE & OENO
ISLANDS
(to UK)

A   R   G   E   N   T   I   N   A

*Lord Howe Island
(to Australia)*

*Kermadec Island
(to NZ)*

*Juan Fernandez Island
(to Chile)*

URUGUAY

NEW
ZEALAND

*Chatham Island
(to NZ)*

P A C I F I C

O C E A N

*Bounty Island
(to NZ)*

*Campbell Island
(to NZ)*

FALKLAND ISLANDS
(to UK)

Macquarie Island (to Australia)

CHILE

ABBREVIATIONS: AFGH. Afghanistan, ALB. Albania, AUT. Austria,
AZ. or AZERB. Azerbaijan, BELG. Belgium, BELA. Belarus,
B.&H. Bosnia & Herzegovina, BOTS. Botswana, BULG. Bulgaria,
CAMB. Cambodia, C.A.R. Central African Republic, CRO. Croatia,
CZ. REP. Czech Republic (Czechia), DOM. REP. Dominican Republic,
EST. Estonia, HUNG. Hungary, KOS. Kosovo, KYRG. Kyrgyzstan,
LAT. Latvia, LIECH. Liechtenstein, LITH. Lithuania, LUX. Luxembourg,

MACED. Macedonia, MOLD. Moldova, MON. Montenegro,
NETH. Netherlands, PORT. Portugal, ROM. Romania,
RUSS. FED. Russian Federation, S.M. San Marino,
SLVK. Slovakia, SLVN. Slovenia, SWITZ. Switzerland,
THAI. Thailand, TURKMEN. Turkmenistan,
U.A.E. United Arab Emirates, UZBEK. Uzbekistan,
VAT. CITY Vatican City, ZIMB. Zimbabwe.

SOUTH GEORGIA &
SOUTH SANDWICH ISLANDS
(to UK)

5

Antarctic Circle

ANTARCTICA

E   F   G   H

The Physical World

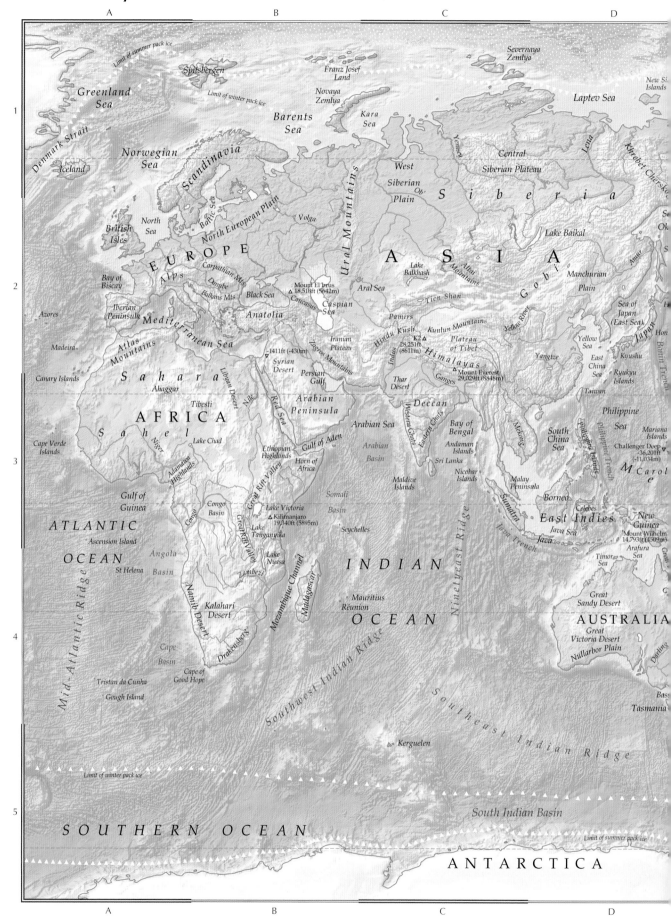

Limit of summer pack ice
Spitsbergen
Franz Josef Land
Severnaya Zemlya
New Si... Islands

Greenland Sea
Limit of winter pack ice
Novaya Zemlya
Kara Sea
Laptev Sea

Denmark Strait
Norwegian Sea
Barents Sea

Iceland
Scandinavia
West Siberian Plain
Ob'
Yenisey
Central Siberian Plateau
Lena
Khrebet Cherskiy

Baltic Sea
Siberia
Se... Ok...

North Sea
British Isles
North European Plain
Volga
Ural Mountains
ASIA
Lake Baikal

EUROPE
Carpathian Mts
Lake Balkhash
Altai Mountains
Gobi
Manchurian Plain
Amur

Bay of Biscay
Alps
Danube
Black Sea
Mount El'brus 18,510ft (5642m)
Caspian Sea
Aral Sea
Tien Shan
Sea of Japan (East Sea)
Japan

Iberian Peninsula
Balkans Mts
Anatolia
Caucasus
Pamirs
Hindu Kush
Kunlun Mountains
Plateau of Tibet
Yellow River
Yellow Sea
Kyushu
Bonin Tre...

Azores
Mediterranean Sea
-1411ft (-430m)
Syrian Desert
Iranian Plateau
Zagros Mountains
Indus
K2 28,251ft (8611m)
Himalayas
Mount Everest 29,029ft (8848m)
Ganges
Yangtze
East China Sea
Ryukyu Islands

Madeira
Atlas Mountains
Persian Gulf
Thar Desert
Taiwan

Canary Islands
Ahaggar
Sahara
Libyan Desert
Nile
Red Sea
Arabian Peninsula
Deccan
Western Ghats
Eastern Ghats
Bay of Bengal
Philippine Sea
Mariana Islands

AFRICA
Tibesti
Sahel
Lake Chad
Niger
Ethiopian Highlands
Gulf of Aden
Arabian Sea
Andaman Islands
Mekong
South China Sea
M Carol...
Challenger Deep -36,201ft (-11,034m)

Cape Verde Islands
Adamawa Highlands
Horn of Africa
Arabian Basin
Sri Lanka
Nicobar Islands
Malay Peninsula
Philippine Trench
Philippine Islands

Gulf of Guinea
Congo Basin
Congo
Great Rift Valley
Maldive Islands
Java Trench
Borneo
Celebes

ATLANTIC
Somali Basin
Lake Victoria
Kilimanjaro 19,340ft (5895m)
Seychelles
East Indies
Java Sea
New Guinea
Mount Wilhelm 14,793ft (4509m)

Ascension Island
Great Rift Valley
Lake Tanganyika
Sumatra
Java
Arafura Sea

OCEAN
Angola Basin
St Helena
Lake Nyasa
Zambezi
INDIAN
Timor Sea

Namib Desert
Mozambique Channel
Madagascar
Mauritius
Réunion
OCEAN
AUSTRALIA
Great Sandy Desert

Cape Basin
Kalahari Desert
Ninetyeast Ridge
Great Victoria Desert
Nullarbor Plain
Darling

Tristan da Cunha
Drakensberg
Cape of Good Hope
Bass...
Tasmania

Mid-Atlantic Ridge
Gough Island
Cape Basin

Southwest Indian Ridge
Southeast Indian Ridge
Kerguelen

Limit of winter pack ice
South Indian Basin

Limit of summer pack ice

SOUTHERN OCEAN

ANTARCTICA

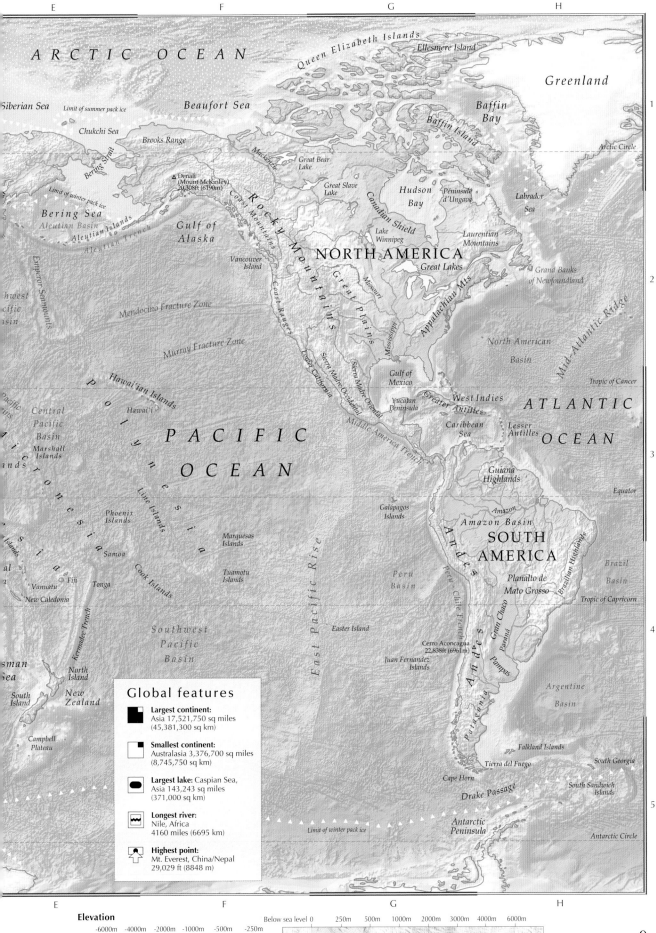

Standard Time Zones

The numbers at the top of the map indicate how many hours each time zone is ahead or behind Coordinated Universal Time (UTC). The row of clocks indicate the time in each zone when it is 12:00 noon UTC.

TIME ZONES

Because Earth is a rotating sphere, the Sun shines on only half of its surface at any one time. Thus, it is simultaneously morning, evening, and night time in different parts of the world. Because of these disparities, each country or part of a country adheres to a local time. A region of the Earth's surface within which a single local time is used is called a time zone.

COORDINATED UNIVERSAL TIME (UTC)

Coordinated Universal Time (UTC) is a reference by which the local time in each time zone is set. UTC is a successor to, and closely approximates, Greenwich Mean Time (GMT). However, UTC is based on an atomic clock, whereas GMT is determined by the Sun's position in the sky relative to the 0° longitudinal meridian, which runs through Greenwich, UK.

THE INTERNATIONAL DATELINE

The International Dateline is an imaginary line from pole to pole that roughly corresponds to the 180° longitudinal meridian. It is an arbitrary marker between calendar days. The dateline is needed because of the use of local times around the world rather than a single universal time.

The
WORLD
ATLAS

THE MAPS IN THIS ATLAS ARE ARRANGED CONTINENT BY CONTINENT, STARTING FROM THE INTERNATIONAL DATE LINE, AND MOVING EASTWARD. THE MAPS PROVIDE A UNIQUE VIEW OF TODAY'S WORLD, COMBINING TRADITIONAL CARTOGRAPHIC TECHNIQUES WITH THE LATEST REMOTE-SENSED AND DIGITAL TECHNOLOGY.

North & Central America

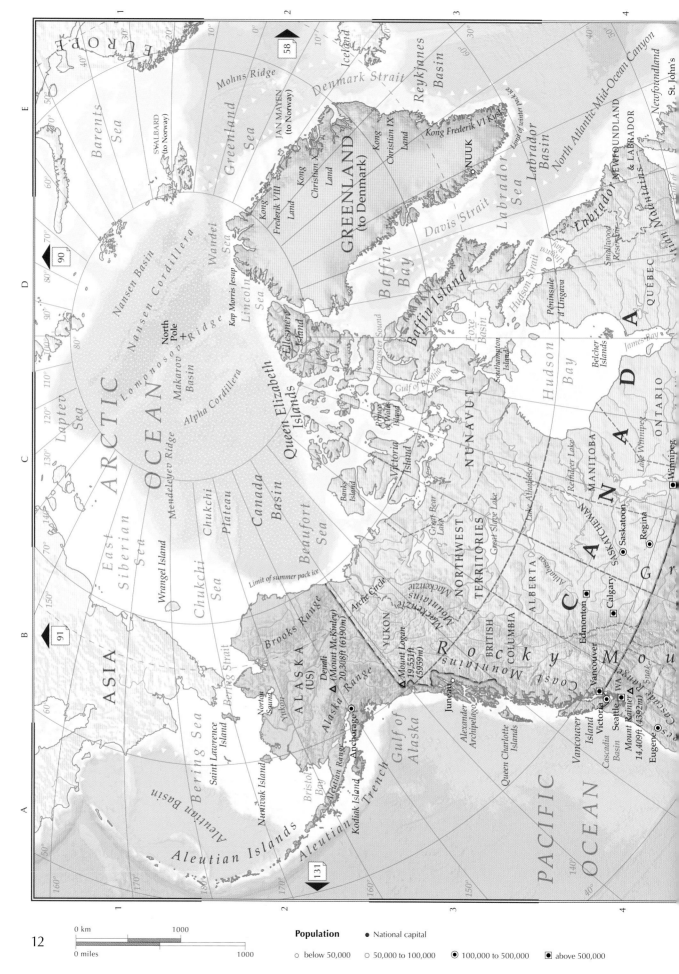

0 km   1000

0 miles   1000

Population   ● National capital

○ below 50,000   ○ 50,000 to 100,000   ◉ 100,000 to 500,000   ▣ above 500,000

Map labels

EUROPE

Barents Sea

SVALBARD (to Norway)

Mohns Ridge

Greenland Sea

JAN MAYEN (to Norway)

Iceland

Denmark Strait

Reykjanes Basin

Kong Frederik VIII Land

Kong Christian X Land

Kong Christian IX Land

Kong Frederik VI Kyst

GREENLAND (to Denmark)

NUUK

North Atlantic Mid-Ocean Canyon

Newfoundland

St. John's

Labrador Basin

Labrador Sea

Kap Morris Jesup

Lincoln Sea

Wandel Sea

Nansen Basin

Nansen Cordillera

North Pole

Makarov Basin

Lomonosov Ridge

Alpha Cordillera

Ellesmere Island

Queen Elizabeth Islands

Baffin Bay

Davis Strait

Baffin Island

Smallwood Reservoir

Péninsule d'Ungava

QUÉBEC

NEWFOUNDLAND & LABRADOR

Ungava Bay

Hudson Strait

Foxe Basin

Southampton Island

Belcher Islands

James Bay

ONTARIO

ARCTIC OCEAN

Mendeleyev Ridge

East Siberian Sea

Chukchi Plateau

Chukchi Sea

Canada Basin

Banks Island

Victoria Island

Prince of Wales Island

Lancaster Sound

Gulf of Boothia

NUNAVUT

Hudson Bay

Winnipeg

MANITOBA

Lake Winnipeg

Reindeer Lake

Laptev Sea

Wrangel Island

Limit of summer pack ice

Beaufort Sea

Mackenzie Mountains

NORTHWEST TERRITORIES

Great Bear Lake

Great Slave Lake

Lake Athabasca

SASKATCHEWAN

Saskatoon

Regina

ASIA

Bering Sea

Saint Lawrence Island

Bering Strait

Nunivak Island

Norton Sound

Yukon

ALASKA (US)

Brooks Range

Arctic Circle

Denali (Mount McKinley) 20,308ft (6190m)

Alaska Range

YUKON

Mount Logan 19,551ft (5959m)

Mackenzie

Rocky Mountains

BRITISH COLUMBIA

ALBERTA

Edmonton

Calgary

CANADA

Aleutian Basin

Bristol Bay

Aleutian Range

Kodiak Island

Gulf of Alaska

Aleutian Trench

Juneau

Alexander Archipelago

Queen Charlotte Islands

Coast Mountains

Cascadia Basin

Vancouver Island

Vancouver

Victoria

Seattle WA

Mount Rainier 14,409ft (4392m)

Eugene

Cascade Range

Snake

Aleutian Islands

PACIFIC OCEAN

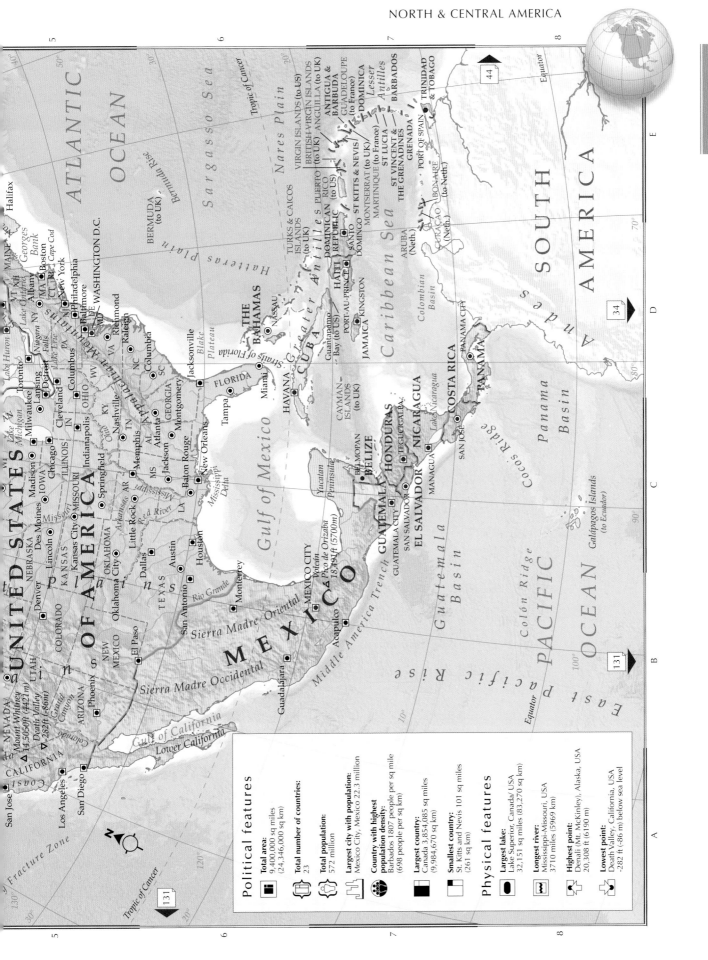

Political features

Total area:
9,400,000 sq miles
(24,346,000 sq km)

Total number of countries:
23

Total population:
572 million

Largest city with population:
Mexico City, Mexico 22.3 million

Country with highest population density:
Barbados 1807 people per sq mile
(698 people per sq km)

Largest country:
Canada 3,854,085 sq miles
(9,984,670 sq km)

Smallest country:
St Kitts and Nevis 101 sq miles
(261 sq km)

Physical features

Largest lake:
Lake Superior, Canada/USA
32,151 sq miles (83,270 sq km)

Longest river:
Mississippi-Missouri, USA
3710 miles (5969 km)

Highest point:
Denali (Mt. McKinley), Alaska, USA
20,308 ft (6190 m)

Lowest point:
Death Valley, California, USA
-282 ft (-86 m) below sea level

Western Canada & Alaska

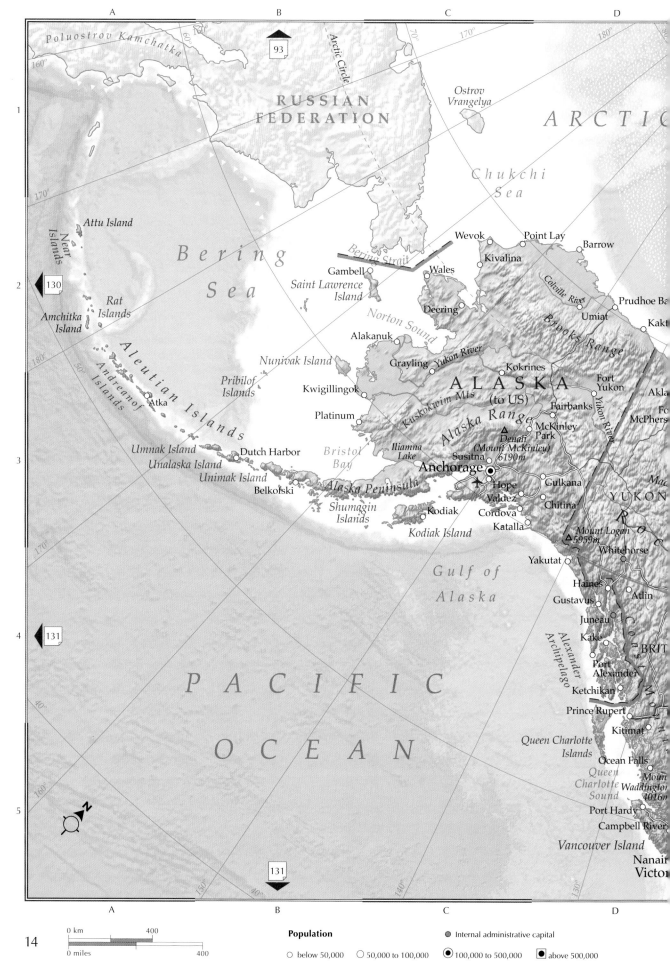

Poluostrov Kamchatka

93

RUSSIAN FEDERATION

Arctic Circle

Ostrov Vrangelya

ARCTIC

Chukchi Sea

Bering Strait

Wevok • Point Lay
Barrow
Kivalina

Colville River

Prudhoe Ba

Bering Sea

Gambell
Wales
Deering

Umiat

Kakt

Brooks Range

Alakanuk

Saint Lawrence Island

Norton Sound

130

Attu Island

Near Islands

Rat Islands

Amchitka Island

Grayling
Yukon River

Kokrines

A L A S K A
(to US)

Fort Yukon

Akla

Nunivak Island

Fairbanks

Fo
McPhers

Pribilof Islands

Kwigillingok

Kuskokwim Mts

McKinley Park
△ Denali
(Mount McKinley)
6190m

Yukon River

Aleutian Islands

Andreanof Islands

Atka

Platinum

Alaska Range

Iliamna
Lake

Susitna
Anchorage

Umnak Island
Unalaska Island

Dutch Harbor

Bristol Bay

Hope
Valdez

Gulkana

Chitina

Mac

YUKON

RO

Unimak Island

Belkofski

Alaska Peninsula

Kodiak
Cordova
Katalla

△ Mount Logan
5959m

Whitehorse

Shumagin Islands

Kodiak Island

Yakutat

Haines

Gustavus

Juneau

Atlin

Gulf of Alaska

Kake

BRIT

P A C I F I C

Alexander Archipelago

Port Alexander

Ketchikan

Prince Rupert

Kitimat

O C E A N

Queen Charlotte Islands

Ocean Falls

Queen Charlotte Sound

Moun Waddington 4016m

Port Hardy

131

Campbell River

Vancouver Island

Nanair

Victor

131

0 km   400

0 miles   400

Population

◯ below 50,000   ◯ 50,000 to 100,000   ◉ 100,000 to 500,000   ◼ above 500,000

⬤ Internal administrative capital

E · F · G · H

1

Alert

GREENLAND
(to Denmark)

Knud Rasmussen Land

Ellesmere Island

Axel Heiberg Island

Ellef Ringnes Island
Isachsen

Amund Ringnes Island

Prince Patrick Island

Mould Bay

Melville Island

Bathurst Island

Cornwallis Island

Devon Island

Resolute
(Qausuittuq)

Lancaster Sound

Baffin Bay

Arctic Circle

60

2

OCEAN

Queen Elizabeth Islands

Viscount Melville Sound

Banks Island

M'Clintock Channel

Prince of Wales Island

Somerset Island

Brodeur Peninsula

Baffin Island

Davis Strait

eaufort Sea

s Harbour
(Ikaahuk)

yaktuk

Paulatuk

Amundsen Gulf

Holman

Victoria Island

Boothia Peninsula

Gulf of Boothia

Igloolik

Cumberland Sound

ort
Good Hope
(Rádeyilikóé)

Kugluktuk
(Coppermine)

Cambridge Bay
(Ikaluktutiak)

Gjoa Haven
(Uqsuqtuuq)

King William Island

Kugaaruk
(Pelly Bay)

Melville Peninsula

Foxe Basin

Nettilling Lake

Amadjuak Lake

Iqaluit
(Frobisher Bay)

60

3

Mackenzie

Great Bear Lake

Echo Bay

Burnside

Back

N U N A V U T

Garry Lake

Baker Lake

Repulse Bay

Southampton Island

Coral Harbour
(Salliq)

Hudson Strait

Péninsule d'Ungava

QUÉBEC

N O R T H W E S T
T E R R I T O R I E S

ten

Edzo
Fort Simpson

Yellowknife
Reliance

Lutselk'e
(Snowdrift)

Great Slave Lake

Rankin Inlet

Whale Cove
(Tikiarjuaq)

Arviat

Coats Island

Mansel Island

70

4

ort Providence

Fort Liard

Hay River

Fort Smith

Dubawnt

Lake Athabasca

Reindeer Lake

Churchill

Hudson Bay

Belcher Islands

16

Fort Nelson

Nelson

James Bay

MBIA

Fort St. John

Fort Vermilion

C A N A D A

Wollaston Lake

Southern Indian Lake

50

A L B E R T A

Grande Prairie

Fort McMurray

Buffalo Narrows

Lynn Lake

Thompson

ONTARIO

nce George

Athabasca

SASKATCHEWAN

Flin Flon

Lake Winnipeg

5

Athabasca

North Saskatchewan

The Pas

M A N I T O B A

Edmonton

Mount Robson 3954m

Leduc

Prince Albert

Saskatoon

Saskatchewan

Red Deer

Kindersley

Yorkton

Lake Manitoba

Winnipeg

Lake of the Woods

Lake Superior

Lake Huron

Kamloops

Calgary

Regina

Qu'Appelle

Brandon

Kelowna

Cranbrook

Medicine Hat

Weyburn

Melita

Lake Michigan

ouver

Lethbridge

Milk River

Estevan

23

U N I T E D · S T A T E S · O F · A M E R I C A

E · F · G · H

Elevation

-6000m · -4000m · -2000m · -1000m · -500m · -250m

Below sea level 0 · 250m · 500m · 1000m · 2000m · 3000m · 4000m · 6000m

-19,658ft · -13,124ft · -6562ft · -3281ft · -1640ft · -820ft · -328ft/-100m · 0

820ft · 1640ft · 3281ft · 6562ft · 9843ft · 13,124ft · 19,685ft

Eastern Canada

NORTHWEST TERRITORIES

NUNAVUT

15

SASKATCHEWAN

MANITOBA

Churchill

Southern Indian Lake

Nelson

Hayes

15

Cedar Lake

Lake Winnipeg

Lake Winnipegosis

Lake Manitoba

C

Severn

Fort Severn

Peawanuk

Wintisk

Sandy Lake

Attawapiskat

Attawapiskat

A

Albany

Fort Albany

O N T A R I O

N

Hudson Bay

Ottawa Islands

Coats Island

Mansel Island

Inukjuak
(Port Harrison)

Belcher Islands

James Bay

Akimiski Island

Moosonee

Moose

Ivujivik

Charles Island

Péninsu d' Unga

Lac Min

Bie

Q U

Eastmain

Rivière de Rupert

Mistass

Chibougamau

Réservoi Gouin

Lac Seul

Armstrong

Kenora   Dryden

Lake of the Woods

Lake Nipigon

Longlac

Hearst

Kapuskasing

Cochrane

Amos

Rouyn-Noranda

Val-d'Or

Harricana

Red River

Fort Frances   Atikokan

Rainy Lake

Nipigon

Thunder Bay

Tip Top Mountain
△640m

Marathon

Timmins

Foleyet

Lake Superior

NORTH DAKOTA

MINNESOTA

23

Wawa

Kirkland Lake

Sault Ste.Marie

Sudbury

North Bay

SOUTH DAKOTA

MICHIGAN

Manitoulin Island

Georgian Bay

Pembroke

Gatineau

Hull

OTTAWA

UNITED STATES

WISCONSIN

Lake Michigan

Lake Huron

Midland

Peterborough

Brampton

Kitchener

Hamilton

Sarnia

Oshawa

Toronto

St.Catharines

London

Kings

Lake Ont

Niagara Falls

NEW YOR

NEBRASKA

OF AMERICA

IOWA

Mississippi River

18

Windsor

Leamington

Lake Erie

ILLINOIS

INDIANA

OHIO

PENNSYLVANIA

0 km   300

0 miles   300

Population   ● National capital   ● Internal administrative capital

○ below 50,000   ○ 50,000 to 100,000   ◉ 100,000 to 500,000   ◼ above 500,000

Baffin Island

Resolution Island

Button Islands

Akpatok Island

Ungava Bay

ujjuaq

Rivière à la Baleine

Caniapiscau

Labrador Sea

Nain

Hopedale

Makkovik

Cape Harrison

Cartwright

NEWFOUNDLAND & LABRADOR

Schefferville

Smallwood Reservoir

Lake Melville

Churchill

St.Anthony

servoir de niapiscau

E C

D

A

Gagnon

Réservoir Manicouagan

Laurentian Mountains

Havre-St-Pierre

Strait of Belle Isle

Gander

Grand Falls

St.John's

Sept-Îles

Île d'Anticosti

Newfoundland

Baie-Comeau

St.Lawrence

Péninsule de Gaspé

Gaspé

Îles de la Madeleine

Corner Brook

Gulf of St. Lawrence

Channel-Port aux Basques

Cape Race

Chicoutimi

Matane

Rimouski

Cabot Strait

ST PIERRE & MIQUELON (to France)

ière

Rivière-du-Loup

Edmundston

Bathurst

PRINCE EDWARD ISLAND

Sydney

Glace Bay

Tuque

NEW BRUNSWICK

Charlesbourg

Moncton

Charlottetown

Cape Breton Island

St-Georges

Oromocto

Amherst

Québec

Fredericton

New Glasgow

Trois-Rivières

Drummondville

Saint John

Truro

NOVA SCOTIA

ntréal

MAINE

Dartmouth

Sherbrooke

Halifax

Sable Island

Liverpool

Yarmouth

NEW HAMPSHIRE

ATLANTIC

Cape Cod

SSACHUSETTS

OCEAN

NNECTICUT

RHODE ISLAND

Elevation

						Below sea level 0	250m	500m	1000m	2000m	3000m	4000m	6000m
-6000m	-4000m	-2000m	-1000m	-500m	-250m								
-19,658ft	-13,124ft	-6562ft	-3281ft	-1640ft	-820ft	-328ft/-100m 0	820ft	1640ft	3281ft	6562ft	9843ft	13,124ft	19,685ft

USA: The Northeast

Population ● National capital ● Internal administrative capital
○ below 50,000 ○ 50,000 to 100,000 ⊙ 100,000 to 500,000 ◼ above 500,000

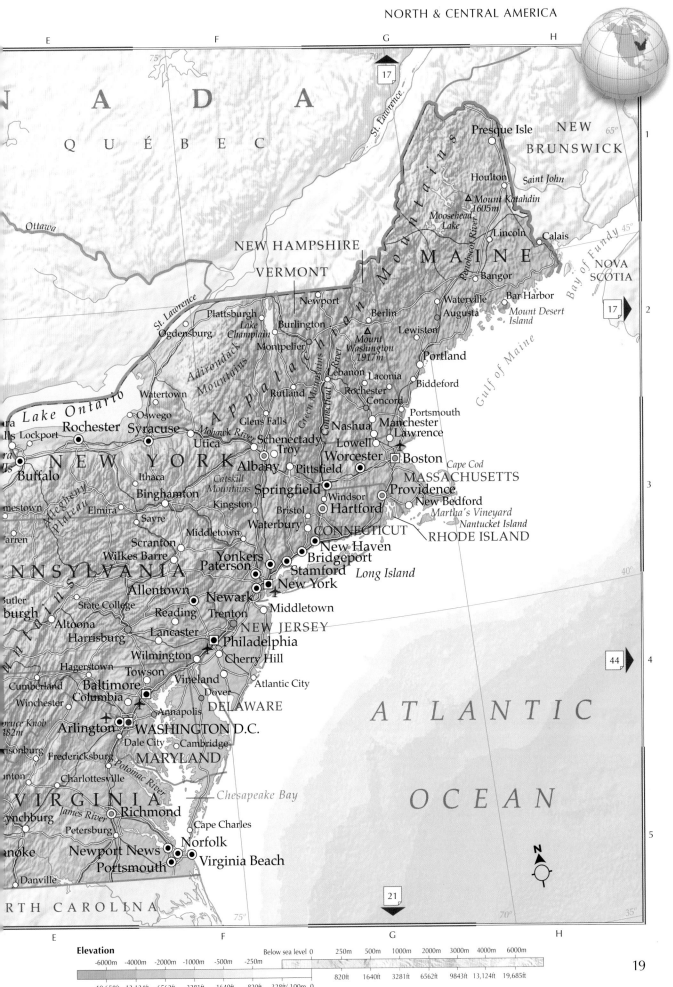

E   F   G   H

17

75°   70°

St. Lawrence

C A N A D A

Ottawa

Q U É B E C

65°

Presque Isle   **NEW BRUNSWICK**

Houlton   Saint John

△ *Mount Katahdin*
1605m

Moosehead Lake

Lincoln   Calais

NOVA SCOTIA

45°

NEW HAMPSHIRE

VERMONT

M A I N E

Bangor

Bay of Fundy

17

Newport

Plattsburgh

St. Lawrence

Burlington

Berlin

Waterville

Augusta

Bar Harbor

Mount Desert Island

Gulf of Maine

Ogdensburg

Lake Champlain

Montpelier

Mount Washington
1917m

Lewiston

2

Adirondack Mountains

Connecticut River

Lebanon

Laconia

Portland

Green Mountains

Rochester

Biddeford

Watertown

Rutland

Concord

Portsmouth

Lake Ontario

Oswego

Glens Falls

Nashua

Manchester

Lockport

Rochester

Syracuse

Mohawk River

Schenectady

Lowell

Lawrence

Buffalo

NEW YORK

Utica

Albany

Troy

Worcester

Boston

Cape Cod

Ithaca

Pittsfield

3

Binghamton

Catskill Mountains

Springfield

Windsor

Providence

MASSACHUSETTS

Allegheny Plateau

Elmira

Sayre

Kingston

Bristol

Hartford

New Bedford

Martha's Vineyard

Nantucket Island

RHODE ISLAND

Jamestown

Middletown

Waterbury

CONNECTICUT

Warren

Scranton

New Haven

40°

PENNSYLVANIA

Wilkes Barre

Yonkers

Paterson

Bridgeport

Stamford

Long Island

Allentown

New York

Butler

State College

Reading

Newark

burgh

Altoona

Lancaster

Trenton

Middletown

NEW JERSEY

Harrisburg

Wilmington

Philadelphia

Cherry Hill

4

Hagerstown

Towson

Vineland

Atlantic City

Cumberland

Baltimore

Columbia

Dover

DELAWARE

Winchester

Annapolis

Bruce Knob
882m

Arlington

WASHINGTON D.C.

A T L A N T I C

Harrisonburg

Dale City

Cambridge

Fredericksburg

MARYLAND

Potomac River

nton

Charlottesville

Chesapeake Bay

V I R G I N I A

James River

Richmond

O C E A N

Lynchburg

Petersburg

Cape Charles

Roanoke

Newport News

Norfolk

5

Danville

Portsmouth

Virginia Beach

N

RTH C A R O L I N A

21

75°   70°

35°

E   F   G   H

Elevation

Below sea level 0   250m   500m   1000m   2000m   3000m   4000m   6000m

-6000m   -4000m   -2000m   -1000m   -500m   -250m

820ft   1640ft   3281ft   6562ft   9843ft 13,124ft 19,685ft

-19,658ft -13,124ft -6562ft -3281ft -1640ft   -820ft   -328ft/-100m 0

USA: The Southeast

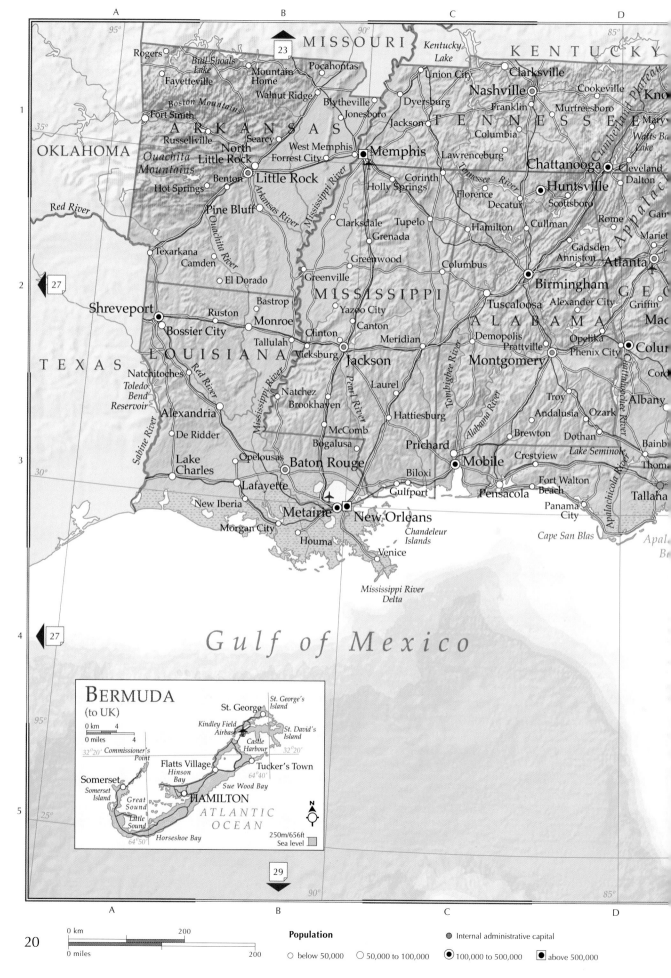

MISSOURI

Kentucky Lake

KENTUCKY

Rogers
Bull Shoals Lake
Mountain Home
Pocahontas
Union City
Clarksville
Cookeville
Knox

Fayetteville
Walnut Ridge
Blytheville
Dyersburg
Nashville
Murfreesboro
Mary

Fort Smith
Boston Mountains
Jonesboro
Jackson
Franklin
Columbia
TENNESSEE
Watts Ba
Lake

OKLAHOMA
Russellville
Searcy
West Memphis
Memphis
Lawrenceburg
Chattanooga
Cleveland

Ouachita Mountains
North Little Rock
Forrest City
Corinth
Florence
Decatur
Scottsboro
Huntsville
Dalton

Red River
Benton
Little Rock
Holly Springs
Hamilton
Cullman
Rome
Gad

Hot Springs
Pine Bluff
Clarksdale
Tupelo
Gadsden
Anniston
Atlanta
Mariet

Texarkana
Camden
El Dorado
Arkansas River
Greenwood
Grenada
Columbus
Birmingham
GEO

Shreveport
Bastrop
MISSISSIPPI
Tuscaloosa
Alexander City
Griffin
Mac

Bossier City
Ruston
Monroe
Yazoo City
Canton
ALABAMA
Opelika
Colu

TEXAS
Tallulah
Clinton
Meridian
Demopolis
Prattville
Phenix City
Cord

LOUISIANA
Vicksburg
Jackson
Montgomery
Albany

Natchitoches
Red River
Natchez
Laurel
Troy
Andalusia
Ozark

Toledo Bend Reservoir
Brookhaven
Hattiesburg
Brewton
Dothan
Bainb

Alexandria
McComb
Prichard
Crestview
Lake Seminole
Thom

De Ridder
Bogalusa
Mobile
Fort Walton Beach
Panama City
Talla

Lake Charles
Opelousas
Baton Rouge
Biloxi
Pensacola
Cape San Blas
Apal

Lafayette
Metairie
Gulfport
Be

New Iberia
New Orleans

Morgan City
Houma
Chandeleur Islands

Venice

Mississippi River Delta

Gulf of Mexico

BERMUDA
(to UK)

0 km 4
0 miles 4

Commissioner's Point
St. George
St. George's Island

Kindley Field Airbase
St. David's Island

Castle Harbour

Flatts Village
Tucker's Town

Somerset
Hinson Bay

Somerset Island
Sue Wood Bay

Great Sound
HAMILTON

Little Sound
ATLANTIC OCEAN

Horseshoe Bay

250m/656ft
Sea level

N

20

0 km 200
0 miles 200

Population

○ below 50,000   ○ 50,000 to 100,000   ◉ 100,000 to 500,000   ■ above 500,000

● Internal administrative capital

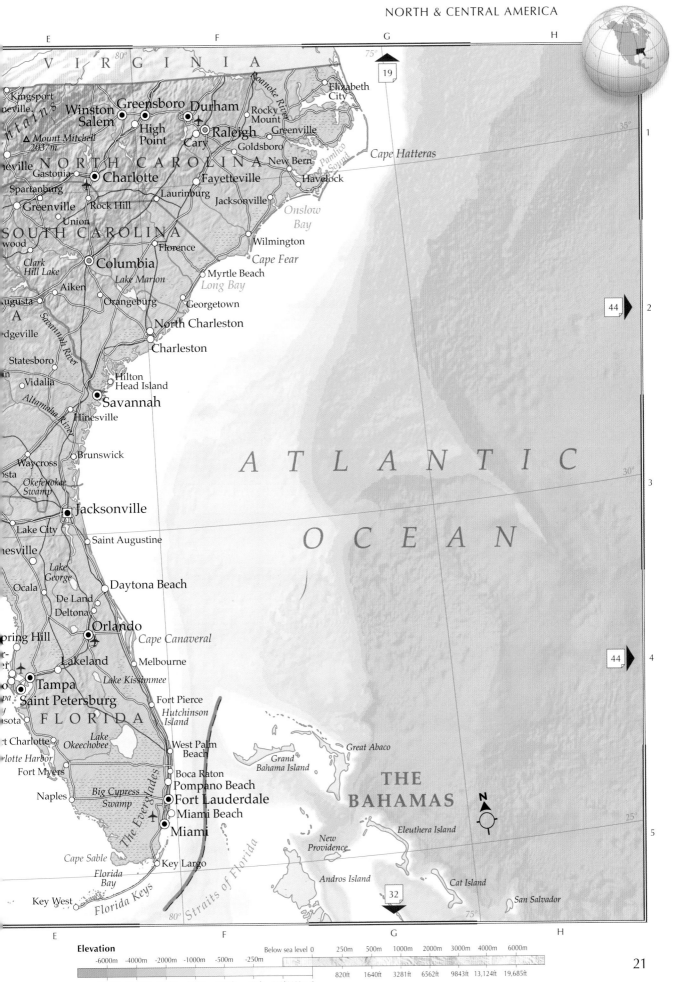

Elevation

| -6000m | -4000m | -2000m | -1000m | -500m | -250m | Below sea level 0 | 250m | 500m | 1000m | 2000m | 3000m | 4000m | 6000m |

-19,658ft -13,124ft -6562ft -3281ft -1640ft -820ft -328ft/-100m 0 820ft 1640ft 3281ft 6562ft 9843ft 13,124ft 19,685ft

USA: Central States

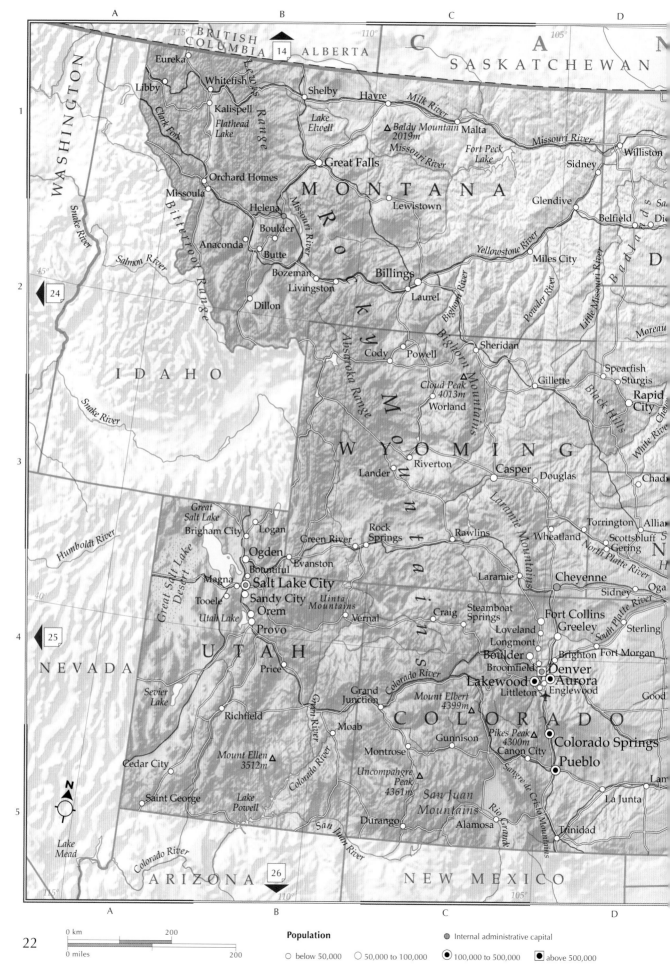

Population

0 km   200
0 miles   200

○ below 50,000   ○ 50,000 to 100,000   ◉ 100,000 to 500,000   ■ above 500,000

● Internal administrative capital

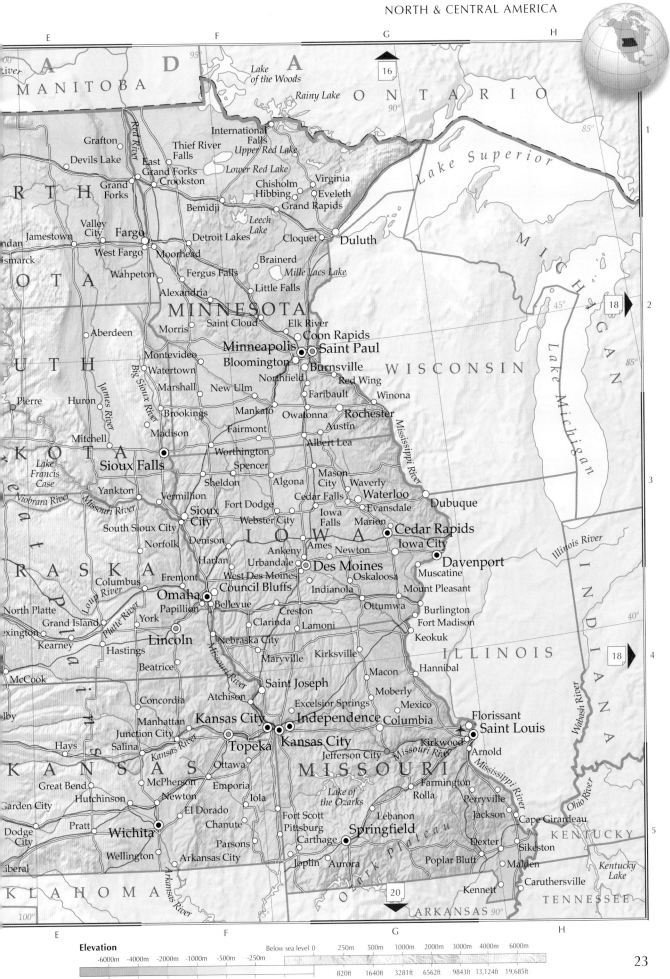

Elevation

							Below sea level 0	250m	500m	1000m	2000m	3000m	4000m	6000m

-6000m   -4000m   -2000m   -1000m   -500m   -250m

-19,658ft -13,124ft -6562ft -3281ft -1640ft -820ft -328ft/-100m 0

820ft   1640ft   3281ft   6562ft   9843ft 13,124ft 19,685ft

USA: The West

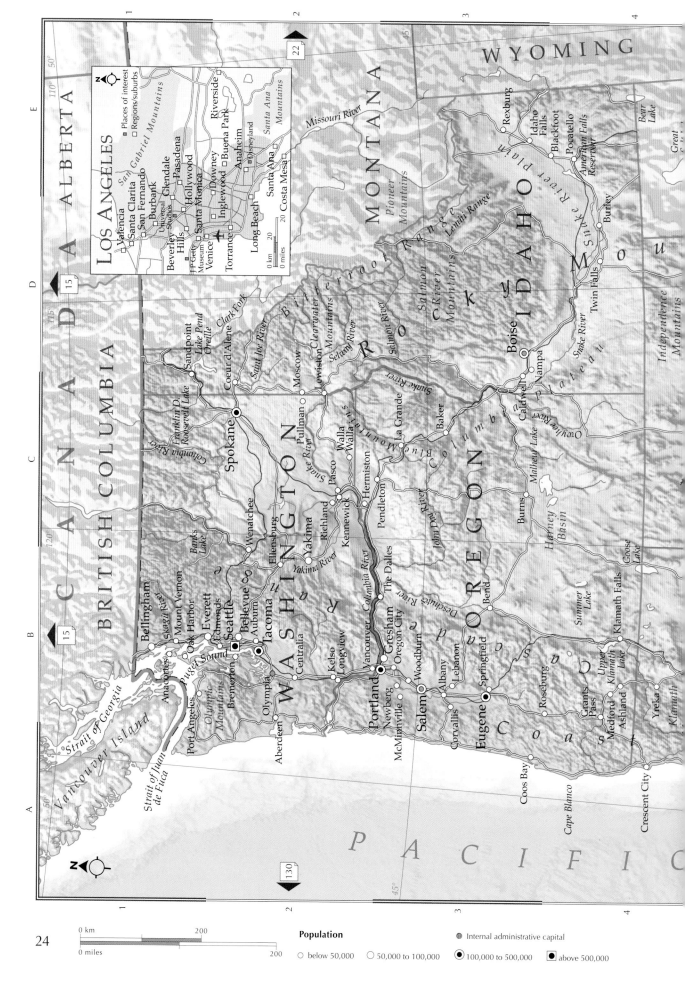

LOS ANGELES

Places of interest
Regions/suburbs

Valencia
Santa Clarita
San Fernando
San Gabriel Mountains
Universal Studios
Burbank
Glendale
Pasadena
Hollywood
Beverley Hills
Getty Museum
Santa Monica
Venice
Downey
Inglewood
Torrance
Buena Park
Anaheim
Disneyland
Riverside
Santa Ana Mountains
Santa Ana
Costa Mesa
Long Beach

0 km   20
0 miles   20

WYOMING

MONTANA

CANADA
ALBERTA
BRITISH COLUMBIA

Missouri River

Rexburg
Idaho Falls
Blackfoot
Pocatello
American Falls Reservoir
Bear Lake
Great

IDAHO

Pioneer Mountains
Salmon River Mountains
Lemhi Range
Burley
Twin Falls

ROCKY Mo u

Bitterroot Range
Clearwater Mountains
Selway River
Salmon River
Snake River Plain
Snake River
Independence

Sandpoint
Lake Pend Oreille
Clark Fork
Saint Joe River
Moscow
Lewiston
Pullman
Coeur d'Alene

Franklin D. Roosevelt Lake
Columbia River

Spokane

Walla Walla
Snake River

Boise
Nampa
Caldwell

La Grande
Baker
Owyhee River
Malheur Lake

WASHINGTON

Banks Lake
Wenatchee
Ellensburg
Yakima
Yakima River
Richland
Kennewick
Pasco
Hermiston
Pendleton

Blue Mountains

Columbia

OREGON

John Day River
Burns
Harney Basin

Bellingham
Skagit River
Mount Vernon
Oak Harbor
Everett
Edmonds
Bellevue
Seattle
Auburn
Tacoma
Olympia

Puget Sound

Anacortes
Bremerton
Port Angeles
Olympic Mountains
Aberdeen

Centralia
Kelso
Longview

Vancouver
Columbia River
The Dalles
Gresham
Oregon City
Portland
Newberg
McMinnville
Woodburn
Salem
Albany
Lebanon
Corvallis
Springfield
Eugene

Deschutes River
Bend

Summer Lake
Klamath Falls
Goose Lake

Roseburg
Grants Pass
Medford
Ashland
Upper Klamath Lake
Yreka
Klamath

Coast

Strait of Georgia
Vancouver Island
Strait of Juan de Fuca

Coos Bay
Cape Blanco
Crescent City

PACIFIC

24

0 km   200
0 miles   200

Population

Internal administrative capital

○ below 50,000
○ 50,000 to 100,000
◉ 100,000 to 500,000
◼ above 500,000

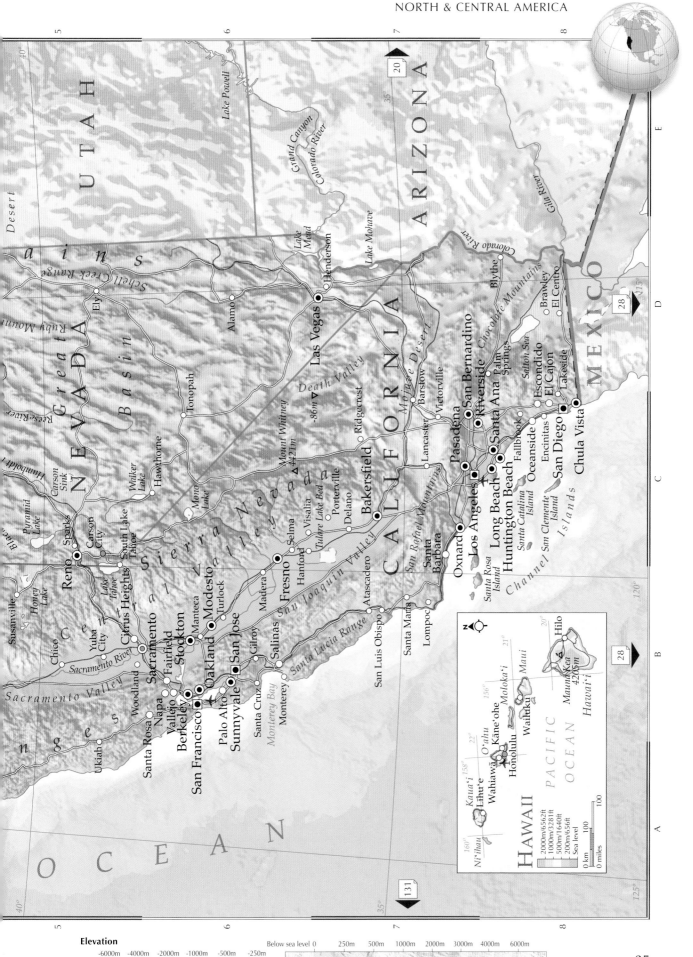

Elevation

-6000m	-4000m	-2000m	-1000m	-500m	-250m	Below sea level 0	250m	500m	1000m	2000m	3000m	4000m	6000m

| -19,658ft | -13,124ft | -6562ft | -3281ft | -1640ft | -820ft | -328ft/-100m | 0 | | 820ft | 1640ft | 3281ft | 6562ft | 9843ft | 13,124ft | 19,685ft |

25

USA: The Southwest

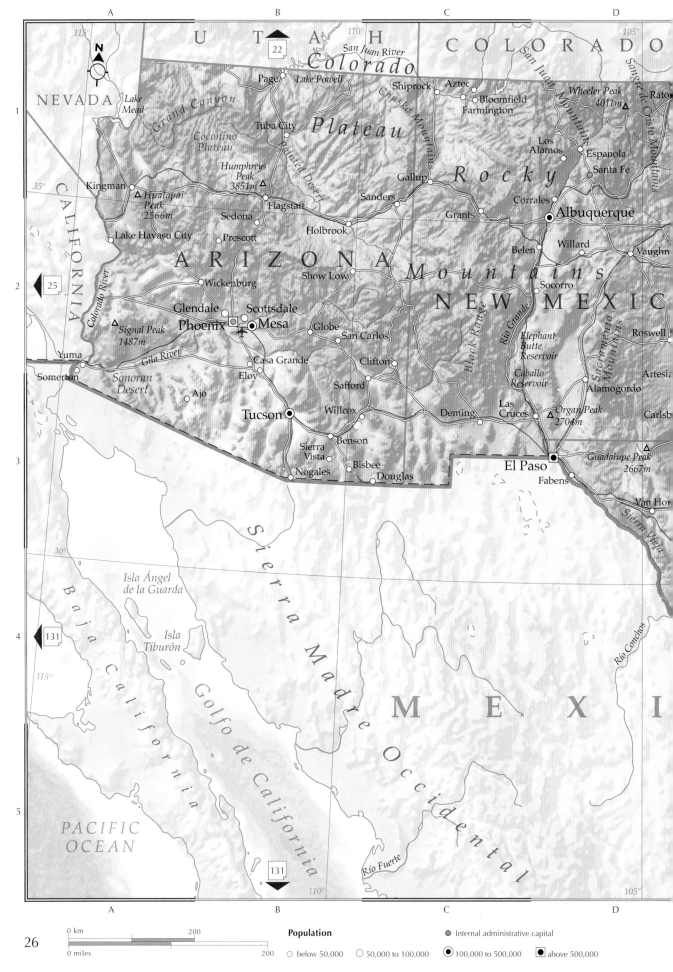

0 km   200

0 miles   200

Population

○ below 50,000   ○ 50,000 to 100,000   ◉ 100,000 to 500,000   ■ above 500,000

● Internal administrative capital

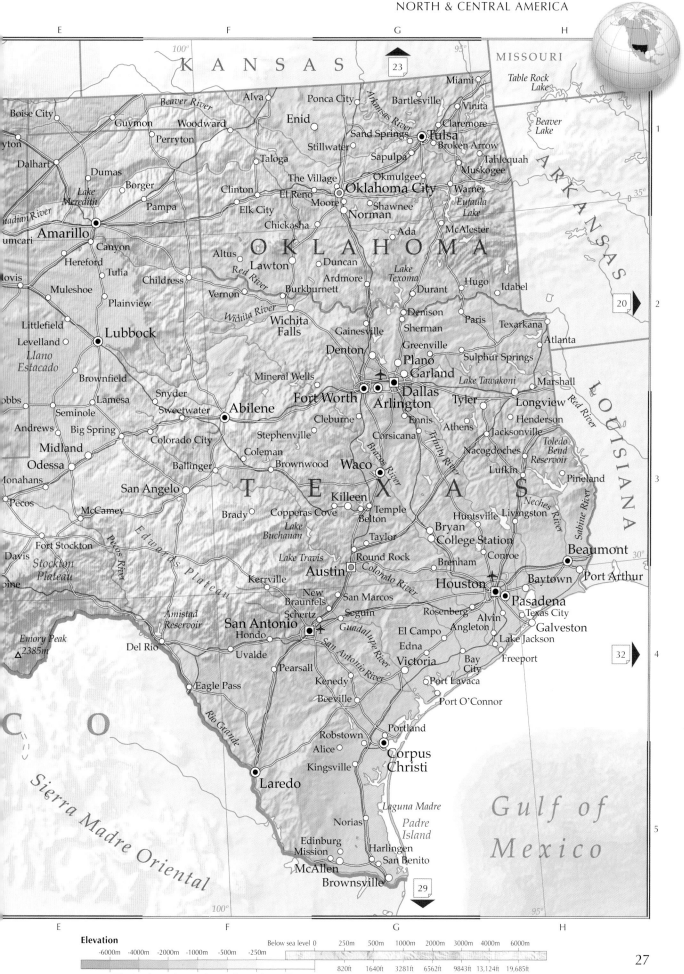

KANSAS

MISSOURI

Table Rock
Lake

Beaver
Lake

ARKANSAS

Boise City
Guymon
Woodward
Alva
Ponca City
Bartlesville
Miami

Perryton
Enid
Sand Springs
Claremore
Vinita

Dalhart
Dumas
Borger
Pampa
Stillwater
Tulsa
Broken Arrow

Clinton
El Reno
The Village
Okmulgee
Sapulpa
Tahlequah
Muskogee

Amarillo
Canyon
Elk City
Oklahoma City
Moore
Norman
Shawnee
Warner
McAlester

OKLAHOMA

Hereford
Tulia
Childress
Altus
Lawton
Duncan
Ada
Eufaula
Lake

Muleshoe
Plainview
Vernon
Burkburnett
Ardmore
Durant
Hugo
Idabel

Littlefield
Levelland
Wichita
Falls
Gainesville
Denison
Sherman
Paris
Texarkana
Atlanta

Lubbock
Mineral Wells
Denton
Plano
Garland
Greenville
Sulphur Springs
Marshall

Brownfield
Snyder
Fort Worth
Dallas
Arlington
Tyler
Longview
Lake Tawakoni

Lamesa
Sweetwater
Abilene
Cleburne
Ennis
Athens
Henderson
Jacksonville

Big Spring
Colorado City
Stephenville
Corsicana
Nacogdoches
Toledo
Bend
Reservoir

Midland
Coleman
Brownwood
Waco
Lufkin
Pineland

Odessa
Ballinger
TEXAS
Killeen
Bryan
Livingston
Neches River

Monahans
San Angelo
Brady
Copperas Cove
Temple
Belton
College Station
Huntsville
Beaumont

Pecos
McCamey
Lake
Buchanan
Taylor
Conroe
Baytown
Port Arthur

Fort Stockton
Edwards Plateau
Lake Travis
Round Rock
Brenham
Houston
Pasadena

Davis
Stockton
Plateau
Kerrville
Austin
Colorado River
Rosenberg
Texas City
Galveston

Amistad
Reservoir
New
Braunfels
San Marcos
Seguin
Alvin
Lake Jackson

Emory Peak
2385m
San Antonio
Hondo
El Campo
Angleton
Freeport

Del Rio
Uvalde
Edna
Bay
City

Pearsall
Guadalupe River
Victoria
Port Lavaca

Eagle Pass
Kenedy
San Antonio River
Port O'Connor

Beeville

Rio Grande
Robstown
Portland

Alice
Corpus
Christi

Sierra Madre Oriental
Laredo
Kingsville
Laguna Madre
Padre
Island
Gulf of
Mexico

Norias

Edinburg
Mission
Harlingen
San Benito

McAllen
Brownsville

LOUISIANA

Sabine River

Elevation

Below sea level 0   250m   500m   1000m   2000m   3000m   4000m   6000m

-6000m -4000m -2000m -1000m -500m -250m

820ft 1640ft 3281ft 6562ft 9843ft 13,124ft 19,685ft

-19,658ft -13,124ft -6562ft -3281ft -1640ft -820ft -328ft/-100m 0

Mexico

CALIFORNIA

ARIZONA

NEW MEXICO

UNITED STATES O

Colorado River

26

Tijuana
Rosarito
Mexicali
San Luis Río Colorado
Ensenada

Ciudad Juárez

Pecos River

Desierto de Altar

Nogales
Agua Prieta
Samalayuca

Rio Bravo del Norte

Río Grande

Villa Acu

Cananea

Caborca
Magdalena
Cumpas

Nuevo
Casas Grandes
El Sueco
Ojinaga
Boquillas

Río Bavispe

Golfo

Isla Ángel
de la Guarda

San Pedro
de la Cueva
El Sáuz
San Miguel

130

Bahía Sebastían Vizcatno

Chihuahua

Río Conchos

Nueva Ros
Sab

Hermosillo

Isla
Tiburón

Cuauhtémoc
Delicias
Ciudad Camargo

Isla Cedros

Guerrero Negro

Guaymas
Empalme
Esperanza

San Francisco
del Oro
Jiménez

Monclo

San Ignacio

Ciudad
Obregón
Navojoá

Hidalgo del Parral

Santa Barbara

San Ped
Par

Sierra San Pedro Mártir

Baja California

Huatabampo

Gómez Palacio
Torreón
Ciudad Lerdo

San Blas

Loreto

Guasave
Guamúchil

Los Mochis

Matamoros

M E X

Sierra de la Giganta

Culiacán

Miguel Auza
Juan Alda

Isla Magdalena

Navolato

Sierra Madre Occidental

Río Gra

Isla Santa Margarita

Bahía
de La
Paz

El Dorado

Durango
Fresnillo

La Paz

Tropic of Cancer

Santa Genoveva
2406m

Miraflores

Mazatlán

Zacatecas
Guadalupe
Villanueva
Aguascalientes
Jalpa

Escuinapa

Acaponeta
Tuxpan

Lagos de More
Yahualica

Isla San
Juanito

Islas Marías

Tepic

Tequila

Lago
Chaj

Isla María Madre
Isla María Magdalena

Isla María
Cleofas

Guadalajara

130

Puerto Vallarta

Tlaquepaque
Zamora de Hida

Zap
Tuxpa

Isla San Benedicto

Ciudad Guzmán

Colima

Isla Roca Partida

Isla Socorro

Manzanillo
Tecomán

Agu

Isla Clarión

Islas Revillagigedo
(to Mexico)

Lázaro Cárd

PACIFIC OCEAN

N

131

28

0 km 300
0 miles 300

Population ● National capital

○ below 50,000 ○ 50,000 to 100,000 ◉ 100,000 to 500,000 ◼ above 500,000

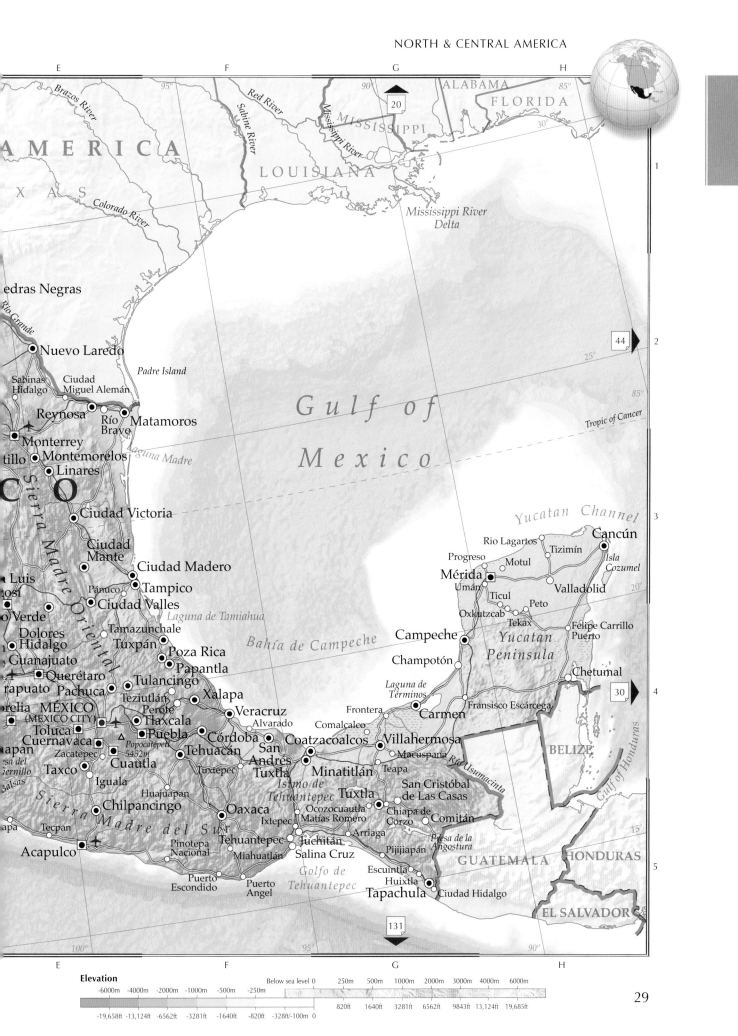

E   F   G   H

95°   90°   85°

ALABAMA

20

30°

FLORIDA

MISSISSIPPI

Mississippi River

LOUISIANA

1

Brazos River

Red River

Sabine River

Mississippi River Delta

Colorado River

A M E R I C A

T E X A S

edras Negras

Rio Grande

44

2

25°

85°

Nuevo Laredo

Padre Island

G u l f   o f

Sabinas Hidalgo

Ciudad Miguel Alemán

Reynosa

Río Bravo

Matamoros

M e x i c o

Tropic of Cancer

Monterrey

Laguna Madre

tillo

Montemorelos

Linares

C O

Sierra Madre Oriental

Ciudad Victoria

3

Yucatan Channel

20°

Ciudad Mante

Rio Lagartos

Tizimín

Cancún

Progreso

Isla Cozumel

Ciudad Madero

Motul

Mérida

Luis

Umán

Valladolid

osí

Pánuco

Tampico

Ciudad Valles

Ticul

Peto

o Verde

Laguna de Tamiahua

Oxkutzcab

Tekax

Dolores Hidalgo

Tamazunchale

Bahía de Campeche

Campeche

Yucatan Peninsula

Félipe Carrillo Puerto

Guanajuato

Túxpán

Poza Rica

Champotón

rapuato

Querétaro

Papantla

Chetumal

4

Pachuca

Tulancingo

Laguna de Términos

30

relia

Teziutlán

Xalapa

Frontera

Fransisco Escárcega

MÉXICO

Perote

Veracruz

Carmen

BELIZE

(MEXICO CITY)

Tlaxcala

Alvarado

Comalcalco

Villahermosa

Gulf of Honduras

Toluca

Cuernavaca

Puebla

Córdoba

Coatzacoalcos

Macuspana

apan

Popocatepetl 5452m

San Andrés Tuxtla

Minatitlán

Teapa

Río Usumacinta

Zacatepec

Cuautla

Tehuacán

Taxco

Tuxtepec

Istmo de Tehuantepec

San Cristóbal de Las Casas

Iguala

Huajuapan

Tuxtla

Balsas

Sierra

Chilpancingo

Madre del Sur

Oaxaca

Ocozocuautla

Chiapa de Corzo

Comitán

apa

Ixtepec

Matías Romero

Tecpan

Tehuantepec

Juchitán

Arriaga

Presa de la Angostura

15°

5

Acapulco

Pinotepa Nacional

Miahuatlán

Salina Cruz

Pijijiapán

HONDURAS

Puerto Escondido

Puerto Angel

Golfo de Tehuantepec

Escuintla

GUATEMALA

Huixtla

Tapachula

Ciudad Hidalgo

EL SALVADOR

131

100°

95°

90°

E   F   G   H

Elevation

Below sea level 0   250m   500m   1000m   2000m   3000m   4000m   6000m

-6000m   -4000m   -2000m   -1000m   -500m   -250m

820ft   1640ft   3281ft   6562ft   9843ft   13,124ft   19,685ft

-19,658ft -13,124ft -6562ft -3281ft -1640ft -820ft -328ft/-100m 0

Central America

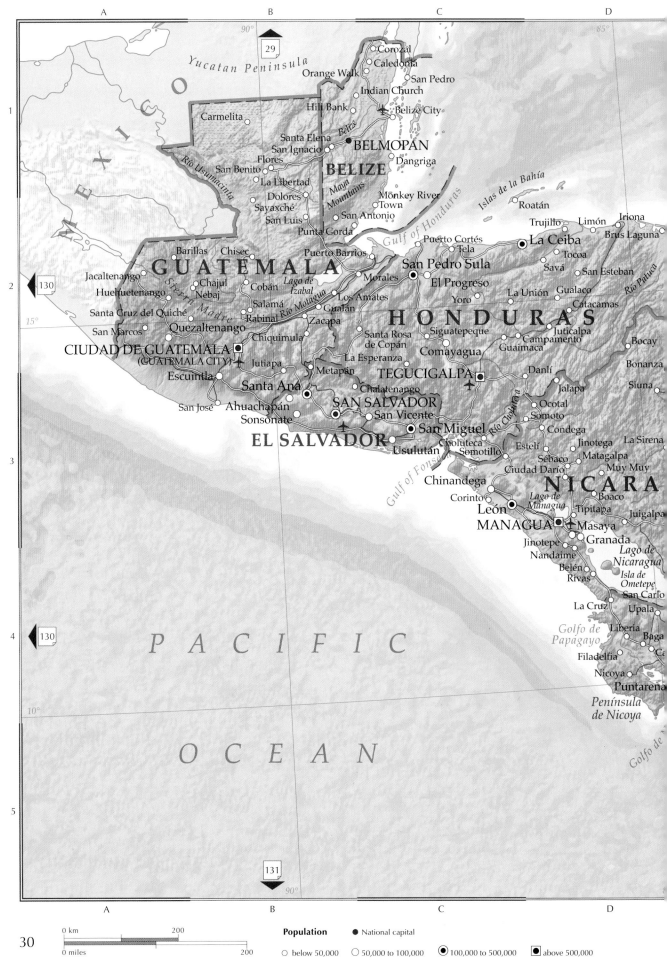

Yucatan Peninsula

MEXICO

Corozal
Caledonia
Orange Walk
San Pedro
Indian Church
Hill Bank
Belize City
Carmelita
Santa Elena
San Ignacio
BELMOPAN
Flores
Dangriga
San Benito
La Libertad
Río Usumacinta
Dolores
Maya
Sayaxché
Mountains
Monkey River
San Luis
San Antonio
Town
Punta Gorda

BELIZE

Gulf of Honduras
Islas de la Bahía
Roatán
Puerto Cortés
Trujillo
Limón
Iriona
Tela
Brus Laguna

Barillas
Chisec
Puerto Barrios
San Pedro Sula
La Ceiba

GUATEMALA
Morales
Tocoa
Savá
San Esteban

Jacaltenango
Chajul
Cobán
Lago de
El Progreso
La Unión
Gualaco
Río Patuca
Nebaj
Izabal
Yoro
Catacamas
Huehuetenango
Salamá
Los Amates
Santa Cruz del Quiché
Rabinal
Río Motagua
Gualán
HONDURAS
Siguatepeque
Campamento
Juticalpa
Bocay
San Marcos
Zacapa
Santa Rosa
Guaimaca
Quezaltenango
Chiquimula
de Copán
Comayagua
Bonanza
La Esperanza
Danlí
Siuna
CIUDAD DE GUATEMALA
La Esperanza
TEGUCIGALPA
Jalapa
(GUATEMALA CITY)
Jutiapa
Metapán
Chalatenango
Ocotal
Condega
Escuintla
Santa Ana
SAN SALVADOR
San Vicente
Somoto
La Sirena
San José
Ahuachapán
Sonsonate
San Miguel
Estelí
Jinotega
Matagalpa
EL SALVADOR
Usulután
Choluteca
Sébaco
Muy Muy
Somotillo
Ciudad Darío
NICARA
Chinandega
Boaco
Corinto
Lago de
Tipitapa
Juigalpa
Managua
León
MANAGUA
Masaya
Jinotepe
Granada
Nandaime
Lago de
Belén
Nicaragua
Rivas
Isla de
Ometepe
San Carlo
La Cruz
Upala
Golfo de
Liberia
Papagayo
Baga
Filadelfia
Nicoya
Puntarena
Península
de Nicoya

PACIFIC

OCEAN

Golfo de

0 km   200

0 miles   200

Population ● National capital

○ below 50,000   ○ 50,000 to 100,000   ◉ 100,000 to 500,000   ◼ above 500,000

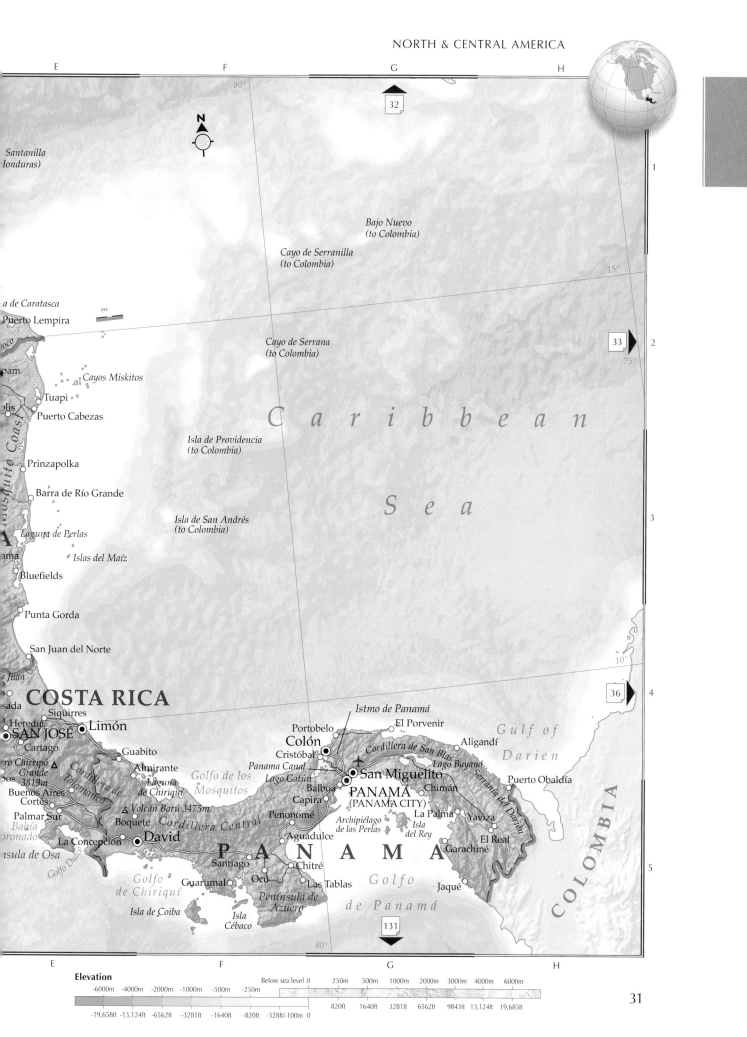

E   F   G   H

N

Santanilla
(Honduras)

1

Bajo Nuevo
(to Colombia)

Cayo de Serranilla
(to Colombia)

15°

a de Caratasca
Puerto Lempira

Cayo de Serrana
(to Colombia)

33   2

75°

oco

pam

lis   Tuapi
Puerto Cabezas

Cayos Miskitos

C a r i b b e a n

Prinzapolka

Isla de Providencia
(to Colombia)

Barra de Río Grande

S e a

Laguna de Perlas

3

ama

Islas del Maíz

Bluefields

Isla de San Andrés
(to Colombia)

Punta Gorda

San Juan del Norte

10°

n Juan

COSTA RICA

Istmo de Panamá

36   4

sada

Siquirres

El Porvenir

Gulf of

Heredia   Limón

Portobelo

Aligandí

Darien

SAN JOSÉ

Colón

Cartago

Cristóbal

Cordillera de San Blas

Guabito

Panama Canal

Lago Bayano

ro Chiripó   Almirante

Golfo de los

Lago Gatún

San Miguelito

Serranía del Darién

Grande   *Cordillera de*   *Laguna*

Mosquitos

Balboa

PANAMÁ

Chimán

Puerto Obaldía

3819m   *Talamanca*   *de Chiriquí*

Capira

(PANAMA CITY)

Buenos Aires

Volcán Barú 3475m

Penonomé

Archipiélago

Isla

La Palma   Yaviza

Cortés

Boquete   *Cordillera Central*

de las Perlas

del Rey

Palmar Sur

David

P A N A M Á

Aguadulce

El Real

Bahía   La Concepción

Garachiné

ronado

Santiago

Chitré

Golfo

sula de Osa

Golfo Dulce

Guarumal   Ocú

Las Tablas

de Panamá

Jaqué

5

Golfo
de Chiriquí

Peninsula de
Azuero

de Panamá

COLOMBIA

Isla de Coiba

Isla
Cébaco

80°

131

Elevation

-6000m -4000m -2000m -1000m -500m -250m   Below sea level 0   250m 500m 1000m 2000m 3000m 4000m 6000m

-19,658ft -13,124ft -6562ft -3281ft -1640ft -820ft -328ft/-100m 0   820ft 1640ft 3281ft 6562ft 9843ft 13,124ft 19,685ft

The Caribbean

N

UNITED STATES OF AMERICA

Gulf of Mexico

Tropic of Cancer

Grand Bahama Island

Freeport

Marsh Harbour

Great Abaco

Bimini Islands

Northeast Providence Channel

Berry Islands

Nicholls Town

NASSAU

Eleuthera Island

The Everglades

Florida Keys

Straits of Florida

New Providence

Rock Sound

Andros Town

Andros Island

Exuma Cays

Exuma Sound

Cat Island

San Salvador

Cay Sal

THE BAHAMAS

George Town

Rum Cay

LA HABANA (HAVANA)

Guanabacoa

Cárdenas

Anguilla Cays

Great Exuma Island

Long Island

Artemisa

Matanzas

Sagua la Grande

Archipiélago de Camagüey

Clarence Town

Crooked Island

Pinar del Río

Consolación del Sur

Santa Clara

Ragged Island Range

Crooked Island Passage

La Fé

Cienfuegos

Placetas

Acklins Island

Mayaguana Passage

Mayagı

Nueva Gerona

Cayo Largo

Sancti Spíritus

Morón

Ciego de Ávila

Caicos Passa

Isla de la Juventud

Bahía de Cochinos

C U B A

Little Inagua

Archipiélago de los Canarreos

Camagüey

Nuevitas

Holguín

Lake Rosa

Matthew Town

Great Ina

Archipiélago de los Jardines de la Reina

Las Tunas

Manzanillo

Bayamo

Palma Soriano

Guantánamo

Windward Passage

Ca

Haïti

Little Cayman

Cayman Brac

Santiago de Cuba

Guantánamo Bay (to US)

GEORGE TOWN

Grand Cayman

G

NAVASSA ISLAND (to US)

Gonaïves

Île de la Gonâve

HA

CAYMAN ISLANDS (to UK)

r

e

Jérémie

PORT-AU-PRINCE

Montego Bay

a

Jamaica Channel

Cayes

Jacm

Spanish Town

t

Portmore

KINGSTON

e

JAMAICA

r

Pedro Cays

HONDURAS

C

a

r

i

b

b

e

a

n

JAMAICA

Montego Bay

Lucea

Falmouth

Discovery Bay

St Ann's Bay

Caribbean Sea

The Cockpit Country

Ocho Rios

Cambridge

Annotto Bay

Buff Bay

Savanna-La-Mar

Christiana

Ewarton

Port Antonio

Mandeville

Spanish Town

Blue Mountain Peak △2258m

Black River

May Pen

Old Harbour

KINGSTON

Portmore

Morant Bay

N

Portland Bight

Caribbean Sea

0 km   20

0 miles   20

2000m/6562ft
1000m/3281ft
500m/1640ft
200m/656ft
Sea level

NICARAGUA

COSTA RICA

COLOMBI

0 km   200

0 miles   200

Population   ● National capital

○ below 50,000   ○ 50,000 to 100,000   ◉ 100,000 to 500,000   ◙ above 500,000

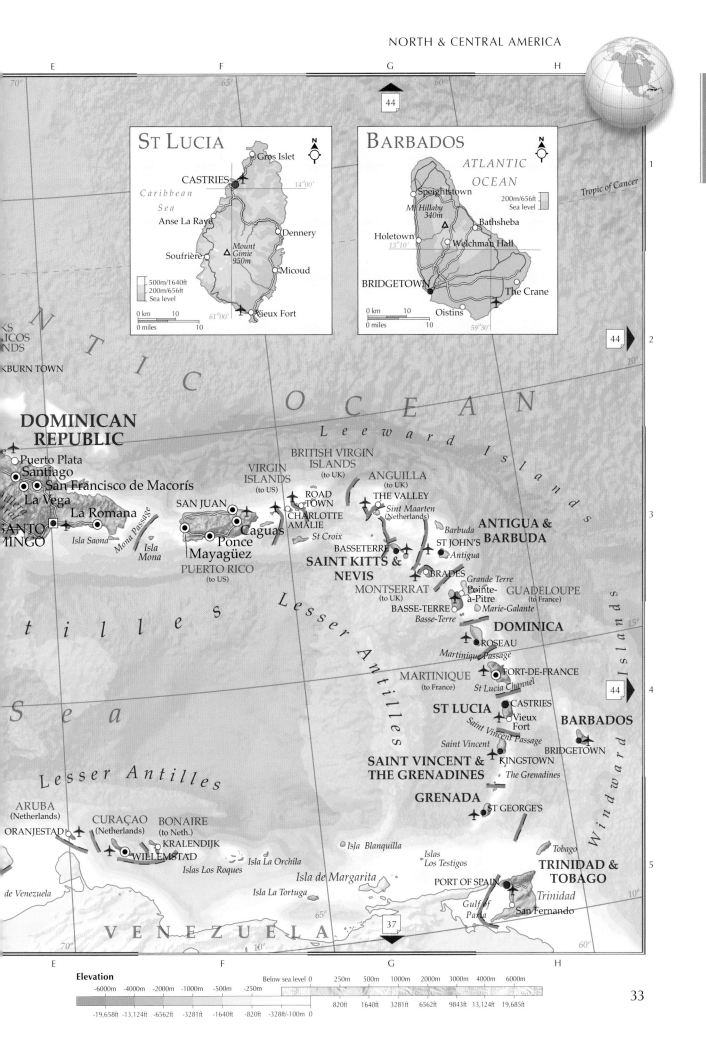

ST LUCIA

N

Gros Islet

CASTRIES

Caribbean Sea

Anse La Raye

Dennery

Soufrière

△ Mount Gimie 950m

Micoud

14°00'

61°00'

500m/1640ft
200m/656ft
Sea level

0 km   10
0 miles   10

Vieux Fort

BARBADOS

N

ATLANTIC OCEAN

Speightstown

Mt Hillaby 340m △

Bathsheba

Holetown

Welchman Hall

13°10'

BRIDGETOWN

The Crane

Oistins

59°30'

200m/656ft
Sea level

0 km   10
0 miles   10

44

Tropic of Cancer

44

20°

A N T I C   O C E A N

L e e w a r d   I s l a n d s

DOMINICAN REPUBLIC

Puerto Plata

Santiago

San Francisco de Macorís

La Vega

SANTO DOMINGO

La Romana

Isla Saona

Mona Passage

Isla Mona

SAN JUAN

Caguas

Ponce

Mayagüez

PUERTO RICO (to US)

VIRGIN ISLANDS (to US)

BRITISH VIRGIN ISLANDS (to UK)

ROAD TOWN

CHARLOTTE AMALIE

St Croix

ANGUILLA (to UK)

THE VALLEY

Sint Maarten (Netherlands)

Barbuda

ANTIGUA & BARBUDA

ST JOHN'S

Antigua

BASSETERRE

SAINT KITTS & NEVIS

BRADES

MONTSERRAT (to UK)

Grande Terre

Pointe-à-Pitre

GUADELOUPE (to France)

BASSE-TERRE

Basse-Terre

Marie-Galante

DOMINICA

ROSEAU

Martinique Passage

MARTINIQUE (to France)

FORT-DE-FRANCE

St Lucia Channel

ST LUCIA

CASTRIES

Vieux Fort

Saint Vincent Passage

Saint Vincent

SAINT VINCENT & THE GRENADINES

KINGSTOWN

The Grenadines

GRENADA

ST GEORGE'S

BARBADOS

BRIDGETOWN

W i n d w a r d   I s l a n d s

15°

L e e w a r d   I s l a n d s

L e s s e r   A n t i l l e s

Lesser Antilles

ARUBA (Netherlands)

ORANJESTAD

CURAÇAO (Netherlands)

BONAIRE (to Neth.)

KRALENDIJK

WILLEMSTAD

Isla La Orchila

Islas Los Roques

de Venezuela

Isla Blanquilla

Islas Los Testigos

Isla de Margarita

Isla La Tortuga

Tobago

TRINIDAD & TOBAGO

PORT OF SPAIN

Trinidad

Gulf of Paria

San Fernando

V E N E Z U E L A

70°

65°

60°

10°

37

44

TKS ICOS NDS

KBURN TOWN

1

2

3

4

5

E   F   G   H

Elevation

| -6000m | -4000m | -2000m | -1000m | -500m | -250m | | Below sea level 0 | 250m | 500m | 1000m | 2000m | 3000m | 4000m | 6000m |

-19,658ft -13,124ft -6562ft -3281ft -1640ft -820ft -328ft/-100m 0   820ft 1640ft 3281ft 6562ft 9843ft 13,124ft 19,685ft

South America

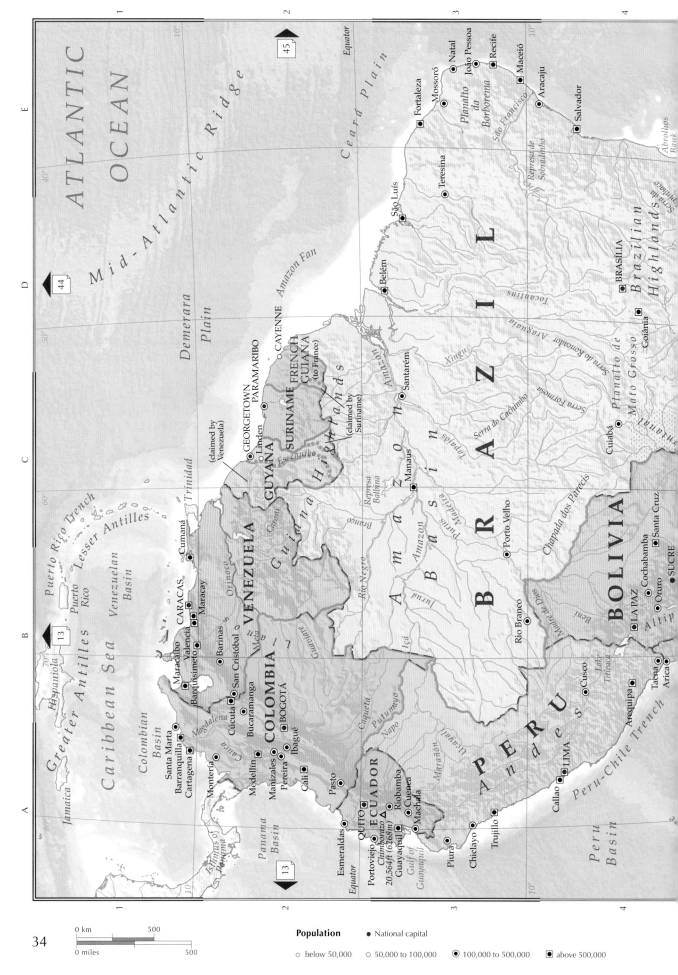

ATLANTIC OCEAN

Mid-Atlantic Ridge

Demerara Plain

Ceara Plain

Amazon Fan

Equator

Natal
João Pessoa
Recife
Maceió
Mossoró
Aracaju
Fortaleza
Salvador
Planalto da Borborema
São Francisco
Represa de Sobradinho
Abrolhos Bank
Teresina
São Luís
BRASÍLIA
Brazilian Highlands
Serra do Espinhaço
Belém
Tocantins
Goiânia
Planalto de Mato Grosso
Xingu
Santarém
Cuiabá
Araguaia
Serra do Roncador
Serra Formosa
Serra do Cachimbo
B R A Z I L
A m a z o n   B a s i n
Manaus
Represa Balbina
Tapajós
Madeira
Purus
Amazon
Porto Velho
Chapada dos Parecis
BOLIVIA
Santa Cruz
Cochabamba
LA PAZ
Oruro
SUCRE
Altip...
Madre de Dios
Beni
Rio Branco
Rio Negro
Branco
Juruá
Içá
CAYENNE
FRENCH GUIANA (to France)
PARAMARIBO
SURINAME
GEORGETOWN
Linden
GUYANA
Essequibo
Guiana Highlands (claimed by Surinam)
(claimed by Venezuela)
Trinidad
Cumaná
Maracay
CARACAS
Valencia
VENEZUELA
Orinoco
Caroni
Barinas
San Cristóbal
Barquisimeto
Maracaibo
Guaviare
Meta
Caquetá
Putumayo
Napo
Ucayali
Marañón
Colombian Basin
Santa Marta
Barranquilla
Cartagena
Montería
Medellín
Manizales
Pereira
Cali
Cúcuta
Bucaramanga
BOGOTÁ
Ibagué
COLOMBIA
Magdalena
Cauca
Pasto
ECUADOR
QUITO
Esmeraldas
Portoviejo
Chimborazo (6268m) 20,564ft △
Riobamba
Cuenca
Machala
Guayaquil
Gulf of Guayaquil
Piura
Chiclayo
Trujillo
P E R U
A n d e s
Cusco
Lake Titicaca
Arequipa
Tacna
Arica
Callao
LIMA
Peru-Chile Trench
Peru Basin
Panama Basin
Isthmus of Panama
Caribbean Sea
Greater Antilles
Lesser Antilles
Puerto Rico Trench
Puerto Rico
Venezuelan Basin
Jamaica
Hispaniola
Equator

34

0 km   500
0 miles   500

Population   ● National capital

○ below 50,000   ◎ 50,000 to 100,000   ◉ 100,000 to 500,000   ■ above 500,000

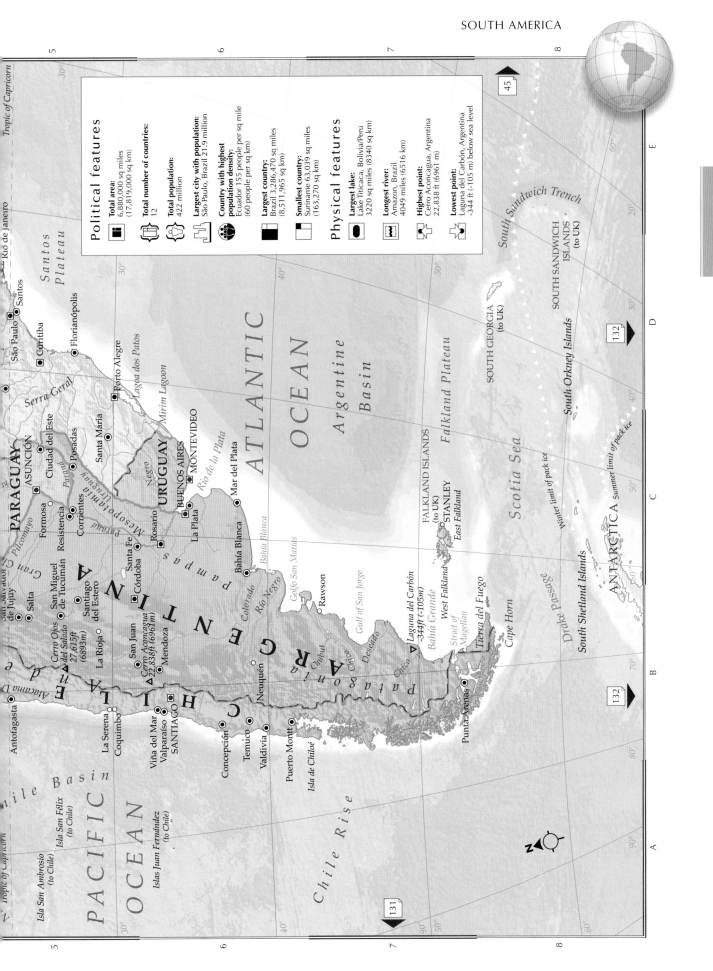

Political features

Total area:
6,880,000 sq miles (17,819,000 sq km)

Total number of countries:
12

Total population:
422 million

Largest city with population:
São Paulo, Brazil 21.9 million

Country with highest population density:
Ecuador 155 people per sq mile (60 people per sq km)

Largest country:
Brazil 3,286,470 sq miles (8,511,965 sq km)

Smallest country:
Suriname 63,039 sq miles (163,270 sq km)

Physical features

Largest lake:
Lake Titicaca, Bolivia/Peru 3220 sq miles (8340 sq km)

Longest river:
Amazon, Brazil 4049 miles (6516 km)

Highest point:
Cerro Aconcagua, Argentina 22,838 ft (6961 m)

Lowest point:
Laguna del Carbón, Argentina -344 ft (-105 m) below sea level

Northern South America

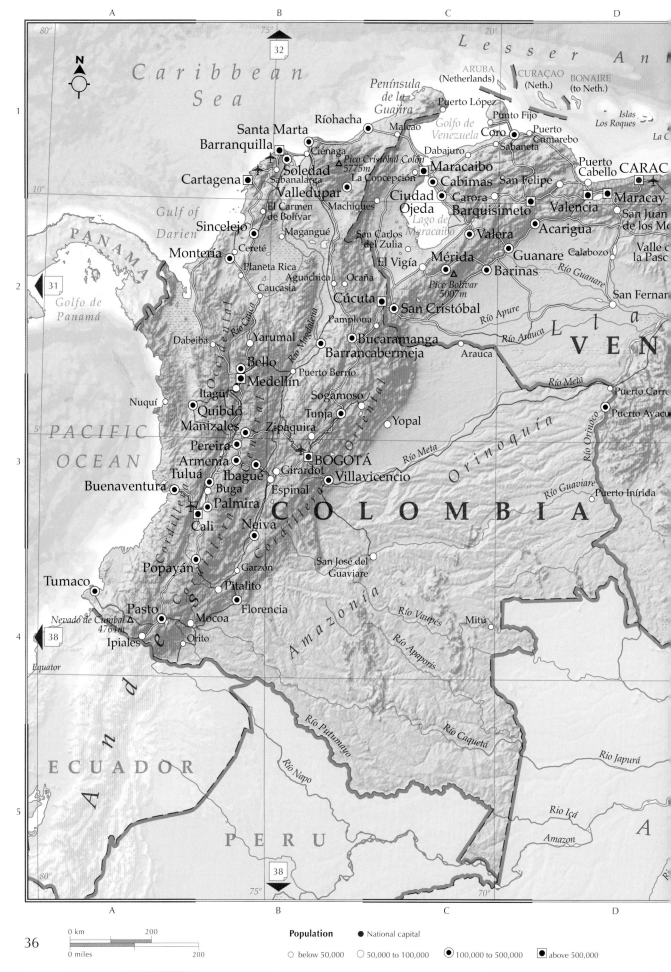

Caribbean Sea

Península de la Guajira

ARUBA (Netherlands)
CURAÇAO (Neth.)
BONAIRE (to Neth.)

L e s s e r   A n t

Santa Marta
Barranquilla
Cartagena
Soledad
Sabanalarga
Ciénaga
Valledupar
El Carmen de Bolívar
Sincelejo
Montería
Cereté
Planeta Rica
Magangué
Aguachica
Caucasia

Ríohacha
Maicao
Pico Cristóbal Colón 5775m
La Concepción
Machiques

Puerto López
Punto Fijo
Coro
Dabajuro
Sabaneta
San Felipe
Maracaibo
Cabimas
Ciudad Ojeda
Carora
Barquisimeto
Lago del Maracaibo
San Carlos del Zulia
El Vigía
Mérida
Pico Bolívar 5007m

Puerto Cumarebo
Puerto Cabello
CARAC
Maracay
Valencia
Acarigua
Valera
Guanare
Barinas
Calabozo
San Juan de los Mo
Valle de la Pasc

Golfo de Venezuela

Islas Los Roques
La C

Gulf of Darien

PANAMA

Golfo de Panamá

Ocaña
Cúcuta
Pamplona
San Cristóbal
Bucaramanga
Barrancabermeja
Arauca

San Fernan

Río Apure
Río Arauca
Río Meta
V E N

PACIFIC OCEAN

Nuquí
Dabeiba
Yarumal
Bello
Medellín
Itagüí
Quibdó
Manizales
Pereira
Armenia
Tuluá
Buga
Ibagué
Palmira
Buenaventura
Cali
Neiva
Popayán
Garzón
Pitalito
Florencia

Puerto Berrío
Sogamoso
Tunja
Zipaquirá
Yopal
BOGOTÁ
Girardot
Villavicencio
Espinal

San José del Guaviare

Río Meta
Río Orinoco
Puerto Carre
Puerto Ayacu
Puerto Inírida

Orinoquía

C O L O M B I A

Tumaco
Nevado de Cumbal 4764m
Pasto
Mocoa
Ipiales
Orito

A n d e s
Cordillera Occidental
Cordillera Central
Cordillera Oriental
Río Cauca
Río Magdalena

Equator

A m a z o n i a

Río Putumayo
Río Napo
Río Vaupés
Río Apaporis
Río Caquetá
Mitú

Río Japurá

E C U A D O R

P E R U

Río Içá
Amazon

0 km 200
0 miles 200

Population   ● National capital
○ below 50,000   ○ 50,000 to 100,000   ◉ 100,000 to 500,000   ▣ above 500,000

ATLANTIC

OCEAN

SAINT VINCENT &
THE GRENADINES

BARBADOS

GRENADA

Isla Blanquilla
Isla de
Margarita

Islas Los Testigos

Tobago

La Asunción

Carúpano

Güiria

Cariaco
Gulf of
Paria

TRINIDAD &
TOBAGO

Trinidad

Puerto La Cruz

Barcelona

The Serpent's Mouth

San Mateo

Anaco

Maturín

Cantaura

Tucupita

El Tigre

Río Orinoco

Ciudad Guayana

Upata

Ciudad
Bolívar

Embalse de Guri

U E L A

El Callao

Matthews
Ridge

Charity

Spring Garden

El Dorado

Parika

Aurora

GEORGETOWN

New
Amsterdam

PARAMARIBO

Salto
Angel

Peters Mine

Bartica

Rockstone

Totness

Nieuw Amsterdam

St-Laurent-du-Maroni

Río Caura

Cuyuni River

Linden

Sinnamary

Río Paragua

Nieuw
Nickerie

Kaaimanston

Kourou

Kamarang

Orealla

Apoera

Río Caroni

Mount Roraima
2810m

GUYANA

W. J. van
Blommesteinmeer

Maroni River

Grand-
Santi

Montagnes
de la Trinité

CAYENNE

Ouanary

Pakaraima Mountains

Kurupukari

SURINAME

Juliana Top
1230m

Montagne
Tortue

St-Georges

(Venezuela claims all
of Guyana west of
Essequibo River)

Lethem

Essequibo River

FRENCH
GUIANA
(to France)

Camopi

z
u
i
a
n
a

Orinoco

H
i
g
h
l
a
n
d
s

Courantyne River

Tumuc-Humac Mountains

(claimed by
Suriname)

Acarai Mountains

(claimed by
Suriname)

Equator

Río Negro

B R A Z I L

Amazon

Rio Purus

Amazon

Amazon

Rio Tapajós

zon Basin

Amazon

Elevation

-6000m -4000m -2000m -1000m -500m -250m

Below sea level 0   250m 500m 1000m 2000m 3000m 4000m 6000m

-19,658ft -13,124ft -6562ft -3281ft -1640ft -820ft -328ft/-100m 0

820ft 1640ft 3281ft 6562ft 9843ft 13,124ft 19,685ft

Western South America

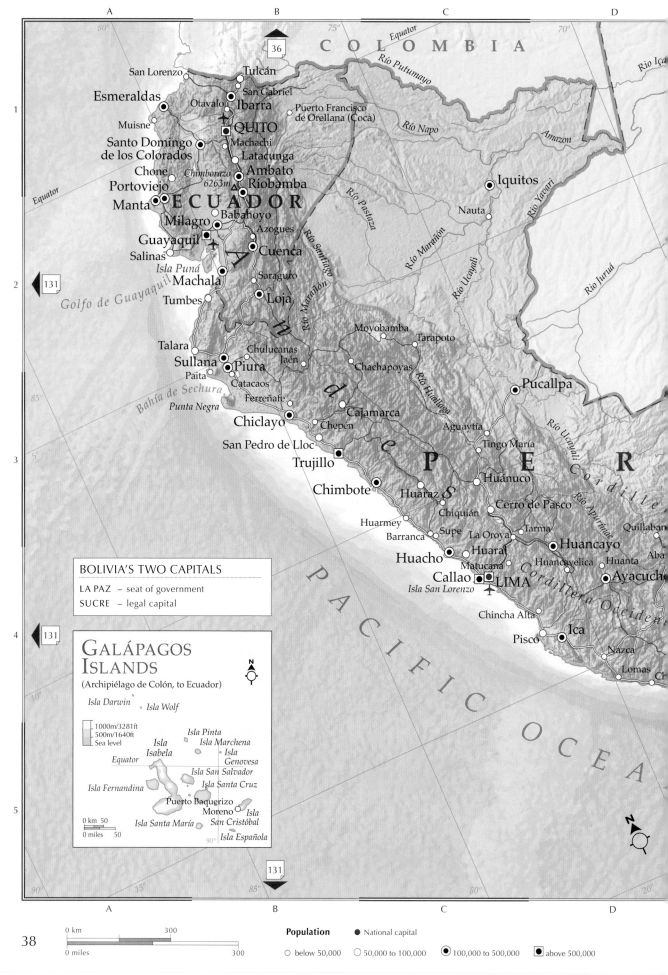

COLOMBIA

Equator

Río Putumayo

San Lorenzo
Tulcán
Esmeraldas
San Gabriel
Otavalo
Ibarra
Muisne
Puerto Francisco
de Orellana (Coca)
Río Napo
QUITO
Machachi
Santo Domingo
de los Colorados
Latacunga
Chone
Ambato
Chimborazo
6263m
Riobamba
Portoviejo
ECUADOR
Río Pastaza
Iquitos
Manta
Babahoyo
Nauta
Milagro
Azogues
Milagro
Río Santiago
Guayaquil
Cuenca
Río Marañón
Río Ucayali
Río Yavari
Salinas
Isla Puná
Saraguro
Río Iurua
Machala
Loja
Golfo de Guayaquil
Tumbes

Amazon
Río Içá

Talara
Chulucanas
Jaén
Moyobamba
Tarapoto
Sullana
Piura
Chachapoyas
Paita
Catacaos
Pucallpa
Bahía de Sechura
Ferreñafe
Cajamarca
Río Huallaga
Río Ucayali
Punta Negra
Chiclayo
Chepen
Aguaytía
San Pedro de Lloc
Tingo María
Trujillo
Huánuco
PERU
Chimbote
Huaraz
Cerro de Pasco
Río Apurímac
Quillabam
Chiquián
Huarmey
Barranca
Supe
La Oroya
Tarma
Huancayo
Huanta
Aba
Huaral
Huancavelica
Huacho
Matucana
Callao
LIMA
Ayacucho
Isla San Lorenzo
Chincha Alta
Ica
Pisco
Nazca
Lomas
Ch

PACIFIC OCEAN

BOLIVIA'S TWO CAPITALS

LA PAZ — seat of government
SUCRE — legal capital

GALÁPAGOS ISLANDS

(Archipiélago de Colón, to Ecuador)

N

Isla Darwin
Isla Wolf

1000m/3281ft
500m/1640ft
Sea level

Isla Pinta
Isla Marchena
Isla
Isabela
Isla
Genovesa
Equator
Isla San Salvador
Isla Fernandina
Isla Santa Cruz
Puerto Baquerizo
Moreno
Isla
0 km 50
Isla Santa María
San Cristóbal
0 miles 50
Isla Española
N

0 km 300
0 miles 300

Population   ● National capital

○ below 50,000   ○ 50,000 to 100,000   ◉ 100,000 to 500,000   ▣ above 500,000

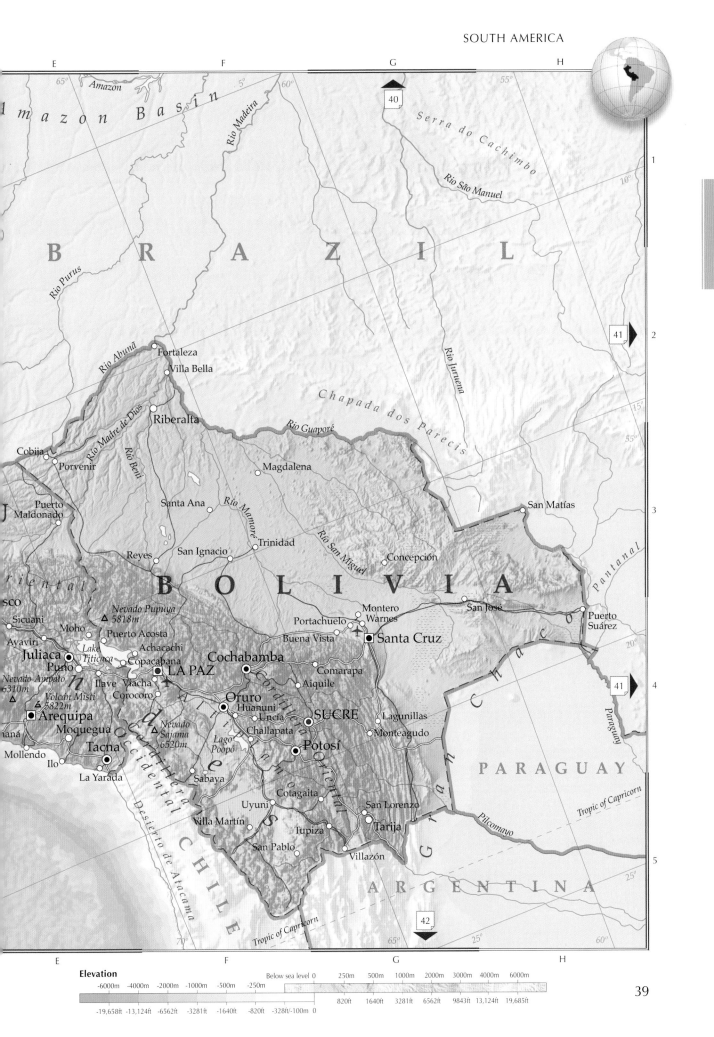

E · F · G · H

Amazon

65° *5°* *60°* *55°*

40

Serra do Cachimbo

1

Rio São Manuel

A m a z o n   B a s i n

10°

Rio Madeira

B   R   A   Z   I   L

41

2

Rio Purus

Rio Iauena

Rio Abunã

Fortaleza
Villa Bella

Chapada dos Parecis

15°

Riberalta

Rio Madre de Dios
Rio Guaporé

55°

Cobija
Porvenir

Rio Beni

Magdalena

San Matías

3

Puerto
Maldonado

Santa Ana

Rio Mamoré

Rio San Miguel

Pantanal

Reyes

San Ignacio

Trinidad

Concepción

Puerto
Suárez

SCO

B   O   L   I   V   I   A

Sicuani

Nevado Pupuya
5818m

Montero
Warnes

San José

Moho

Puerto Acosta

Portachuelo

Ayaviri

Achacachi

Buena Vista

Santa Cruz

20°

Juliaca

Lake
Titicaca

Copacabana

Cochabamba

Puno

Comarapa

Nevado Ampato
6310m

Ilave
Viacha

LA PAZ

Aiquile

4

Corocoro

Volcán Misti
5822m

Huanuni

SUCRE

Lagunillas

Arequipa

Oruro

Uncía

Moquegua

Nevado
Sajama
6520m

Challapata

Monteagudo

Mollendo

Tacna

Lago
Poopó

Potosí

P   A   R   A   G   U   A   Y

Ilo

La Yarada

Sabaya

Cotagaita

San Lorenzo

Tropic of Capricorn

Uyuni

Tarija

Pilcomayo

Villa Martín

Tupiza

25°

San Pablo

Villazón

5

A R G E N T I N A

25°

42

Tropic of Capricorn

70° 65° 60°

E · F · G · H

Elevation

Below sea level 0   250m 500m 1000m 2000m 3000m 4000m 6000m

-6000m -4000m -2000m -1000m -500m -250m

820ft 1640ft 3281ft 6562ft 9843ft 13,124ft 19,685ft

-19,658ft -13,124ft -6562ft -3281ft -1640ft -820ft -328ft/-100m 0

39

Brazil

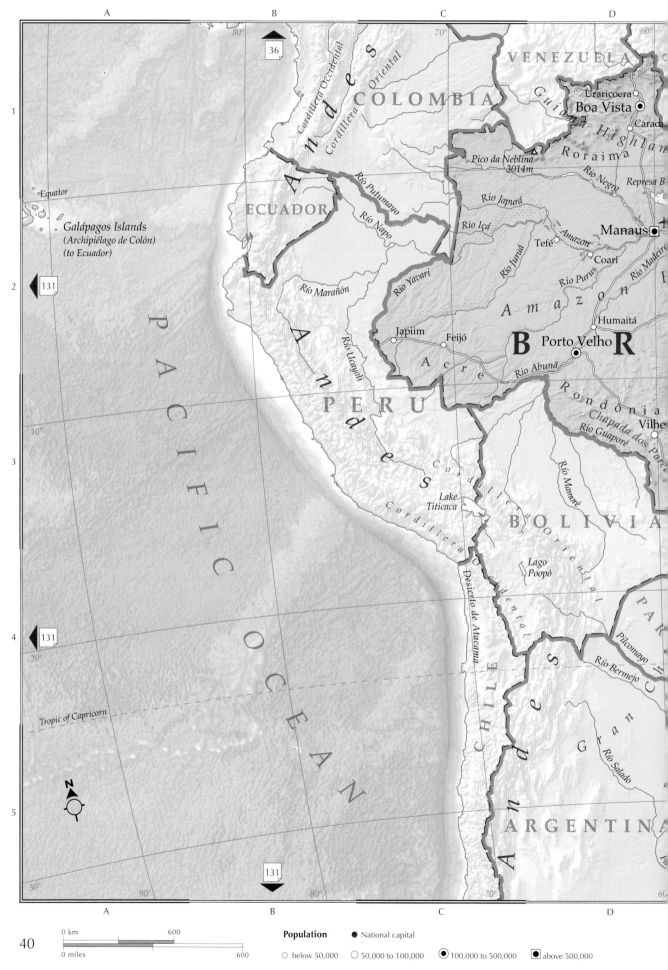

A

B

C

D

36

80°

VENEZUELA

Uraricoera

Boa Vista

Caraco

COLOMBIA

Cordillera Occidental

Cordillera Oriental

Guiana Highlan

Roraima

1

70°

60°

Pico da Neblina
3014m

Rio Negro

Represa B

Río Putumayo

Rio Japurá

Equator

ECUADOR

Río Napo

Rio Içá

Manaus

Galápagos Islands
(Archipiélago de Colón)
(to Ecuador)

A
n
d
e
s

Tefé

Amazon

Coari

Rio Madeira

Río Marañón

Rio Yavari

Rio Juruá

Rio Purus

131

2

A
m
a
z
o
n

Humaitá

Japiim

Feijó

B Porto Velho

R

Río Ucayali

A
c
r
e

Rio Abunã

Rondônia

Chapada dos Pare

PERU

A
n
d
e
s

Río Guaporé

Vilhe

10°

3

Cordillera
Oriental

Rio Mamoré

Lake
Titicaca

BOLIVIA

PACIFIC OCEAN

Cordillera Occidental

Lago
Poopó

PA

Pilcomayo

131

4

Desierto de Atacama

A
n
d
e
s

Río Bermejo

C

20°

CHILE

Gran

Tropic of Capricorn

G

Río Salado

N

5

A
n
d
e
s

ARGENTINA

30°

90°

80°

70°

60

A

B

C

D

40

0 km   600

0 miles   600

Population   ● National capital

○ below 50,000   ○ 50,000 to 100,000   ◉ 100,000 to 500,000   ▣ above 500,000

ATLANTIC OCEAN

44

FRENCH
GUIANA
(to France)

SURINAME

*Tumuc-Humac
Mountains*

Mouths of the Amazon

Ilha Caviana de Fora

Macapá

Baía de Marajó

Equator

*Ilha
de Marajó*   Belém

Baía de São Marcos

Santarém   Altamira

Amazon

Amapá

São Luís   Parnaíba

Camocim

enquer

Itaituba

*Represa de
Tucuruí*   Bacabal   *Piripiri*   Fortaleza

Marabá   Imperatriz   Teresina   Mossoró

Atol das Rocas

*San Fernando de Noronha
(to Brazil)*

Rio Xingu

Maranhão   *Ceará*   Assu   *Cabo de São Roque*

Carolina   Floriano   Juazeiro do Norte   Natal   45

Balsas   Picos   *Rio Grande do Norte*

Serra do Cachimbo   *Piauí*   João Pessoa
Campina Grande

B R A Z I L   *Paraíba*

Pernambuco   Recife

Palmas do
Tocantins   *Represa de Sobradinho*   Juazeiro   Maceió

Rio Tocantins   *Alagoas*

Serra dos Gradaús   *Tocantins*   *Rio São Francisco*   Aracaju
Estância

Serra Formosa   Taguatinga   *Chapada
Diamantina*   Feira de Santana

Goiás   B a h i a   Salvador

Planalto   *Baía de Todos os Santos*

Cuiabá   *Rio Araguaia*   Janaúba   Itabuna

Anápolis   BRASÍLIA   *Central*   Vitória da Conquista

donópolis   Goiânia   Montes Claros   *Canavieiras*

Jataí   *Araçuai*

Mato Grosso   Araguari   *Minas Gerais*

do Sul   Uberlândia   Governador Valadares

Campo Grande   Uberaba   *Espírito
Santo*

quidauana   Ribeirão Preto   Belo Horizonte

idente Prudente   Marília   Divinópolis   Vitória

Juiz de Fora

Londrina   Campinas   Campos dos Goytacazes

Maringá   *Paraná*   Nova
Iguaçu   Rio de Janeiro

*Represa
de Itaipú*   São Paulo   Santos

*Saltos do
Iguaçu*   *Rio Iguaçu*   Ponta Grossa

araná   Curitiba

Joinville

Santa Catarina   Blumenau

Florianópolis

Passo Fundo

ta Maria   Canoas

do Sul   Porto Alegre

Bagé   *Lagoa dos Patos*

o Negro   Rio Grande

URUGUAY   *Mirim Lagoon*

Tropic of Capricorn

45

45

ATLANTIC OCEAN

Elevation

-6000m -4000m -2000m -1000m -500m -250m   Below sea level 0   250m 500m 1000m 2000m 3000m 4000m 6000m

-19,658ft -13,124ft -6562ft -3281ft -1640ft -820ft -328ft/-100m 0   820ft 1640ft 3281ft 6562ft 9843ft 13,124ft 19,685ft

Southern South America

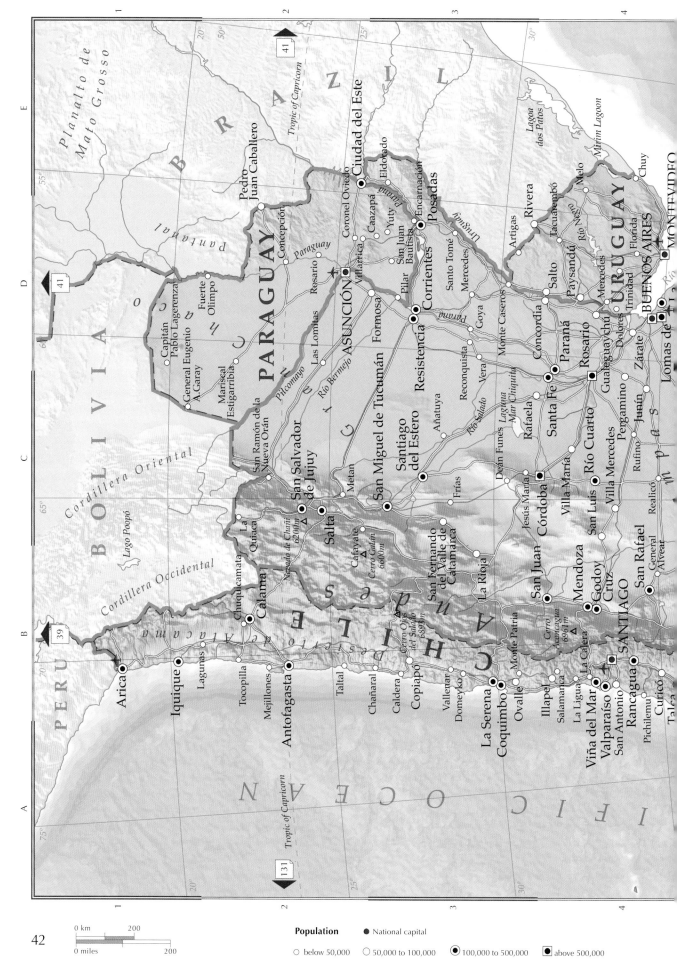

0 km   200

0 miles   200

Population   ● National capital

○ below 50,000   ◎ 50,000 to 100,000   ◉ 100,000 to 500,000   ■ above 500,000

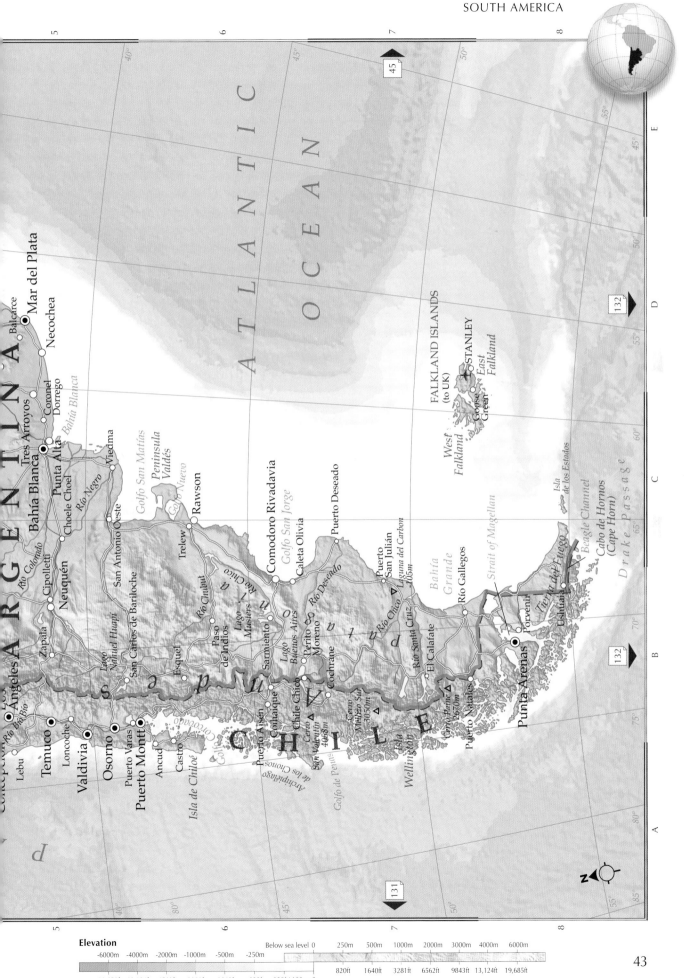

ARGENTINA

Mar del Plata
Balcarce
Necochea
Coronel
Dorrego
Tres Arroyos
Bahía Blanca
Punta Alta
Choele Choel
Río Colorado
Río Negro
Viedma
Bahía Blanca
Neuquén
Cipolletti
Zapala
San Antonio Oeste
Golfo San Matías
Península Valdés
Golfo Nuevo
Rawson
Trelew

Los Ángeles
Río Bío Bío
Lebu
Temuco
Loncoche
Valdivia
Osorno
Puerto Varas
Puerto Montt
Ancud
Castro
Isla de Chiloé

San Carlos de Bariloche
Lago Nahuel Huapi
Paso de Indios
Esquel
Río Chubut

Sarmiento
Lago Musters
Río Chico
Lago Buenos Aires
Perito Moreno
Cochrane

Comodoro Rivadavia
Caleta Olivia
Golfo San Jorge
Puerto Deseado
Río Deseado

Puerto San Julián
Laguna del Carbón
-105m
Río Chico
Río Santa Cruz
El Calafate
Río Gallegos

Bahía Grande

Strait of Magellan

FALKLAND ISLANDS
(to UK)
STANLEY
East Falkland
West Falkland
Goose Green

Isla de los Estados
Beagle Channel
Cabo de Hornos
(Cape Horn)
Drake Passage

Tierra del Fuego
Ushuaia
Porvenir
Punta Arenas
Puerto Natales

Cerro Murallón 3050m
Cerro San Valentín 4058m
Cerro Paine 2670m

Coihaique
Chile Chico
Puerto Aisén
Golfo Corcovado
Golfo de Penas
Archipiélago de los Chonos
Isla Wellington

CHILE

d

45

132

132

131

N

Elevation

| -6000m | -4000m | -2000m | -1000m | -500m | -250m | Below sea level 0 | 250m | 500m | 1000m | 2000m | 3000m | 4000m | 6000m |

-19,658ft -13,124ft -6562ft -3281ft -1640ft -820ft -328ft/-100m 0   820ft 1640ft 3281ft 6562ft 9843ft 13,124ft 19,685ft

The Atlantic Ocean

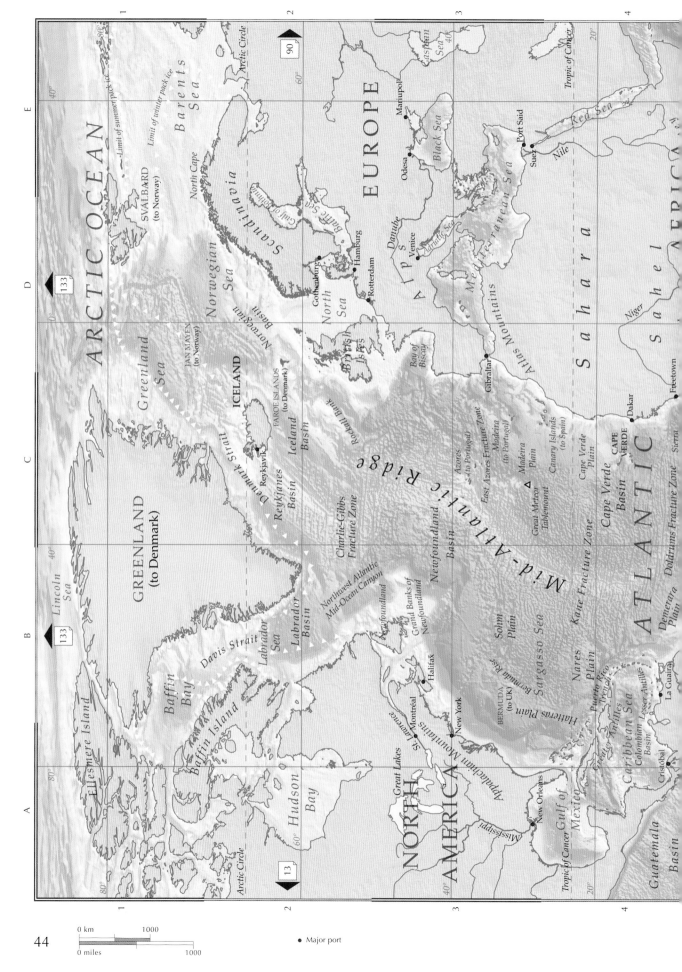

ARCTIC OCEAN

Lincoln Sea

Limit of summer pack ice
Limit of winter pack ice

Barents Sea

Arctic Circle

SVALBARD (to Norway)

North Cape

Scandinavia

Greenland Sea

JAN MAYEN (to Norway)

Norwegian Sea

Norwegian Basin

GREENLAND (to Denmark)

ICELAND

Denmark Strait

Reykjavik

Reykjanes Basin

FAROE ISLANDS (to Denmark)

Iceland Basin

Rockall Bank

British Isles

Bay of Biscay

Gulf of Bothnia

Baltic Sea

Gothenburg

North Sea

Hamburg

Rotterdam

EUROPE

Alps

Danube

Venice

Adriatic Sea

Mediterranean Sea

Black Sea

Mariupol'

Odesa

Caspian Sea

Port Said

Suez

Red Sea

Nile

Tropic of Cancer

Gibraltar

Atlas Mountains

Sahara

Niger

Sahel

AFRICA

Dakar

Freetown

Sierra

Azores (to Portugal)

East Azores Fracture Zone

Madeira (to Portugal)

Madeira Plain

Canary Islands (to Spain)

Great Meteor Tablemount

Cape Verde Plain

CAPE VERDE

Cape Verde Basin

ATLANTIC

Kane Fracture Zone

Doldrums Fracture Zone

Mid-Atlantic Ridge

Charlie-Gibbs Fracture Zone

Newfoundland Basin

Northwest Atlantic Mid-Ocean Canyon

Labrador Sea

Labrador Basin

Davis Strait

Baffin Bay

Baffin Island

Ellesmere Island

Hudson Bay

Great Lakes

St. Lawrence

Montreal

Appalachian Mountains

New York

Halifax

Newfoundland

Grand Banks of Newfoundland

Sohm Plain

Bermuda Rise

BERMUDA (to UK)

Hatteras Plain

Sargasso Sea

Nares Plain

Puerto Rico Trench

La Guaira

Lesser Antilles

Greater Antilles

Caribbean Sea

Colombian Basin

Cristobal

Demerara Plain

NORTH AMERICA

New Orleans

Gulf of Mexico

Mississippi

Tropic of Cancer

Guatemala Basin

Arctic Circle

• Major port

44

0 km   1000

0 miles   1000

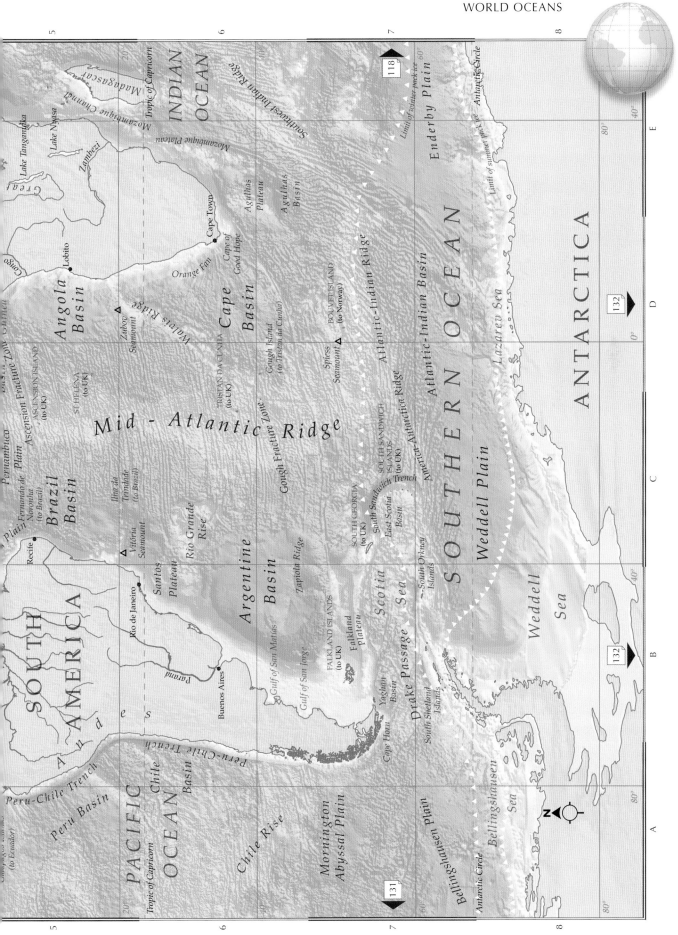

INDIAN OCEAN

Madagascar

Lake Tanganyika
Lake Nyasa

Zambezi

Mozambique Channel

Mozambique Plateau

Southwest Indian Ridge

Limit of winter pack ice

Enderby Plain

Antarctic Circle

118

Great
Rift Valley

Cape Town

Cape of
Good Hope

Agulhas
Plateau

Agulhas
Basin

Limit of summer pack ice

Orange Fan

Congo

Lobito

Angola
Basin

Walvis Ridge

Zubov
Seamount

Cape
Basin

Atlantic–Indian Ridge

BOUVET ISLAND
(to Norway)

Gough Island
(to Tristan da Cunha)

Spiess
Seamount

Atlantic–Indian Basin

SOUTHERN OCEAN

ANTARCTICA

Lazarev Sea

132

ST HELENA
(to UK)

TRISTAN DA CUNHA
(to UK)

ASCENSION ISLAND
(to UK)

Ascension Fracture Zone

Mid - Atlantic Ridge

Gough Fracture Zone

America–Antarctica Ridge

SOUTH SANDWICH
ISLANDS
(to UK)

Weddell Plain

Pernambuco
Plain

Fernando de
Noronha
(to Brazil)

Brazil
Basin

Ilha da
Trindade
(to Brazil)

Recife

Vitória
Seamount

Rio Grande
Rise

Santos
Plateau

SOUTH GEORGIA
(to UK)

South Sandwich Trench

East Scotia
Basin

South Orkney
Islands

Zapiola Ridge

SOUTH
AMERICA

Rio de Janeiro

Argentine
Basin

Scotia
Sea

Weddell
Sea

132

Paraná

Buenos Aires

Gulf of San Matías

Gulf of San Jorge

FALKLAND ISLANDS
(to UK)

Falkland
Plateau

Drake Passage

South Orkney
Islands

Andes

Yaghan
Basin

South Shetland
Islands

Cape Horn

Bellingshausen
Sea

Peru–Chile Trench

Chile
Basin

Chile Trench

Mornington
Abyssal Plain

Bellingshausen Plain

Antarctic Circle

N

131

Galápagos Islands
(to Ecuador)

Peru–Chile Trench

PACIFIC
OCEAN

Tropic of Capricorn

Chile Rise

Peru Basin

Elevation

-6000m	-4000m	-2000m	-1000m	-500m	-250m	-100m	0
-19,658ft	-13,124ft	-6562ft	-3281ft	-1640ft	-820ft	-328ft/-100m	0

Africa

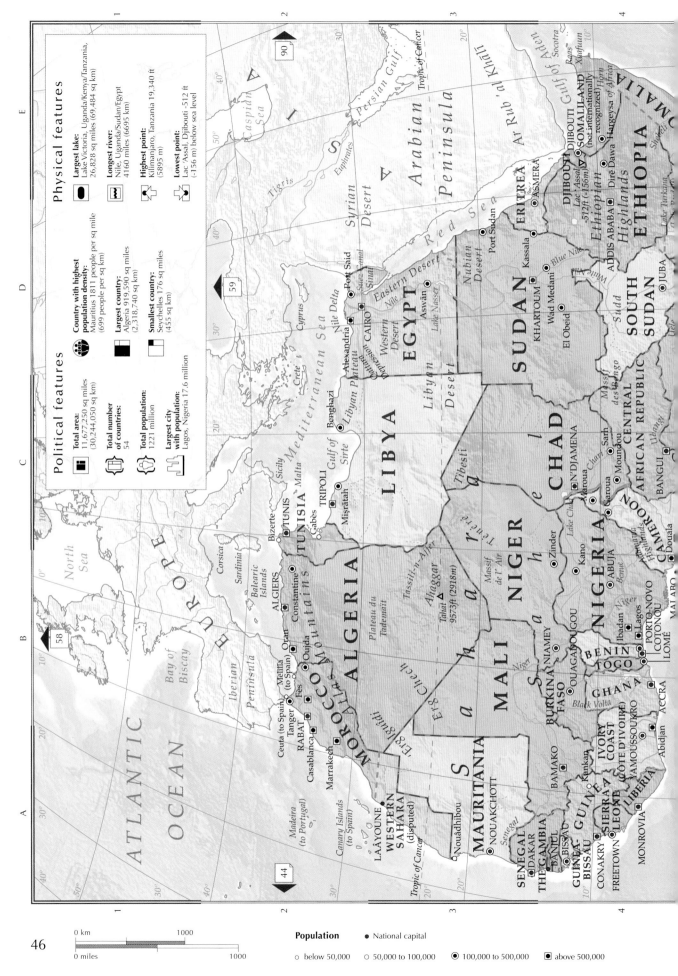

Northwest Africa

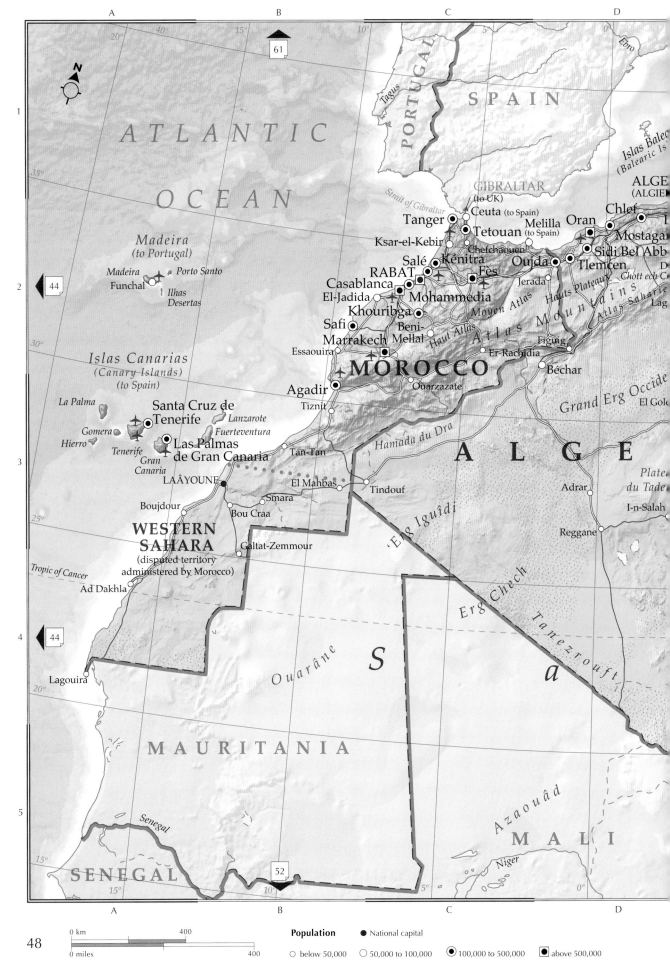

ATLANTIC

OCEAN

SPAIN

PORTUGAL

Tagus

Ebro

Islas Balea
(Balearic Is

GIBRALTAR
(to UK)
Ceuta (to Spain)
Tanger
Tetouan
Ksar-el-Kebir
Melilla
(to Spain)
Oran
Chlef
ALGE
(ALGIE
Mostaga
Sidi Bel Abb

Chefchaouen
Salé
RABAT
Kénitra
Fès
Oujda
Tlemcen
Casablanca
Mohammedia
El-Jadida
Jerada
Moyen Atlas
Hauts Plateaux
Chott ech
Lag
Khouribga
Beni-
Mellal
Safi
Haut Atlas
Atlas Mountains
Essaouira
Marrakech
MOROCCO
Er-Rachidia
Figuig
Béchar
Ouarzazate
Agadir
Grand Erg Occide
El Gol
Tiznit

Madeira
(to Portugal)
Madeira
Porto Santo
Funchal
Ilhas
Desertas

Islas Canarias
(Canary Islands)
(to Spain)

La Palma
Lanzarote
Santa Cruz de
Tenerife
Fuerteventura
Gomera
Hierro
Tenerife
Las Palmas
de Gran Canaria
Gran
Canaria
Tan-Tan
Hamáda du Dra
ALGE
Adrar
Plate
du Tade
I-n-Salah
Reggane

LAÂYOUNE
El Mahbas
Tindouf
Smara
Boujdour
Bou Craa

**WESTERN
SAHARA**
(disputed territory
administered by Morocco)
Galtat-Zemmour

Tropic of Cancer
Ad Dakhla

Erg Iguîdi
Erg Chech
Tanezrouft

Lagouira

S

a

Ouarâne

MAURITANIA

Azaouâd

MALI

Senegal
Niger

SENEGAL

0 km   400
0 miles   400

Population   ● National capital
○ below 50,000   ○ 50,000 to 100,000   ◉ 100,000 to 500,000   ■ above 500,000

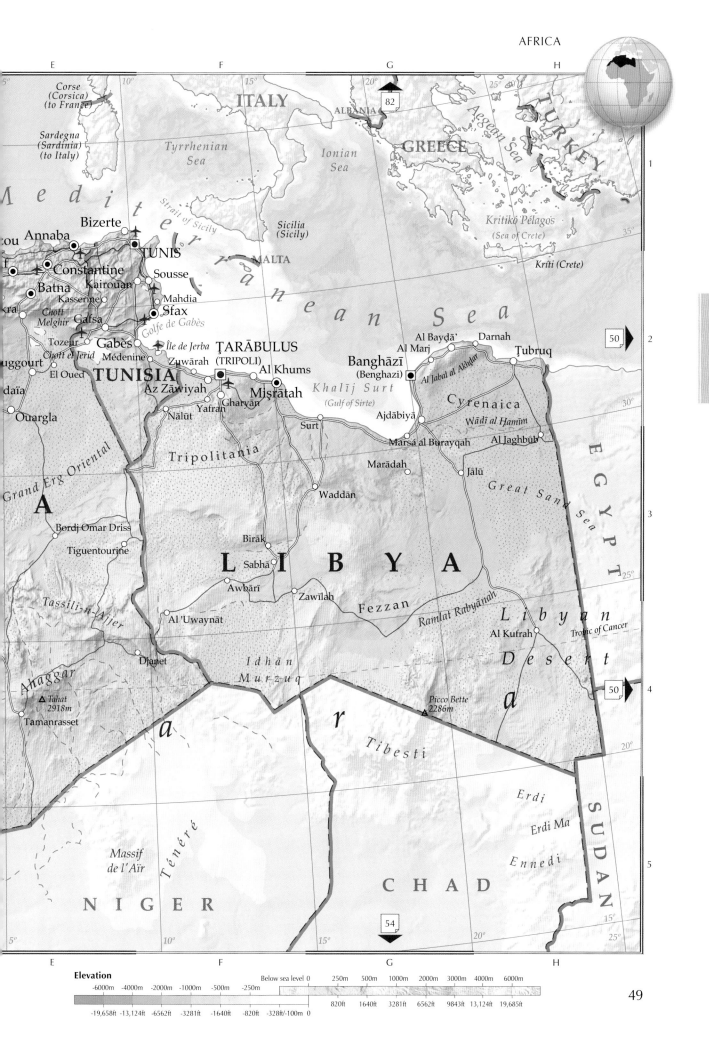

Corse
(Corsica)
(to France)

ITALY

ALBANIA

82

GREECE

TURKEY

Sardegna
(Sardinia)
(to Italy)

Tyrrhenian
Sea

Ionian
Sea

Aegean
Sea

Kritikó Pélagos
(Sea of Crete)

35°

Bizerte

Annaba

TUNIS

Sousse

MALTA

Sicilia
(Sicily)

Strait of Sicily

Kríti (Crete)

Constantine

Kairouan

Batna

Kasserine

Mahdia

Sfax

Al Baydā'

Darnah

50

2

Gafsa

Chott
Melghir

Golfe de Gabès

Al Marj

Ţubruq

Tozeur

Gabès

Île de Jerba

ŢARĀBULUS

Banghāzī

Chott el Jerid

Médenine

Zuwārah

(TRIPOLI)

(Benghazi)

Al Jabal al Akhdar

El Oued

TUNISIA

Az Zāwiyah

Al Khums

Khalīj Surt

Cyrenaica

30°

daïa

Yafran

Mişrātah

(Gulf of Sirte)

Ouargla

Nālūt

Gharyān

Ajdābiyā

Wādī al Ḩamīm

Surt

Al Jaghbūb

Tripolitania

Marsá al Burayqah

Marādah

Jālū

Waddān

Grand Erg Oriental

A

Great Sand Sea

EGYPT

3

Bordj Omar Driss

Birāk

Tiguentourine

L

Sabhā

I

B

Y

A

25°

Awbārī

Zawīlah

Fezzan

Ramlat Rabyānah

Libyan

Tassili-n-Ajjer

Al 'Uwaynāt

Al Kufrah

Tropic of Cancer

Djanet

Idhān

Desert

50

4

Ahaggar

Murzuq

a

20°

△ Tahat
2918m

Picco Bette
2286m

△

Tamanrasset

a

r

Tibesti

Erdi

Massif
de l'Aïr

Ténéré

Erdi Ma

Ennedi

SUDAN

5

N I G E R

C H A D

15°

54

5°

10°

15°

20°

25°

Elevation

-6000m	-4000m	-2000m	-1000m	-500m	-250m	Below sea level 0	250m	500m	1000m	2000m	3000m	4000m	6000m
-19,658ft	-13,124ft	-6562ft	-3281ft	-1640ft	-820ft	-328ft/-100m 0	820ft	1640ft	3281ft	6562ft	9843ft	13,124ft	19,685ft

Northeast Africa

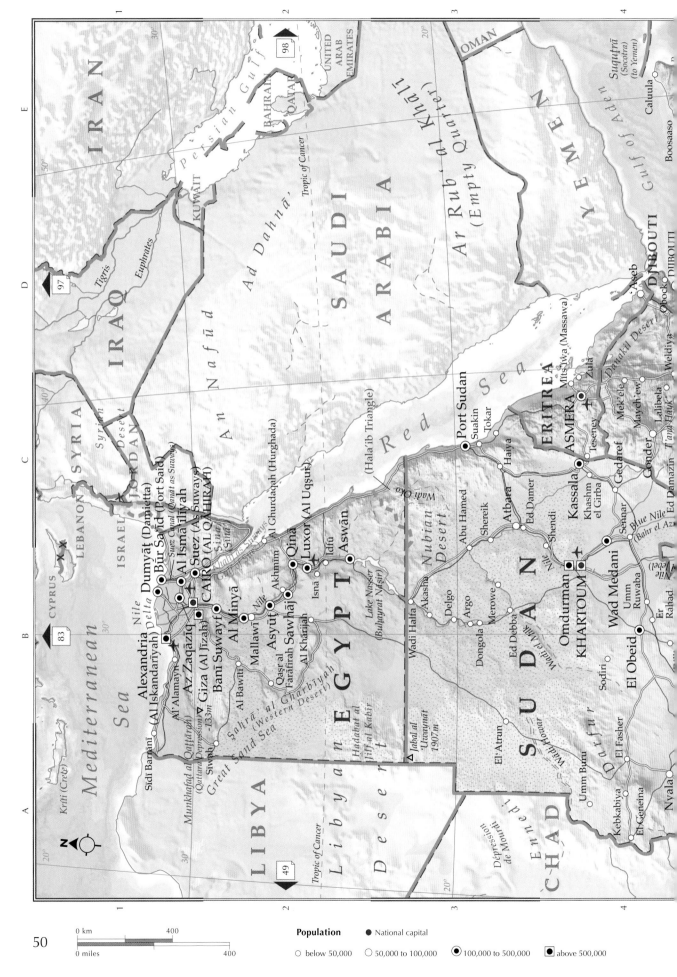

IRAN

IRAQ

SYRIA

LEBANON

CYPRUS

ISRAEL

JORDAN

Krtti (Crete)

Mediterranean Sea

Tigris

Euphrates

Syrian Desert

KUWAIT

Persian Gulf

BAHRAIN

QATAR

UNITED ARAB EMIRATES

OMAN

An Nafūd

Ad Dahnā'

SAUDI ARABIA

Ar Rub' al Khālī (Empty Quarter)

Tropic of Cancer

YEMEN

Gulf of Aden

Suqutrā (Socotra) (to Yemen)

Caluula

Boosaaso

DJIBOUTI

Aseb

Obock

DJIBOUTI

Danakil Desert

Weldiya

Mits'iwa (Massawa)

Zula

Mek'ele

Maych'ew

Lalibela

Tana Häyk'

ERITREA

ASMERA

Teseney

Gedaret

Gonder

Ed Damazin

Red Sea

Port Sudan

Suakin

Tokar

(Hala'ib Triangle)

Wadi Oko

Haiya

Abu Hamed

Shereik

Atbara

Ed Damer

Kassala

Khashm el Girba

Sennar

Blue Nile (Bahr el Azr)

Nubian Desert

Akasha

Delgo

Argo

Merowe

Dongola

Ed Debba

Shendi

Omdurman

KHARTOUM

Wad Medani

Umm Ruwaba

El Obeid

SUDAN

Wadi Halfa

Lake Nasser (Buḥayrat Nāṣir)

Nile

Wadi el Milk

Wadi Howar

Sodiri

Er

Nile (Jebel)

Jabal al 'Uwaynāt 1907m

El'Atrun

Umm Buru

Kebkabiya

El Fasher

Darfur

El Geneina

Nyala

CHAD

Ennedi

Dépression de Mourdi

Alexandria (Al Iskandarīyah)

Dumyāṭ (Damietta)

Būr Sa'īd (Port Said)

Al Ismā'īlīyah

Suez Canal (Qanāt as Suways)

Suez (As Suways)

CAIRO (AL QĀHIRAH)

Nile Delta

Az Zaqāzīq

Giza (Al Jīzah)

Banī Suwayf

Al Minyā

Mallawī

Asyūṭ

Qasr al Farāfirah

Sawhāj

Akhmīm

Al Khārijah

Qinā

Luxor (Al Uqṣur)

Idfū

Isnā

Aswān

Al Ghurdaqah (Hurghada)

Sinai

Gulf of Suez

Gulf of Aqaba

EGYPT

Ṣaḥrā' al Gharbīyah (Western Desert)

Hadabat al Jilf al Kabīr

Great Sand Sea

Al Bawīṭī

Siwah

Munkhafad al Qaṭṭārah (Qattara Depression) -133m

At Alamayn

Sidi Barrāni

Tropic of Cancer

Libyan Desert

LIBYA

N

83

49

97

98

0 km 400

0 miles 400

Population ● National capital

○ below 50,000   ○ 50,000 to 100,000   ◉ 100,000 to 500,000   ■ above 500,000

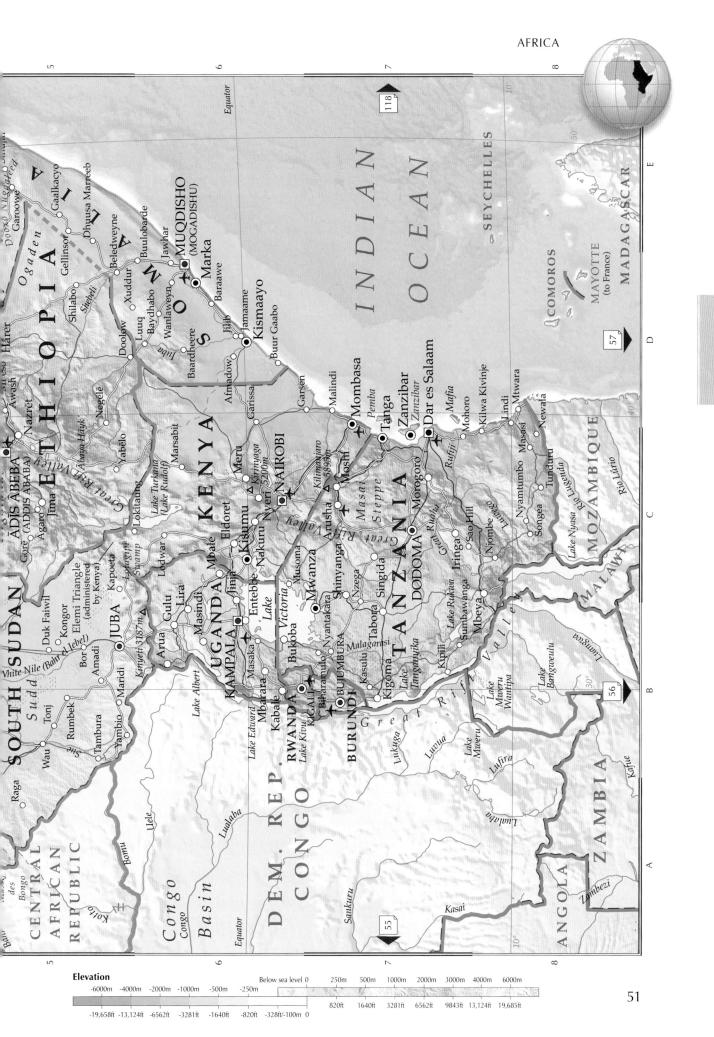

57

SEYCHELLES

INDIAN

OCEAN

COMOROS

MAYOTTE
(to France)

MADAGASCAR

ETHIOPIA

Ogaden

Doxo Nugaaleed

Garoowe

Gaalkacyo

Dhuusa Marreeb

Gellinsor

Beledweyne

Buulobarde

Jawhar

MUQDISHO
(MOGADISHU)

Marka

Baraawe

Jamaame

Kismaayo

Buur Gaabo

S
O
M
A
L
I
A

Shilabo

Xuddur

Baydhabo

Wanlaweyn

Jiliib

Luuq

Doolow

Baardheere

Afmadow

Garissa

Garsen

Malindi

Mombasa

Pemba

Tanga

Zanzibar
Zanzibar

Dar es Salaam

Mafia

Mohoro

Kilwa Kivinje

Lindi

Mtwara

Newala

Masasi

Tunduru

Songea

Rio Lúrio

MOZAMBIQUE

Shebeli

Nazrēt

Awash

Harer

Ago so Hāyk'

KENYA

Marsabit

Meru

Kirinyaga
5200m

NAIROBI

Nyeri

Moshi

Kilimanjaro
5895m

Arusha

Masai
Steppe

Morogoro

DODOMA

Great Ruaha

Sao Hill

Iringa

Njombe

Nyamtumbo

Lake Nyasa
Rio Lugenda

MALAWI

Lake Malombe

ADĪS ĀBEBA
(ADDIS ABABA)

Jima

Agaro

Gorē

Negēlē

Abelo

Lake Turkana
(Lake Rudolf)

Lokitaung

Lodwar

Eldoret

Nakuru

Kisumu

Mbale

Jinja

Entebbe

KAMPALA

Masaka

Bukoba

Musoma

Mwanza

Nyantakata

Nzega

Shinyanga

Singida

Tabora

Kipili

Sumbawanga

Mbeya

Great Rift Valley

Lake Rukwa

Njombe

T A N Z A N I A

Great Rift Valley

Lake Victoria

JUBA

SOUTH SUDAN

Bor

Amadi

Maridi

Yambio

Tambura

Rumbek

Tonj

Wau

Raga

Bau

des
Bongo

Sudd

White Nile (Bahr el Jebel)

She

Kapoeta

Lotagipi
Swamp

Kinyeti 3187m

Arua

Gulu

Lira

Masindi

UGANDA

Mbarara

Kabale

Kigali

RWANDA

Lake Kivu

BUJUMBURA

BURUNDI

Biharamulo

Kasulu

Kigoma

Lake
Tanganyika

Malagarasi

Lukuga

Luvua

Lake Mweru

Lake Mweru
Wantipa

Lufira

Lake
Bangweulu

Luapula

ZAMBIA

ANGOLA

Zambezi

Kafue

Elemi Triangle
(administered
by Kenya)

CENTRAL
AFRICAN
REPUBLIC

Kotto

Uele

Bomu

Congo

DEM. REP.
CONGO

Congo
Basin

Lualaba

Lomami

Lualaba

Sankuru

Kasai

Kasai

Lake Albert

Lake Edward

Lake
Edward

Duk Faiwil

Kongor

Shilabo

Baydhabo

Equator

Equator

55

56

51

Elevation

Below sea level 0   250m   500m 1000m 2000m 3000m 4000m 6000m

-6000m -4000m -2000m -1000m -500m -250m

-19,658ft -13,124ft -6562ft -3281ft -1640ft -820ft -328ft/-100m 0

820ft   1640ft   3281ft   6562ft   9843ft 13,124ft 19,685ft

West Africa

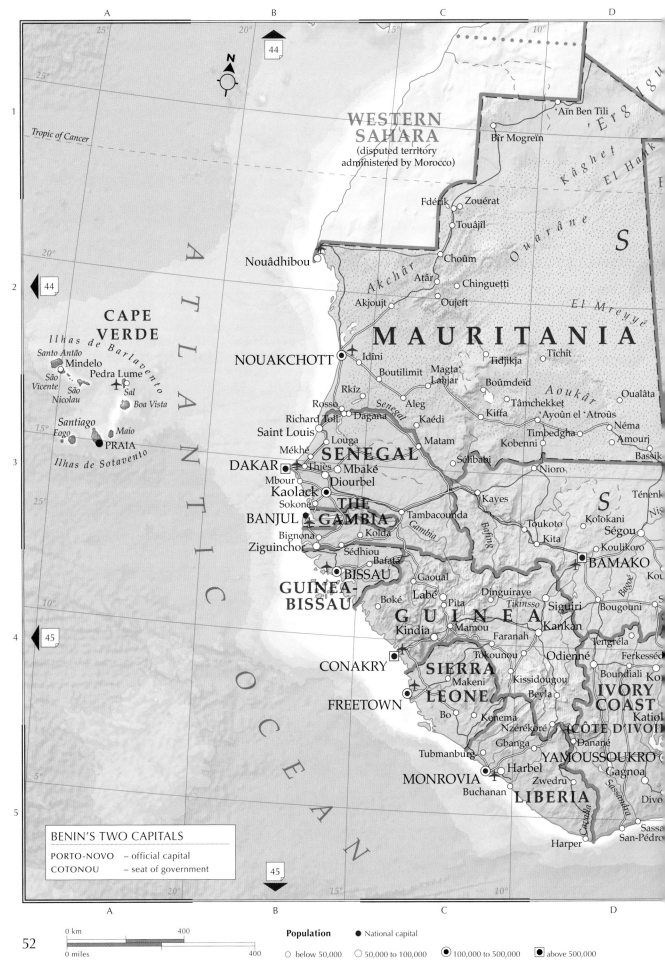

A · B · C · D

WESTERN
SAHARA
(disputed territory
administered by Morocco)

Aïn Ben Tili
Bîr Mogreïn

Tropic of Cancer

Fdérik · Zouérat
Touâjîl

Nouâdhibou
Choûm
Atâr · Chinguetti
Akjoujt · Oujeft

M A U R I T A N I A

CAPE
VERDE

Ilhas de Barlavento

Santo Antão
Mindelo
São
Vicente
São
Nicolau
Pedra Lume
Sal
Boa Vista

NOUAKCHOTT
Idîni
Boutilimit · Magta
Lahjar
Tidjikja · Tîchît
Rkîz
Boûmdeïd · Oualâta
Rosso
Aleg
Tâmchekket
'Ayoûn el 'Atroûs
Richard Toll · Dagana
Saint Louis
Kaédi · Kiffa
Kobenni · Timbedgha · Néma
Amourj

Santiago
Fogo
Maio
PRAIA

Ilhas de Sotavento

Mékhé
Louga
DAKAR
Thiès · Mbaké
Matam
Sélibabi
Nioro · Bassik

Mbour
Diourbel
Kayes
Tének
Kaolack
Sokone
THE
Tambacounda
Kolokani · Nig
BANJUL · GAMBIA
Toukoto
Ségou
Bignona
Kolda
Kita
Koulikoro
Ziguinchor
Sédhiou
BAMAKO
Bafatá
Gaoual
Kou
BISSAU
GUINEA-
BISSAU
Boké · Labé · Pita
Dinguiraye
Tikinsso · Siguiri
Bougouni
Kindia
Mamou
Kankan
Faranah
Tengréla
CONAKRY
Tokounou
Odienné · Ferkessé
SIERRA
Kissidougou
Boundiali · Ko
Makeni
Beyla
IVORY
FREETOWN
LEONE
COAST
Bo
Kenema
Katiol
Nzérékoré
CÔTE D'IVOI
Gbanga
Danané
Tubmanburg
YAMOUSSOUKRO
Harbel
Gagnoa
MONROVIA
Zwedru
Divo
Buchanan
LIBERIA
Sassa
San-Pédro
Harper

A T L A N T I C   O C E A N

S

S

BENIN'S TWO CAPITALS
PORTO-NOVO – official capital
COTONOU   – seat of government

0 km   400

0 miles   400

Population   ● National capital

○ below 50,000   ○ 50,000 to 100,000   ◉ 100,000 to 500,000   ▣ above 500,000

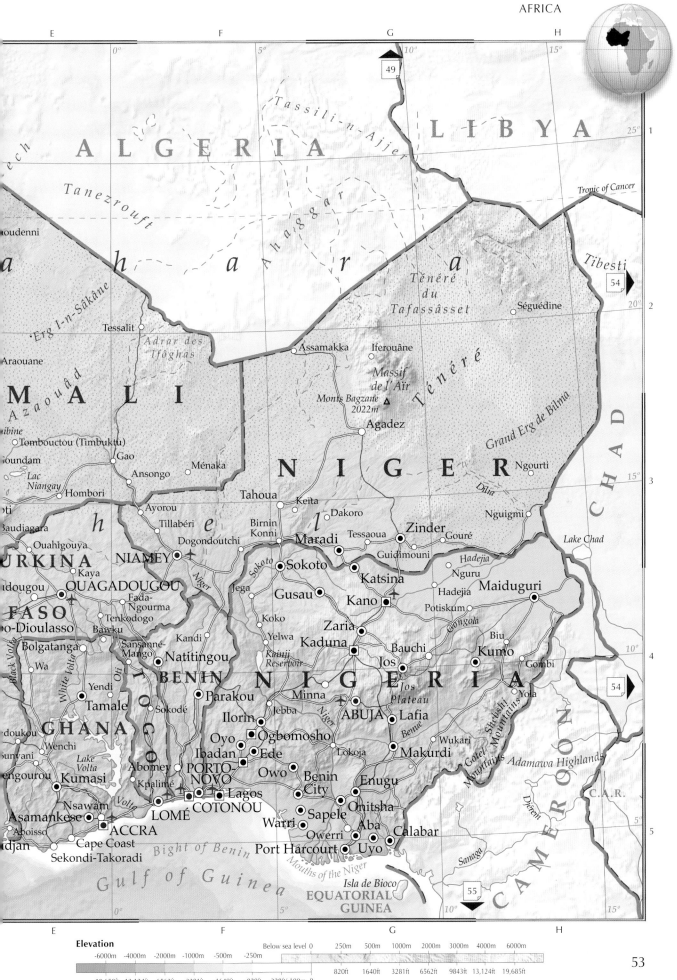

AFRICA

E   F   G   H

0°   5°   10°   15°   25° 1

49

Tropic of Cancer

LIBYA

ALGERIA

Tassili-n-Ajjer

Tanezrouft

Ahaggar

Tibesti

54

Ténéré
du
Tafassâsset

Séguédine

20° 2

Saharã

Assamakka

Iferouâne

Massif
de l'Aïr

Ténéré

Grand Erg de Bilma

CHAD

Moudenni

Erg I-n-Sâkâne

Tessalit

Araouane

Adrar des
Ifôghas

Azaouâd

Monts Bagzane
2022m

Agadez

Ngourti

MALI

ibine

Tombouctou (Timbuktu)

Gao

Ansongo

Ménaka

N I G E R

Dîlia

Nguigmi

15° 3

Goundam

Lac
Niangay

Hombori

Tahoua

Keïta

Dakoro

Lake Chad

oti

Ayorou

Birnin
Konni

Tessaoua

Zinder

Gouré

Hadejia

audiagara

Dogondoutchi

Maradi

Nguru

Ouahigouya

h

e

l

Guidimouni

Hadejia

URKINA

Tillabéri

Sokoto

Sokoto

Katsina

Maiduguri

Kaya

NIAMEY

Jega

Gusau

Kano

Potiskum

idougou

OUAGADOUGOU

Fada-
Ngourma

Koko

Zaria

Biu

Gombi

FASO

Tenkodogo

Yelwa

Kaduna

Bauchi

Kumo

o-Dioulasso

Bawku

Sansanné-
Mango

Kandi

Kainji
Reservoir

Jos

Yola

Bolgatanga

Oti

Natitingou

Jos
Plateau

54

Wa

Yendi

BENIN

N I G E R I A

Shebshi
Mountains

CAMEROON

Tamale

Parakou

Minna

Adamawa Highlands

doukou

GHANA

Sokodé

Ilorin

Jebba

ABUJA

Lafia

Wukari

Gotel
Mountains

Benue

Makurdi

C.A.R.

Wenchi

Oyo

Ogbomosho

Lokoja

unyani

Lake
Volta

Ibadan

Ede

Owo

Benin
City

Enugu

Djérem

engourou

Kumasi

Abomey

PORTO
NOVO

Owo

Onitsha

Nsawam

Kpalimé

Lagos

Sapele

Asamankese

LOMÉ

COTONOU

Warri

Aba

Calabar

Sanaga

Aboisso

ACCRA

Owerri

Uyo

idjan

Cape Coast

Bight of Benin

Port Harcourt

Sekondi-Takoradi

Gulf of Guinea

Mouths of the Niger

Isla de Bioco

55

EQUATORIAL
GUINEA

0°   5°   10°   15°   5

E   F   G   H

Elevation

Below sea level 0   250m 500m 1000m 2000m 3000m 4000m 6000m

-6000m -4000m -2000m -1000m -500m -250m

820ft 1640ft 3281ft 6562ft 9843ft 13,124ft 19,685ft

-19,658ft -13,124ft -6562ft -3281ft -1640ft -820ft -328ft/-100m 0

53

Central Africa

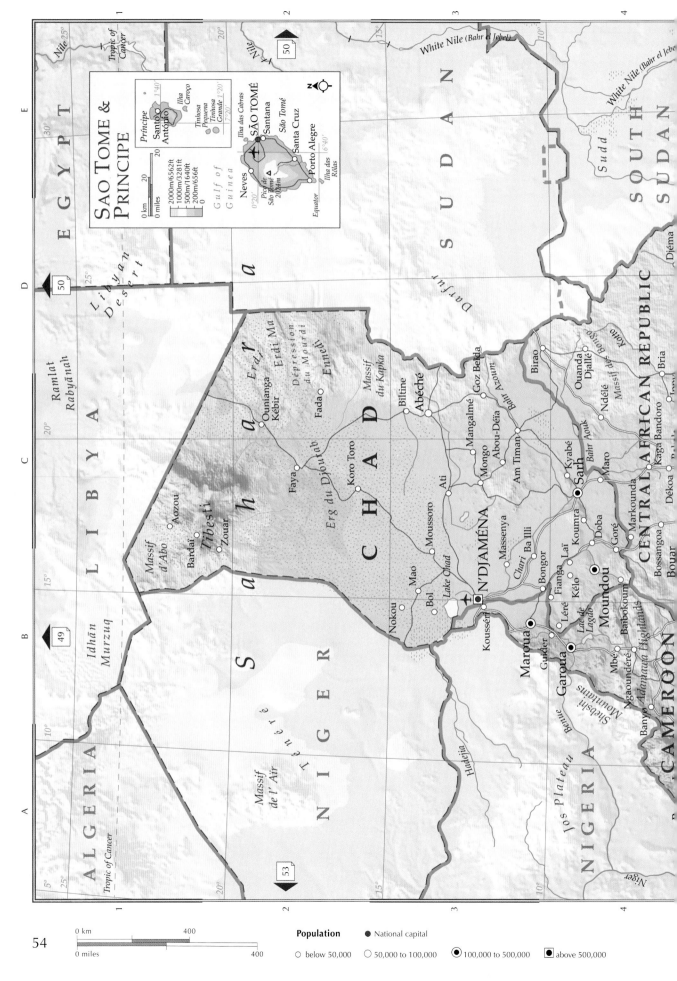

Sao Tome & Principe

Principe
Santo António
Ilha Caroço
Tinhosa Pequena
Tinhosa Grande

SÃO TOMÉ
Ilha das Cabras
Santana
São Tomé
Santa Cruz
Porto Alegre
Ilha das Rôlas

Neves
Pico de São Tomé 2024m

Gulf of Guinea

Equator

2000m/6562ft
1000m/3281ft
500m/1640ft
200m/656ft
0

EGYPT

Nile
Tropic of Cancer
Libyan Desert
Ramlat Rabyānah

ALGERIA
Tropic of Cancer

LIBYA
Idhān Murzuq

Massif d'Abo
Aozou
Bardaï
Tibesti
Zouar

NIGER
Ténéré
Massif de l'Aïr

Sahara

Erg du Djourab
Faya
Koro Toro
Ati
Moussoro
Mao
Bol
Nokou
Lake Chad

S U D A N
Darfur
White Nile (Bahr el Jebel)

SOUTH SUDAN
White Nile (Bahr el Jebel)
Sudd

Ennedi
Erdi Ma
Dépression du Mourdi
Erdi
Ounianga Kébir
Fada
Massif du Kapka
Biltine
Abéché

C H A D

Goz Beïda
Mangalmé
Abou-Déïa
Mongo
Am Timan
Massenya

Birao
Ouanda Djallé
Ndélé
Massif des Bongo
Bria
Bamingui Bangoran

CENTRAL AFRICAN REPUBLIC
Bossangoa
Bouar
Kaga Bandoro
Dékoa

Bahr Azoum
Kyabé
Sarh
Maro
Bahr Aouk
Koumra
Doba
Goré
Markounda

N'DJAMÉNA
Chari
Bongor
Fianga
Léré
Lac de Léré
Lai
Kélo
Moundou
Baïbokoum

Koussér
Maroua
Guider
Garoua
Mbé
Ngaoundéré

CAMEROON
Shebshi Mountains
Bénoué
Banyo Adamaua Highlands
Jos Plateau

NIGERIA
Hadejia
Niger

Djéma
Koto

Population

● National capital
○ below 50,000
◉ 50,000 to 100,000
◉ 100,000 to 500,000
■ above 500,000

54

0 km 400
0 miles 400

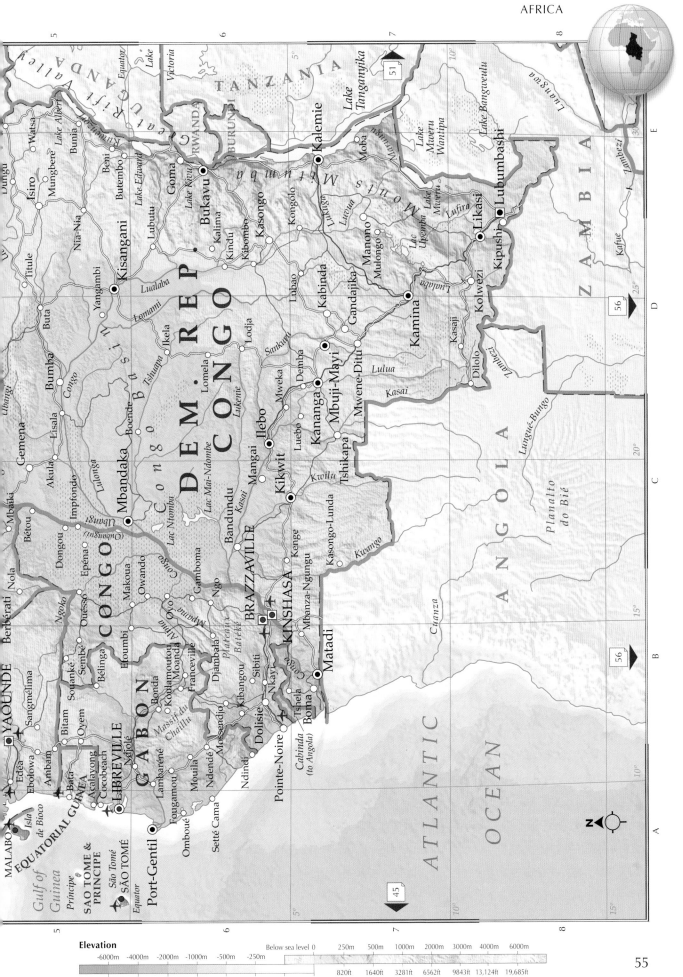

5

6

7

8

Great Rift Valley
Equator
Lake
Victoria
TANZANIA
Lake Tanganyika
51

RWANDA
BURUNDI
Kalemie
Moba
Lake Mweru Wantipa
Lake Bangweulu
Luangwa
E
30°

Watsa
Lake Albert
Ruwenzori
Bunia
Beni
Butembo
Lake Edward
Goma
Bukavu
Lake Kivu
Kongolo
Lukuga
Luvua
Lufira
Likasi
Lubumbashi
ZAMBIA
Kafue
Zambezi

Dungu
Isiro
Mungbere
Nia-Nia
Kisangani
Lubutu
Kalima
Kindu
Kibombo
Kasongo
Manono
Lac Upemba
Mulongo
Kipushi
Kolwezi
Lualaba
Mts.
Mitumba
D
25°
56

Titule
Yangambi
Lualaba
Lomami
Ikela
Lubao
Kabinda
Gandajika
Kamina
Kasaji
Dilolo
Zambezi

Buta
DEM. REP. CONGO
Tshuapa
Lomela
Lodja
Sankuru
Demba
Mwene-Ditu
Lulua
Kasai
Lungué-Bungo
Planalto do Bié
C
20°

Bumba
Lisala
Lukenie
Mweka
Mbuji-Mayi
Kananga
Tshikapa

Gemena
Akula
Lac Mai-Ndombe
Lac Ntomba
Lulonga
Congo Basin
Mbandaka
Boende
Bandundu
Mangai
Kasai
Ilebo
Luebo
Kikwit
Kwilu
Kasongo-Lunda
ANGOLA
15°

Mbaïki
Bétou
Impfondo
Dongou
Ubangi
(Oubangui)
Ubangi
Kenge
Mbanza-Ngungu
Kwango
Cuanza

Berbérati
Nola
Ngoko
Epéna
Ouésso
CONGO
Makoua
Owando
Gamboma
Ngo
Plateaux Batéké
BRAZZAVILLE
KINSHASA
Matadi

Eloumbi
Bélinga
Sembé
Souanké
Bonda
Koulamoutou
Moanda
Franceville
Djambala
Kibangou
Sibiti
Nkayi
Tshela
Boma
Congo

YAOUNDÉ
Edéa
Ebolowa
Ambam
Bitam
Oyem
Mpoua
Oyo
Alima
Mpama
Dolisie
Pointe-Noire
Cabinda (to Angola)

MALABO
EQUATORIAL GUINEA
Bata
Acalayong
Cocobeach
LIBREVILLE
Ndjolé
GABON
Lambaréné
Mouila
Ndendé
Mossendjo
Ndindi

Isla de Bioco
Gulf of Guinea
Principe
Fougamou
Massif du Chaillu

SAO TOME & PRINCIPE
São Tomé
SÃO TOMÉ
Equator
Port-Gentil
Omboué
Setté Cama

ATLANTIC
OCEAN
N
45

A
B
C
D
E

5
6
7
8

Elevation

Below sea level 0   250m   500m   1000m   2000m   3000m   4000m   6000m

-6000m   -4000m   -2000m   -1000m   -500m   -250m

-19,658ft -13,124ft -6562ft -3281ft -1640ft -820ft -328ft/-100m 0

820ft   1640ft   3281ft   6562ft   9843ft 13,124ft 19,685ft

55

Southern Africa

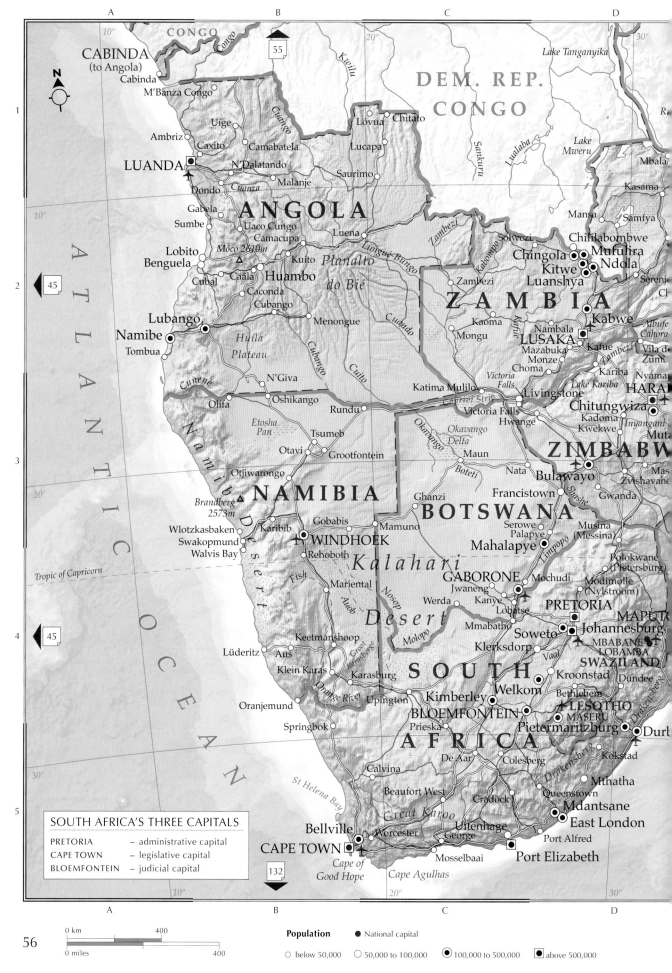

CONGO

CABINDA
(to Angola)
Cabinda
M'Banza Congo

DEM. REP.
CONGO

Lake Tanganyika

Uíge
Lóvua
Chitato
Ambriz
Caxito
Camabatela
Lucapa
LUANDA
N'Dalatando
Saurimo
Dondo
Malanje
Cuanza

Mbala

Kasama

Mansa
Samfya

Gabela

Sumbe
Uaco Cungo
Camacupa
Luena
Solwezi
Chililabombwe
Chingola
Mufulira
Kitwe
Ndola
Luanshya

ANGOLA

Lobito
Benguela
Cubal
Caála
Huambo
Kuito
Planalto
do Bié
Zambezi

Môco 2620m
Caconda
Cubango

ZAMBIA
Kaoma
Kabwe
Lubango
Menongue
Mongu
Nambala
LUSAKA
Namibe
Tombua
Huíla
Plateau
Cubango
Cuito
Mazabuka
Kafue
Monze
Choma
Kariba

N'Giva
Katima Mulilo
Victoria
Falls
Livingstone
HARA
Olifa
Oshikango
Rundu
Caprivi Strip
Victoria Falls
Hwange
Chitungwiza
Kadoma
Kwekwe
Etosha
Pan
Okavango
Delta
Tsumeb
Okavango
ZIMBABW
Otavi
Grootfontein
Maun
Nata
Bulawayo
Otjiwarongo
Boteti
Francistown
Gwanda

NAMIBIA
Ghanzi
BOTSWANA
Serowe
Palapye
Musina
(Messina)
Mahalapye
Limpopo
Polokwane
(Pietersburg)

Brandberg
2573m
Gobabis
Mamuno
Karibib
Wlotzkasbaken
Swakopmund
Walvis Bay
WINDHOEK
Rehoboth
Kalahari
GABORONE
Mochudi
Modimolle
(Nylstroom)

Tropic of Capricorn
Fish
Mariental
Desert
Jwaneng
Werda
Kanye
Lobatse
PRETORIA
MAPUT
Auob
Nosop
Mmabatho
Johannesburg
Lüderitz
Aus
Keetmanshoop
Molopo
Soweto
MBABANE
LOBAMBA
SWAZILAND
Klein Karas
Karasburg
SOUTH
Klerksdorp
Vaal
Kroonstad
Dundee
Oranjemund
Upington
Kimberley
Welkom
Bethlehem
LESOTHO
Orange River
Prieska
BLOEMFONTEIN
MASERU
Pietermaritzburg
Durb
Springbok
AFRICA
Kokstad
De Aar
Colesberg
Calvina
Mthatha
Queenstown
Beaufort West
Cradock
Mdantsane
St Helena Bay
Great Karoo
Uitenhage
East London
Bellville
Worcester
George
Port Alfred
CAPE TOWN
Mosselbaai
Port Elizabeth
Cape of
Good Hope
Cape Agulhas

SOUTH AFRICA'S THREE CAPITALS

PRETORIA — administrative capital
CAPE TOWN — legislative capital
BLOEMFONTEIN — judicial capital

0 km 400
0 miles 400

Population • National capital

○ below 50,000 ○ 50,000 to 100,000 ◉ 100,000 to 500,000 ◼ above 500,000

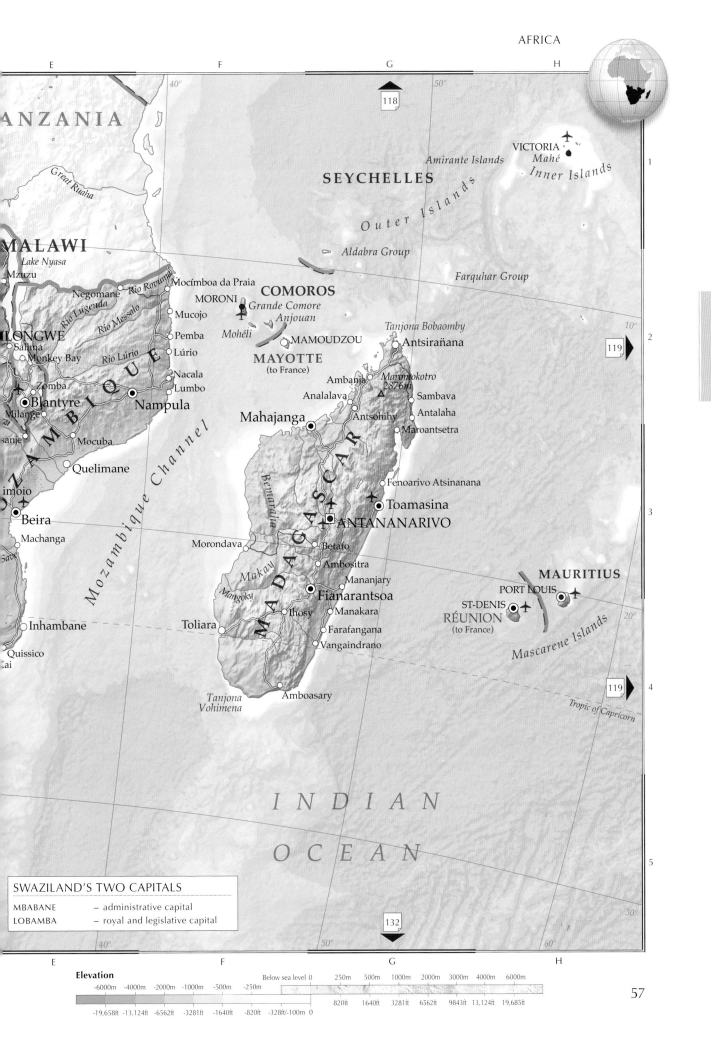

E F G H

118

SEYCHELLES

Amirante Islands
VICTORIA
Mahé
Inner Islands

Outer Islands

Aldabra Group

Farquhar Group

1

ANZANIA

Great Ruaha

MALAWI
Lake Nyasa
Mzuzu

Negomane
Rio Rovuma
Mocímboa da Praia
Rio Lugenda
Mucojo
Rio Messalo
MORONI
Grande Comore
COMOROS
Anjouan

10°

LONGWE
Rio Lúrio
Pemba
Mohéli
MAMOUDZOU
Tanjona Bobaomby

119

Salima
Lúrio
MAYOTTE
(to France)
Antsirañana

Monkey Bay
Nacala
Ambanja
Maromokotro 2876m
Zomba
Lumbo
Analalava
Antsohihy
Sambava
Milange
Blantyre
Nampula
Mahajanga
Antalaha
2
sanje
Mocuba
Maroantsetra

Quelimane

3

imoio
Machanga
Fenoarivo Atsinanana

Beira
Morondava
Toamasina
Betafo
ANTANANARIVO

Save
Ambositra
Mananjary
MAURITIUS
PORT LOUIS
ST-DENIS
Inhambane
Toliara
Ihosy
Manakara
Fianarantsoa
RÉUNION
(to France)
20°
Quissico
Farafangana
Vangaindrano
Mascarene Islands
ai

Tanjona Vohimena
Amboasary
119
4

Tropic of Capricorn

I N D I A N

O C E A N
5

30°

SWAZILAND'S TWO CAPITALS

MBABANE   – administrative capital
LOBAMBA   – royal and legislative capital

E F G H

Elevation

-6000m -4000m -2000m -1000m -500m -250m | Below sea level 0 | 250m 500m 1000m 2000m 3000m 4000m 6000m

-19,658ft -13,124ft -6562ft -3281ft -1640ft -820ft -328ft/-100m 0 | | 820ft 1640ft 3281ft 6562ft 9843ft 13,124ft 19,685ft

57

Europe

Political features

Total area:
4,809,200 sq miles
(12,456,000 sq km)

Total number of countries:
44

Total population:
724 million

Largest city with population:
Moscow, European Russia 17.1 million

Country with highest population density:
Monaco 50,667 people per sq mile
(19,487 people per sq km)

Largest country:
European Russia 1,527,341 sq miles
(3,955,818 sq km)

Smallest country:
Vatican City, Italy 0.17 sq miles
(0.44 sq km)

Physical features

Largest lake:
Lake Lagoda, European Russia
7100 sq miles (18,390 sq km)

Longest river:
Volga, European Russia
2290 miles (3688 km)

Highest point:
El'brus, Caucasus, European Russia
18,510ft (5642 m)

Lowest point:
Volga Delta, Caspian Sea, European
Russia -92ft (-28m) below sea level

Limit of winter pack ice

Reykjanes Ridge

REYKJAVÍK
ICELAND
Vatnajökull
Arctic Circle

Iceland Basin

Faroe-Iceland Ridge

FAROE ISLANDS
(to Denmark)

Norwegian Basin

Norwegian Sea

Hatton Ridge

Faroe-Shetland Trough

Shetland Islands

Trondheim

Rockall Bank

Outer Hebrides

Orkney Islands

Bergen

Rockall Trough

British Isles

Glasgow
Edinburgh

OSLO

Stavanger

North Sea

Gothenburg
Aalborg
Jönk

Ireland
Belfast
Isle of Man

UNITED KINGDOM

IRELAND
DUBLIN
Liverpool
Manchester

DENMARK
Odense
Jutland
COPENH
Mal

Porcupine Plain

Celtic Sea

Birmingham

Cardiff
LONDON

Britain

NETHERLANDS
THE
HAGUE
AMSTERDAM
Rotterdam

Hamburg

N

Celtic Shelf

English Channel

Channel Islands

le Havre

BELGIUM
BRUSSELS
Liège

Hanover

Düsseldorf
BERLIN

ATLANTIC

Azores-Biscay Rise

Charcot Seamounts

Biscay Plain

Rennes

Seine

PARIS

Nantes

Loire

Orléans

LUXEMBOURG
LUXEMBOURG

Bonn

Frankfurt
am Main

GERMANY

Wro

PRAG

OCEAN

Iberian Plain

Bay of Biscay

A Coruña

Bordeaux

FRANCE

Garonne

Strasbourg

Stuttgart

Zurich

BERN
SWITZERLAND

Lyon
Massif
Central

Rhône

Munich

CZECH REPU
(CZECHIA

BRAT

VIENNA
Salzburg

AUSTRI

Galicia Bank

Bilbao

Cordillera Cantábrica

Porto

Mont Blanc
15,774ft
(4807m)

Innsbruck

LIECH

SLOVEN

PORTUGAL

Iberian

Zaragoza

Duero

Pyrenees

Toulouse

Nice

Milan

Turin

Venice

LJUBLJANA

Trieste
CRO

Tagus
Plain

LISBON

Tagus

MADRID

Ebro

ANDORRA

Marseille

MONACO

Pisa

Bologna

SAN
MARINO

Adriatic S

Horseshoe Seamounts

SPAIN

Peninsula

Guadalquivir

Barcelona

Valencia

ITALY

Apennines

SAR
Ma

VATICAN CITY
ROME

Madeira
(to Portugal)

Seville

Málaga

Palma

Balearic Islands

Corsica

Sardinia

Naples

Bari

GIBRALTAR
(to UK)
Ceuta
(to Spain)

Melilla
(to Spain)

Algerian Basin

Tyrrhenian Sea

Cagliari

Cosenza

Canary Islands
(to Spain)

Mediter

Atlas Mountains

AFRICA

Mount Etna
10,922ft
(3340m)

Sicily

Palermo

Catania

MALTA
VALLETTA

Strait of Gibraltar

133

44

44

46

0 km   500

0 miles   500

Population   ● National capital

○ below 50,000   ○ 50,000 to 100,000   ◉ 100,000 to 500,000   ■ above 500,000

Barents Sea

North Cape

Ostrov Kolguyev

Arctic Circle

Ob'

Irtysh

1

● Murmansk

Kola
Peninsula

White
Sea

● Archangel

Northern Dvina

R U S S I A

Perm' ●

133

● Tampere

Lake Onega

Lake Ladoga

90

2

Turku ●
● HELSINKI

● Saint Petersburg

● Vologda

Ufa ●

TALLINN

Kazan' ●

CKHOLM

ESTONIA

Yaroslavl' ●

Nizhniy
Novgorod ●

Ul'yanovsk ●

Orenburg ●

LATVIA

MOSCOW ●

Samara ●

Ural

RIGA ●

LITHUANIA

Vitsyebsk /
Vitebsk ●

Central
Russian
Upland

Volga Uplands

Syr Darya

3

grad ●

Kaunas ●

VILNIUS ●

● MINSK

KALININGRAD
(to Russia)

Babruysk /
Bobruysk ●

Homyel' /
Gomel' ●

● Voronezh

Don

Volga

Ural

Aral Sea

szcz

WARSAW ●

BELARUS

Brest ●

Pripet
Marshes

Dnieper Lowlands

Kharkiv ●

Volgograd ●

Amu Darya

LAND

Brig

● KIEV

Dnieper

Kraków ●

● L'viv

UKRAINE

Dniester

Dnipro
(Dnipropetrovs'k) ●

Donets'k ●

Rostov-na-Donu ●

Astrakhan' ●

Volga Delta
98ft (-28m)
▽

60°

VAKIA

Chernivtsi ●

MOLDOVA

Ural

40°

APEST

CHIŞINĂU ●

Sea of
Azov

Stavropol' ●

Caspian Sea

90

4

GARY

Cluj-Napoca ●

Odesa ●

Crimea

C a u c a s u s

ROMANIA

Braşov ●

Simferopol' ●

El'brus 18,510ft
(5642m) △

BELGRADE

(since 2014 the Ukrainian
territory of Crimea has been
annexed by Russia)

BUCHAREST ●

Danube

Black
Sea

RBIA

OSOVO
(disputed)

BULGARIA

● Varna

Balkan Mountains

RISTINA

SOFIA ●

● Burgas

GRICA

SKOPJE

TURKEY

MACED

ANA

ANIA

Aegean
Sea

Anatolia

Z ā g r o s   M o u n t a i n s

30°

GREECE

ATHENS ●

Piraeus ●

5

Peloponnese

Cyprus

Tigris

Euphrates

50°

96

40°

e a

Irákleio ●

Crete

The North Atlantic

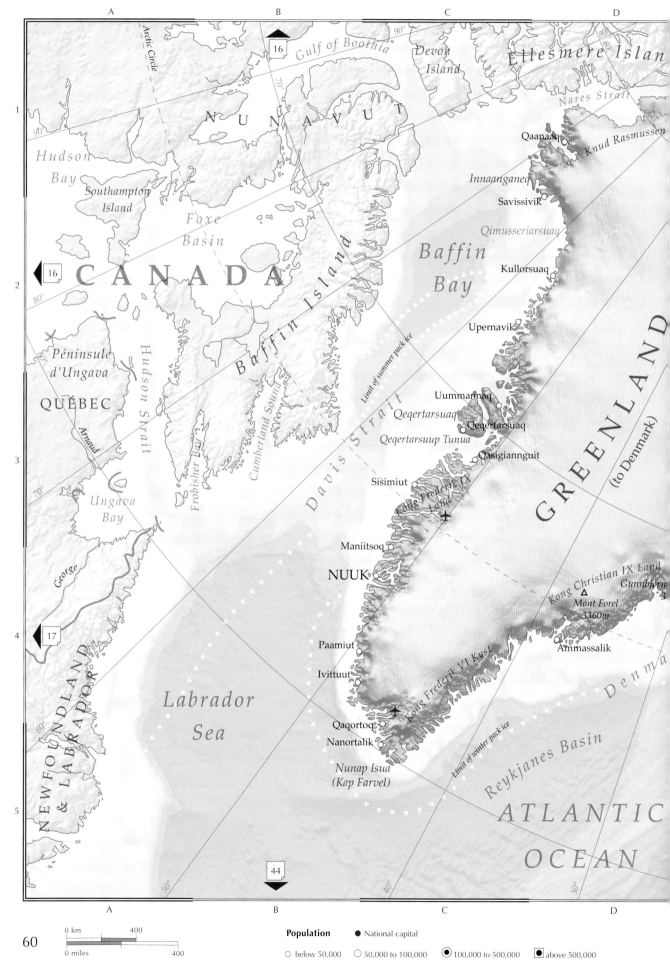

A   B   C   D

Arctic Circle

16

Gulf of Boothia

Devon Island

Ellesmere Islan

Nares Strait

N U N A V U T

Hudson Bay

Southampton Island

Foxe Basin

16

C A N A D A

Qaanaaq

Knud Rasmussen

Innaanganeq

Savissivik

Qimusseriarsuaq

Baffin Island

Baffin Bay

Kullorsuaq

Upernavik

Limit of summer pack ice

Péninsule d'Ungava

QUÉBEC

Hudson Strait

Cumberland Sound

Uummannaq

Qeqertarsuaq

Qeqertarsuup Tunua

Qeqertarsuaq

Qasigiannguit

G R E E N L A N D

(to Denmark)

Arnaud

Davis Strait

Frobisher Bay

Sisimiut

Kong Frederik IX Land

Ungava Bay

Maniitsoq

George

NUUK

Kong Christian IX Land

Gunnbjørn

Mont Forel 3360m

17

N E W F O U N D L A N D   &   L A B R A D O R

Paamiut

Ivittuut

Kong Frederik VI Kyst

Ammassalik

Denma

Labrador Sea

Qaqortoq

Nanortalik

Reykjanes Basin

Nunap Isua (Kap Farvel)

Limit of winter pack ice

A T L A N T I C

44

O C E A N

A   B   C   D

0 km   400
0 miles   400

Population   ● National capital

○ below 50,000   ○ 50,000 to 100,000   ◉ 100,000 to 500,000   ◼ above 500,000

ARCTIC
OCEAN

Lincoln Sea

Kap Morris Jesup

Wandel Sea

SVALBARD
(to Norway)

Independence Fjord

Nord

Kong Frederik VIII Land

Zemlya Frantsa-Iosifa

Kvitøya

Nordaustlandet

Kong Karls Land

Spitsbergen

Barentsøya

Edgeøya

LONGYEARBYEN
Barentsburg

Novaya Zemlya

Barents Sea

88

Storfjorden

Limit of winter pack ice

Greenland Sea

Bjørnøya
(to Norway)

Kong Christian X Land

Petermann Bjerg
2940m

Daneborg

Limit of summer pack ice

Nordkapp
(North Cape)

FINLAND

Kong Oscar Fjord

Mohns Ridge

Ittoqqortoormiit

Kangertittivaq

Kangikajik

JAN MAYEN
(to Norway)

*Norwegian
Sea*

Norwegian Basin

Vestfjorden

Arctic Circle

62

rait

ICELAND

Bolungarvík
Siglufjörður   Raufarhöfn
fjörður
Húsavík
Akureyri
Stykkishólmur   Seyðisfjörður
REYKJAVÍK   Neskaupstaður
Selfoss   Vatnajökull   Djúpivogur
orlákshöfn   Hvannadalshnúkur
urtsey   Vestmannaeyjar   2119m

SWEDEN

Gulf
of
Bothnia

FAROE ISLANDS
(to Denmark)

N

TÓRSHAVN

NORWAY

Shetland
Islands

63

E   F   G   H

Elevation

| Below sea level 0 | 250m | 500m | 1000m | 2000m | 3000m | 4000m | 6000m |

-6000m   -4000m   -2000m   -1000m   -500m   -250m

820ft   1640ft   3281ft   6562ft   9843ft   13,124ft   19,685ft

-19,658ft   -13,124ft   -6562ft   -3281ft   -1640ft   -820ft   -328ft/-100m   0

Scandinavia & Finland

RUSSIA

FINLAND

Barents Sea

ARCTIC OCEAN

Norwegian Sea

Kirkenes
Varangerhalvøya
Varangerfjorden
Tana Bru
Deatnu
Nordkapp
(North Cape)
Mageroya
Porsangerhalvøya
Porsangerfjorden
Soroya
Talvik
Alta
Lakselv
Kvaloya
Ringvassoya
Senja
Andøya
Tromsø
Kvaloya
Vesterålen
Lofoten
Vestfjorden
Sortfjorden
Harstad
Hinnøya
Narvik
Bodø
Fauske
Mo i Rana
Mosjøen
Vega
Namsos
Steinkjer
Frøya
Foldereid

Finnmarksvidda
Kautokeino
Karasjok
Karigasniemi
Váljohka
Inari
Inarijärvi
Kaamanen
Ivalo
Saariselkä
Sodankylä
Kittilä
Sattanen
Ounasjoki
Muonio
Muonionjoki
Kolari
Tornionjoki
Visttasjohka
Kebnekaise
2117m
Kiruna
Torneträsk
Gällivare
Malmberget
Skalka
Jokkmokk
Arvidsjaur
Skellefteälven
Storuman
Vilhelmina
Dorotea
Kaprberga
vattnet
Börgefjell

Lapland

Kaaresuvanto

Kuusamo
Pudasjärvi
Kemijärvi
Rovaniemi
Tjolmoy
Kemi
Tornio
Haparanda
Kalix
Luleå
Luleälven
Boden
Piteå
Skellefteå
Storuman
Lycksele
Ångerman
Umeälven

Suomussalmi
Kuhmo
Sotkamo
Kajaani (Kajana)
Oulujärvi
Iijoki
Oulujoki
Oulu (Uleåborg)
Hailuoto
Kempele
Raahe
Kokkola
(Karleby)
Kivalo

Population
- ● National capital
- ○ below 50,000
- ○ 50,000 to 100,000
- ◉ 100,000 to 500,000
- ▣ above 500,000

0 km 200
0 miles 200

RUSSIA

BELARUS

ESTONIA

LATVIA

LITHUANIA

Western Dvina

Lake Peipus

Gulf of Finland

Gulf of Finland

Gulf of Riga

Hiiumaa

Saaremaa

KALININGRAD
(to Russia)

Neman

Courland Lagoon

Gulf of Danzig

76

Wisła

P O L A N D

Oder

G E R M A N Y

Elbe

Weser

Ems

72

Ladozhskoye
Ozero

Varkaus
Hankasi

Äänekoski
Saimaa
Imatra
Lappeenranta
(Villmanstrand)

Seinäjoki
Keuruu
Jyväskylä
Niisijärvi
Piihlava
Joutseno
Lahti (Lahtis)
Kouvola
Kotka

Tampere (Tammerfors)
Nokia
HELSINKI
Hyvinkää
Porvoo (Borgå)
Kirkkonummi
Vantaa
Espoo
Salo

Hämeenlinna
(Tavastehus)
Salo
Hanko
(Hangö)

(Vasa)
Närpes (Närpiö)
Pori
(Björneborg)
Rauma
Turku
(Åbo)

Kankaanpää

Åland
Ålands Hav

STOCKHOLM
Uppsala
Norrtälje
Täby

Härnösand
Sundsvall
Hudiksvall
Söderhamn
Gävle
Sandviken
Tierp
Sala

S W E D E N

Kramfors
Timrå
Ange
Liusnan
Ljusdal
Bollnäs
Rättvik
Leksand
Falun
Avesta
Sala
Sollentuna
Mölnlycke
Södertälje
Nyköping

Norrköping
Linköping

Svenstavik
Rätan
Sveg
Idre
Mora
Malung
Borlänge
Ludvika
Nora
Askersund
Motala
Hjälmaren
Hjälmaren
Mjölby

Gotland
Visby

Borgholm
Öland
Kalmar
Karlskrona

Jönköping

Oskarshamn
Växjö

Åndalsnes
Dombås
Ringebu
Gol
Gjøvik
Hamar
Mjøsa
Lillehammer
Hamar

N O R W A Y

Roros
Glåma
Eidfjord
Geilo
Hønefoss
Horten
Kongsberg
Porsgrunn
Arendal
Setesdal
Evje
Moi
Lyngdal
Liknes
Kristiansand

Jotunheimen
Galdhøpiggen
2472m

Lillestrøm
OSLO
Ski
Moss
Sarpsborg
Halden
Mellerud
Åmål
Lidköping
Grums

Sandvika
Drammen
Fredrikstad
Strömstad
Uddevalla
Trollhättan

Karlstad
Filipstad
Säffle
Vänern
Vättern

Örebro
Västerås

Göteborg
(Gothenburg)
Borås
Mölndal
Kungsbacka
Varberg
Halmstad
Laholm
Ljungby
Helsingborg
Lund
Malmö
Landskrona
Kristianstad
Hanöbukten

Ronne
Bornholm

DENMARK
KØBENHAVN
Copenhagen

Hjørring
Aalborg
Holstebro
Viborg
Randers
Aarhus
Hobro
Ding Skoven
173m
Jylland

Ringkøbing Fjord
Varde
Esbjerg
Rømø
Kolding
Odense
Fyn
Slagelse
Nykøbing
Korsør
Sjælland
Storebælt
Møn
Falster
Lolland

Ålesund
Hardangervidda
Hauges
Haugesund
Skudnes
Stavanger
Sandnes
Bergen

*Hardanger-
fjorden*

K a t t e g a t
Læsø
Fyn
Aarhus

S k a g e r r a k

N o r t h
S e a

B a l t i c   S e a

67

63

Elevation

Below sea level 0	250m	500m	1000m	2000m	3000m	4000m	6000m

-6000m -4000m -2000m -1000m -500m -250m

-19,658ft -13,124ft -6562ft -3281ft -1640ft -820ft -328ft/-100m 0

820ft 1640ft 3281ft 6562ft 9843ft 13,124ft 19,685ft

The Low Countries

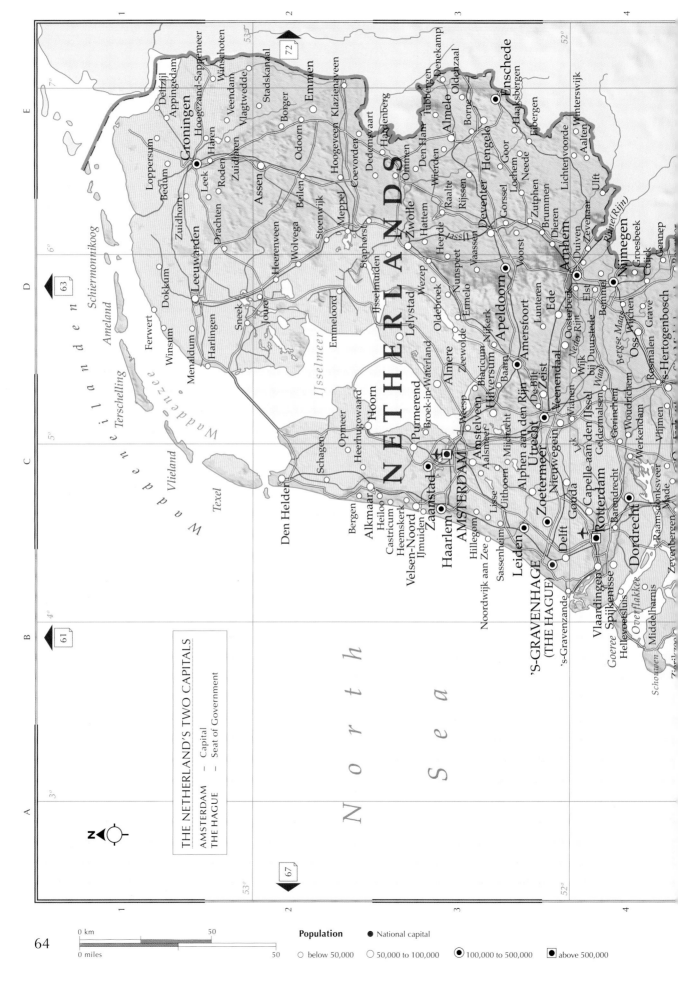

THE NETHERLAND'S TWO CAPITALS

AMSTERDAM – Capital
THE HAGUE – Seat of Government

NETHERLANDS

Population

- ● National capital
- ○ below 50,000
- ○ 50,000 to 100,000
- ◉ 100,000 to 500,000
- ▣ above 500,000

0 km   50
0 miles   50

Elevation

	Below sea level 0	250m	500m	1000m	2000m	3000m	4000m	6000m
-6000m -4000m -2000m -1000m -500m -250m								
-19,658ft -13,124ft -6562ft -3281ft -1640ft -820ft -328ft/-100m 0		820ft	1640ft	3281ft	6562ft	9843ft	13,124ft	19,685ft

The British Isles

North Sea

ATLANTIC

OCEAN

SCOTLAND

Unst
Yell
Fetlar
Mainland
Lerwick
Shetland Islands

Fair Isle

Sanday
Kirkwall
Mainland
Hoy
Orkney Islands
John o'Groats

Thurso
Ben Hope
927m △
Ullapool

Moray Firth
Elgin
Fraserburgh
Peterhead
Aberdeen

Dee
Inverness
Loch Ness
Aviemore
Grampian Mountains
Spey

Montrose
Arbroath
St Andrews
Dundee
Forfar
Perth
Tay
Firth of Forth
Dunfermline
Stirling
Edinburgh
Galashiels
Hawick
Berwick-upon-Tweed
Newcastle upon Tyne

North West Highlands
Ben Nevis
1343m △
Fort William
Loch
Lomond
Forth

Glasgow
Paisley
Greenock
Hamilton
East Kilbride
Clyde
Kilmarnock
Prestwick
Ayr

Isle of Lewis
Stornoway
Harris
North Uist
South Uist
Barra
Outer Hebrides
St Kilda

Mallaig
Isle of Skye
Stromeferry
Rhum
Eigg
Coll
Tiree
Isle of Mull
Oban
Firth of Lorn
Jura
Islay
Kintyre
Isle of Arran
Inner Hebrides

The Minch
The Little Minch

66

0 km   100
0 miles   100

Population
● National capital
● Internal administrative capital
○ below 50,000
○ 50,000 to 100,000
◉ 100,000 to 500,000
▣ above 500,000

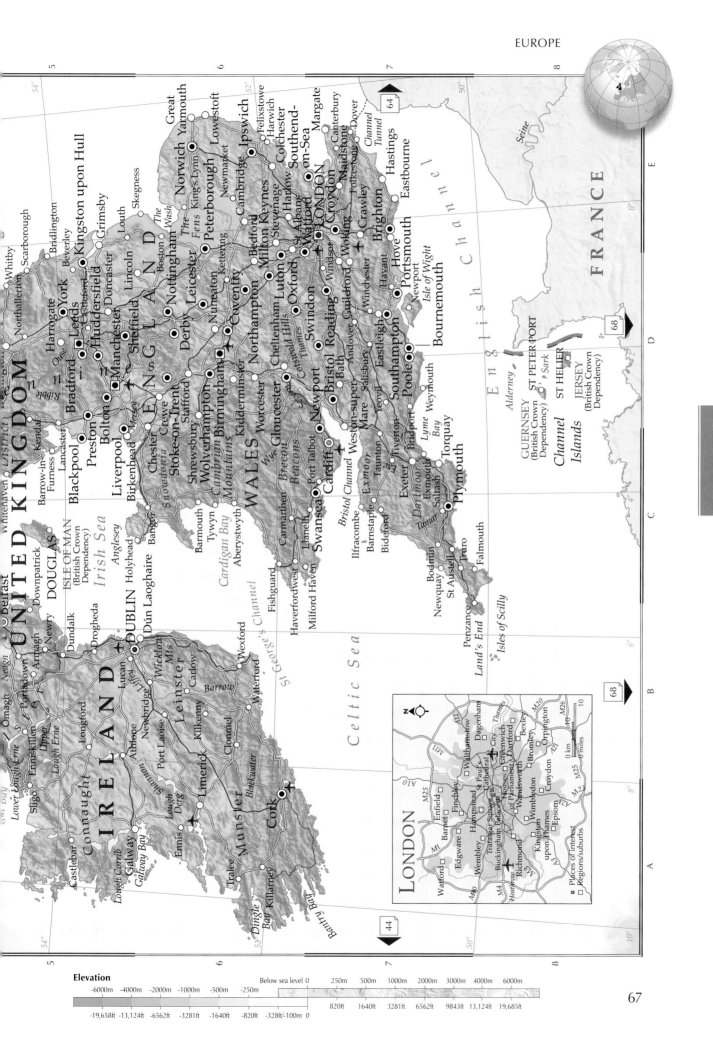

FRANCE

UNITED KINGDOM

ENGLAND

WALES

IRELAND

Munster

Leinster

Connaught

LONDON

English Channel

Celtic Sea

Irish Sea

Cardigan Bay

ISLE OF MAN
(British Crown Dependency)
DOUGLAS

GUERNSEY
(British Crown Dependency)
ST PETER PORT

JERSEY
(British Crown Dependency)
ST HELIER

Channel Islands

St George's Channel

Great Yarmouth, Lowestoft, Ipswich, Felixstowe, Harwich, Colchester, Southend-on-Sea, Margate, Canterbury, Dover, Folkestone, Hastings, Eastbourne, Maidstone, Crawley, Brighton, Hove, Portsmouth, Isle of Wight, Newport, Bournemouth, Poole, Weymouth, Torquay, Plymouth, Falmouth, Truro, Penzance, Land's End, Isles of Scilly, Newquay, Bodmin, St Austell, Bideford, Barnstaple, Ilfracombe, Exeter, Exmouth, Saltash, Tamar, Dartmoor, Exmoor, Taunton, Bridport, Lyme Bay, Yeovil, Bath, Bristol, Reading, Swindon, Oxford, Luton, Milton Keynes, Bedford, Northampton, Coventry, Birmingham, Wolverhampton, Kidderminster, Worcester, Gloucester, Cheltenham, Newport, Cardiff, Swansea, Port Talbot, Llanelli, Carmarthen, Barmouth, Tywyn, Aberystwyth, Fishguard, Haverfordwest, Milford Haven, Bangor, Holyhead, Anglesey, Chester, Crewe, Stoke-on-Trent, Stafford, Shrewsbury, Derby, Nottingham, Leicester, Kettering, Peterborough, Cambridge, Newmarket, Norwich, King's Lynn, Boston, Skegness, Grimsby, Lincoln, Doncaster, Sheffield, Nottingham, Huddersfield, Bradford, Leeds, York, Harrogate, Castleford, Kingston upon Hull, Beverley, Bridlington, Scarborough, Whitby, Northallerton, Kendal, Lancaster, Barrow-in-Furness, Blackpool, Preston, Bolton, Manchester, Liverpool, Birkenhead

DUBLIN, Dún Laoghaire, Lucan, Drogheda, Dundalk, Newry, Armagh, Portadown, Omagh, Enniskillen, Longford, Athlone, Castlebar, Sligo, Galway, Ennis, Limerick, Tralee, Dingle Bay, Killarney, Bantry Bay, Cork, Clonmel, Kilkenny, Carlow, Port Laoise, Newbridge, Wexford, Waterford, Belfast, Downpatrick, Holywood

Cotswold Hills, *Brecon Beacons*, *Cambrian Mountains*, *Snowdonia*, *The Wash*, *The Fens*, *Wicklow Mts*, *Shannon*, *Lough Derg*, *Lough Corrib*, *Lough Ree*, *Blackwater*, *Barrow*, *Wye*, *Usk*, *Severn*, *Thames*, *Ribble*, *Ouse*, *Mersey*, *Trent*, *Tamar*, *Exe*, *Upper Lough Erne*, *Lower Lough Erne*, *Lough Neagh*, *Lough*, *Seine*

Windsor, Guildford, Woking, Watford, St Albans, Stevenage, Harlow, Croydon, Havant, Winchester, Eastleigh, Salisbury, Andover, Weston-super-Mare, Bristol Channel

Channel Tunnel

LONDON inset:
N, M25, M1, M11, A10, A12, M20, M26, M25, M2, A2, A20, M23, A3, M3, M4, M40, A40, A1, A23
Watford, Enfield, Barnet, Finchley, Edgware, Wembley, Hampstead, Walthamstow, Dagenham, Bexley, Greenwich, Dartford, Bromley, Orpington, Croydon, Epsom, Wimbledon, Wandsworth, Richmond, Kingston upon Thames, Heathrow, Thames, City, St Paul's Cathedral, Trafalgar Square, Houses of Parliament, Buckingham Palace
0 km 10 / 0 miles 10
□ Places of interest
□ Regions/suburbs

Elevation

| -6000m | -4000m | -2000m | -1000m | -500m | -250m | | Below sea level 0 | 250m | 500m | 1000m | 2000m | 3000m | 4000m | 6000m |

| -19,658ft | -13,124ft | -6562ft | -3281ft | -1640ft | -820ft | -328ft/-100m | 0 | | 820ft | 1640ft | 3281ft | 6562ft | 9843ft | 13,124ft | 19,685ft |

France, Andorra & Monaco

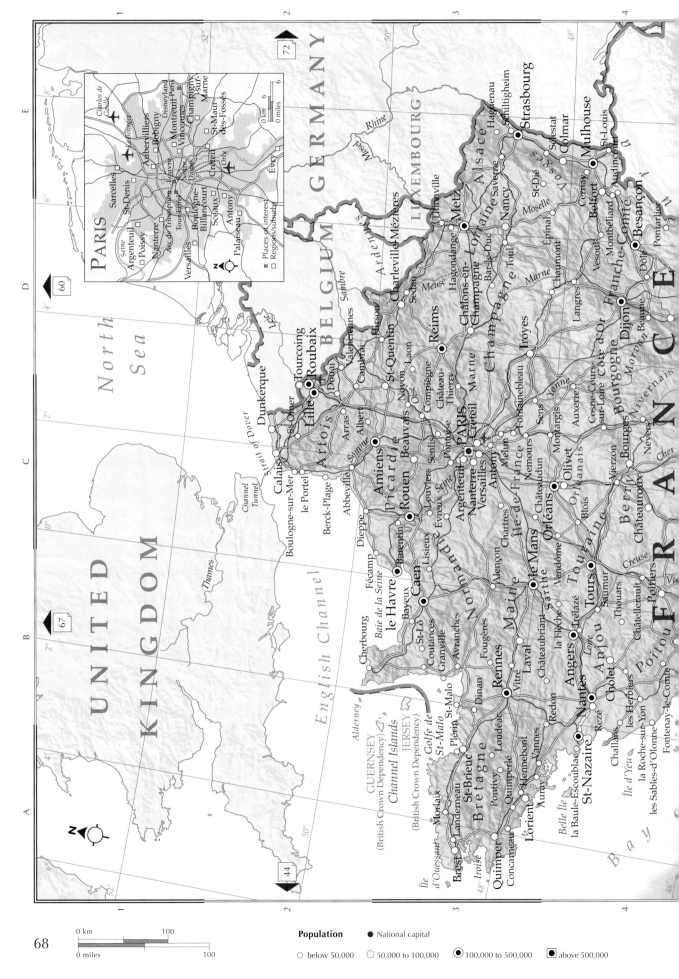

Population • National capital

○ below 50,000 ○ 50,000 to 100,000 ◉ 100,000 to 500,000 ▣ above 500,000

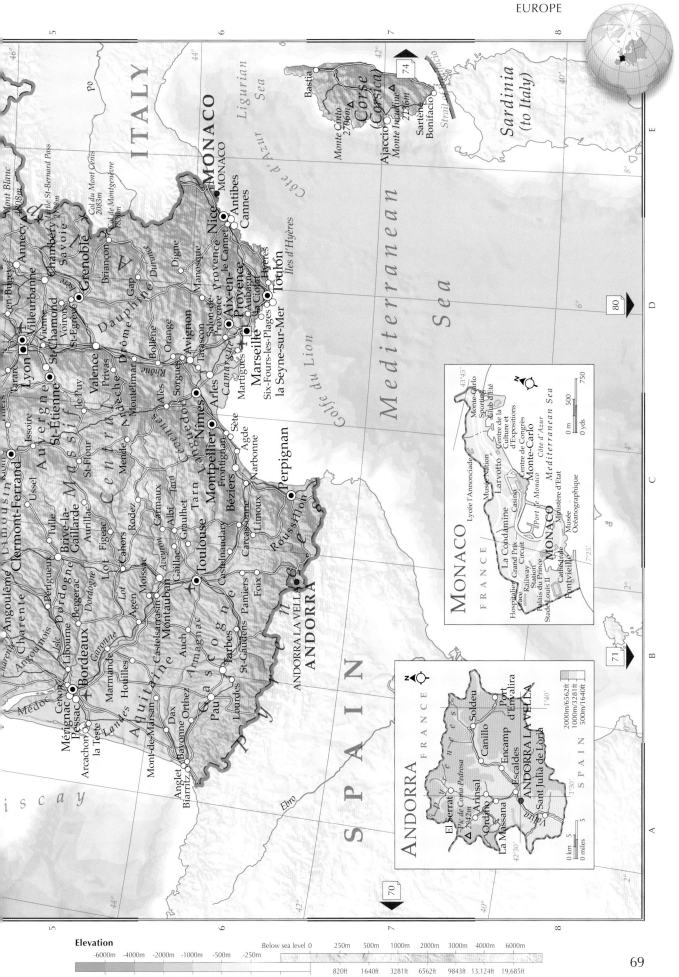

EUROPE

Mont Blanc 4808m

Little St-Bernard Pass 2188m

Col du Mont Cenis 2083m

Col de Montgenèvre 1850m

ITALY

Po

Annecy

Savoie

Chambéry

Grenoble

Voiron

St-Egreve

en-Bugey

Villeurbanne

Vienne

St-Chamond

Lyon

Tarare

Issoire

Ussel

Brive-la-Gaillarde

Périgueux

Tulle

Bergerac

Libourne

Bordeaux

Cenon

Mérignac Pessac

Médoc

Arcachon

la Teste

Landes

Mont-de-Marsan

Dax

Anglet

Bayonne

Biarritz

Orthez

Marmande

Agen

Moissac

Montauban

Castelsarrasin

Auch

St-Gaudens

Tarbes

Pau

Lourdes

Toulouse

Gaillac

Carmaux

Albi

Rodez

Figeac

Cahors

Aurillac

St-Flour

Mende

Pamiers

Foix

Castelnaudary

Carcassonne

Limoux

Béziers

Narbonne

Agde

Sète

Montpellier

Frontignan

Nîmes

Arles

Avignon

Sorgues

Orange

Bollène

Montélimar

Privas

Valence

le Puy

St-Étienne

Gap

Digne

Durance

Briançon

Manosque

Salon-de-Provence

Aix-en-Provence

Martigues

Marseille

Six-Fours-les-Plages

la Seyne-sur-Mer

Toulon

Hyères

Îles d'Hyères

la Ciotat

Aubagne

Tarascon

ANDORRA LA VELLA

ANDORRA

Perpignan

ANTIBES

Nice

Cannes

MONACO

MONACO

Côte d'Azur

Ligurian Sea

Mediterranean Sea

Golfe du Lion

Bastia

Monte Cinto 2706m

Corse (Corsica)

Monte Incudine 2136m

Sartène

Ajaccio

Bonifacio

Strait of Bonifacio

Sardinia (to Italy)

Ebro

Biscay

SPAIN

ITALY

MONACO

MONACO

FRANCE

Lycée l'Annonciade

Musée Nation

Larvotto

La Condamine

Hospital Grand Prix Circuit

Grace

Palais du Prince

Stade Louis II

Monte-Carlo Sporting Club d'Été

Centre de la Culture et d'Expositions

Centre de Congrès

Casino

Port de Monaco

MONACO

Cathédrale

Pontvieille

Ministère d'État

Musée Océanographique

Côte d'Azur

Mediterranean Sea

0 m   500   750

0 yds

ANDORRA

FRANCE

Soldeu

El Serrat

Pic de Coma Pedrosa 2942m

Ordino

Arinsal

La Massana

Canillo

Encamp

Escaldes

ANDORRA LA VELLA

Port d'Envalira

Sant Julia de Loria

SPAIN

Pyrénées

Valira

2000m/6562ft
1000m/3281ft
500m/1640ft

0 km   5
0 miles   5

N

71

70

74

80

Elevation

| Below sea level 0 | 250m | 500m | 1000m | 2000m | 3000m | 4000m | 6000m |

-6000m -4000m -2000m -1000m -500m -250m

-19,658ft -13,124ft -6562ft -3281ft -1640ft -820ft -328ft/-100m 0

820ft   1640ft   3281ft   6562ft   9843ft 13,124ft 19,685ft

Spain & Portugal

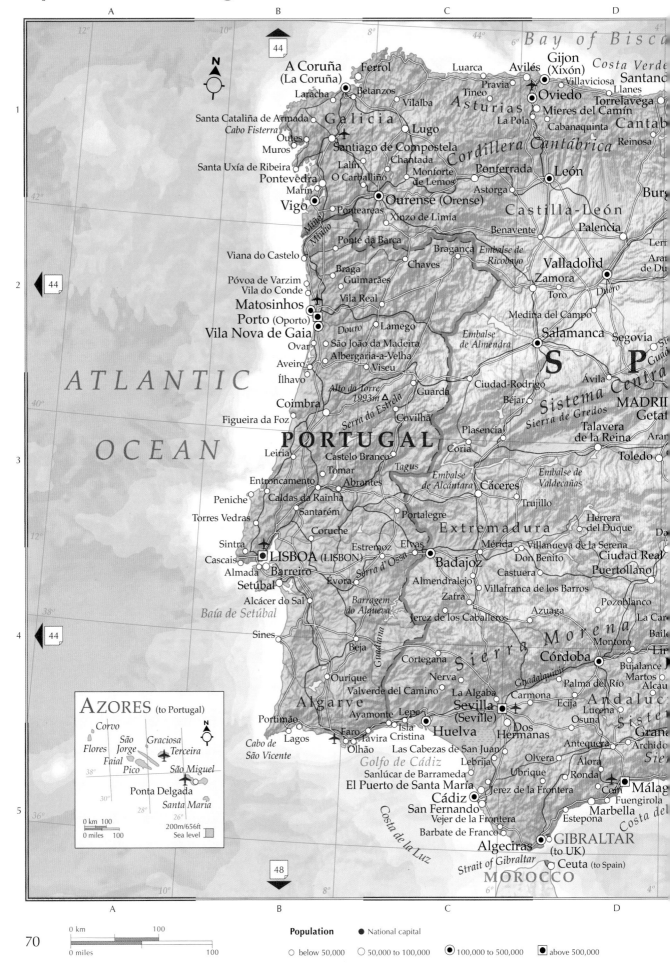

ATLANTIC

OCEAN

Bay of Biscay

Costa Verde

Galicia

A Coruña (La Coruña)
Ferrol
Laracha
Betanzos
Luarca
Avilés
Gijon (Xixón)
Villaviciosa
Santand
Santa Cataliña de Armada
Cabo Fisterra
Vilalba
Pravia
Tineo
Oviedo
Llanes
Cantá
Outes
Muros
Santiago de Compostela
Lugo
La Pola
Mieres del Camín
Cabanaquinta
Santa Uxía de Ribeira
Lalín
Chantada
Monforte de Lemos
Ponferrada
León
Reinosa
Pontevedra
O Carballiño
Astorga
Cordillera Cantábrica
Burg
Marín
Ponteareas
Ourense (Orense)
Benavente
Palencia
Vigo
Xinzo de Limia
Castilla-León
Lerr
Ponte da Barca
Bragança
Embalse de Ricobayo
Valladolid
Aran de Du
Viana do Castelo
Chaves
Zamora
Póvoa de Varzim
Braga
Guimarães
Toro
Duero
Vila do Conde
Vila Real
Medina del Campo
Matosinhos
Salamanca
Segovia
Porto (Oporto)
Douro
Lamego
Embalse de Almendra
S
P
Vila Nova de Gaia
São João da Madeira
Guad
Ovar
Albergaria-a-Velha
Ciudad-Rodrigo
Ávila
Aveiro
Viseu
Sistema Central
MADRI
Ílhavo
Béjar
Getaf
Alto da Torre 1993m
Guarda
Sierra de Gredos
Coimbra
Covilhã
Talavera de la Reina
Figueira da Foz
Plasencia
de la Reina
Aran

PORTUGAL
Serra da Estrela
Coria

Leiria
Castelo Branco
Tagus
Toledo
Tomar
Embalse de Valdecañas
Entroncamento
Abrantes
Embalse de Alcántara
Cáceres
Peniche
Caldas da Rainha
Portalegre
Trujillo
Herrera del Duque
Torres Vedras
Santarém
Coruche
Mérida
Villanueva de la Serena
Extremadura
Sintra
Estremoz
Elvas
Don Benito
Ciudad Real
Cascais
Serra d'Ossa
Badajoz
Castuera
Puertollano
Almada
LISBOA (LISBON)
Barreiro
Évora
Almendralejo
Villafranca de los Barros
Setúbal
Zafra
Pozoblanco
La Car
Alcácer do Sal
Barragem do Alqueva
Jerez de los Caballeros
Azuaga
Baía de Setúbal

Sines
Guadiana
Sierra
Morena
Córdoba
Li
Beja
Montoro
Baile
Cortegana
Guadalquivir
Bujalance
Martos
Ourique
Nerva
Palma del Río
Alcau
Valverde del Camino
La Algaba
Carmona
Andaluc
Algarve
Ayamonte
Lepe
Sevilla (Seville)
Ecija
Lucena
Sist
Portimão
Isla
Dos Hermanas
Osuna
Gran
Cabo de São Vicente
Lagos
Faro
Tavira
Cristina
Huelva
Antequera
Sier
Olhão
Las Cabezas de San Juan
Olvera
Álora
Golfo de Cádiz
Lebrija
Ubrique
Ronda
Com
Málag
Sanlúcar de Barrameda
Jerez de la Frontera
Fuengirola
El Puerto de Santa María
Marbella
Cádiz
Estepona
Costa del
San Fernando
Vejer de la Frontera
Barbate de Franco
GIBRALTAR (to UK)
Algeciras
Ceuta (to Spain)
Costa de la Luz
Strait of Gibraltar
MOROCCO

AZORES (to Portugal)

Corvo
Flores
São Jorge
Graciosa
Terceira
Faial
Pico
São Miguel
Ponta Delgada
Santa Maria

0 km 100
0 miles 100

200m/656ft
Sea level

Population ● National capital

○ below 50,000 ○ 50,000 to 100,000 ◉ 100,000 to 500,000 ◼ above 500,000

0 km 100
0 miles 100

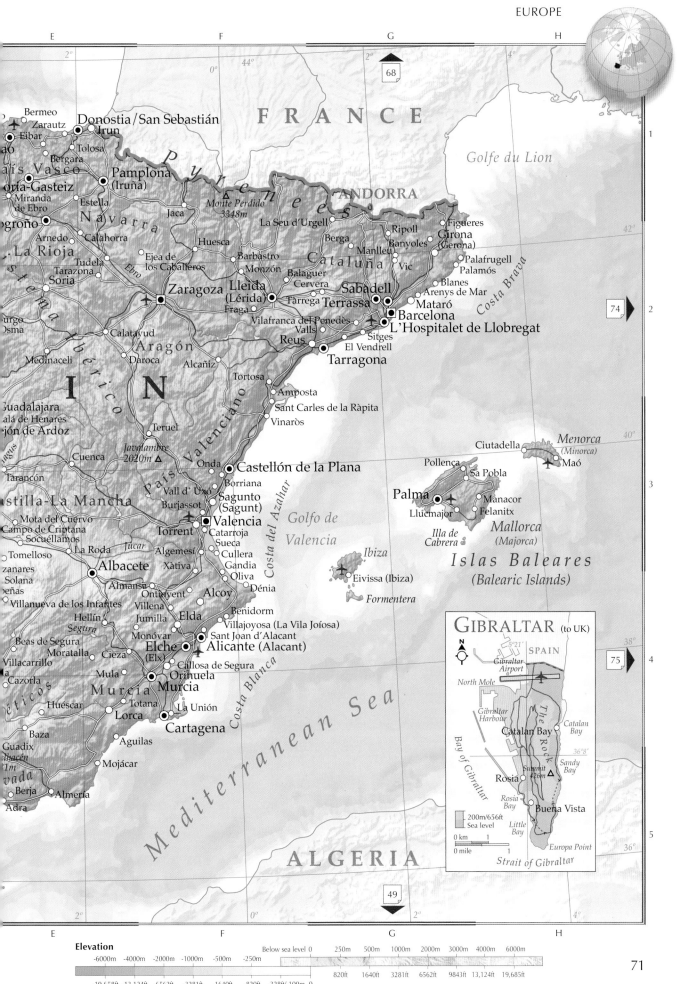

E   F   G   H

68

F R A N C E

Golfe du Lion

Bermeo
Zarautz
Donostia / San Sebastián
Eibar
Irun
Tolosa
Bergara
País Vasco
ria-Gasteiz
Pamplona
(Iruña)
goño
Miranda
de Ebro
Estella
Navarra
Jaca
Monte Perdido
3348m
La Seu d'Urgell
ANDORRA
Arnedo
Calahorra
Huesca
Berga
Ripoll
Figueres
La Rioja
Tudela
Ejea de
los Caballeros
Barbastro
Banyoles
Girona
(Gerona)
Tarazona
Soria
Monzón
Cataluña
Vic
Manlleu
Palafrugell
Palamós
burgo
Osma
Medinaceli
Zaragoza
Lleida
(Lérida)
Balaguer
Cervera
Tàrrega
Sabadell
Arenys de Mar
Blanes
Costa Brava
Calatayud
Aragón
Fraga
Vilafranca del Penedès
Terrassa
Mataró
Daroca
Alcañiz
Valls
Barcelona
L'Hospitalet de Llobregat
I   N
Reus
Sitges
El Vendrell
Tortosa
Tarragona
Teruel
Amposta
Sant Carles de la Ràpita
Javalambre
2020m
Vinaròs
Ciutadella
Menorca
(Minorca)
Cuenca
Onda
Castellón de la Plana
Pollença
Sa Pobla
Maó
Tarancón
Vall d'Uxó
Borriana
Sagunto
(Sagunt)
Palma
Manacor
Felanitx
stilla-La Mancha
Burjassot
Valencia
Torrent
Catarroja
Costa del Azahar
Golfo de
Valencia
Lluchmajor
Illa de
Cabrera
Mallorca
(Majorca)
Mota del Cuervo
Campo de Criptana
Socuéllamos
Sueca
Algemesí
Cullera
Gandia
Oliva
Islas Baleares
(Balearic Islands)
Tomelloso
La Roda
Xàtiva
Dénia
Ibiza
zanares
Solana
Almansa
Albacete
Ontinyent
Eivissa (Ibiza)
peñas
Villanueva de los Infantes
Villena
Jumilla
Alcoy
Elda
Benidorm
Villajoyosa (La Vila Joíosa)
Formentera
Hellín
Monóvar
Sant Joan d'Alacant
Beas de Segura
Moratalla
Cieza
Elche
(Elx)
Alicante (Alacant)
Villacarrillo
Mula
Callosa de Segura
Cazorla
Murcia
Orihuela
Murcia
ticos
Huéscar
Totana
La Unión
Baza
Lorca
Cartagena
Guadix
Aguilas
hacén
m
Mojácar
vada
Berja
Almería
Adra

Mediterranean Sea

ALGERIA

49

74

75

Elevation

-6000m -4000m -2000m -1000m -500m -250m

Below sea level 0   250m 500m 1000m 2000m 3000m 4000m 6000m

-19,658ft -13,124ft -6562ft -3281ft -1640ft -820ft -328ft/-100m 0

820ft 1640ft 3281ft 6562ft 9843ft 13,124ft 19,685ft

Germany & The Alpine States

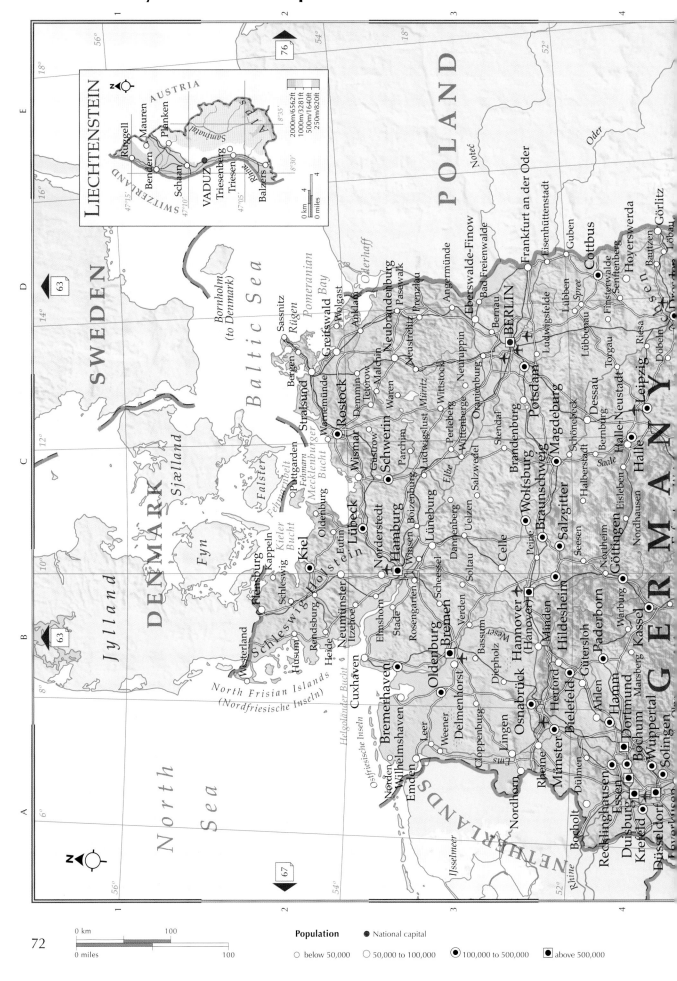

LIECHTENSTEIN

AUSTRIA

SWITZERLAND

Ruggell
Mauren
Planken
Benderm
Schaan
VADUZ
Triesenberg
Triesen
Balzers

Samtnatal

Rhine

2000m/6562ft
1000m/3281ft
500m/1640ft
250m/820ft

0 km   4
0 miles   4

POLAND

Oder

Noteć

SWEDEN

Bornholm
(to Denmark)

Baltic Sea

Pomeranian

Greifswald Bay
Oderhaff

Sassnitz
Rügen
Bergen
Stralsund
Warnemünde
Rostock
Wismar

Anklam
Wolgast

Neubrandenburg
Pasewalk
Prenzlau
Demmin
Malchin
Teterow
Waren
Neustrelitz
Müritz
Wittstock
Neuruppin
Oranienburg

Angermünde
Eberswalde-Finow
Bad Freienwalde
Bernau

Frankfurt an der Oder
Eisenhüttenstadt

Cottbus
Guben
Finsterwalde
Forst
Senftenberg
Hoyerswerda
Bautzen
Görlitz
Löbau

BERLIN
Potsdam
Ludwigsfelde
Lübben
Spree
Lübbenau
Torgau
Riesa
Döbeln
S a c h s e n
Leipzig

Brandenburg
Magdeburg
Schönebeck
Dessau
Bernburg
Halberstadt
Halle-Neustadt
Eisleben
Halle
Saale

Wolfsburg
Braunschweig
Salzgitter
Seesen
Nordhausen

DENMARK

Sjælland
Falster
Fyn
Jylland

Kiel
Kieler Bucht
Fehmarn
Fehmarnbelt
Puttgarden
Mecklenburger Bucht

Flensburg
Kappeln
Schleswig
S c h l e s w i g - H o l s t e i n
Rendsburg
Husum
Heide
Neumünster
Lübeck
Eutin
Oldenburg
Güstrow
Schwerin
Parchim
Ludwigslust
Boizenburg
Dömitz
Perleberg
Wittenberge
Salzwedel
Stendal
Gardelegen

Norderstedt
Hamburg
Itzehoe
Elmshorn
Wedel
Lüneburg
Uelzen
Dannenberg

North Frisian Islands
(Nordfriesische Inseln)
Westerland

Helgoländer Bucht
Ostfriesische Inseln
Norden
Emden

Cuxhaven
Stade
Scheessel
Rosengarten
Soltau
Celle
Peine
Hannover
(Hanover)

Bremerhaven
Wilhelmshaven
Bremen
Verden
Bassum
Diepholz
Weser
Minden
Hildesheim
Göttingen
Northeim

Oldenburg
Delmenhorst
Weener
Leer
Cloppenburg
Lingen
Nordhorn

Ems

Osnabrück
Rheine
Bielefeld
Herford
Gütersloh
Paderborn
Warburg
Kassel
Marsberg

Münster
Ahlen
Hamm
Dülmen

NETHERLANDS

Ijsselmeer
Rhine
Bocholt
Recklinghausen
Duisburg
Essen
Krefeld
Bochum
Dortmund
Düsseldorf
Solingen
Wuppertal

G E R M A N Y

North
Sea

Population
● National capital
○ below 50,000
◎ 50,000 to 100,000
◉ 100,000 to 500,000
■ above 500,000

0 km   100
0 miles   100

72

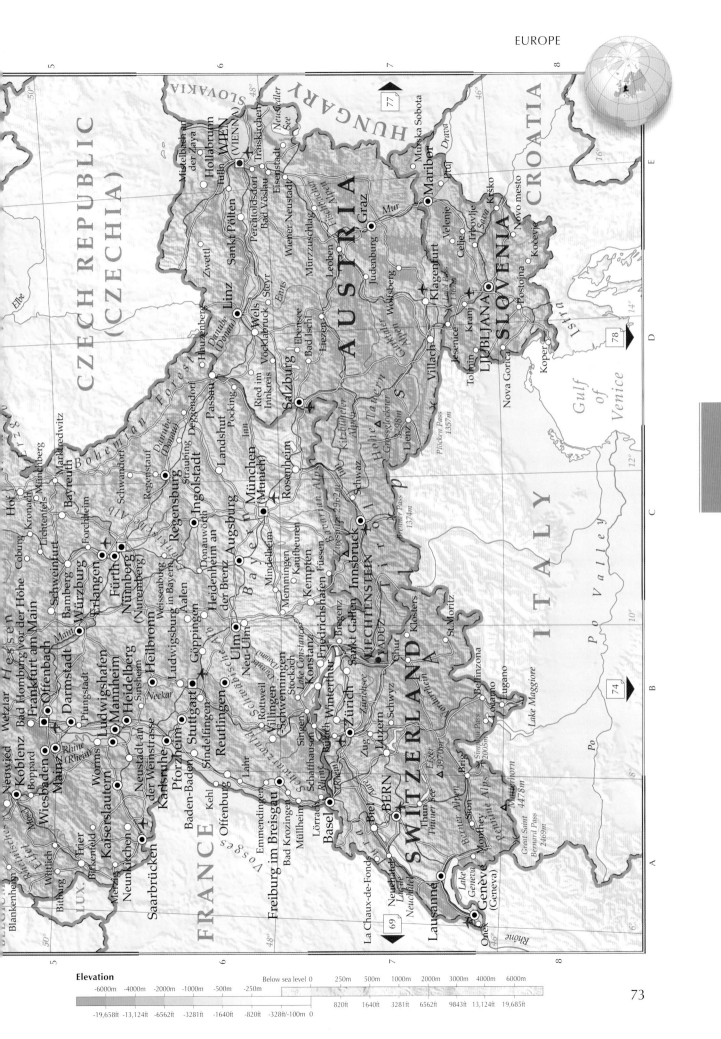

Elevation

-6000m -4000m -2000m -1000m -500m -250m

-19,658ft -13,124ft -6562ft -3281ft -1640ft -820ft -328ft/-100m 0

Below sea level 0   250m   500m   1000m   2000m   3000m   4000m   6000m

820ft   1640ft   3281ft   6562ft   9843ft 13,124ft 19,685ft

Italy

SLOVAKIA

HUNGARY

Drava

ITALY

AUSTRIA

SLOVENIA

CROATIA

BOSNIA & HERZEGOVINA

Sava

GERMANY

SWITZERLAND

LIECHTENSTEIN

FRANCE

MONACO

Adriatic Sea

Dalmacija

Istra

SAN MARINO

Doganaz
Serravalle
Fiorina
Gualdicciolo
Borgo Maggiore
SAN MARINO
Chiesanuova
Cailungo
Monte Titano 739m
Murata
Montegiardino
Faetano
ITALY

Appennino

500m/1640ft
200m/656ft
100m/328ft

0 km   2
0 miles   2

Tarvisio
Udine
Cortina d'Ampezzo
Gemona del Friuli
Pordenone
Montfalcone
Portogruaro
Trieste
Gulf of Venice
Foci del Po

Bressanone
Merano
Bolzano
Dolomitiche
Alpi
Trento
Bassano del Grappa
Treviso
Mestre
Venezia (Venice)
Chioggia

Edolo
Arco
Lago di Garda
Vicenza
Padova
Monselice
Ostiglia
Rovigo
Adige
Ferrara
Comacchio
Ravenna

Como
Lago di Como
Bergamo
Brescia
Verona
Mantova
Cremona
Carpi
Modena
Imola
Forlì
Cesena
Rimini
SAN MARINO
SAN MARINO
Pesaro
Fano
Falconara Marittima
Ancona
Civitanova Marche
Fermo
Ascoli Piceno
Giulianova
Teramo
Pescara
Ortona
Chieti

Monza
Milano (Milan)
Pavia
Piacenza
Parma
Reggio nell'Emilia
Bologna
Faenza
Sansepolcro
Arezzo
Perugia
Foligno
Todi
Terni
L'Aquila
Avezzano
Tivoli

Novara
Vercelli
Varese
Castéggio
Genova (Genoa)
Carrara
Massa
Viareggio
Pisa
Lucca
Pistoia
Prato
Firenze (Florence)
Siena
Livorno
Cecina
Grosseto
Viterbo
Civitavecchia
VATICAN CITY

Aosta
Gran Paradiso 4061m
Torino (Turin)
Asti
Alessandria
Savona
La Spezia
Finale Ligure
Imperia
San Remo
Golfo di Genova

Mont Blanc 4808m
Little St-Bernard Pass 2188m
Great Saint Bernard Pass 2469m
Susa
Rivoli
Moncalieri
Savigliano
Cuneo
Mondovi
Ventimiglia
Piemonte
Appennino

Brenner Pass 1374m
Inn
Rhône
Rhine
Lake Geneva
Lake Constance
Lago Maggiore
Lake Maggiore

Lombardia
EMILIA
Po
Adda
Ticino
Valli

Marche
Umbro-Marchigiano
Appennino
Lago Trasimeno
Toscana
Chianti
Arno

Ligurian Sea

Corse (Corsica) (to France)
Archipelago Toscano
Isola d'Elba
Portoferraio
Piombino
Orbetello

Strait of Bonifacio

74

Population   ● National capital

○ below 50,000   ○ 50,000 to 100,000   ◉ 100,000 to 500,000   ■ above 500,000

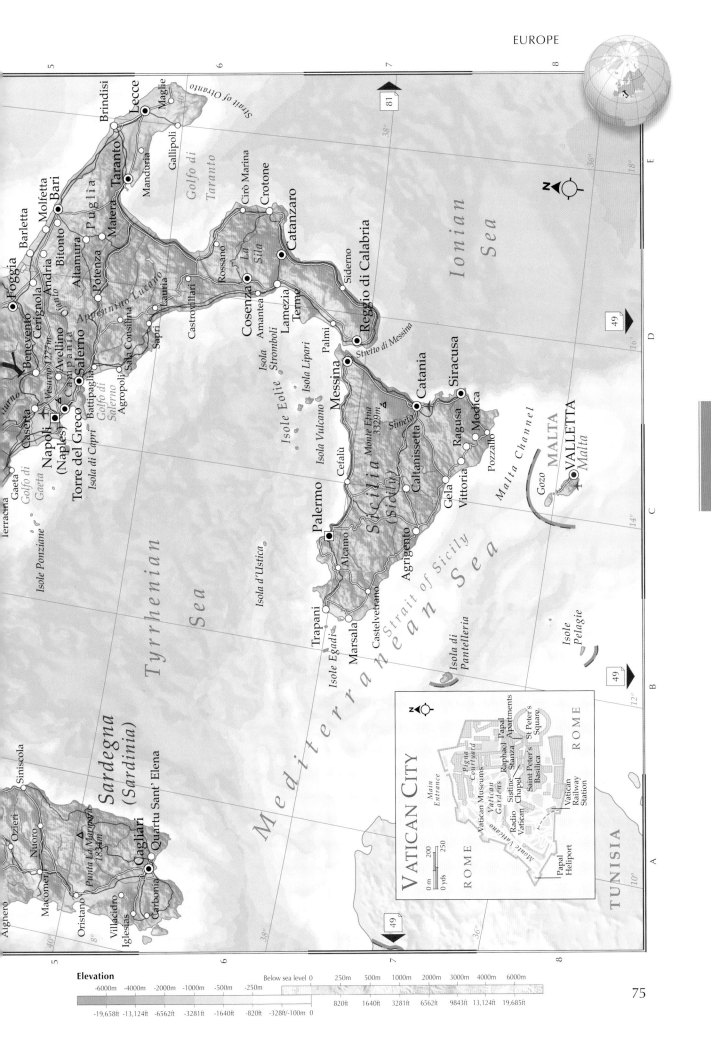

Brindisi
Lecce
Maglie
Strait of Otranto
Taranto
Bari
Molfetta
Bitonto
Barletta
Andria
Cerignola
Foggia
Benevento
Avellino
Salerno
Caserta
Napoli (Naples)
Torre del Greco
Battipaglia
Agropoli
Sala Consilina
Sapri
Lauria
Castrovillari
Cosenza
Amantea
Lamezia Terme
Palmi
Reggio di Calabria
Stretto di Messina
Messina
Catanzaro
Crotone
Cirò Marina
Rossano
Sidemo
Matera
Manduria
Gallipoli
Altamura
Potenza

Puglia
Campania
Appennino Lucano
Vesuvio 1277m
Golfo di Gaeta
Gaeta
Terracina
Isole Ponziane

Sardegna (Sardinia)
Siniscola
Ozieri
Nuoro
Macomer
Oristano
Villacidro
Iglesias
Carbonia
Cagliari
Quartu Sant'Elena
Punta La Marmora 1834m

Tyrrhenian Sea

Isola di Capri
Golfo di Salerno

Isola d'Ustica

Isole Eolie
Isola Stromboli
Isola Lipari
Isola Vulcano
Cefalù

Palermo
Alcamo
Trapani
Isole Egadi
Marsala
Castelvetrano

Sicilia (Sicily)
Agrigento
Caltanissetta
Gela
Vittoria
Ragusa
Modica
Pozzallo
Catania
Siracusa
Monte Etna 3329m
Simeto

Strait of Sicily

Mediterranean Sea

Isola di Pantelleria

Malta Channel
Gozo
MALTA
VALLETTA
Malta
Isole Pelagie

Ionian Sea

Golfo di Taranto

TUNISIA

Vatican City

81
49
49
49

N
W E
S

VATICAN CITY
Main Entrance
Vatican Gardens
Radio Vatican
Vatican Museums
Pigna Courtyard
Sistine Chapel
Raphael Stanza
Papal Apartments
St Peter's Square
Saint Peter's Basilica
Vatican Railway Station
Papal Heliport
Monte Vaticano
ROME
ROME

0 m   200
0 yds   250

Elevation

Below sea level	0	250m	500m	1000m	2000m	3000m	4000m	6000m

-6000m   -4000m   -2000m   -1000m   -500m   -250m

-19,658ft   -13,124ft   -6562ft   -3281ft   -1640ft   -820ft   -328ft/-100m   0

820ft   1640ft   3281ft   6562ft   9843ft   13,124ft   19,685ft

Central Europe

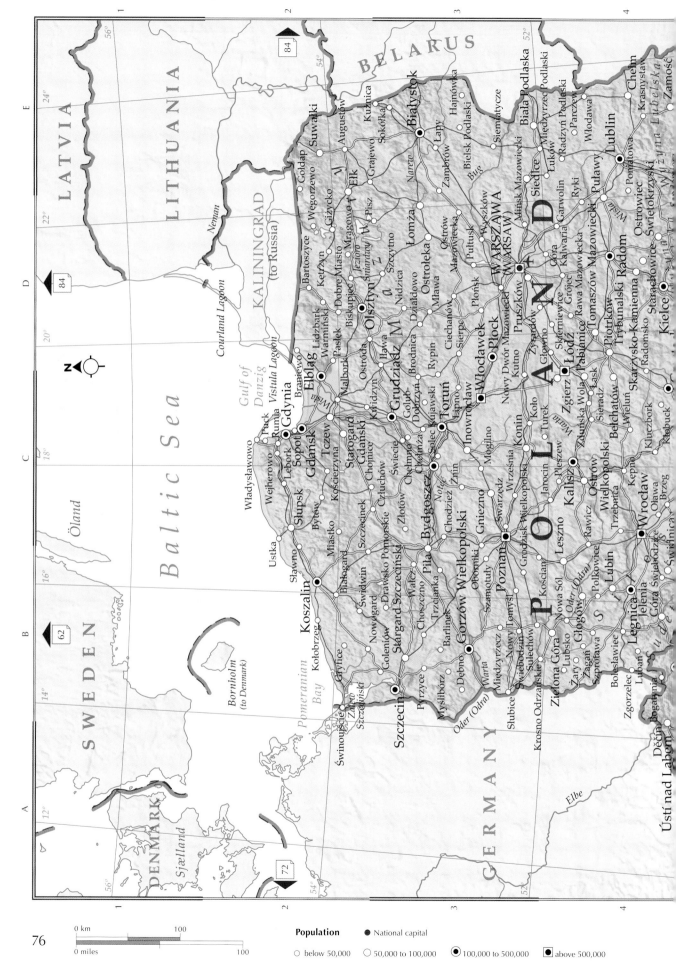

SWEDEN

DENMARK

Sjælland

Öland

LATVIA

LITHUANIA

KALININGRAD
(to Russia)

BELARUS

POLAND

GERMANY

Baltic Sea

Bornholm
(to Denmark)

Neman

Courland Lagoon

*Gulf of
Danzig*

Vistula Lagoon

*Pomeranian
Bay*

Oder (Odra)

Elbe

Władysławowo
Puck
Rumia
Gdynia
Sopot
Gdańsk
Tczew
Wejherowo
Lębork
Kościerzyna
Starogard
Gdański
Chojnice
Świecie
Chełmno
Chełmża
Grudziądz
Kwidzyn
Malbork
Elbląg
Braniewo
Frombork
Pasłęk
Lidzbark
Warmiński
Bartoszyce
Kętrzyn
Gołdap
Węgorzewo
Goldap
Suwałki
Augustów
Sokółka
Kuźnica
Grajewo
Białystok
Łapy
Bielsk Podlaski
Hajnówka
Siemiatycze
Biała Podlaska
Międzyrzec Podlaski
Radzyń Podlaski
Parczew
Włodawa
Chełm
Krasnystaw
Zamość
Lublin
Puławy
Ryki
Garwolin
Kozienice
Radom
Ostrowiec
Świętokrzyski
Starachowice
Skarżysko-Kamienna
Kielce
Ostrów
Mazowiecka
Maków
Mazowiecki
Pułtusk
Płońsk
Ciechanów
Mława
Działdowo
Nidzica
Szczytno
Pisz
Giżycko
Mrągowo
Ruciane
Iława
Brodnica
Rypin
Lipno
Golub
Dobrzyń
Toruń
Solec
Kujawski
Włocławek
Kutno
Koło
Konin
Turek
Września
Jarocin
Pleszew
Kalisz
Ostrów
Wielkopolski
Krotoszyn
Rawicz
Trzebnica
Oleśnica
Wrocław
Oława
Brzeg
Strzelin
Świdnica
Dzierżoniów
Góra
Świebodzice
Legnica
Jelenia
Góra
Lubań
Bolesławiec
Zgorzelec
Zary
Żagań
Głogów
Lubsko
Zielona Góra
Sprotawa
Nowa Sól
Polkowice
Lubin
Nowy Tomyśl
Grodzisk
Wielkopolski
Leszno
Gostyń
Kościan
Śrem
Środa
Wielkopolska
Poznań
Swarzędz
Szamotuły
Oborniki
Wągrowiec
Gniezno
Żnin
Mogilno
Inowrocław
Bydgoszcz
Nakło
Chodzież
Piła
Wałcz
Trzcianka
Czarnków
Wronki
Międzychód
Gorzów Wielkopolski
Sulęcin
Świebodzin
Międzyrzecz
Słubice
Krosno Odrzańskie
Krosno Odrzański
Świebodzice
Sulechów
Kostrzyn
Dębno
Myślibórz
Pyrzyce
Gryfice
Kołobrzeg
Koszalin
Darłowo
Sławno
Słupsk
Ustka
Białogard
Świdwin
Drawsko
Pomorskie
Szczecinek
Złotów
Miastko
Bytów
Człuchów
Tuchola
Nowogard
Goleniów
Stargard
Szczeciński
Police
Świnoujście
Szczecin

Słubice
Wschowa
Krotoszyn
Milicz

Bielsk Podlaski
Łomża
Zambrów
Ostrołęka
Ostrów

Dobre Miasto
Bisztynek
Olsztyn
Biskupiec
Ostróda

Warta

Notec

Wisla

Bug

Narew

Oder (Odra)

MAZOWSZE
WARMIA
MAZURY

WARSZAWA
WARSAW

Grodzisk
Mazowiecki
Żyrardów
Pruszków
Piaseczno
Grójec
Góra
Kalwaria
Rawa
Mazowiecka
Skierniewice
Łowicz
Głowno
Zgierz
Łódź
Pabianice
Zduńska
Wola
Sieradz
Łask
Bełchatów
Wieluń
Wieruszów
Kępno
Kluczbork
Namysłów
Olesno
Ostrzeszów
Syców
Piotrków
Trybunalski
Tomaszów
Mazowiecki
Opoczno
Radomsko

Łuków
Mińsk Mazowiecki
Siedlce
Sokołów
Podlaski
Węgrów

Ełk
Olecko

Bielsk

Czersk

Kętrzyn

Śrem

SUDETY
SUDETES
JIZERA

Děčín
Ústí nad Labem
Bogatynia

Międzyrzec
Podlaski

Pisz

Population

● National capital
○ below 50,000
◯ 50,000 to 100,000
◉ 100,000 to 500,000
▣ above 500,000

0 km ———— 100
0 miles ———— 100

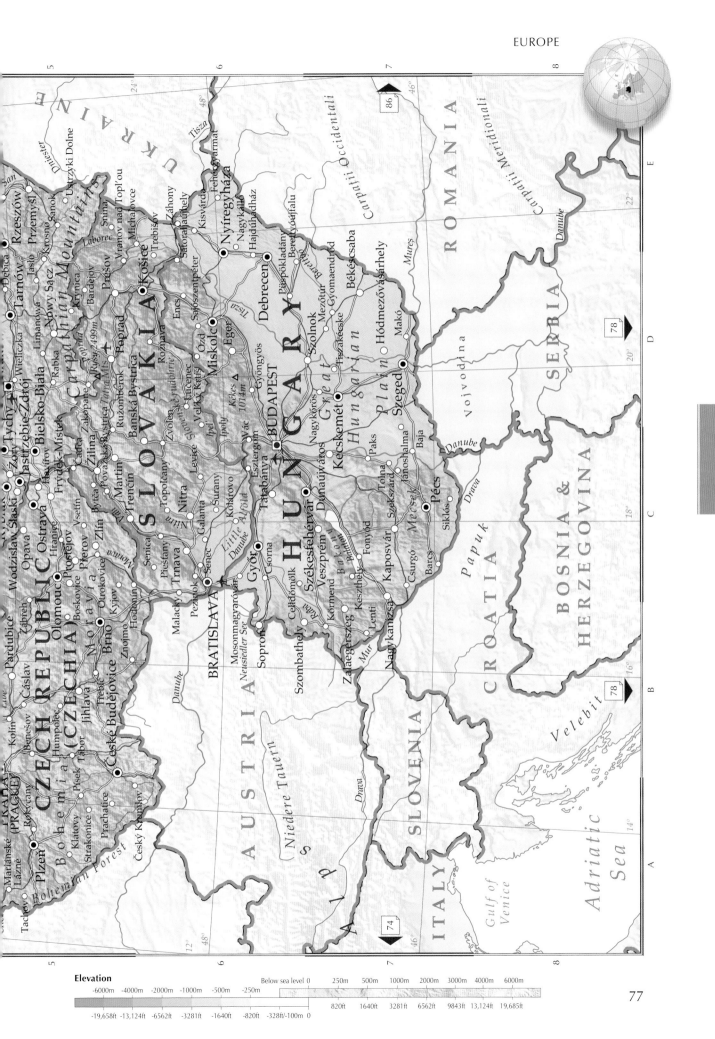

Elevation

| Below sea level | 0 | 250m | 500m | 1000m | 2000m | 3000m | 4000m | 6000m |

-6000m -4000m -2000m -1000m -500m -250m

820ft 1640ft 3281ft 6562ft 9843ft 13,124ft 19,685ft

-19,658ft -13,124ft -6562ft -3281ft -1640ft -820ft -328ft/-100m 0

Southeast Europe

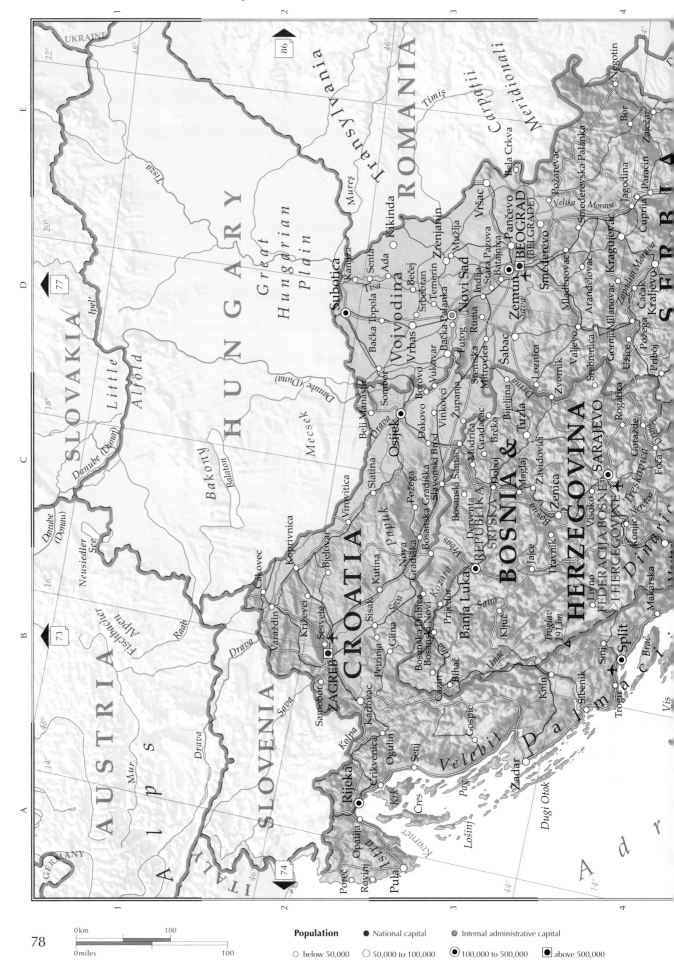

UKRAINE

SLOVAKIA

AUSTRIA

Alps
Fischbacher
Alpen

GERMANY

ITALY

SLOVENIA

HUNGARY

Great
Hungarian
Plain

Little
Alföld

Bakony

Mecsek

Balaton
Neusiedler
See

Danube (Donau)
Danube (Duna)

Tisza
Ipel'
Mur
Raab
Drava
Mur

Transylvania

ROMANIA

Carpaţii Meridionali

Timiş
Mureş

Negotin
Bor
Zaječar
Kikinda
Bela Crkva
Vršac
Požarevac
Smederevska Palanka
Jagodina
Ćuprija
Paraćin
Subotica
Kanjiža
Senta
Ada
Bečej
Zrenjanin
Mužlja
Velika Morava
Pančevo
BEOGRAD
(BELGRADE)
Smederevo
Mladenovac
Aranđelovac
Gornji Milanovac
Kragujevac
Zapadna Morava
Čačak
Kraljevo
Tisa
Ipola
Srbobran
Temerin
Novi Sad
Indija
Stara Pazova
Batajnica
Zemun
Bačka Palanka
Ruma
Sremska Mitrovica
Valjevo
Užice
Požega
Pribo
Bačka Topola
Vrbas
Futog
Šabac
Loznica
Srebrenica
SERBIA
Sombor
Beli Manastir
Vukovar
Županja
Zvornik
Rogatica
Goražde
Osijek
Đakovo
Vinkovci
Derventa
Brčko
Tuzla
Sarajevo
Foča
Beli Manastir
Virovitica
Slatina
Požega
Nova Gradiška
Bosanska Gradiška
Slavonski Brod
Modriča
Gradačac
Doboj
Maglaj
Zavidovići
Zenica
Visoko
REPUBLIKA
SRPSKA
BOSNIA &
HERZEGOVINA
Tešanj
Trepča
Konjic
Foča
Koprivnica
Bjelovar
Papuk
Kozara
Bosanski Samac
Banja Luka
FEDERACIJA BOSNE
I HERCEGOVINE
Livno
Jajce
Travnik
Kolašin
Dinaric
Čakovec
Varaždin
Križevci
Sesvete
CROATIA
Sisak
Kutina
Bosanska Dubica
Bosanski Novi
Prijedor
Ključ
Troglav
1913m
Sinj
Makarska
Samobor
ZAGREB
Karlovac
Glina
Petrinja
Cazin
Bihać
Una
Sana
Vrbas
Bosna
Drina
Unac
Knin
Šibenik
Trogir
Split
Brač
Sava
Kolpa
Crikvenica
Ogulin
Senj
Gospić
Velebit
Zadar
Vis
Rijeka
Opatija
Krk
Cres
Pag
Dugi Otok
Lošinj
Poreč
Rovinj
Pula
Istria
Krmen
Palma
Adriatic
Drava

Population

● National capital   ● Internal administrative capital

○ below 50,000   ○ 50,000 to 100,000   ◉ 100,000 to 500,000   ◼ above 500,000

0km   100
0miles   100

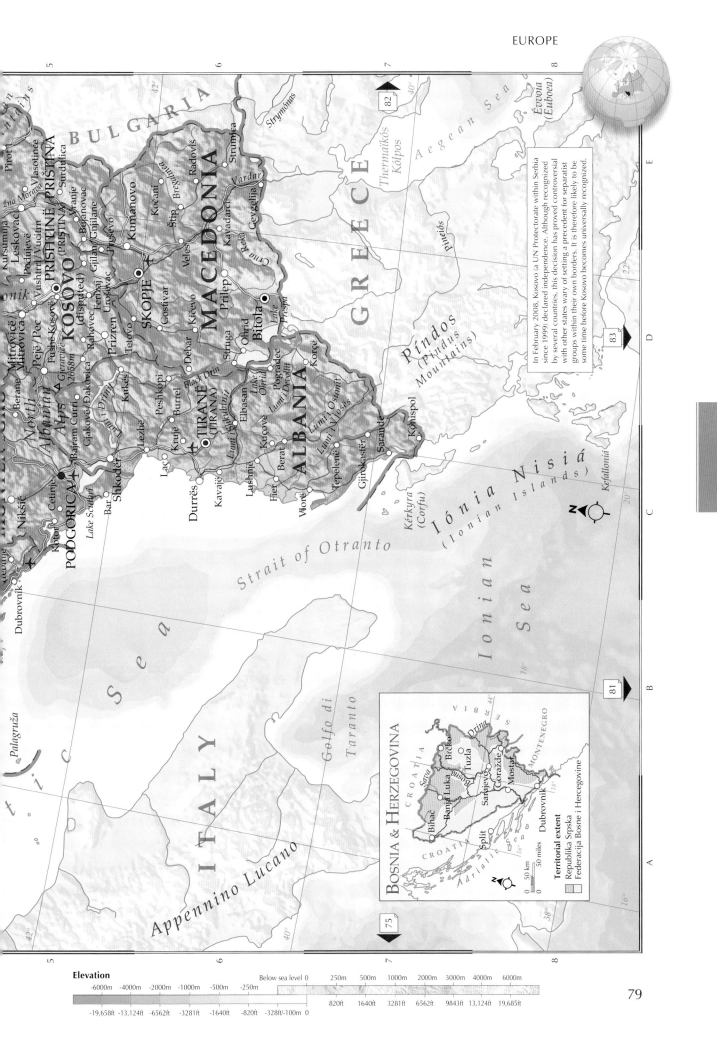

In February 2008, Kosovo (a UN Protectorate within Serbia since 1999) declared independence. Although recognized by several countries, this decision has proved controversial with other states wary of setting a precedent for separatist groups within their own borders. It is therefore likely to be some time before Kosovo becomes universally recognized.

BULGARIA

GREECE

Aegean Sea

Thermaïkós Kólpos

Strymónas

Évvoia (Euboea)

MACEDONIA

KOSOVO
PRISHTINË PRIŠTINA
(PRISTINA)

Pirot
Nksuma
žna Morava
Vlasotince
Leskovac
Podujevě
Vushtrri/Vučitrn
Surdulica
Vranje
Bujanovac
Gnjilane
Gjilan/
Ferizaj/
Uroševac
Preševo
Kumanovo
Kočani
Štip
Bregalnica
Radoviš
Strumica
Vardar
Crna Reka
Kavadarci
Gevgelija

SKOPJE
Gostivar
Veles
Prilep
Bitola
Lake Prespa

Mitrovicë
Mitrovica
Berane
Pejë / Pec
Fushë Kosovë
Rahovec/
Orahovac
Prizren
Djakovica
Tetovo
Debar
Kičevo
Ohrid
Struga
Lake Ohrid
Pogradec
Lumi i Devollit
Korçë

Pindos
(Pindus
Mountains)

North
Albanian
Alps
2658m
Glaniani
Bajram Curri
Gjakovë/
Kukës
Peshkopi
Burrel
Black Drin
Elbasan
Lumi i Shkumbinit
Kuçovë
Berat
Lumi i Osumit

ALBANIA
TIRANË (TIRANA)
Lezhë
Lac
Krujë
Kavajë
Lushnjë
Fier
Vlorë
Tepelenë
Gjirokastër
Sarandë
Lumi i Vjosës
Kelcyrë

Niksić
Cetinje
PODGORICA
Kotor
Bar
Shkodër
Lake Scutari
Durrës
Kônispol

Dubrovnik
Palagruža

Kérkyra
(Corfu)

Ióna Nisiá
(Ionian Islands)
Kefalloniá

Strait of Otranto

Ionian
Sea

Sea

Adriatic
Sea

ITALY
Appennino Lucano
Golfo di Taranto

N

BOSNIA & HERZEGOVINA

CROATIA
SERBIA
Brčko
Tuzla
Banja Luka
Bihać
Sava
Drina
Bosna
Goražde
Sarajevo
Mostar
MONTENEGRO
Dubrovnik
Split
CROATIA
Adriatic Sea

N

0   50 km
0   50 miles

Territorial extent
Republika Srpska
Federacija Bosne i Hercegovine

Elevation

-6000m	-4000m	-2000m	-1000m	-500m	-250m				

Below sea level 0   250m   500m   1000m   2000m   3000m   4000m   6000m

-19,658ft -13,124ft -6562ft -3281ft -1640ft -820ft -328ft/-100m 0

820ft   1640ft   3281ft   6562ft   9843ft 13,124ft 19,685ft

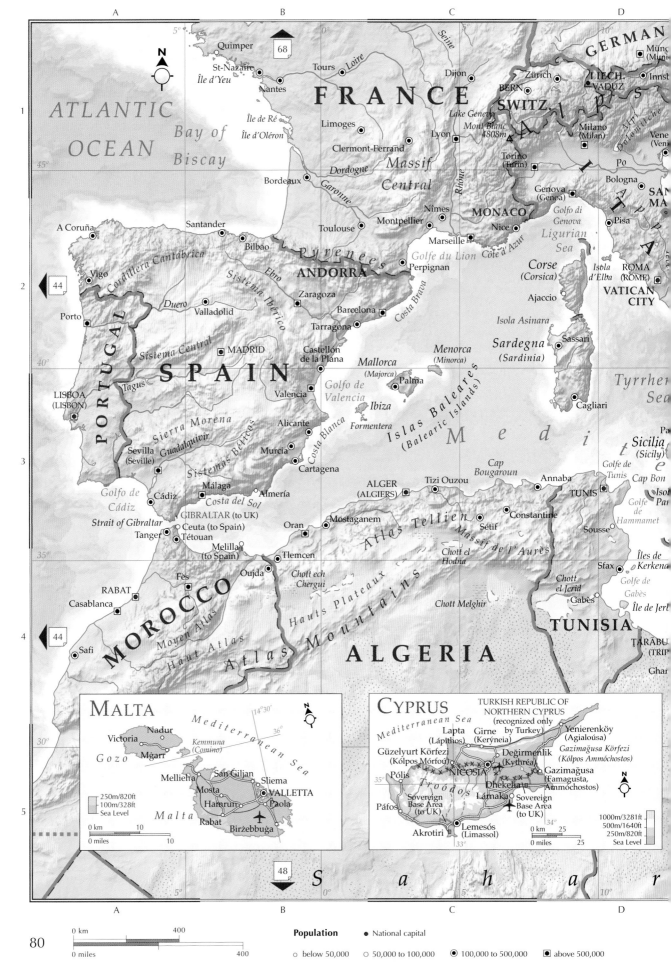

MALTA

Mediterranean Sea

Victoria
Nadur
Mġarr
Gozo
Kemmuna
(Comino)

Mellieħa
San Ġiljan
Mosta
Sliema
Ħamrun
VALLETTA
Paola
Malta
Rabat
Birżebbuġa

250m/820ft
100m/328ft
Sea Level

0 km   10
0 miles   10

CYPRUS

Mediterranean Sea

TURKISH REPUBLIC OF
NORTHERN CYPRUS
(recognized only
by Turkey)

Lapta
(Lápithos)
Girne
(Kerýneia)
Yenierenköy
(Agialoúsa)

Güzelyurt Körfezi
(Kólpos Mórfou)
Değirmenlik
(Kythréa)
Gazimağusa Körfezi
(Kólpos Ammóchostos)

NICOSIA
Pólis
Gazimağusa
(Famagusta,
Ammóchostos)
Troodos
Dhekelia
Páfos
Sovereign
Base Area
(to UK)
Lárnaka
Sovereign
Base Area
(to UK)
Lemesós
(Limassol)
Akrotiri

1000m/3281ft
500m/1640ft
250m/820ft
Sea Level

0 km   25
0 miles   25

Population   ● National capital

○ below 50,000   ◎ 50,000 to 100,000   ◉ 100,000 to 500,000   ■ above 500,000

0 km   400
0 miles   400

E | F | G | H

SLOVAKIA
WIEN
(NNA)

Danube

HUNGARY

Tisza

Carpathian Mountains

Satu Mare

BUDAPEST

Bâlti

UKRAINE

86

*Kakhovs'ka
Vodoskhovyshche*

Târgu Mureş

Great
Hungarian
Plain

MOLD.

CHIŞINĂU

ZAGREB

Nistru

Odesa

Dnieper

Berdyans'k

Sea of Azov

CROATIA

Novi Sad

ROMANIA

Carpaţii Meridonali

Galaţi

*Kryms'kyy
Pivostrov
(Crimea)*

Kerch

RUSS.

Sava

BOSNIA
& HERZ.

BEOGRAD
(BELGRADE)

BUCUREŞTI
(BUCHAREST)

Danube

Constanţa

Sevastopol'

Novorossiysk

SARAJEVO

SERBIA

BULGARIA

Varna

*Black
Sea*

(since 2014 the Ukrainian
territory of Crimea has
been annexed by Russia)

MON.

PRISHTINË

KOSOVO
(disputed)

SOFIYA
(SOFIA)

Balkan Mountains

Burgas

95

PODGORICA

SKOPJE
(SKOPJE)

MACED.

*Rhodope
Mountains*

Edirne

*İstanbul
Boğazı
(Bosporus)*

İstanbul

Zonguldak

Küre Dağları

Samsun

Ordu

TIRANË
(TIRANA)

ALBANIA

*Pindos
(Pindus)
Mts*

Thessaloníki
(Salonica)

*Marmara
Denizi*

Bursa

ANKARA

Kızıl Irmak

Bari

Límnos

Lárisa

Balıkesir

TURKEY

Kayseri

(Naples)

*Strait of
Otranto*

GREECE

*Aegean
Sea*

Chíos

İzmir

*Tuz
Gölü*

Vesuvio 1277m
Lecce

*Golfo di
Taranto*

Kefallonia

ATHÍNA
(ATHENS)

Sámos

*Dodekánisa
(Dodecanese)*

Ionian

Kýthira

Zákynthos

*Kykládes
(Cyclades)*

*Mirtóo
Pelagos*

*Dodekánisa
(Dodecanese)*

Antalya

Toros Dağları

Gaziantep

Adana

Euphrates

Catanzaro

Monte Etna
329m

Catania

Sea

*Krikikó Pélagos
(Sea of Crete)*

Ródos
(Rhodes)

*Antalya
Körfezi*

İskenderun Körfezi

Halab
(Aleppo)

Siracusa

Kárpathos

NICOSIA

35°

LLETTA
TA

Irákleio

*Kríti
(Crete)*

CYPRUS

Lemesós
(Limassol)

Lárnaka

SYRIA

LEBANON

BEYROUTH
(BEIRUT)

DIMASHQ
(DAMASCUS)

97

Mediterranean

Darnah

Hefa
(Haifa)

ISRAEL

Tel Aviv-Yafo

AMMAN

Banghāzī
(Benghazi)

Ţubruq

JERUSALEM

Gaza

Dead Sea

JORDAN

Mişrātah

Alexandria
(Al Iskandarīyah)

*Nile
Delta*

Būr Sa'īd
(Port Said)

*Qanāt as Suways
(Suez Canal)*

In 1974 Turkey occupied the northern part
of Cyprus while Greek Cypriots remained in
control of the south. Cyprus was effectively
partitioned and a UN buffer zone currently
divides the two areas. In 1983 the north of
the island proclaimed itself the Turkish
Republic of North Cyprus. It was only
recognized by Turkey.

CAIRO
(AL QĀHIRAH)

Giza
(Al Jīzah)

Suez
(As Suways)

Al 'Aqabah

Elat

*Sinai
(Sinā')*

**SAUDI
ARABIA**

*Munkhafaḍ al Qaṭṭārah
(Qattara Depression)*

*Libyan
Plateau*

*Great
Sand
Sea*

EGYPT

Nile

*Ṣaḥrā' ash Sharqīyah
(Eastern Desert)*

Khalīj as Suways (Gulf of Suez)

*Red
Sea*

LIBYA

*Libyan
Desert*

50

E | F | G | H

Elevation

						Below sea level 0	250m	500m	1000m	2000m	3000m	4000m	6000m
-6000m	-4000m	-2000m	-1000m	-500m	-250m								
-19,658ft	-13,124ft	-6562ft	-3281ft	-1640ft	-820ft	-328ft/-100m 0	820ft	1640ft	3281ft	6562ft	9843ft	13,124ft	19,685ft

81

Bulgaria & Greece

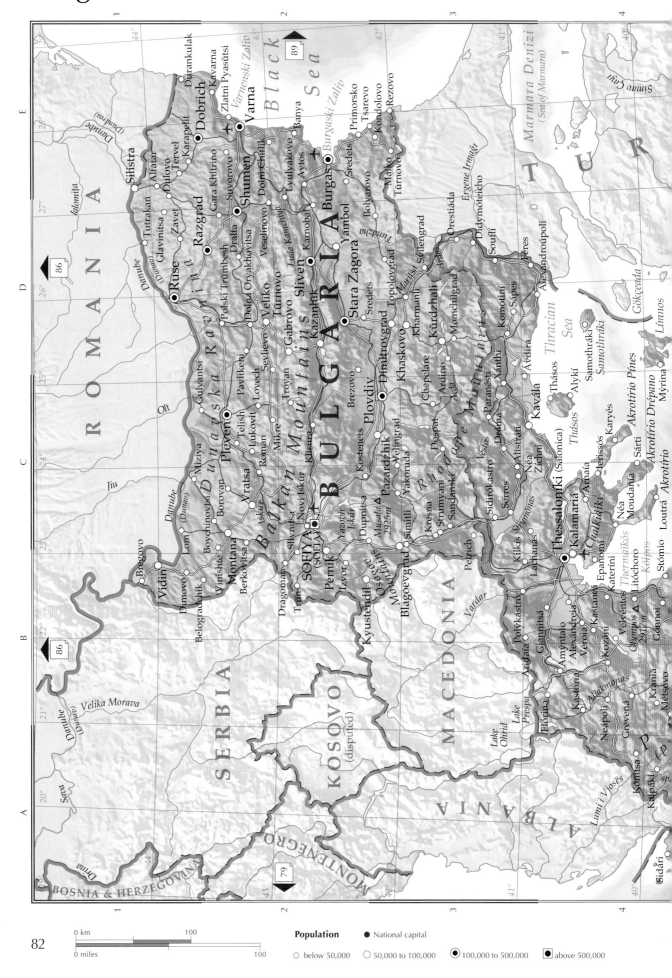

Population

● National capital

○ below 50,000   ○ 50,000 to 100,000   ◉ 100,000 to 500,000   ◼ above 500,000

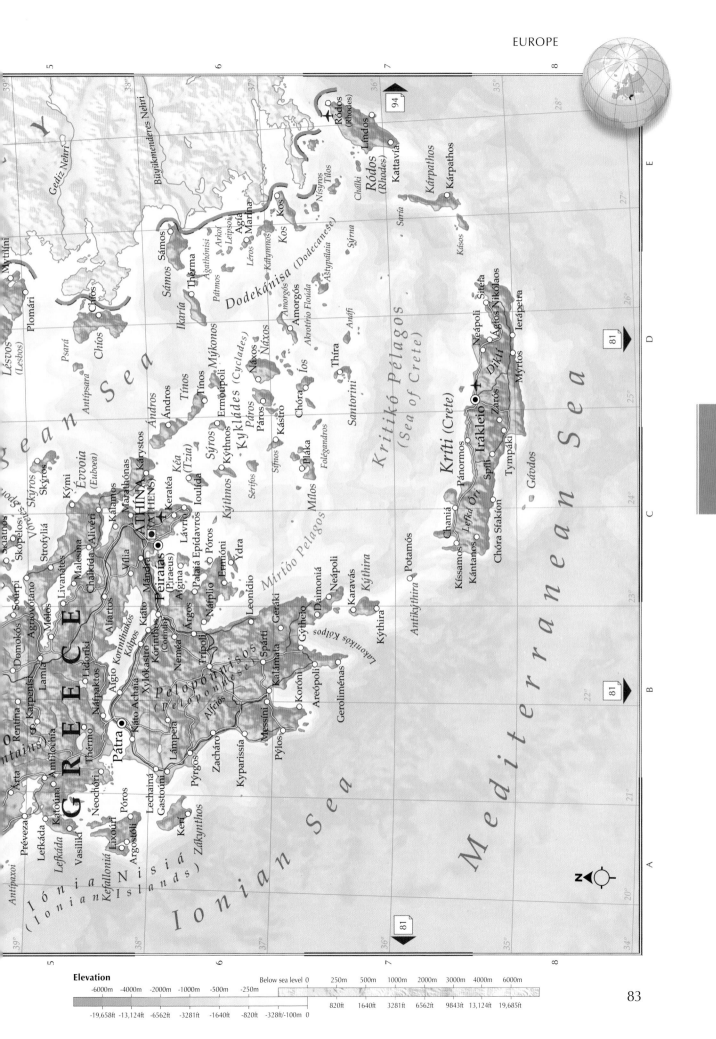

Elevation

-6000m	-4000m	-2000m	-1000m	-500m	-250m
-19,658ft	-13,124ft	-6562ft	-3281ft	-1640ft	-820ft

Below sea level 0   250m   500m   1000m   2000m   3000m   4000m   6000m

820ft   1640ft   3281ft   6562ft   9843ft 13,124ft 19,685ft

-328ft/-100m 0

83

The Baltic States & Belarus

SWEDEN

FINLAND

RUSSIA

ESTONIA

LATVIA

LITHUANIA

Gulf of Finland

Narva Bay

Baltic Sea

Gulf of Riga

Gotland

Courland Lagoon

Lake Peipus

Lake Pskov

TALLINN

Paldiski
Keila
Maardu
Loksa
Aegviidu
Raasiku
Rakvere
Tapa
Rakke
Kohtla-Järve
Kunda
Sillamäe
Narva
Risti
Rapla
Paide
Haapsalu
Lihula
Pärnu-Jaagupi
Viljandi
Puurmani
Tartu
Otepää
Võnnu
Rõpina
Kallaste
Palamuse
Mõisaküla
Võru
Sindi
Pärnu
Audru
Uulu
Kilingi-Nõmme
Kõmme
Staicele
Rõngu
Tõrva
Valga
Valka
Smiltene
Ape
Alūksne
Balvi
Viļaka
Rūjiena
Rugāji
Aloja
Burtnieks
Valmiera
Gulbene
Jaunpiebalga
Lubāns
Varakļāni
Kārsava
Ludza
Rēzekne
Malta
Spogi
Vormsi
Kärdla
Hiiumaa
Emmaste
Orissaare
Saaremaa
Kuressaare
Saäre
Väinameri
Suur Väin
Kihnu
Ruhnu
Ainaži
Salacgrīva
Mērsrags
Saulkrasti
Cēsis
Gauja
Munamägi
318m
RIGA
Jūrmala
Jelgava
Iecava
Bauska
Madona
Jēkabpils
Pļaviņas
Līvāni
Western Dvina
Obeliai
Rokiškis
Nereta
Viesīte
Aizkraukle
Pasvalys
Biržai
Pakruojis
Radviliškis
Panevėžys
Subačius
Naujamiestis
Dotnuva
Kolkasrags
Kolka
Roja
Talsi
Engures Ezers
Engure
Tukums
Brocēni
Saldus
Kandava
Usmas Ezers
Ugāle
Venta
Kuldīga
Durbe
Skrunda
Ventspils
Mazirbe
Pavilosta
Grobiņa
Skuodas
Mažeikiai
Telšiai
Papilė
Joniškis
Šiauliai
Kelmė
Raseiniai
Skaudvilė
Jurbarkas
Plungė
Gargždai
Šiaulė
Liepāja
Rucava
Kretinga
Priekulė
Klaipėda
Šilutė
Taurage
Neman
Nida
Zelenogradsk
Gvardeysk
Chernyakhovsk
Gusev
KALININGRAD (to Russia)
Primorsk
Pionerskiy
Mamonovo
Kaliningrad
Bagrationovsk
Zheleznodorozhnyy
Zhelezdorozhnyy

Emajõgi
Žemaičių Aukštumas
Neman

Population

● National capital

○ below 50,000   ○ 50,000 to 100,000   ◉ 100,000 to 500,000   ▪ above 500,000

0 km   100
0 miles   100

84

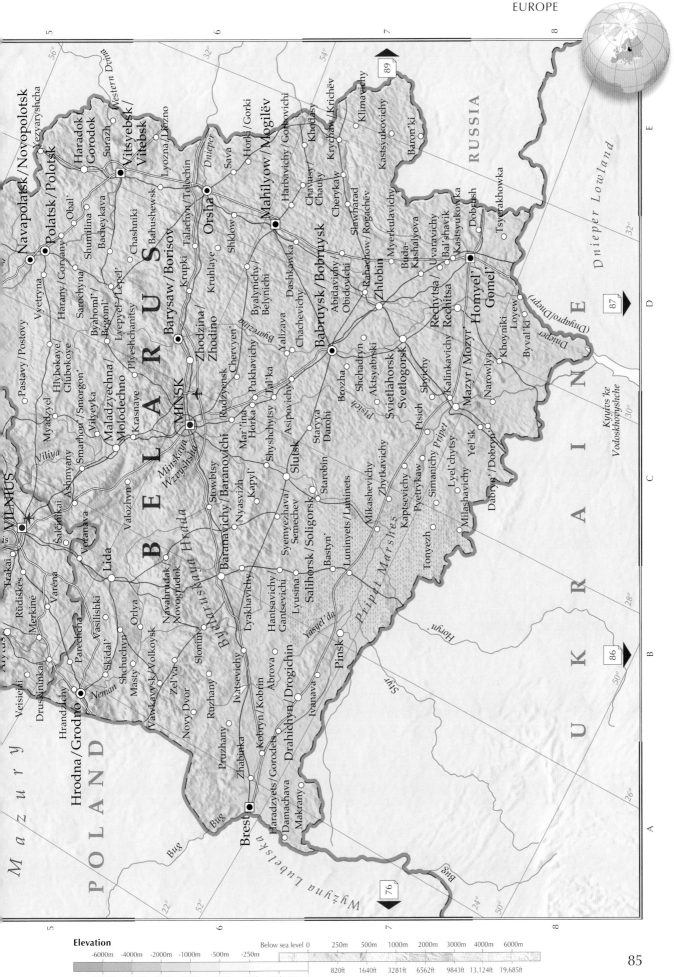

Elevation

| Below sea level 0 | 250m | 500m | 1000m | 2000m | 3000m | 4000m | 6000m |

| -6000m | -4000m | -2000m | -1000m | -500m | -250m |

| 820ft | 1640ft | 3281ft | 6562ft | 9843ft | 13,124ft | 19,685ft |

| -19,658ft | -13,124ft | -6562ft | -3281ft | -1640ft | -820ft | -328ft/-100m | 0 |

Ukraine, Moldova & Romania

POLAND

BELARUS

Małopolska

Wyżyna Lubelska

Pripet

Pripet Marshes

Kovel'
Sarny
Olevs'k
Korosten
Volodymyr-Volyns'kyy
Novovolyns'k
Kivertsi
Luts'k
Rivne
Sokal'
Dubno
Novohrad-Volyns'kyy
Radomys
Zhovkva
Chervonohrad
Slavuta
Shepetivka
Yavoriv
Kremenets'
Polonne
Zhytom
L'viv
Zolochiv
Izyaslav
Berdy
Horodok
Zbarazh
Starokostyantyniv

Carpathian Mountains

Tatra Mountains

77

SLOVAKIA

Slovenské Rudohorie

Sambir
Drohobych
Boryslav
Khodoriv
Berezhany
Ternopil'
Khmel'nyts'kyy
Koz
Stryy
Zhydachiv
Vinnytsya
Kalush
Chortkiv
Dolyna
Ivano-Frankivs'k
Zhmerynka
Uzhhorod
Nadvirna
Kam'yanets'-Podil's'kyy
Lypov
Mukacheve
Kolomyya
Chernivtsi
Mohyliv-Podil's'kyy
Berehove
Darabani
Soroca
Vynohradiv
Khust
Hora Hoverla 2061m
Negresti-Oaş
Satu Mare
Rădăuţi
Dorohoi
Balţi
Botoşani
Rîbniţa
Carei
Baia Mare
Solca
Marghita
Baia Sprie
Borşa
Suceava
MOLDOVA
Şimleu Silvaniei
Zalău
Năsăud
Fălticeni
Călăraşi
Oradea
Dej
Bistriţa
Târgu-Neamţ
Paşcani
Orhei
Aleşd
Toplita
Bicaz
Iaşi
Ungheni
Străseni
Beiuş
Cluj-Napoca
Reghin
Roman
CHIŞINĂU
Salonta
Ineu
Turda
Ludus
Gheorgheni
Piatra-Neamţ
(KISHINEV)
Tig
Transylvania
Reghin
Bacău
Hinceşti
Curtici
Muntii Apuseni
Abrud
Aiud
Târgu Mureş
Miercurea-Ciuc
Vaslui
Tirasp
Arad
Cristuru Secuiesc
Târgu Ocna
Sânnicolau Mare
Lipova
Alba Iulia
Medias
Rupea
Bârlad
Basara
Jimbolia
Deva
ROMANIA
Comrat
Ciadîr-Lu
Timişoara
78
Făgăraş
Târgu Secuiesc
Adjud
Artsyz
Lugoj
Hunedoara
Sibiu
Codlea
Sfântu Gheorghe
Cahul
Taraclia
Oţelu Roşu
Cisnădie
Vârful Moldoveanu 2544m
Focşani
Tecuci
Bolhrad
Bocşa
Hateg
Câmpulung
Braşov
Reni
Reşiţa
Petroşani
Carpaţii Meridionali
Râşnov
Ozero Yalpuh
Kili
Anina
Târgu Jiu
Călimăneşti
Sinaia
Râmnicu Sărat
Galaţi
Oraviţa
Curtea de Arges
Câmpina
Buzău
Braila
Moldova Nouă
Motru
Moreni
Mizil
Macin
Izmayil
Orşova
Râmnicu Vâlcea
Pitesti
Urziceni
Isaccea
Tulcea
Drobeta-Turnu Severin
Strehaia
Titu
Ploieşti
Ţăndărei
Babadag
Filiaşi
Drăgăşani
Târgovişte
Buftea
Lacul Razi
SERBIA
Wallachia
Sloboza
Hârşova
Lacul Sinoie
Craiova
Slatina
Olteniţa
Fetesti
Medgidia
Baileşti
Caracal
BUCUREŞTI
(BUCHAREST)
Călăraşi
Calafat
Roşiori de Vede
Alexandria
Constanţa
Corabia
Olt
Turnu Măgurele
Giurgiu
Techirghiol
82
Zimnicea
Eforie-Sud
Dunavska Ravnina
Mangalia
BULGARIA

0 km 100
0 miles 100

Population ● National capital

○ below 50,000 ○ 50,000 to 100,000 ◉ 100,000 to 500,000 ■ above 500,000

RUSSIA

Srednerusskaya Vozvyshennosť

Don

Horodnya
Snovs'k (Shchors)
Shostka
Hlukhiv
Chernihiv
Krolevets'
Konotop
Bakhmach
Nizhyn
Romny
Sumy
Nosivka
Prylyky
Lebedyn
Yahotyn
Pyryatyn
Okhtyrka
Brovary
Zolochiv
Derhachi
Vasyl'kiv
Hrebinka
Lubny
Myrhorod
Lyubotyn
Kharkiv
Kaniv
Bila Tserkva
Merefa
Kup"yans'k
Bohuslav
Zolotonosha
Hlobyne
Poltava
Starobil's'k
Horodyshche
Cherkasy
Donets
Izyum
Kreminna
Smila
Chyhyryn
Kremenchuts'ke Vodoskhovyshche
Rubizhne
Shpola
Syeverodonets'k
Tal'ne
Oleksandrivka
Svitlovods'k
Kremenchuk
Slov"yans'k
Znam"yanka
Dniprodzherzhyns'ke Vodoskhovyshche
Kramators'k
Lysychans'k
Mala Vyska
Oleksandriya
Zolote
Luhans'k
Zhovti Vody
Kam"yans'ke (Dniprodzherzhyns'k)
Novomoskovs'k
Kostyantynivka
Kadiyivka (Stakhanov)
Sorokyne (Krasnodon)
Kropyvnyts'kyy (Kirovohrad)
P"yatykhatky
Dnipro (Dnipropetrovs'k)
Pavlohrad
Horlivka
Yenakiyeve
Khrustal'nyy (Krasnyy Luch)
Dolyns'ka
Synel'nykove
Makiyivka
Pervomays'k
Bobrynets'
Kryvyy Rih
Pokrovs'ke
Chystyakove (Torez)
Amvrosiyivka
Arbuzynka
Inhulets'
Donets'k
Novyy Buh
Pokrov (Ordzhonikidze)
Nikopol
Zaporizhzhya
Orikhiv
Dokuchayevs'k
Voznesens'k
Kam"yanka-Dniprovs'ka
Marhanets'
Volnovakha
Dniprorudne
Polohy
Don
Tokmak
Novoazovs'k
Kakhovs'ka Vodoskhovyshche
Molochans'k
Mariupol'
Mykolaiv
Kakhovka
Melitopol'
Gulf of Taganrog
Yeya
Zhovtneve
Dnieper (Dnipro)
Kherson
Yakymivka
Prymors'k
Berdyans'k
Ochakiv
Oleshky (Tsyurupyns'k)
Odesa
Hola Prystan'
Chaplynka
Novotroyits'ke
Heniches'k
88
Chornomors'k (Illichivs'k)
Kalanchak
Armyans'k
Sea of Azov
Yany Kapu (Krasnoperekops'k)
RUSSIA
Karkinits'ka Zatoka
Rozdol'ne
Dzhankoy
Nyzhn'ohirs'kyy
Kerch Strait
Chornomors'ke
Krasnohvardiys'ke
Zatoka Syvash
Kerch
Kuban'
Yevpatoriya
Kryms'kyy Pivostriv (Crimea)
Lenine
Saky
Feodosiya
Simferopol'
(since 2014 the Ukrainian territory of Crimea has been annexed by Russia)
Bakhchysaray
Kryms'ki Hory
Sevastopol'
Alushta
Yalta
Alupka

Black Sea

94

Elevation

| -6000m | -4000m | -2000m | -1000m | -500m | -250m | Below sea level 0 | 250m | 500m | 1000m | 2000m | 3000m | 4000m | 6000m |

-19,658ft -13,124ft -6562ft -3281ft -1640ft -820ft -328ft/-100m 0   820ft 1640ft 3281ft 6562ft 9843ft 13,124ft 19,685ft

European Russia

0 km 300
0 miles 300

Population ● National capital

○ below 50,000   ○ 50,000 to 100,000   ◉ 100,000 to 500,000   ◼ above 500,000

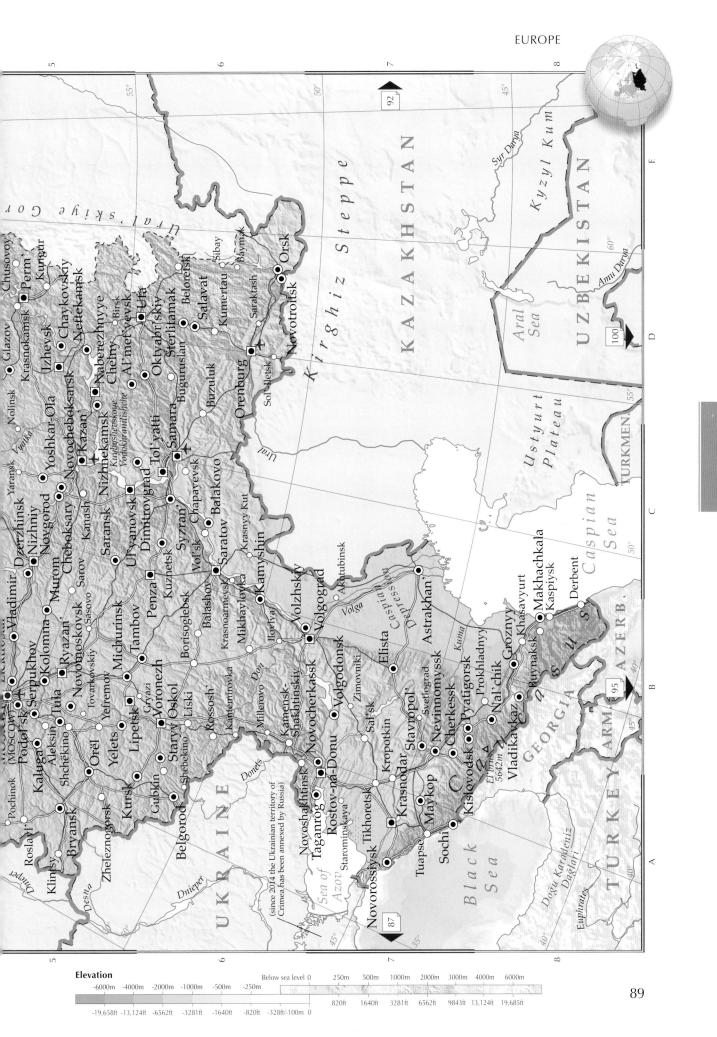

Elevation

						Below sea level 0	250m	500m	1000m	2000m	3000m	4000m	6000m
-6000m	-4000m	-2000m	-1000m	-500m	-250m								
-19,658ft	-13,124ft	-6562ft	-3281ft	-1640ft	-820ft	-328ft/-100m 0		820ft	1640ft	3281ft	6562ft	9843ft	13,124ft 19,685ft

(since 2014 the Ukrainian territory of Crimea has been annexed by Russia)

North & West Asia

133

Franz Josef Land

A R C T I C

Severnaya Ze

Ostrov Komsomolets

Ostrov Oktyabr'skoy Revolyutsii

Ostrov Bol'shevik

Summer limit of pack ice

Norwegian
Sea   North Cape

Winter limit of pack ice

Barents
Sea

Novaya Zemlya

East Novaya Zemlya Trench

Kara Sea

Poluostrov Taymy

North Siber

Kheta

Ostrov
Kolguyev

Poluostrov
Yamal

Central
Siberian
Plateau

Noril'sk

Murmansk
Kola
Peninsula

Arctic Circle

59

White Sea

R   U   S   S   i

Lake
Onega

Northern
Dvina

Ob'

West Siberian
Plain

Ob'

Kureyka

Lower Tunguska

Yenisey

Stony Tunguska

Angara

Chulym

Gulf of Bothnia

Lake Ladoga

Saint Petersburg
Yaroslavl'

Vologda

Nizhniy
Novgorod

Perm'

Ural Mountains

Yekaterinburg

Irtysh

Irtysh

Ishim

Tomsk

Krasnoyarsk

Irkt

Baltic Sea

MOSCOW

Kazan'

Ufa

Chelyabinsk

Omsk

Novosibirsk

Novokuznetsk

Kaliningrad

Central
Russian
Upland

Ul'yanovsk

Samara

Volga

ASTANA

Sayanskiy Khrebet

KALININGRAD
(to Russia)

Voronezh

Saratov

Orenburg

Ural'sk

Karagandy

Semipalatinsk

Altai Mountains

Volgograd

Ural

Kirghiz
Steppe

Kazakh Uplands

Ozero
Zaysan

(since 2014 the Ukrainian
territory of Crimea has
been annexed by Russia)

Volga

Don

Danube

Rostov-na-Donu

Stavropol'

Astrakhan'

KAZAKHSTAN

Aral'sk

Syr Darya

Lake
Balkhash

Ili

E U R O P E

Black Sea

El'brus
18,510ft
(5642m)

Caucasus

Caspian

Aktau

Ustyurt
Plateau

Aral
Sea

Kyzyl
Kum

Kyzylorda

Taraz

Almaty

Tien

Shan

Jengish Chokusu/Tömür Feng
24,406ft (7443m)

Istanbul

Küre Dağları

Sea

UZBEKISTAN

Dasoguz

Amu Darya

BISHKEK

TASHKENT

KYRGYZSTAN

ANKARA

GEORGIA
ARMENIA

YEREVAN

TBILISI

BAKU

AZERB.

TURKMENISTAN

Garagum

DUSHANBE

TAJIKISTAN

TURKEY

Lake
Van

Tabriz

ASHGABAT

Kunlun Mountains

Adana

Gaziantep

Mosul

TEHRAN

Hindu Kush

KABUL

Jalalabad

Aleppo

Qom

IRAN

Isfahan

Herat

Khyber Pass

Himalayas

CYPRUS

81

SYRIA

IRAQ

DAMASCUS

Tigris

BAGHDAD

Euphrates

Syrian
Desert

Iranian
Plateau

AFGHANISTAN

Mediterranean Sea

BEIRUT
LEBANON
ISRAEL
JERUSALEM

AMMAN

Basra

Zagros Mountains

Shiraz

Zahedan

Thar Desert

Ganges

Dead Sea
-1411ft
(- 430m)

JORDAN

KUWAIT
KUWAIT

Persian Gulf

Bandar-e 'Abbas

Indus Fan

An Nafud

MANAMA

Dubai

Murray Ridge

Ganges Fan

SAUDI
ARABIA

BAHRAIN

RIYADH

QATAR

DOHA
U.A.E.

Gulf of Oman

MUSCAT

Tropic of Cancer

Jedda

ABU
DHABI

Sur

Nile

Red Sea

Arabian
Peninsula

OMAN

Murray Ridge

At Ta'if

Ar Rub' al Khali

Arabian
Sea

Bay of
Bengal

A F R I C A

N

47

SANAA

YEMEN

Ta'izz

Aden

Gulf of Aden

Socotra
(to Yemen)

0 km   800

0 miles   800

Population   • National capital

○ below 50,000   ○ 50,000 to 100,000   ◉ 100,000 to 500,000   ◼ above 500,000

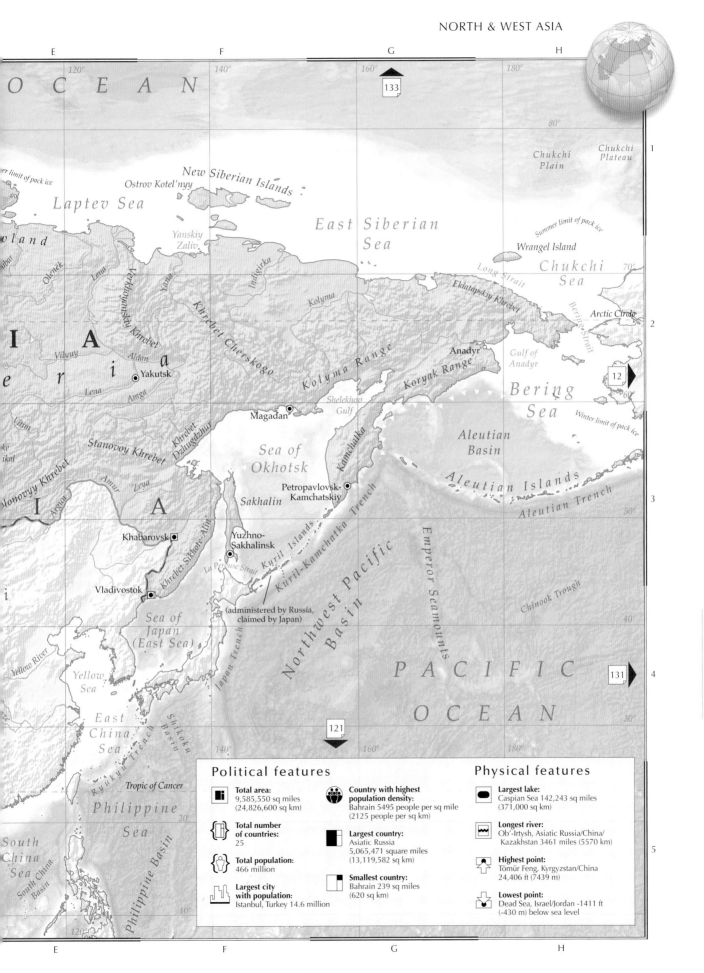

E 120° F 140° G 160° H 180°

O C E A N

133

80°

1

Upper limit of pack ice

Ostrov Kotel'nyy

New Siberian Islands

Laptev Sea

Chukchi Plain

Chukchi Plateau

East Siberian Sea

Yanskiy Zaliv

Summer limit of pack ice

Wrangel Island

Long Strait

Chukchi Sea

Ekiatapskiy Khrebet

70°

Bering Strait

Arctic Circle

2

S I B E R I A

Olenëk

Lena

Verkhoyanskiy Khrebet

Vilyuy

Aldan

● Yakutsk

Lena

Amga

Indigirka

Kolyma

Khrebet Cherskogo

Kolyma Range

Koryak Range

● Anadyr

Gulf of Anadyr

12

60°

Bering Sea

Vitim

Aldan

Stanovoy Khrebet

Golonovyy Khrebet

Khrebet Dzhugdzhur

Shelekhov Gulf

Aleutian Basin

Winter limit of pack ice

Magadan ●

Amur

Zeya

Kamchatka

Sea of Okhotsk

Petropavlovsk-Kamchatskiy ●

Aleutian Islands

Aleutian Trench

50°

3

Khabarovsk ◉

Sakhalin

Yuzhno-Sakhalinsk ◉

Khrebet Sikhote-Alin'

Kuril Islands

La Pérouse Strait

Kuril-Kamchatka Trench

Northwest Pacific Basin

Emperor Seamounts

Chinook Trough

40°

Vladivostok ◉

(administered by Russia, claimed by Japan)

Sea of Japan (East Sea)

Japan Trench

P A C I F I C

131

4

Yellow River

Yellow Sea

East China Sea

Shikoku Basin

O C E A N

30°

121

F 140° G 160° H 180°

Tropic of Cancer

20°

Philippine Sea

Ryukyu Trench

South China Sea

South China Basin

Philippine Basin

10°

120° E

5

Political features

📊 **Total area:**
9,585,550 sq miles
(24,826,600 sq km)

🔢 **Total number of countries:**
25

👥 **Total population:**
466 million

🏙 **Largest city with population:**
Istanbul, Turkey 14.6 million

👪 **Country with highest population density:**
Bahrain 5495 people per sq mile
(2125 people per sq km)

⬛ **Largest country:**
Asiatic Russia
5,065,471 square miles
(13,119,582 sq km)

⬜ **Smallest country:**
Bahrain 239 sq miles
(620 sq km)

Physical features

⬛ **Largest lake:**
Caspian Sea 142,243 sq miles
(371,000 sq km)

〰 **Longest river:**
Ob'-Irtysh, Asiatic Russia/China/Kazakhstan 3461 miles (5570 km)

⛰ **Highest point:**
Tömür Feng, Kyrgyzstan/China
24,406 ft (7439 m)

🔻 **Lowest point:**
Dead Sea, Israel/Jordan -1411 ft
(-430 m) below sea level

E F G H

Russia & Kazakhstan

Arctic Circle

SVALBARD
(to Norway)

Zemlya Fran
Iosifa

NETH.

DENMARK

GERMANY

SWEDEN

NORWAY

Baltic Sea

Gulf of Bothnia

FINLAND

Gulf of Finland

Nordkapp
(North Cape)

Barents
Sea

Novaya Zemlya

Karskoye More

Ostrov Belyy

Diksc

KALININGRAD
(to Russia)

Kaliningrad

POLAND

LITH.

LAT.

EST.

Murmansk

Kandalaksha

Kol'skiy
Poluostrov

Beloye More

Ostrov
Kolguyev

Nar'yan-Mar

Pechora

Poluostrov Yamal

Obskaya Guba

Ta

Sankt-Peterburg

Pskov

Velikiy Novgorod

BELARUS

Ladozhskoye
Ozero

Petrozavodsk

Onezhskoye
Ozero

Severnaya
Dvina

Severodvinsk

Arkhangel'sk

UKRAINE

Smolensk

Cherepovets

MOSKVA
(MOSCOW)

Tver

Vologda

Vel'sk

Kotlas

Ukhta

Vorkuta

Salekhard

Nadym

Ob'

Noril

Igarka

Bryansk

Tula

Yaroslavl'

Kineshma

Vladimir

Syktyvkar

MOLDOVA

(since 2014
the Ukrainian
territory of
Crimea has been
annexed by
Russia)

Belgorod

Ryazan'

Nizhniy Novgorod

Kirov

Solikamsk

Nyagan'

Zapadno-

Yenisey

Tambov

Kazan'

Glazov

Perm'

Serov

Khanty-Mansiysk

Sibirskaya

Voronezh

Penza

Ul'yanovsk

Izhevsk

Lesnoy

Surgut

Nizhnevartovsk

Sea of
Azov

Mikhaylovka

Saratov

Tol'yatti

Naberezhnyye
Chelny

Yekaterinburg

Ravnina

Rostov-na-
Donu

Balakovo

Samara

Ufa

Tyumen'

Tobol'sk

R

U

Krasnodar

Volgograd

Sterlitamak

Chelyabinsk

Ob'

Chulym

Sochi

Stavropol'

Ural'sk

Orenburg

Ishim

El'brus
5642m

Ural

Ural
(Zhayyq)

Magnitogorsk

Orsk

Ishim

Irtysh

Petropavlovsk

Tobol

Nal'chik

Astrakhan'

Omsk

Seversk

Tomsk

Vladikavkaz

Groznyy

Aktobe
(Aktyubinsk)

Alga

Rudnyy

Kostanay

Krasnoy

GEORGIA

Makhachkala

Atyrau

Emba

Kokshetau

Novosibirsk

ARM.

AZERBAIJAN

Fort-Shevchenko

Shalkar

Atbasar

Shchuchinsk

Kemerov

Aktau

Zhanaozen

KAZAKHSTAN

ASTANA

Pavlodar

Barnaul

Novokuznetsk

Ustyurt
Plateau

Aral
Sea

Aral'sk

Ayteke Bi

Temirtau

Saran'

Karagandy

Semey

Aba

Ridder

Zyryanovsk

K

Caspian Sea

Syr Darya

Zhosaly

Zhezkazgan

Kazakhskiy
Melkosopochnik

Shar

Ust'-Kamenogorsk

Gora Belukha
4506m

Kyzyl Kum

Kyzylorda

Balkhash

Ayagoz

Ozero
Zaysan

Altai Mountain

Turkistan

Kentau

Karatau

Ozero
Balkhash

Taldykorgan

Tokeli

IRAN

TURKMENISTAN

Amu Darya

UZBEKISTAN

Arys'

Shymkent

Shu

Taraz

Kirghiz Range

Almaty
(Alma-Ata)

Tien Shan

CHINA

TAJIKISTAN

AFGHANISTAN

KYRGYZSTAN

0 km   600

0 miles   600

Population   ● National capital

○ below 50,000   ○ 50,000 to 100,000   ◉ 100,000 to 500,000   ◼ above 500,000

ALASKA
(to US)

Chukchi
Sea

14

OCEAN

Ostrov
Komsomolets

Severnaya
Zemlya

Ostrov Oktyabr'skoy Revolyutsii

Novosibirskiye
Ostrova

Ostrov
Novaya Sibir'

Vostochno-Sibirskoye
More

Proliv Longa

Ostrov Vrangelya

Ekvyvatapskiy Khrebet

Anadyrskiy
Zaliv

Bering
Sea

Pevek

Ambarchik
Cherskiy

Anadyr'

Koryakskoye Nagor'ye

Ostrov Karaginskiy

Ossora

Zaliv
Shelikhova

More
Laptevykh

Ostrov Kotel'nyy

Ostrov Bol'shoy
Lyakhovskiy

Alazeya

Indigirka

Kolyma

Ust'-Kamchatsk

Vulkan
Klyucheyskaya
Sopka 4688m

Atlasovo

Ostrov Taymyr

Ozero
Taymyr

Ust'-Olenëk

Tiksi

Kazach'ye

Khrebet Cherskogo

Adycha

Susuman

Atka

Magadan

Poluostrov
Kamchatka

Mil'kovo

Petropavlovsk-
Kamchatskiy

Yana

Verkhoyanskiy Khrebet

-Sibirskaya Nizmennost'

Kotuy

Anabar

Olenëk

Olenëk

Srednesibirskoye
Ploskogor'ye

Lena

Khrebet

Aldan

Yakutsk

Vilyuy

Amga

Nyurba

Okhotsk

Khrebet Dzhugdzhur

Okhotskoye
More

Pervyy Kuril'skiy Proliv

Ostrov
Paramushir

Lena

Suntar

Olëkminsk

Aldan

Shantarskiye
Ostrova

Mirnyy

Olëkma

Chunya

S   I   B   I   R   '
(SIBERIA)

Neryungri

Bodaybo

Olëkma

Ostrov Sakhalin

Ostrov Urup

Ostrov Iturup

Kuril'sk

130

R S S I A

Ust'-Ilimsk

Vitim

Tynda

Skovorodino

Komsomol'sk-
na-Amure

Amur

Khrebet Sikhote-Alin'

Kuril'skiye Ostrova (Kuril Islands)

Yuzhno-Sakhalinsk

Bratsk

Ust'-Kut

Ozero
Baykal

Yablonovyy Khrebet

Shilka

Amur

Svobodnyy

Khabarovsk

La Pérouse
Strait

Tulun

Khor

(administered by
Russia, claimed
by Japan)

Usol'ye-Sibirskoye

Angarsk

Chita

Blagoveshchensk

Birobidzhan

Bikin

Irkutsk

Ulan-Ude

Olovyannaya

Krasnokamensk

CHINA

Ussuriysk

Vostochnyy Sayan

Kyakhta

Zabaykal'sk

Vladivostok

Nakhodka

JAPAN

MONGOLIA

G   o   b   i

N

NORTH
KOREA

106

Sea of
Japan
(East Sea)

Elevation

Below sea level 0   250m   500m   1000m   2000m   3000m   4000m   6000m

-6000m -4000m -2000m -1000m -500m -250m

-19,658ft -13,124ft -6562ft -3281ft -1640ft -820ft -328ft/-100m 0

820ft   1640ft   3281ft   6562ft   9843ft 13,124ft 19,685ft

Turkey & The Caucasus

ROMANIA

Dacul Sinoie

Danube

BULGARIA

Maritsa

Varnenski Zaliv

Burgaski Zaliv

UKRAINE

Kryms'kyy Pivostriv (Crimea)

(since 2014 the Ukrainian territory of Crimea has been annexed by Russia)

B l a c k   S e a

82

Kırklareli
Edirne
Çorlu
Tekirdağ

İstanbul Boğazı (Bosporus)

Zonguldak
İstanbul
İzmit
Adapazarı
Bandırma
Yalova
İznik Gölü
Bursa
Bilecik

Marmara Denizi (Sea of Marmara)

Çanakkale

Çanakkale Boğazı (Dardanelles)

Balıkesir
Edremit
Ayvalık

Lésvos

Akhisar
Simav
Gediz
Manisa
Uşak
İzmir
Ödemiş
Alaşehir

Chío
Sámos

Menemen
Gediz Nehri
Bozüyük
Eskişehir
Kütahya

ANKARA

Polatlı
Kırıkkale

T U R K

Kulu
Hırfanlı Barajı
Tuz Gölü

İnebolu
Cide
Bartın
Küre Dağları
Devrek
Karabük
Kastamonu
Kargı
Çerkeş
Bolu
Gerede
Çankırı
Kalecik
Kızıl Irmak

Sinop
Gerze
Bafra
Samsun
Ü

Merzifon
Çorum
Alaca
Tokat
Yıldızeli
Sorgun
Siv
Şarkışla
Boğazlıyan
Bünyan
İncesu
Kayseri
Gürün

Cenik Dağları
Or

Cihanbeyli
Akşehir
Nevşehir
Aksaray
Niğde
Göksun
Gü

A n a t o l i a

Aydın
Nazilli
Büyükmenderes Nehri
Denizli
Dinar
Burdur
Beyşehir Gölü
Konya

Sámos
Söke
Milas
Tavas
Muğla
Burdur Gölü
Isparta
Suğla Gölü
Ereğli
Karaman

GR

83

Bodrum
Marmaris
Dalaman
Fethiye
Kaş
Finike

Dodekánisa (Dodecánese)

Ródos (Rhodes)

Kárpathos

Antalya

Manavgat
Alanya
Mut
Silifke
Anamur

Antalya Körfezi

T o r o s   D a ğ l a r ı

Kahramanma

Gazi

Ceyhan
Tarsus
Mersin (İçel)
Adana
Osmaniye
İskenderun
Kilis
Antakya
Kırıkhan

CYPRUS

TURKISH REPUBLIC OF NORTHERN CYPRUS
(recognized only by Turkey)

Orantes

LEBANON

M e d i t e r r a n e a n

S e a

50

0 km   200
0 miles   200

Population   ● National capital

○ below 50,000   ○ 50,000 to 100,000   ◉ 100,000 to 500,000   ▣ above 500,000

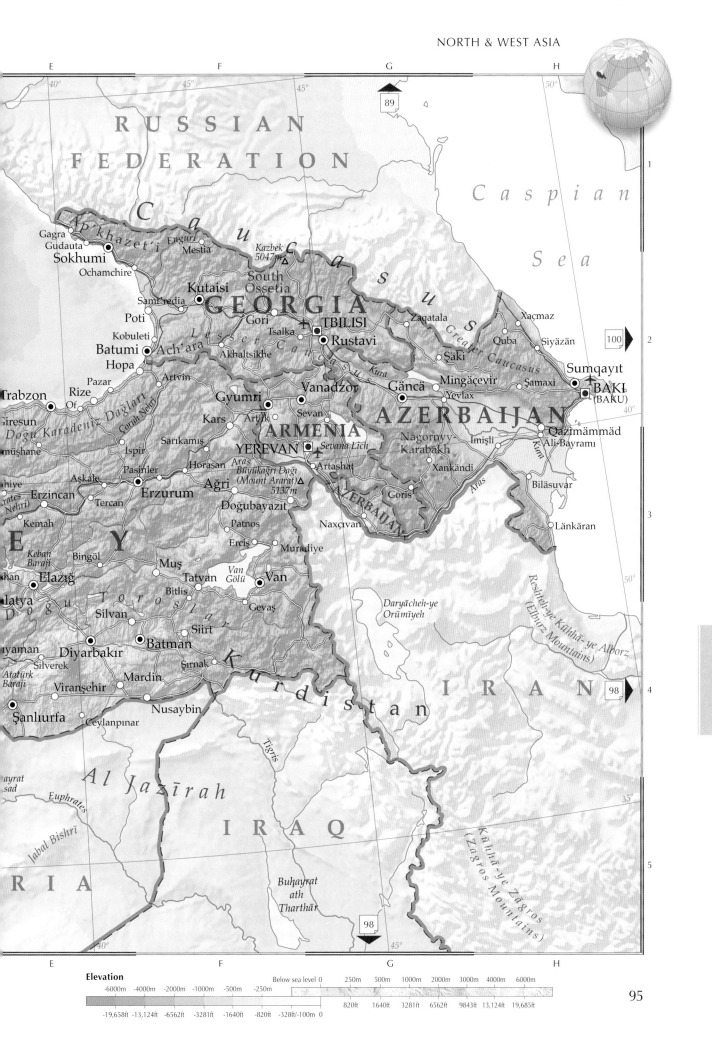

RUSSIAN
FEDERATION

Caspian

Sea

Caucasus

Gagra
Gudauta
Sokhumi
Ochamchire
AP'khazet'i
Enguri
Mestia
Kazbek
5047m
South
Ossetia
Kutaisi
Samtredia
GEORGIA
Gori
Tsalka
TBILISI
Rustavi
Zaqatala
Xaçmaz
Quba
Siyäzän
Poti
Kobuleti
Batumi
Hopa
Ach'ara
Akhaltsikhe
Lesser Caucasus
Şäki
Greater Caucasus
Şamaxı
Sumqayıt
Mingäçevir
Kura
Yevlax
Trabzon
Pazar
Rize
Of
Artvin
Vanadzor
Gäncä
AZERBAIJAN
BAKI
(BAKU)
Giresun
mūshane
Doğu Karadeniz Dağları
Çoruh Nehri
İspir
Gyumri
Kars
Artık
Sevan
Sevana Lich
ARMENIA
YEREVAN
Nagornyy-
Karabakh
İmişli
Qazımämmäd
Äli-Bayramı
Sarıkamış
Horasan
Aras
Artashat
*Büyükağrı Dağı
(Mount Ararat)*
5137m
AZERBAIJAN
Goris
Xankändi
Biläsuvar
Kura
Aras
Aşkale
Pasinler
Ağrı
Doğubayazıt
Patnos
Naxçıvan
Erzincan
Tercan
Erzurum
Länkäran
Kemah
*Euphrates
Nehri*
E Y
Bingöl
Muş
Ercis
Muradiye
Van
Gölü
Van
*Daryācheh-ye
Orūmīyeh*
Reshteh-ye Kūhhā-ye Alborz
(Elburz Mountains)
Elazığ
Keban
Baraji
Doğu Toroslar
Silvan
Bitlis
Tatvan
Gevaş
Diyarbakır
Batman
Siirt
Şırnak
Silverek
Mardin
Kurdistan
IRAN
98
ıyaman
Atatürk
Baraji
Viranşehir
Nusaybin
Tigris
Şanlıurfa
Ceylanpınar
Al Jazīrah
IRAQ
RIA
Euphrates
Jabal Bishrī
ayrat
sad
*Buhayrat
ath
Tharthār*
98
Kūhhā-ye Zāgros
(Zagros Mountains)

89
100
98

Elevation

| Below sea level 0 | 250m | 500m | 1000m | 2000m | 3000m | 4000m | 6000m |

-6000m -4000m -2000m -1000m -500m -250m

820ft 1640ft 3281ft 6562ft 9843ft 13,124ft 19,685ft

-19,658ft -13,124ft -6562ft -3281ft -1640ft -820ft -328ft/-100m 0

The Near East

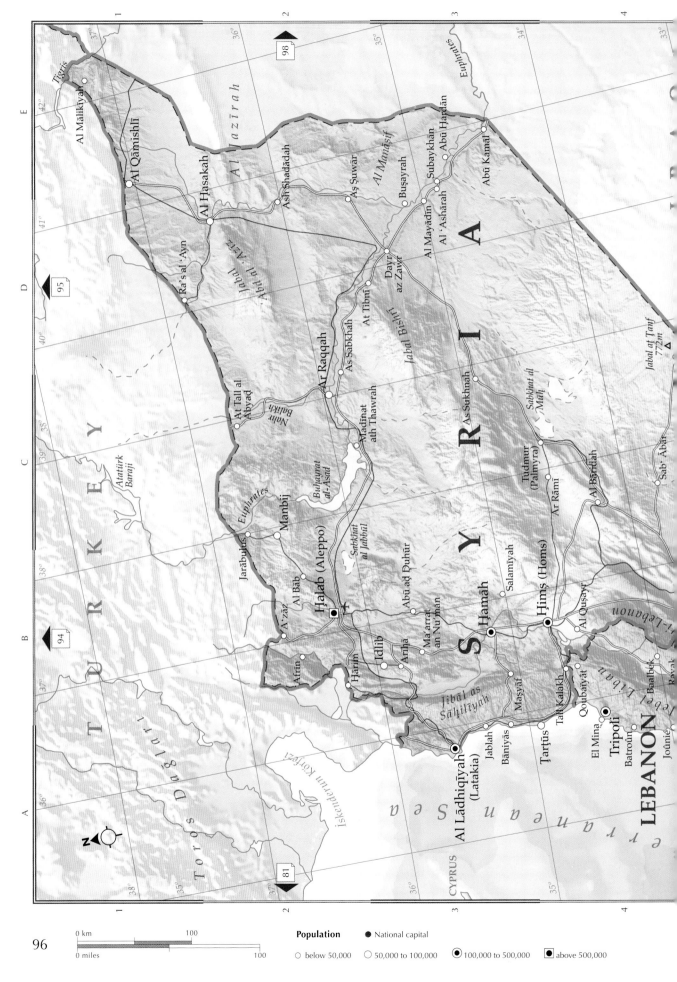

0 km   100

0 miles   100

Population   ● National capital

○ below 50,000   ○ 50,000 to 100,000   ◉ 100,000 to 500,000   ▣ above 500,000

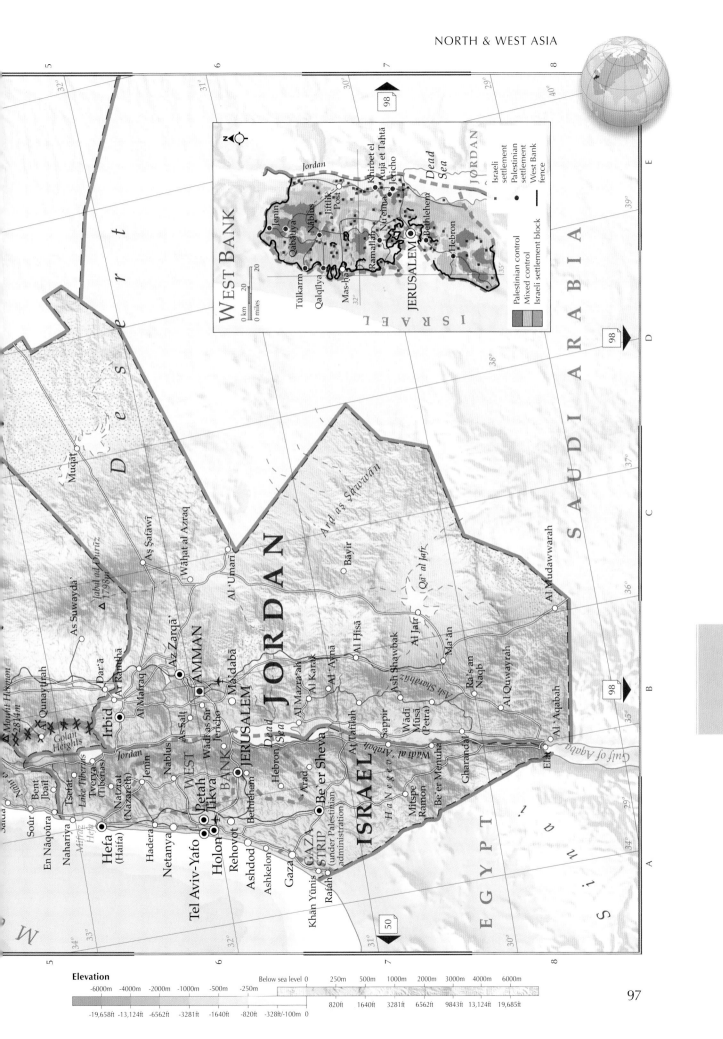

WEST BANK

·	Israeli settlement
●	Palestinian settlement
—	West Bank fence

Palestinian control
Mixed control
Israeli settlement block

Jordan
Khirbet el
Auja et Tahtā
Jericho
Jenin
Jiftlik Post
Nablus
Nu'eima
Qabāṭiya
JERUSALEM
Bethlehem
Ramallāh
Hebron
Tulkarm
Qalqīlya
Mas-ḥa

JORDAN
Dead Sea
ISRAEL

0 km   20
0 miles   20

SAUDI ARABIA

JORDAN

Muqat
As Ṣafāwī
Wāḥat al Azraq
Al 'Umari
Jabal ad Durūz
2814m
As Suwaydā'
Bāyir
Qā' al Jafr
Al Mudawwarah
Ard as Sawwān
As Suwaydā'
Dar'ā
Az Zarqā'
AMMAN
Mā'daba
Al Mafraq
Al Mazra'ah
Al Karak
Al 'Aynā
Ash Shawbak
Al Jafr
Ma'ān
Al Ḥisā
Ash Sharāhiz
Ra's an Naqb
Al Quwayrah
As Salt
Wādī as Sir
Jericho
Al Marah
Al Quwayrah
Sappir
Wādī Mūsā
(Petra)
Al Ḥisā
Gharandal
Al 'Aqabah
Elat
Gulf of Aqaba

JORDAN
Mount Hermon
2814m
Al Qunayṭirah
Golan Heights
Irbid
Ar Ramthā
Jordan
Lake Tiberias
Tverya
(Tiberias)
Naṣrat
(Nazareth)
Jenin
Nablus
WEST
BANK
JERUSALEM
Bethlehem
Dead Sea
Hebron
At Ṭafīlah
Wādī al 'Arabah

Bent
Jbaïl
Ṣaïda
Ṣoûr
En Nâqoûra
Tsefat
Nahariya
Hefa
(Haifa)
Hadera
Netanya
Petaḥ
Tikva
Holon
Tel Aviv-Yafo
Rehovot
Ashdod
Ashkelon
Gaza
Khān Yūnis
Rafah
GAZA
STRIP
(under Palestinian administration)
Be'er Sheva
Arad
ISRAEL
Ha Negev
Mitspe Ramon
Be'er Menuha

EGYPT
Sinai

Elevation

	Below sea level 0	250m	500m	1000m	2000m	3000m	4000m	6000m

-6000m -4000m -2000m -1000m -500m -250m

820ft 1640ft 3281ft 6562ft 9843ft 13,124ft 19,685ft

-19,658ft -13,124ft -6562ft -3281ft -1640ft -820ft -328ft/-100m 0

The Middle East

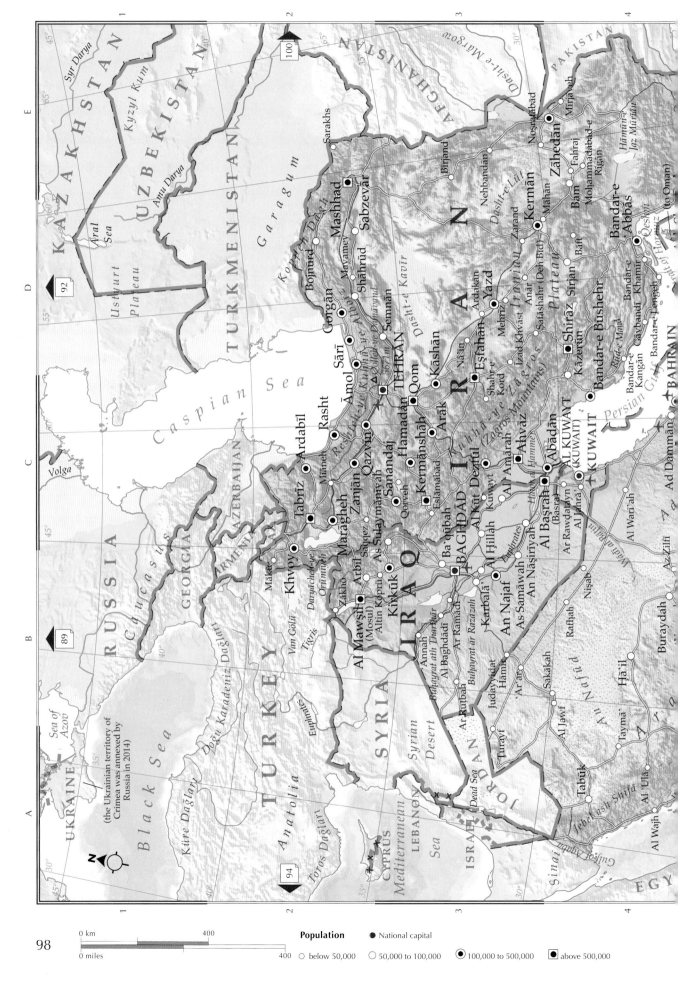

Population ● National capital

0 km	400		
0 miles	400		

○ below 50,000   ○ 50,000 to 100,000   ◉ 100,000 to 500,000   ◼ above 500,000

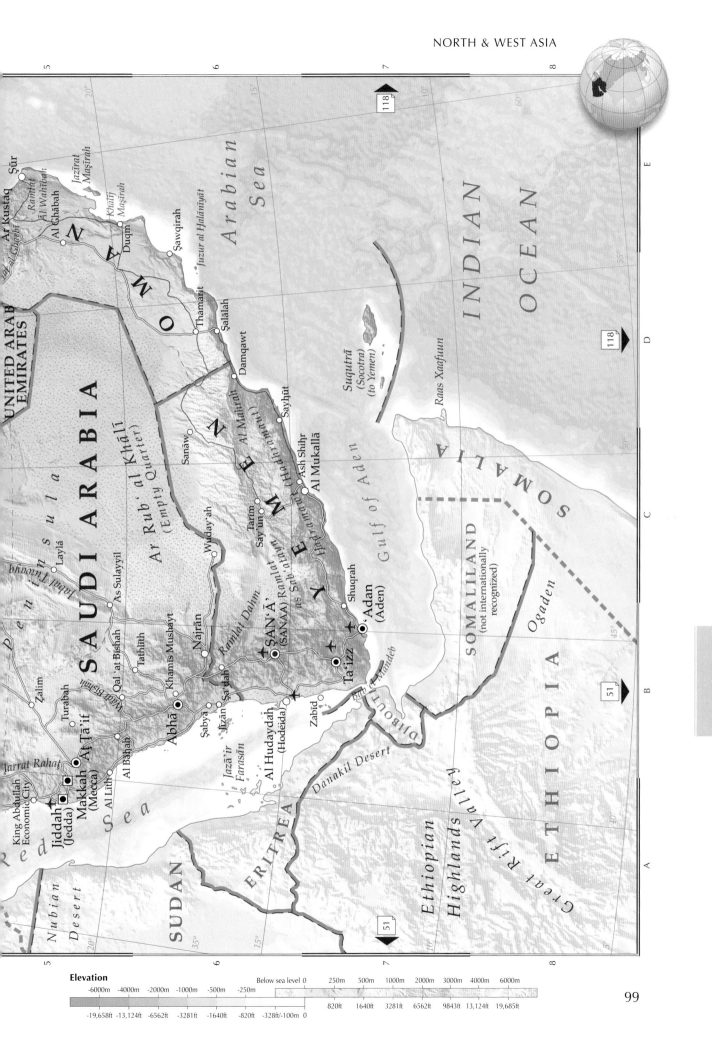

5   6   7   8

UNITED ARAB
EMIRATES

Pen insula

SAUDI ARABIA

O M A N

Şūr
Ar Rustaq
Ramlat
Al Wahîbah
Al Ghabah
Jazīrat
Maṣīrah

Khalīj
Maṣīrah

Duqm
Şawqirah

Jazūr al Halānīyāt

Arabian
Sea

INDIAN

OCEAN

Ar Rub' al Khālī
(Empty Quarter)

Thamarīt
Şalālah

Damqawt
Sanāw

AL MAHRAH

Sayhūt

Suquṭrā
(Socotra)
(to Yemen)

Raas Xaafuun

Jabal Ţuwayq

Laylā

As Sulayyil

Wyday'ah

Tarīm
Sayʾūn

Hadramawt
(Hadramaut)

Ash Shiḥr
Al Mukallā

Gulf of Aden

SOMALIA

Zalim
Turabah

Qal'at Bishah
Tathlīth

Najrān
Khamis Mushayt

Y E M E N

Ramlat as Sab 'atayn

Ramlat Daḥm

Shuqrah

SOMALILAND
(not internationally
recognized)

Ogaden

Jarrat Rahat

Aţ Ţā'if

Abhā
Al Bāḥah

Şabya
Jīzān
Saʿdah

SAN'Ā'
(SANAA)

Ta'izz

Adan
(Aden)

Red Sea

King Abdullah
Economic City
Jiddah
(Jedda)
Maknah
(Mecca)
Al Lith

Zabid

Al Hudaydah
(Hodeida)

Jazā'ir
Farasān

Bab el Mandeb

DJIBOUTI

ETHIOPIA

Nubian
Desert

SUDAN

ERITREA

Danakil Desert

Ethiopian
Highlands

Great Rift Valley

118
51

Elevation

| Below sea level 0 | 250m | 500m | 1000m | 2000m | 3000m | 4000m | 6000m |

-6000m -4000m -2000m -1000m -500m -250m

820ft 1640ft 3281ft 6562ft 9843ft 13,124ft 19,685ft

-19,658ft -13,124ft -6562ft -3281ft -1640ft -820ft -328ft/-100m 0

99

Central Asia

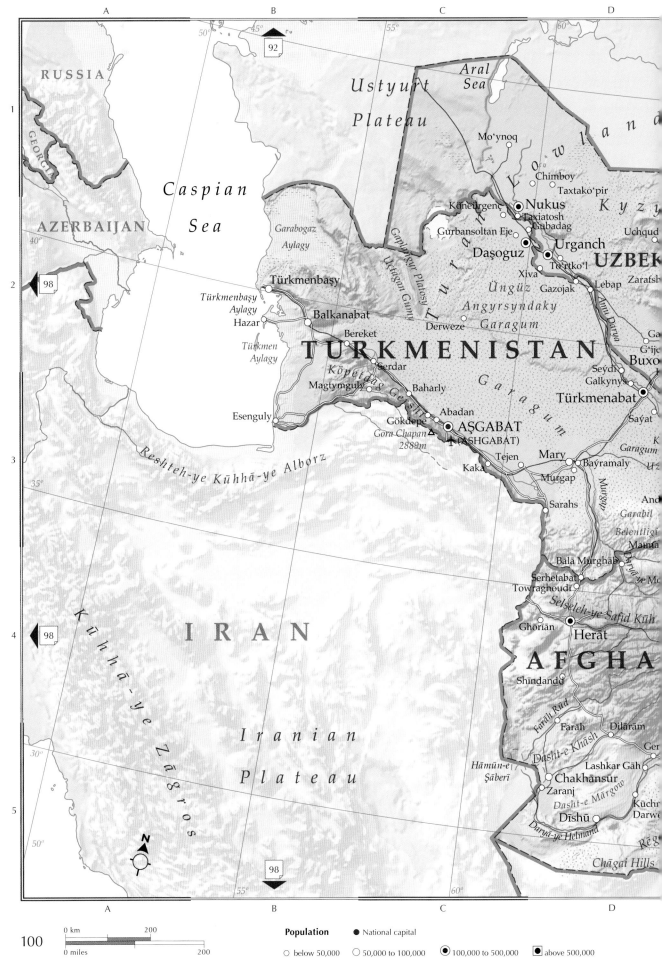

RUSSIA

GEORGIA

AZERBAIJAN

Caspian

Sea

Ustyurt

Plateau

Aral
Sea

Garabogaz
Aylagy

Mo'ynoq

Chimboy

Taxtako'pir

Kyzy

Köneürgenç
Nukus
Taxiatosh
Cubadag

Gurbansoltan Eje
Uchqud

Daşoguz
Urganch
To'rtko'l
UZBEK

Türkmenbaşy
Xiva
Zarafsh

Türkmenbaşy
Aylagy
Gazojak
Lebap

Hazar
Balkanabat
Üngüz
Angyrsyndaky
Garagum

Türkmen
Aylagy
Derweze
Ga

G'ijd

TURKMENISTAN
Buxo

Köpet
Serdar
Seýdi

Magtymguly
Baharly
Garagum
Galkynyş

Esenguly
Abadan
Türkmenabat

Gökdepe
AŞGABAT
Saýat

Gora Chapan
(ASHGABAT)
2889m
Kaka
Tejen
Mary
Bayramaly
Garagum
U

Reshteh-ye Kūhhā-ye Alborz
Murgap
An

Sarahs
Garabil
Belentligi
Maima

IRAN
Bālā Murghāb
Daryā-ye M

Serhetabat

Towraghoudī
Selseleh-ye Safīd Kūh

Ghōriān
Herāt

AFGHA

Kūhhā-ye Zāgros
Shīndandd

Farāh Rūd
Farāh
Dilārām

Iranian
Dasht-e Khāsh
Ger

Hāmūn-e
Şāberī
Lashkar Gāh

Plateau
Chakhānsūr
Küchr

Zaranj
Darw

Dasht-e Mārgow

Dīshū

Daryā-ye Helmand
Rēg

Chāgai Hills

N

0 km   200

0 miles   200

Population   ● National capital

○ below 50,000   ○ 50,000 to 100,000   ◉ 100,000 to 500,000   ◼ above 500,000

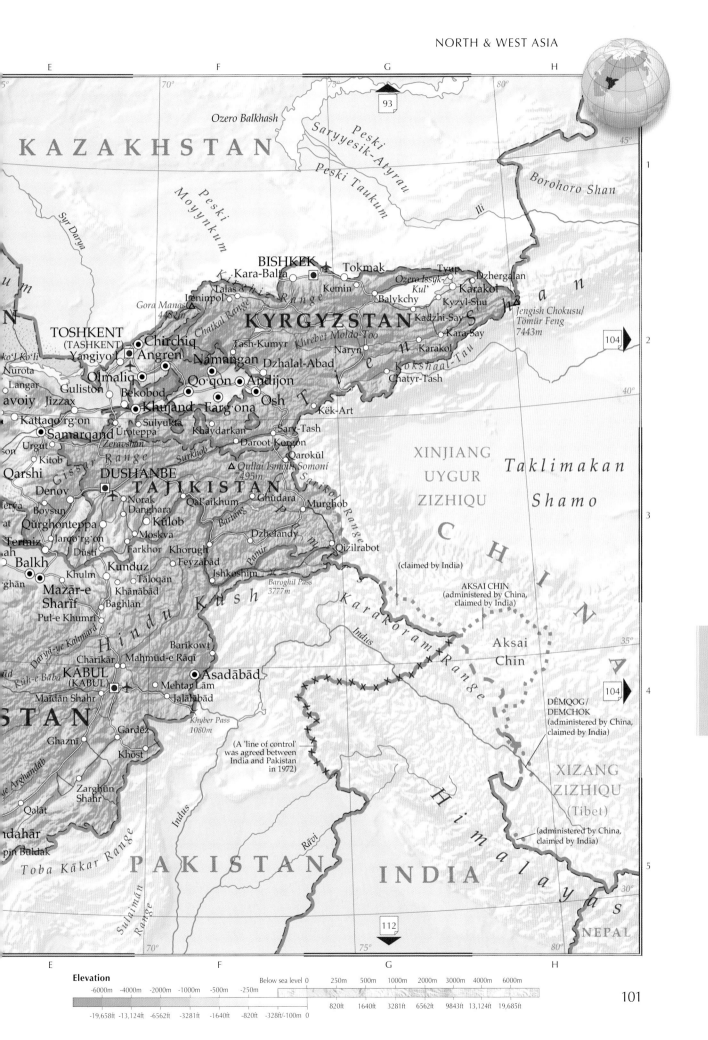

KAZAKHSTAN

Ozero Balkhash

Peski Saryyesik-Atyrau

Peski Taukum

Peski Moyynkum

Syr Darya

Ili

Borohoro Shan

BISHKEK · Tokmak
Kara-Balta
Talas Kemin *Ozero Issyk-* Tyup Dzhergalan
Ireninpol Balykchy *Kul'* Karakol
Gora Manas Kyzyl-Suu *Jengish Chokusu/*
4482m **KYRGYZSTAN** Kadzhi-Say *Tömür Feng*
Kara-Say *7443m*
TOSHKENT *Khrebet Moldo-Too* Karakol
(TASHKENT) Tash-Kumyr Naryn
Yangiyo Chirchiq Namangan Dzhalal-Abad Chatyr-Tash
Angren
Ol maliq Qo'qon Andijon
Nurota Bekobod Osh
Langar Khujand Farg'ona Kök-Art
avoiy Jizzax
Kattaqo'rg'on Samarqand Uroteppa Sulyukta Khaydarkan Sary-Tash
Urgut *Zeravshan* Daroot-Korgon
Kitob *Gissar Range* Surkhob Qarokül
Qarshi **DUSHANBE** △ *Qullai Ismoili Somoni*
Denov Norak *7495m*
Boysun Danghara Qal'aikhum Ghüdara Murghob
Qürghonteppa Kŭlob
Termiz Jarqo'rg'on Moskva Dzhelandy
ah Dusti Farkhor Khorugh Qizilrabot
Balkh Khulm Kunduz Feyzabad
ghan Khānābād Tāloqān Ishkoshim *Baroghil Pass*
Mazār-e Baghlān *3777m*
Sharīf Pul-e Khumrī

Hindu Kush

Daryā-ye Kahmard

Barīkowt
Chārīkār Mahmūd-e Rāqī
KABUL Asadābād
(KABUL)
Maīdān Shahr Mehtar Lām
Kūh-e Bābā Jalālābād

STAN

Ghaznī Gardēz

Khyber Pass (A 'line of control'
1080m was agreed between
Khōst India and Pakistan
in 1972)

Zarghūn
Shahr

ye Arghandāb

Qalāt

ndahār

ṗin Buldak

Toba Kākar Range

PAKISTAN

Sulaimān Range

Indus

Rāvi

INDIA

**XINJIANG
UYGUR
ZIZHIQU**

*Taklimakan
Shamo*

C H I N A

(claimed by India)

AKSAI CHIN
(administered by China,
claimed by India)

Karakoram Range

Aksai
Chin

DÊMQOG/
DEMCHOK
(administered by China,
claimed by India)

**XIZANG
ZIZHIQU**
(Tibet)

(administered by China,
claimed by India)

Himalayas

NEPAL

Elevation

Below sea level 0	250m	500m	1000m	2000m	3000m	4000m	6000m

-6000m -4000m -2000m -1000m -500m -250m

-19,658ft -13,124ft -6562ft -3281ft -1640ft -820ft -328ft/-100m 0

820ft 1640ft 3281ft 6562ft 9843ft 13,124ft 19,685ft

South & East Asia

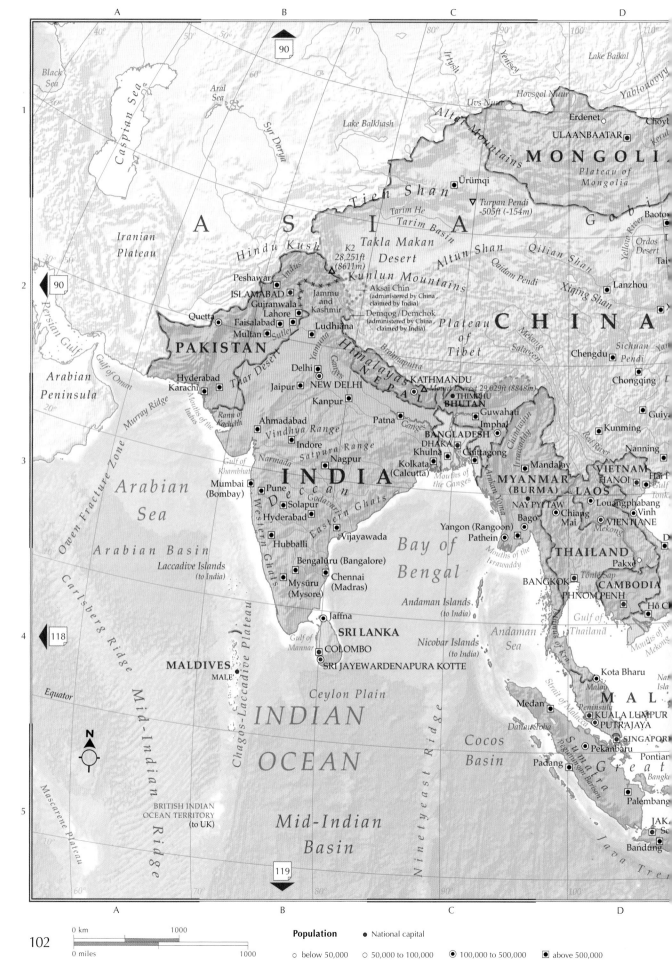

Black Sea

Caspian Sea

Aral Sea

Syr Darya

Lake Balkhash

Irtysh

Yenisey

Hovsgol Nuur

Uvs Nuur

Lake Baikal

Yablonovyy

Altai Mountains

Erdenet

Choyr

ULAANBAATAR

Kerulen

MONGOLIA

Plateau of Mongolia

Gobi

Baotou

Tien Shan

Ürümqi

Turpan Pendi -505ft (-154m)

Iranian Plateau

Hindu Kush

Tarim He

Tarim Basin

Takla Makan Desert

Altun Shan

Qilian Shan

Xiqing Shan

Lanzhou

Ordos Desert

Yellow River

Tai

K2 28,251ft (8611m)

Kunlun Mountains

Aksai Chin (administered by China, claimed by India)

Plateau of Tibet

CHINA

Peshawar

Indus

Demqog/Demchok (administered by China, claimed by India)

Mekong

Salween

Chengdu

Sichuan Pendi

ISLAMABAD

Jammu and Kashmir

Gujranwala

Lahore

Chongqing

Quetta

Faisalabad

Ludhiana

Multan

Sutlej

PAKISTAN

Himalayas

Brahmaputra

Yamuna

Guiy

Persian Gulf

Thar Desert

Delhi

Ganges

NEPAL

KATHMANDU

Mount Everest 29,029ft (8848m)

Kunming

Nanning

Arabian Peninsula

Gulf of Oman

Hyderabad

Jaipur

NEW DELHI

THIMPHU

BHUTAN

Guwahati

Imphal

Karachi

Kanpur

Ganges

Patna

Murray Ridge

Mouths of the Indus

Rann of Kachchh

Ahmadabad

BANGLADESH

DHAKA

Chittagong

Mandalay

VIETNAM

HANOI

Arabian Peninsula

Vindhya Range

Indore

Satpura Range

Khulna

Kolkata (Calcutta)

Chindwin

Hai

Gulf of Tonkin

Gulf of Khambhat

Narmada

Nagpur

Mouths of the Ganges

Irrawaddy

MYANMAR (BURMA)

LAOS

INDIA

Mumbai (Bombay)

Pune

Deccan

Godavari

Solapur

Eastern Ghats

Arakan Yoma

NAY PYI TAW

Louangphabang

Vinh

Arabian Sea

Hyderabad

VIENTIANE

Arabian Basin

Western Ghats

Bago

Chiang Mai

Mekong

Hubballi

Vijayawada

Bay of Bengal

Yangon (Rangoon)

THAILAND

Laccadive Islands (to India)

Bengaluru (Bangalore)

Pathein

Tônlé Sap

Pakxe

Chennai (Madras)

Mouths of the Irrawaddy

BANGKOK

CAMBODIA

Mysuru (Mysore)

Andaman Islands (to India)

PHNOM PENH

D

Hô C

Carlsberg Ridge

Jaffna

SRI LANKA

Gulf of Mannar

Andaman Sea

Nam of Kra

Gulf of Thailand

Mouths of the Mekong

COLOMBO

Nicobar Islands (to India)

MALDIVES

MALE

SRI JAYEWARDENAPURA KOTTE

Kota Bharu

Nat Isla

Chagos-Laccadive Plateau

Ceylon Plain

Medan

Strait of Malacca

MAL

Equator

INDIAN

Malai Peninsula

KUALA LUMPUR

PUTRAJAYA

Mascarene Plateau

Mid-Indian Ridge

OCEAN

Ninetyeast Ridge

Cocos Basin

Pegunungan Barisan

Danau Toba

SINGAPOR

Pekanbaru

Pontian

Padang

Sumatra

Great

Bangk

BRITISH INDIAN OCEAN TERRITORY (to UK)

Palembang

Java Tre

Mid-Indian Basin

JAK Se

Bandung

N

0 km   1000

0 miles   1000

Population ● National capital

○ below 50,000   ◎ 50,000 to 100,000   ◉ 100,000 to 500,000   ■ above 500,000

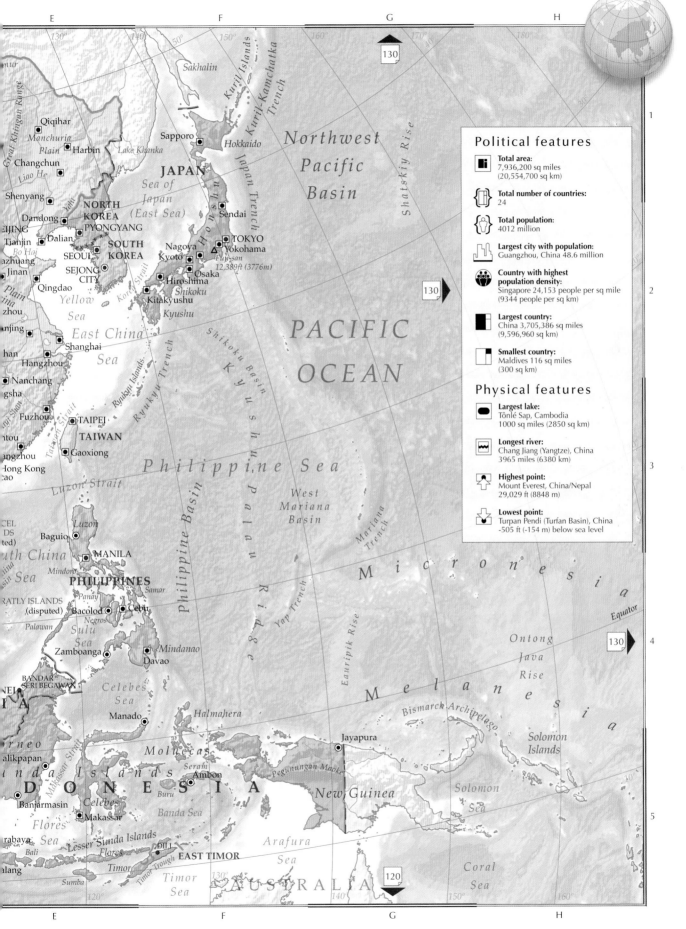

Political features

Total area:
7,936,200 sq miles
(20,554,700 sq km)

Total number of countries:
24

Total population:
4012 million

Largest city with population:
Guangzhou, China 48.6 million

Country with highest population density:
Singapore 24,153 people per sq mile
(9344 people per sq km)

Largest country:
China 3,705,386 sq miles
(9,596,960 sq km)

Smallest country:
Maldives 116 sq miles
(300 sq km)

Physical features

Largest lake:
Tônlé Sap, Cambodia
1000 sq miles (2850 sq km)

Longest river:
Chang Jiang (Yangtze), China
3965 miles (6380 km)

Highest point:
Mount Everest, China/Nepal
29,029 ft (8848 m)

Lowest point:
Turpan Pendi (Turfan Basin), China
-505 ft (-154 m) below sea level

103

Western China & Mongolia

A   70°   75°   B   55°   80°   85°   C   90°   95°   D   100°

R   U   S   S

Yenisey

Zapadnyy Sayan

Hövsgöl Nuur

Kulunda Steppe

1   50°

Uvs Nuur

KAZAKHSTAN

Kazakhskiy

Ulaangom   Mě

Ölgiy

Melkosopochnik

Ozero Zaysan

Altay

Har Us Nuur   Hyargas Nuur   Har Nuur

Hangayn Nuruu   Tsetser

M O N

Ozero Balkhash

45°

2   92

Ulungur Hu

Hovd

Altay   Bayanhongor

Karamay   *Gurbantünggüt Shamo*

70°

Δ Aj Bogd Uul
3802m

G

Ozero Issyk-Kul'

Kuytun   Fukang

Yining   Shihezi   Jimsar

Bo rohoro Shan

Atas Bogd
2695m Δ

KYRGYZSTAN

Ürümqi   Qitai

Tien Shan   Turpan

Hami   Dalain Hc

Δ Jengish Chokusu/Tömür Feng
7443m   Korla   *Bosten Hu*   *Turpan Pendi*

Xingxingxia

3

Kashi   *Tarim He*   *Tarim Basin*   Lop Nur   **GANSU**

TAJIKISTAN   Yengisar

XINJIANG UYGUR   *Qilian Shan*

AFGH.   Shache

Yecheng   *Kuruktag*

ZIZHIQU   Ruoqiang   *Danghe Nanshan*

(claimed by India)   Pishan   *Taklimakan Shamo*   *Altun Shan*   *Qaidam Pendi*   Qinghai

Δ K2
8611m   Moyu   *Shamo*

Hotan   Qira   **C   H**   Δ Dulan

PAKISTAN   *Kunlun Shan*   Golmud

Karakoram Range

Burhan Budai Shan   Anyémaqen

4   112   **JAMMU AND KASHMIR**   **AKSAI CHIN**

Indus

QINGHAI

AKSAI-CHIN (administered by China, claimed by India)   *Bayan Har Sh*

DÊMQOG/DEMCHOK (administered by China, claimed by India)   Rutog   *Qingzang Gaoyuan (Plateau of Tibet)*   *Tongtian He*   Yushu

Gar Xincun   **X I Z A N G**   *Tanggula Shan*   *Mekong*

Zanda   Gozhê   *Siling Co*   Amdo   Qamdo

30°   **Z I Z H I Q U**   *Tangra Yumco*   *Gyaring Co*   Nagqu   Salween

(T i b e t)   *Ngangzê Co*   *Nam Co*   Damxung   *Jinsha Jiang*

5   75°   **N E P A L**   *Braunaputra*   Lhazê   Xigazê   Maizhokunggar   **ARUNACHAL PRADESH (claimed by China)**

Lhasa

Gonggar   *Nyainqêntanglha Shan*

I N D I A   Ganges   Yamuna   Δ Mount Everest
8848m   Gyangzê   *H e r d u i n Shan*

25°   80°   113   85°   **BHUTAN**   **INDIA**   **MYANMAR (BURMA)**

A   B   C   95°   D

0 km   400

0 miles   400

Population   ● National capital   ● Internal administrative capital

○ below 50,000   ○ 50,000 to 100,000   ◉ 100,000 to 500,000   ◼ above 500,000

E   110°   F   115°   120°   G   125°   130°   H   135°

55°

RUSS. FED.

zero Baykal

93

Onon

Shilka

Ergun   Jagdaqi

Amur (Heilong Jiang)

45°

135°

Sühbaatar

HEILONGJIANG

1

Darhan

Erdenet

an

**Hulun Buir
(Hailar)**

Manzhouli

*Hulun
Nur*

*Lake
Khanka*

Onon Gol   Choybalsan

*Menengiyn
Tal*

JILIN

40°

106

2

ULAANBAATAR

Dzuunmod   Öndörhaan

Kerulen

Baruun-Urt

Holin Gol

*Sea of
Japan
(East Sea)*

Saynshand

Tongliao

Xilinhot

Liao He

Erenhot

**Chifeng
(Ulanhad)**

LIAONING

Dalandzadgad

**NORTH
KOREA**

Yin Nuruu

Liaodong Wan

*Korea
Bay*

**SOUTH
KOREA**

35°

130°

3

Ulan Qab (Jining)

BEIJING

Bo Hai

Hohhot

Baotou

TIANJIN

*Huang He
(Yellow River)*

Wuhai
(Haibowan)

*Mu Us
Shadi*

HEBEI

JAPAN

*Tengger
Shamo*

**Yellow
Sea**

NINGXIA

Great Wall of China

SHANDONG

SHANXI

108

ing

Huang He (Yellow River)

JIANGSU

30°

4

N   **A**

HENAN

GANSU

SHAANXI

Han Shui

ANHUI

SHANGHAI SHI

East

HUBEI

ZHEJIANG

China

ICHUAN

Chang Jiang (Yangtze)

Sea

CHONGQING

JIANGXI

*Nansei-shotō
(to Japan)*

25°

125°

HUNAN

FUJIAN

Tropic of Cancer

YUNNAN

GUIZHOU

105°   110°   115°   107   120°   **TAIWAN**

E   F   G   H

Elevation

| | | | | | | | | Below sea level 0 | 250m | 500m | 1000m | 2000m | 3000m | 4000m | 6000m |

-6000m   -4000m   -2000m   -1000m   -500m   -250m

820ft   1640ft   3281ft   6562ft   9843ft   13,124ft   19,685ft

-19,658ft   -13,124ft   -6562ft   -3281ft   -1640ft   -820ft   -328ft/-100m   0

Eastern China & Korea

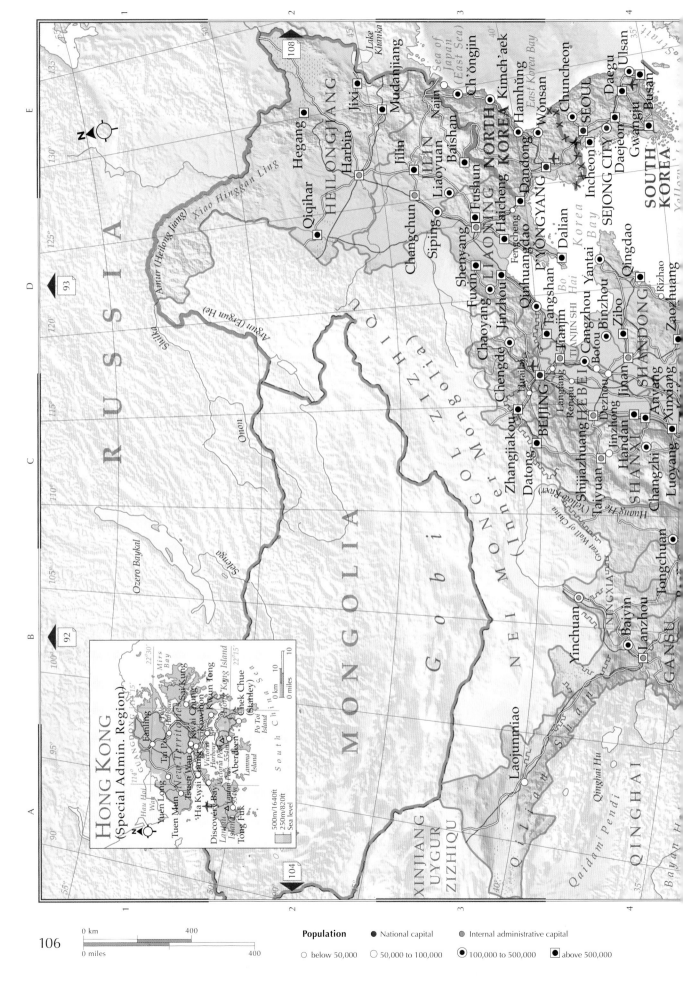

RUSSIA

MONGOLIA

Gobi

Ozero Baykal

Selenga

Onon

Shilka

Amur (Heilong Jiang)

Argun (Ergun He)

Xiao Hinggan Ling

Lake Khanka

Sea of Japan (East Sea)

HEILONGJIANG

Qiqihar
Hegang
Harbin
Jixi
Mudanjiang

JILIN
Changchun
Jilin
Baishan
Liaoyuan
NaJin
Ch'ŏngjin
Kimch'aek
Hamhŭng
East Korea Bay
Wŏnsan
Chuncheon
Ulsan
SEOUL
Daegu
Busan
Incheon
SEJONG CITY
Daejeon
Gwangju
SOUTH KOREA

NORTH KOREA
PYONGYANG
Dandong
Fengcheng
Sinŭiju

LIAONING
Shenyang
Fushun
Fuxin
Haicheng
Siping
Jinzhou
Chaoyang
Qinhuangdao
Dalian
Yantai
Korea Bay
Bo Hai
Qingdao
Rizhao
Zaozhuang

SHANDONG
Zibo
Jinan
Dezhou
Binzhou
Botou
Cangzhou
Tangshan

HEBEI
Chengde
Chaoyang
Zhangjiakou
Datong
BEIJING
TIANJIN SHI
Tianjin
Langfang
Renqiu
Baoding
Shijiazhuang
Xingtai
Handan
Anyang
Xinxiang

SHANXI
Taiyuan
Changzhi
Jinzhong
Yangquan
Luoyang
Tongchuan

NEI MONGOL ZIZHIQU (Inner Mongolia)

NINGXIA
Yinchuan
Baiyin
Lanzhou

GANSU

QINGHAI
Qinghai Hu
Qaidam Pendi
Bayan

XINJIANG UYGUR ZIZHIQU

Qilian Shan

Laojunmiao

Huang He
Great Wall of China

Hau Hai Wan

92
93
108
104

HONG KONG (Special Admin. Region)

GUANGDONG
Fanling
Tuen Mun
Yuen Long
Ta Po
Sai Kung
New Territories
Tsuen Wan
Kwai Chung
Kwun Tong
Kowloon
Victoria Harbour
Victoria Peak 554m
HONG KONG
KYi Island
Chek Chue (Stanley)
Po Toi Island
Aberdeen
Lantau Island
Lamma Island
Discovery Bay
Tong Fuk
Chek Lap Kok
Mirs Bay

500m/1640ft
250m/820ft
Sea level

0 km   10
0 miles   10

Population   ● National capital   ● Internal administrative capital
○ below 50,000   ◐ 50,000 to 100,000   ◉ 100,000 to 500,000   ■ above 500,000

0 km   400
0 miles   400

106

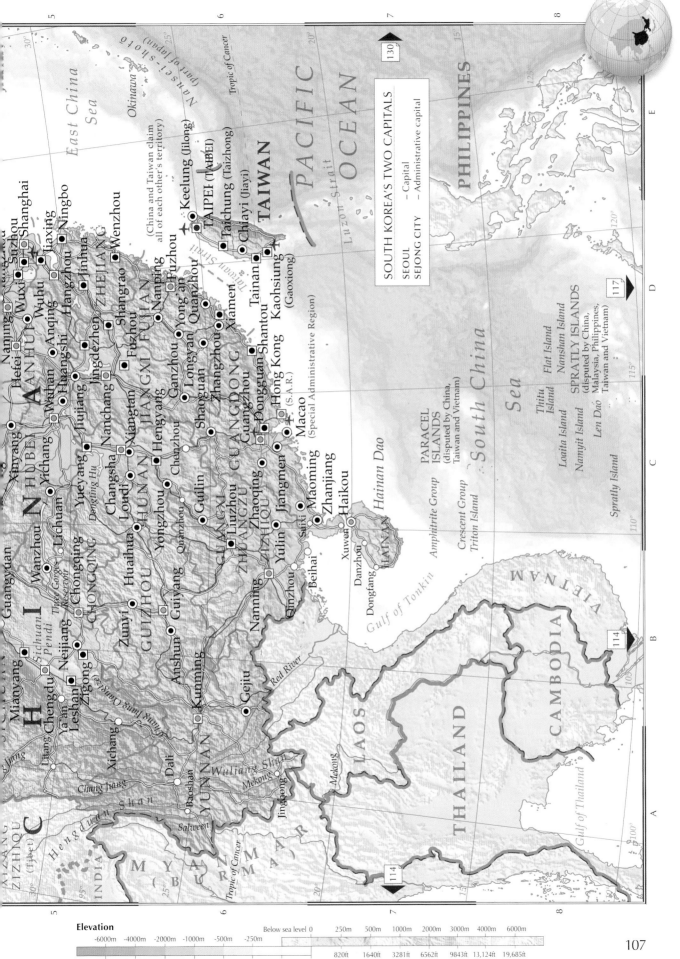

SOUTH KOREA'S TWO CAPITALS

SEOUL – Capital
SEJONG CITY – Administrative capital

PACIFIC OCEAN

PHILIPPINES

East China Sea

Okinawa

Nansei-shoto (part of Japan)

Tropic of Cancer

(China and Taiwan claim
all of each other's territory)

Keelung (Jilong)
TAIPEI (Taibei)
Taichung (Taizhong)
Chiayi (Jiayi)
TAIWAN
Tainan
Kaohsiung (Gaoxiong)

Luzon Strait
Taiwan Strait

Shanghai
Suzhou
Wuxi
Jiaxing
Ningbo
Wenzhou

Nanjing
Hefei
ANHUI
Wuhu
Anqing
Hangzhou
ZHEJIANG
Jinhua
Shangrao

Xinyang
HUBEI
Yichang
Wuhan
Huangshi
Jiujiang
Jingdezhen
JIANGXI
Nanchang
Fuzhou
FUJIAN
Nanping
Fuzhou
Xiamen
Quanzhou
Yong'an
Longyan
Ganzhou

Yiyang
Yueyang
Dongting Hu
Changsha
HUNAN
Loudi
Hengyang
Chenzhou
Quanzhou
Shaoguan
Zhangzhou
Shantou
GUANGDONG
Guangzhou
Dongguan
Hong Kong (S.A.R.)
(Special Administrative Region)
Macao
(Special Administrative Region)

SOUTH KOREA'S TWO CAPITALS

Mianyang
SICHUAN
Sichuan Pendi
Chengdu
Neijiang
Ya'an
Zigong
Leshan
CHONGQING
Chongqing
Guangyuan
Lichuan
The Gorges
Three Gorges Reservoir

GUIZHOU
Zunyi
Anshun
Guiyang

GUANGXI ZHUANGZU
Liuzhou
Yongzhou
Guilin
Yulin
Zhaoqing
Maoming
Zhanjiang
Beihai
Qinzhou
Nanning

Haikou
Danzhou
Dongfang
Hainan Dao
Suixi
Xuwen
LAIZHOU

Gulf of Tonkin

VIETNAM

LAOS

THAILAND

CAMBODIA

South China Sea

PARACEL ISLANDS
(disputed by China,
Taiwan and Vietnam)
Amphitrite Group
Crescent Group
Triton Island

SPRATLY ISLANDS
(disputed by China,
Malaysia, Philippines,
Taiwan and Vietnam)
Flat Island
Nanshan Island
Thitu Island
Loaita Island
Namyit Island
Len Dao
Spratly Island

Baoshan
Dali
YUNNAN
Kunming
Geju
Xichang
Litang
Wuliang Shan
Jingdong
Mekong
Salween
Chang Jiang
Hengduan Shan
XIZANG ZIZHIQU (Tibet)
The Gangs
Jinsha Jiang (Dangtze)

INDIA

MYANMAR (BURMA)

Tropic of Cancer

Red River

Gulf of Thailand

130
117
114
114

Elevation

| Below sea level | 0 | 250m | 500m | 1000m | 2000m | 3000m | 4000m | 6000m |

-6000m -4000m -2000m -1000m -500m -250m

-19,658ft -13,124ft -6562ft -3281ft -1640ft -820ft -328ft/-100m 0

820ft 1640ft 3281ft 6562ft 9843ft 13,124ft 19,685ft

Japan

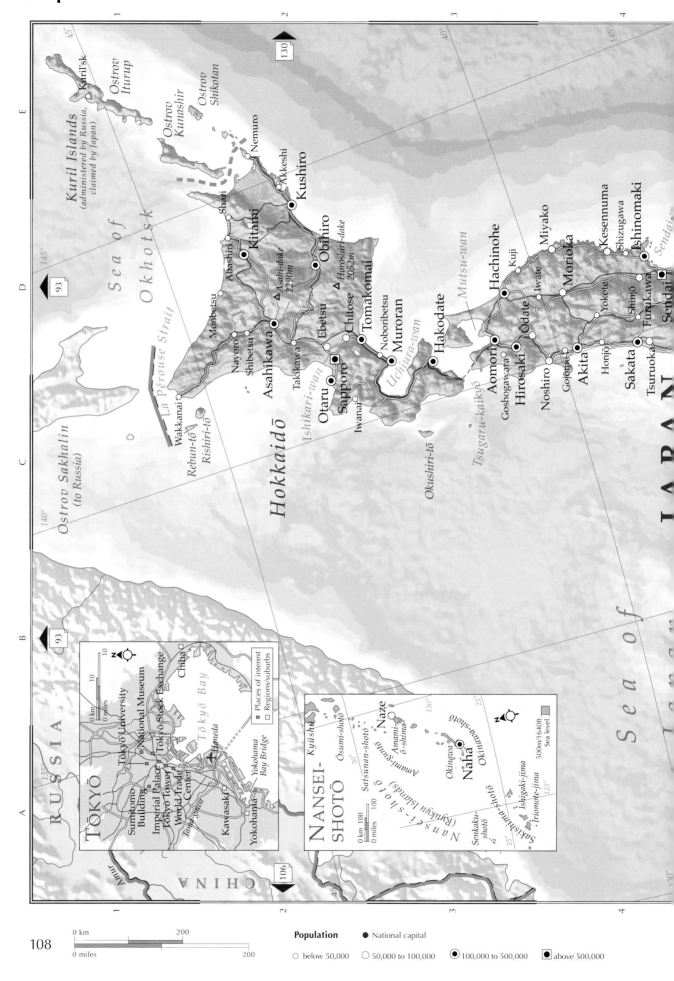

Kuril Islands
(administered by Russia,
claimed by Japan)

Ostrov Iturup
Kuril'sk
Ostrov Kunashir
Ostrov Shikotan

Sea of Okhotsk

Nemuro
Akkeshi
Kushiro
Shari
Kitami
Abashiri
Obihiro
△ Asahi-dake 2290m
△ Horoshiri-dake 2052m
Monbetsu
Nayoro
Shibetsu
Asahikawa
Takkawa
Ebetsu
Chitose
Tomakomai
Noboribetsu
Muroran
Hakodate
Uchiura-wan
Mutsu-wan
Hachinohe
Kuji
Miyako
Morioka
Iwate
Aomori
Odate
Hirosaki
Goshogawara
Yokote
Shinjō
Noshiro
Akita
Honjō
Sakata
Tsuruoka
Furukawa
Kesennuma
Shizugawa
Ishinomaki
Sendai
Gojōme

La Pérouse Strait
Wakkanai
Rebun-tō
Rishiri-tō
Otaru
Sapporo
Iwanai
Okushiri-tō
Ishikari-wan

Hokkaidō

Tsugaru-kaikyō

Ostrov Sakhalin (to Russia)

Sea of Japan

JAPAN

RUSSIA
CHINA
Amur

Sea of Japan

TŌKYŌ

Chiba
Tōkyō Bay
Tokyo University
National Museum
Tōkyō Stock Exchange
Sumitomo Building
Imperial Palace
Tokyo Tower
World Trade Center
Kawasaki
Yokohama
Yokohama Bay Bridge
Honda
Tama-gawa

■ Places of interest
□ Regions/suburbs

0 km   10
0 miles

NANSEI-SHOTŌ

Kyūshū
Naze
Ōsumi-shotō
Satsunan-shotō
Amami-guntō
Amami-ō-shima
Nansei-shotō (Ryūkyū Islands)
Okinawa
Naha
Okinawa-shotō
Senkaku-shotō
Ishigaki-jima
Sakishima-shotō
Iriomote-jima

500m/1640ft
Sea level

0 km   100
0 miles   100

Population

● National capital

○ below 50,000
○ 50,000 to 100,000
◉ 100,000 to 500,000
■ above 500,000

0 km   200
0 miles   200

Honshū

Iwaki
Hitachi
Utsunomiya
Mito
Ōyama
Chōshi
Chiba
Yokohama
Kawagoe
Bōsō-hantō
Kujūkuri-wan
TOKYO
Kawasaki
Maebashi
Sagami-nada
Nagaoka
Nagano
Toyama
Matsumoto
Kōfu
Fuji
Fuji-san
3776m △
Shizuoka
Izu-hantō
Hamamatsu
Toyota
Suruga-wan
Jōetsu
Shinano-gawa
Itoigawa
Hida
Okazaki
Ise-wan
Ise
Takaoka
sanmyaku
Gifu
Ogaki
Nagoya
Ōtsu
Tsu
Owase
Shingū
Kanazawa
Nakatsugawa
Komatsu
Fukui
Tsuruga
Biwa-ko
Kyōto
Osaka
Kōbe
Wakayama
Gobō
Tanabe
Kii-suidō
Toyama-wan
Wakasa-wan
Akashi-shima
Harima-nada
Himeji
Tottori
Okayama
Tokushima
Niihama
Matsuyama
Kōchi
Tosa-wan
Nakamura
Sukumo
Chūgoku-sanchi
Kurashiki
Kure
Shikoku
Yonago
Matsue
Oki-shotō
Dōgo
Dōzen
Gōtsu
Hamada
Hiroshima
Iwakuni
Hōfu
Ube
Ōita
Nobeoka
Miyazaki
Miyakonojō
Kyūshū
Masuda
Iyo-nada
Bungo-suidō
Shibushi-wan
Tanega-shima
Liancourt Rocks
(under South
Korean control)
Nagato
Yamaguchi
Shimonoseki
Kitakyūshū
Fukuoka
Kurume
Ōmuta
Katsushiro
Kumamoto
Satsuma-Sendai
Kagoshima
Yaku-shima
Tsushima
Iki
Sasebo
Nagasaki
Amakusa-nada
Koshikijima-rettō
Kagoshima-wan
Ōsumi-shotō
SOUTH
KOREA
Kō-saki
Korea Strait
Gotō-rettō
East China Sea

(East Sea)

Chūjūniya-san
Izu-shotō
Hachijō-jima
Miyake-jima
Mikura-jima
Nii-jima
Ō-shima
Kōzu-shima

P A C I F I C

O C E A N

130
130
130
106

Elevation

-6000m	-4000m	-2000m	-1000m	-500m	-250m					

Below sea level 0   250m   500m   1000m   2000m   3000m   4000m   6000m

820ft   1640ft   3281ft   6562ft   9843ft   13,124ft   19,685ft

-19,658ft -13,124ft -6562ft -3281ft -1640ft -820ft -328ft/-100m 0

South India & Sri Lanka

Kalyān
112
Mumbai
(Bombay)

N

Pune
Ahmadnagar
Nānded
Jagdalpur

Nizāmābād
Karimnagar

Bārāmati
I N D I A
Vizianagaram

Solāpur
Hyderābād
Seeunderābād
Visākhapat

Sāngli
Kalaburagi (Gulbarga)
Rājahmur

Kolhāpur
Kākin

Arabian
Raīchūr
Krishna
Vijayawāda

Belagāvi
Kurnool
Andhra
Machilīpatna

Panaji
Gadag
Nandyāl
Chīrāla

Sea
Hubballi
Pradesh
Ongole

Tungabhadra Reservoir
Tādpatri
Kāvali

Dāvangere
Anantapur
Nellore

Shivamogga
Cuddapah

Udupi
Bhadrāvati

Tumakūru
Bengalūru
(Bangalore)
Chennai
(Madras)

Mangalūru
(Mangalore)
Mandya
Vellore
Kanchīpuram

Kāsaragod
Krishnagiri
Tiruppattūr

*Amīndīvi
Islands*
Kannur (Cannanore)
Mysūru (Mysore)
Pondicherry

Kozhikode (Calicut)
Erode
Salem
Neyveli

Lakshadweep
(Laccadive Islands)
(to India)
Kavaratti
Island
Coimbatore
Tamil Nādu
Tiruchchirāppalli

Thrissur (Trichūr)
Dindigul

*Nine Degree
Channel*
Kalpeni
Island
Ernākulam
Madurai

Kochi (Cochin)
Jaffna

Minicoy Island
Alappuzha (Alleppey)
Rājapālaiyam
Mannar
SRI LANK

Kollam (Quilon)
Vavuniya

Eight Degree Channel
Thiruvananthapuram
(Trivandrum)
Tuticorin
Trincomalee

Nāgercoil
*Gulf of
Mannar*
Puttalam
Anuradhapura

*Ihavandhippolhu
Atoll*
Matale
Batticaloa

MALDIVES
Negombo
Kandy

COLOMBO

SRI JAYEWARDENAPURA
KOTTE
Ratnapura

Kalutara

Galle

Matara

*Faadhippolhu
Atoll*

*Horsburgh
Atoll*

51
Male' Atoll
Ari Atoll
MAALE (MALE')

Felidhu Atoll

Mulakatholhu

Kolhumadulu

Hadhdhunmathi Atoll

I N D I A N

North Huvadhu Atoll

SRI LANKA'S TWO CAPITALS

COLOMBO – Capital
SRI JAYEWARDENAPURA KOTTE – Administrative capital

Equator

*South Huvadhu
Atoll*

Gan
118

Addu Atoll

0 km   300
0 miles   300

Population   ● National capital

○ below 50,000   ○ 50,000 to 100,000   ◉ 100,000 to 500,000   ◼ above 500,000

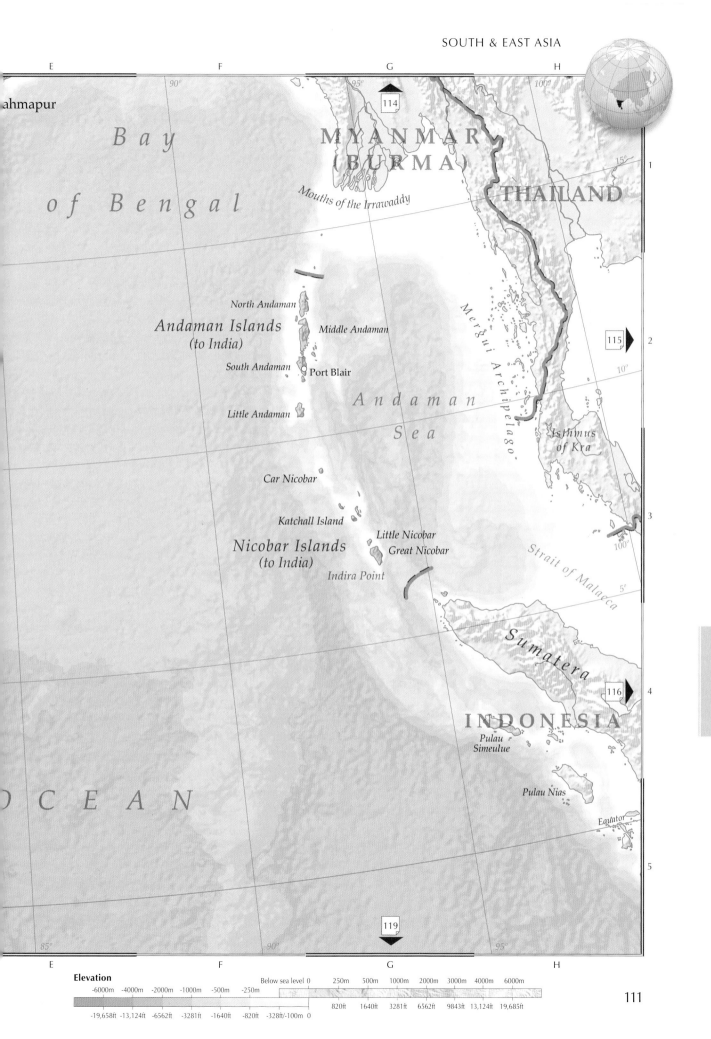

ahmapur

Bay

of Bengal

MYANMAR
(BURMA)

Mouths of the Irrawaddy

THAILAND

North Andaman

Andaman Islands
(to India)

Middle Andaman

South Andaman ○ Port Blair

Mergui Archipelago

*Isthmus
of Kra*

Little Andaman

A n d a m a n

S e a

Car Nicobar

Katchall Island

Strait of Malacca

Nicobar Islands
(to India)

Little Nicobar
Great Nicobar

Indira Point

Sumatera

INDONESIA

*Pulau
Simeulue*

OCEAN

Pulau Nias

Equator

114

115

116

119

Elevation

-6000m -4000m -2000m -1000m -500m -250m   Below sea level 0   250m 500m 1000m 2000m 3000m 4000m 6000m

-19,658ft -13,124ft -6562ft -3281ft -1640ft -820ft -328ft/-100m 0   820ft 1640ft 3281ft 6562ft 9843ft 13,124ft 19,685ft

Northern India, Pakistan & Bangladesh

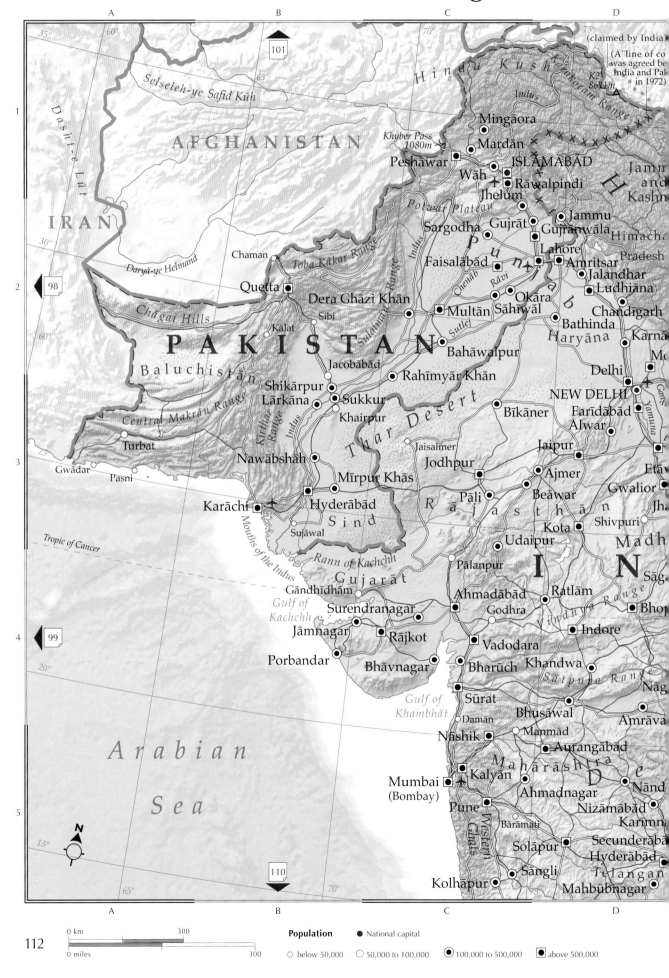

AFGHANISTAN

Selseleh-ye Safid Kūh

Hindu Kush

Karakoram Range

K2 8611m

(claimed by India
(A"line of co
was agreed be
India and Pak
in 1972)

Indus

Dasht-e Lūt

IRAN

Mingāora

Mardān

Khyber Pass
1080m

Peshāwar
Wāh
ISLĀMĀBĀD

Rāwalpindi

Jammu

and

Kashm

Jhelum

Jammu

Potwar Plateau

Himacha

Pradesh

Daryā-ye Helmand

Chaman

Toba Kākar Range

Sargodha
Gujrāt
Gujrānwāla

Lahore

Amritsar

Jalandhar

Quetta

Dera Ghāzi Khan

Faisalābād

Ludhiāna

Sibi

Multan

Okāra
Sāhīwāl

Chandīgarh

Chāgai Hills

Kālat

PAKISTAN

Bahāwalpur

Bathinda

Haryāna

Karna

Baluchistān

Jacobābād

Rahīmyār Khān

Delhi

Mo

Shikārpur

Lārkāna
Sukkur

NEW DELHI

Farīdābād

Turbat

Khairpur

Bīkaner

Alwar

Central Makrān Range

Nawābshāh

Jaisalmer

Jodhpur

Jaipur

Gwādar

Pasni

Mīrpur Khās

Pāli

Ajmer

Beāwar

Gwalior

Jha

Karāchi

Hyderābād

Rā

Kota

Shivpuri

Sind

Udaipur

Madh

Tropic of Cancer

Sujāwal

Rann of Kachchh

Pālanpur

Sāg

Mouths of the Indus

Gujarāt

Gāndhīdhām

Ahmadābād

Ratlām

Bhop

Gulf of
Kachchh

Surendranagar

Godhra

Vindhya Range

Jāmnagar

Rājkot

Indore

Porbandar

Vadodara

Khandwa

Bhāvnagar

Bharūch

Sātpura Range

Nāg

Gulf of
Khambhāt

Sūrat

Daman

Bhusāwal

Amrāva

Nāshik

Manmād

Aurangābād

Arabian

Mumbai
(Bombay)

Kalyān

Maharāshtra

D

Sea

Pune

Ahmadnagar

Nizāmābād

Nānd

Bārāmati

Karīmn

Secunderāba

Solāpur

Hyderābād

Sāngli

Telangan

Kolhāpur

Mahbūbnagar

Western Ghats

0 km   300

0 miles   300

Population ● National capital

○ below 50,000   ○ 50,000 to 100,000   ◉ 100,000 to 500,000   ■ above 500,000

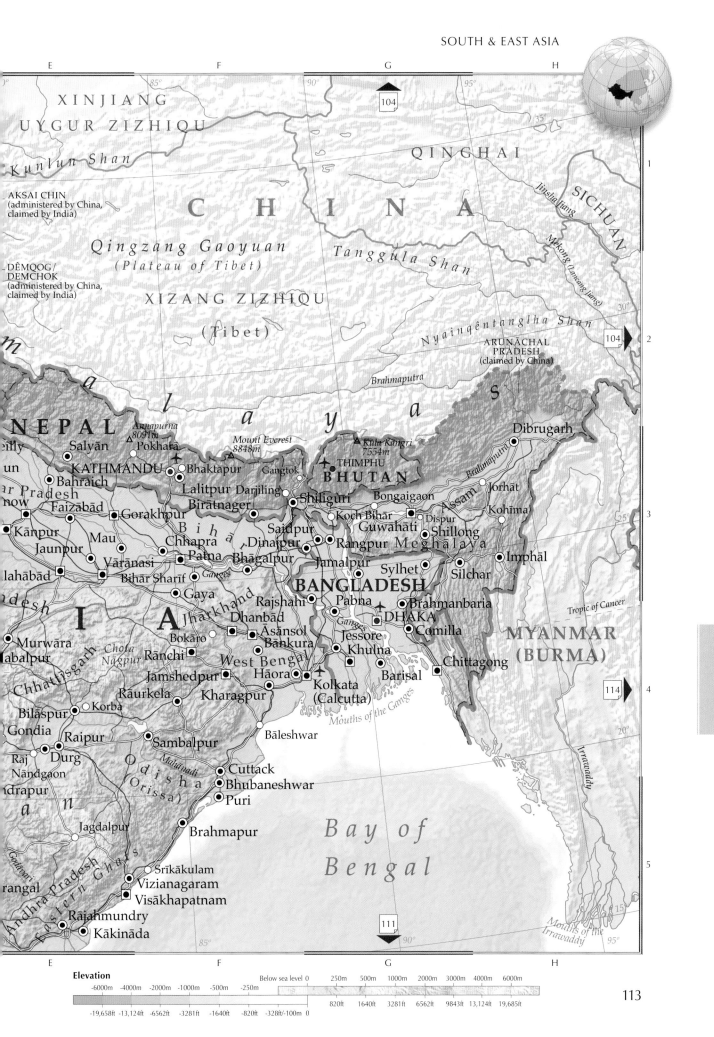

XINJIANG
UYGUR ZIZHIQU

Kunlun Shan

AKSAI CHIN
(administered by China,
claimed by India)

C H I N A

QINGHAI

Jinsha Jiang

SICHUAN

Qingzang Gaoyuan
(Plateau of Tibet)

DÊMQOG /
DEMCHOK
(administered by China,
claimed by India)

XIZANG ZIZHIQU
(Tibet)

Tanggula Shan

Mekong (Lancang Jiang)

Nyainqêntanglha Shan

ARUNÁCHAL
PRADESH
(claimed by China)

Brahmaputra

H   i   m   a   l   a   y   a   s

NEPAL
Salyán
Annapurna
8091m
Pokhara
KATHMANDU
Bhaktapur
Bahraich
Pradesh
now
Faizábad
Gorakhpur
Kánpur
Jaunpur
Mau
Váránasi
lahábad
Bihar Sharíf
I A
Gaya
Murwára
abalpur
Chhattisgarh
Chota
Nágpur
Bilaspur
Gondia
Raj
Nándgaon
drapur
a   n

Mount Everest
8848m
Gangtok
Darjiling
Birátnager
Saidpur
Chhapra
B i h a r
Dinajpur
Patna
Bhagalpur
Ganges
Jharkhand
Dhanbád
Bokáro
Ásánsol
Bánkura
Ránchi
West Bengal
Jamshedpur
Ráurkela
Háora
Kharagpur
Korba
Sambalpur
Raipur
Durg
Báleshwar

Kula Kangri
7554m
THIMPHU
BHUTAN
Shiligúri
Bongaigaon
Koch Bihar
Rangpur
Rajshahi
Pabna
Jessore
Khulna
Kolkata
(Calcutta)
Dispur
Guwáhati
Shillong
Meghalaya
Jamalpur
BANGLADESH
DHAKA
Comilla
Brahmanbaria
Sylhet
Silchar
Barisal

Dibrugarh
Brahmaputra
Assam
Jorhát
Kohíma
Imphál

MYANMAR
(BURMA)

Tropic of Cancer

Mouths of the Ganges

Bay of
Bengal

Chittagong

Irrawaddy

Mahánadi
Odisha
(Orissa)
Cuttack
Bhubaneshwar
Puri
Jagdalpur
Brahmapur
Andhra Pradesh
Eastern Ghats
Srikákulam
Vizianagaram
Visákhapatnam
Rajahmundry
Kákináda
rangal
Godávari

Mouths of the
Irrawaddy

Elevation

| Below sea level 0 | 250m | 500m | 1000m | 2000m | 3000m | 4000m | 6000m |

-6000m -4000m -2000m -1000m -500m -250m

-19,658ft -13,124ft -6562ft -3281ft -1640ft -820ft -328ft/-100m 0

820ft 1640ft 3281ft 6562ft 9843ft 13,124ft 19,685ft

Mainland Southeast Asia

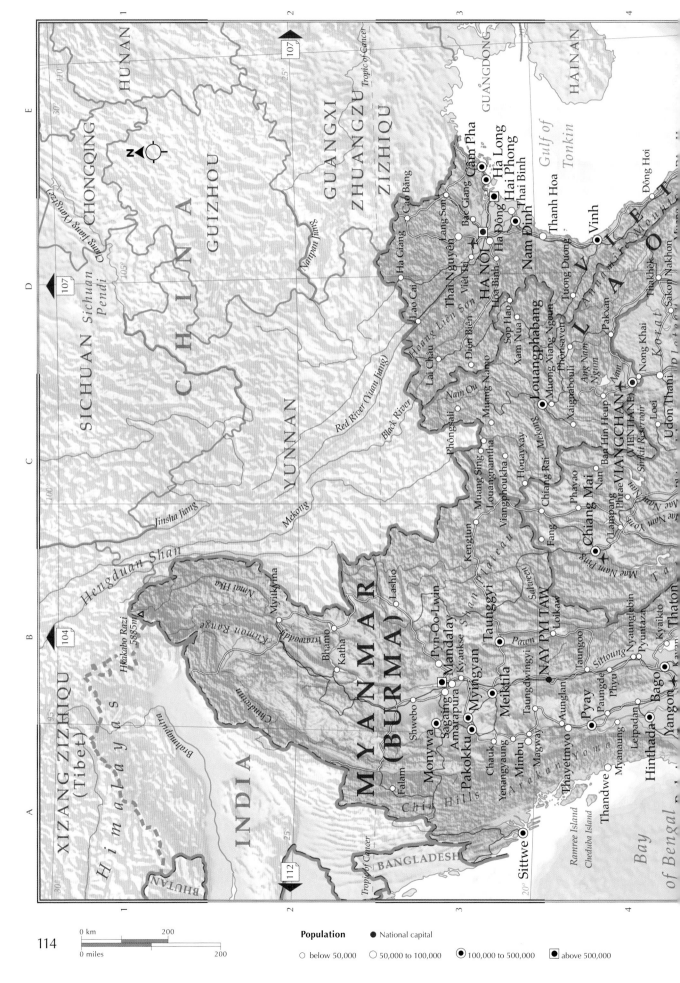

HUNAN

CHONGQING

Chuan Jiang (Yangtze)

SICHUAN Sichuan Pendi

GUIZHOU

GUANGXI ZHUANGZU ZIZHIQU

Nanpan Jiang

GUANGDONG

HAINAN

Gulf of Tonkin

Cam Pha
Ha Long
Hai Phong
Thai Binh
Thanh Hoa
Nam Dinh
Vinh
Dông Hoi

Cao Bang
Bac Giang
Lang Son
Ha Giang
Lao Cai
Thai Nguyen
Viet Tri
HANOI
Ha Dong
Hoa Binh

XIZANG ZIZHIQU (Tibet)

H i m a l a y a s

BHUTAN

INDIA

C H I N A

YUNNAN

Jinsha Jiang

Mekong

Red River (Yuan Jiang)

Black River

Hengduan Shan

Lai Châu
Diên Biên
Sop Hao
Xam Nua
Louangphabang
Muong Xiang Ngeun
Phônsaven
Tuong Duong

Nam Ou
Muang Namo
Muang Sing
Louangnamtha
Houaxay
Houayxay
Viangphoukha
Phôngsali

LAOS

Xaignabouli
Ang Nam Ngum
Pakxan
VIANGCHAN (VIENTIANE)
Nong Khai
Thakhek
Sakon Nakhon

Ko-rat

Chiang Rai
Chiang Mai
Fang
Phayao
Phrae
Nan
Lampang
Udon Thani
Loei
Sirikit Reservoir

Mae Nam Ping

Mae Nam Yom

Mae Nam Nan

Brahmaputra

Hkakabo Razi 5885m

Kumon Range

Nmai Hka

Myitkyina

Bhamo
Katha

Chindwin

Irrawaddy

M Y A N M A R (B U R M A)

Shwebo
Monywa
Sagaing
Amarapura
Mandalay
Pyn-Oo-Lwin
Kyaukse
Myingyan
Meiktila

Lashio
Kengtun

Shan Plateau

Salween

Taunggyi
Pawn
Loikaw

Chauk
Yenangyaung
Minbu
Magway

Pakokku
Taungdwingyi
NAY PYI TAW
Taungoo
Munglan
Sittoung

Falam

Chin Hills

Thayetmyo
Prome
Pyay
Paungde
Phyu
Nyaunglebin

Myanaung
Leitpadan
Pyuntaza
Kyaikto
Bago
Thaton

Hinthada
Yangon

Thandwe
Myanaung

Ramree Island
Cheduba Island

Sittwe

Arakan Yoma

Bay of Bengal

Tropic of Cancer

BANGLADESH

0 km   200
0 miles   200

Population   ● National capital

○ below 50,000   ◔ 50,000 to 100,000   ◉ 100,000 to 500,000   ■ above 500,000

M

Quang Ngai
Quy Nhon
Play Cu
Tônlé Srepok
Cam Ranh
Tuy Hoa
Nha Trang
Da Lat
Phan Rang-
Thap Cham
Phan Thiet
Di-Linh
Khôngxédôn
Pakxe
Attapu
Virôchey
Biên Hoa
Hô Chi Minh
Vung Tau
Champasak
Tônlé
Tônle Kong
Stung Treng
Kâmpóng Trâbêk
Kâmpóng Cham
My Tho
Tra Vinh
Stiêng Sên
Kâmpóng Thom
Kâmpóng Chhnang
Suông
Svay Riêng
Long Xuyên
Soc Trang
Bac Liêu
Con Dao Son
Mouths of the Mekong
Surin
Sisaket
Muang Không
Kralanh
Pursat
Moung Roessei
Kâmpóng Speu
Mekong
Kâmpong
Chuor Phnum
Krâcheh Ôdông
Châu Doc
PHNOM PENH
Can Tho
Rach Gia
Ca Mau
CAMBODIA
Tônlé Sap
Chhot Phnum Dângrêk
Chanthaburi
Reâng Kesei
Battambang
Kâmpôt
Vinh Rach Gia
Buriram
Sâmraông
Sihanoukville
Ubon Ratchathani
Nakhon Ratchasima
Sara Buri
Lop Buri
Chon Buri
Ko Chang
Chanthaburi
Rayong
Pattaya
Samut Prakan
KRUNG THEP
(BANGKOK)
Gulf of Thailand
MALAYSIA
Malay
Peninsula
Ao Krung Thep
Nakhon Pathom
Phetchaburi
Ratchaburi
Ban Hua Hin
Srinagarind Reservoir
Ayutthaya
Narathiwat
Pattani
Songkhla
Yala
Hat Yai
South China Sea
Kepulauan Natuna
(to Indonesia)
Nakhon Si Thammarat
Pak Phanang
Ko Phangan
Ko Samui
Phatthalung
Thale Luang
Chumphon
Lang Suan
Surat Thani
Sichon
Chung Song
Trang
Pulau Pinang
Pulau Langkawi
Ye
Dawei
Mali Kyun
Kadan Kyun
Myeik
Tanintharyi
Daung Kyun
Letsôk-aw Kyun
Lanbi Kyun
Zadetkyi Kyun
Ko Phra Thong
Phang-Nga
Ko Phuket
Phuket
Ko Lanta
Ko Ta Ru Tao
Pulau Langkawi
Strait of Malacca
Sumatera
(Sumatra)
Myeik Archipelago
Ranong
Isthmus of Kra
Bilauktaung Range
INDONESIA
Andaman Sea
Pulau Simeulue
INDIAN OCEAN
North Andaman
Andaman Islands
(to India)
Middle Andaman
South Andaman
Little Andaman
Car Nicobar
Katchall Island
Little Nicobar
Nicobar Islands
(to India)
Great Nicobar

117

116

116

111

Elevation

| | | | | | | | | Below sea level 0 | 250m | 500m | 1000m | 2000m | 3000m | 4000m | 6000m |

-6000m -4000m -2000m -1000m -500m -250m

-19,658ft -13,124ft -6562ft -3281ft -1640ft -820ft -328ft/-100m 0

820ft 1640ft 3281ft 6562ft 9843ft 13,124ft 19,685ft

115

Maritime Southeast Asia

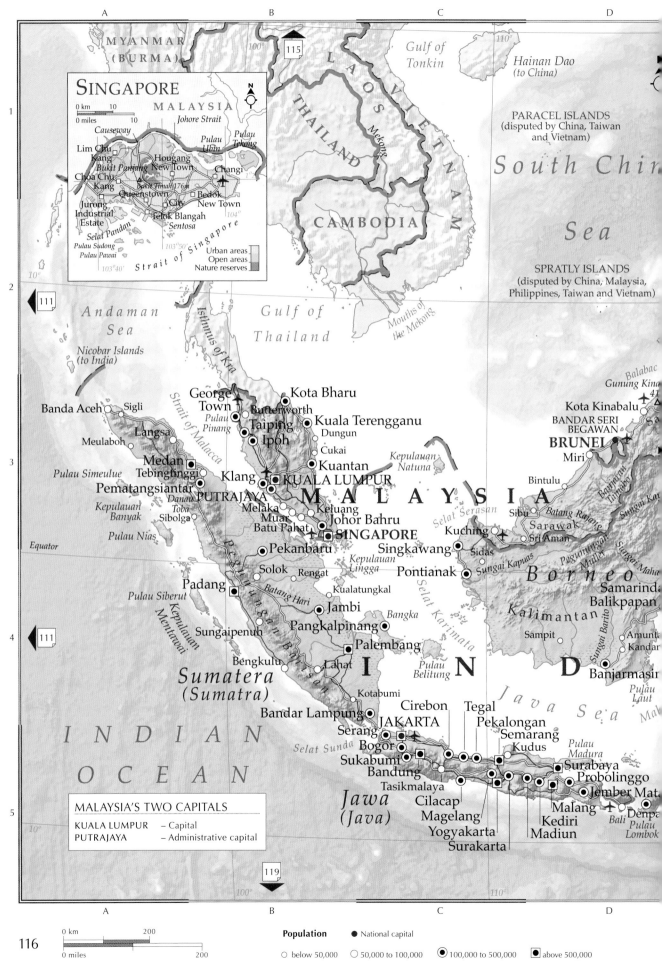

SINGAPORE

MALAYSIA

Johore Strait

Causeway

Lim Chu Kang
Bukit Panjang
Choa Chu Kang
Hougang New Town
Pulau Ubin
Pulau Tekong
Changi
Bukit Timah 176m
Queenstown
City
Bedok New Town
Jurong Industrial Estate
Telok Blangah
Sentosa
Selat Pandan
Pulau Sudong
Pulau Pawai
Strait of Singapore

Urban areas
Open areas
Nature reserves

MYANMAR (BURMA)

LAOS

VIETNAM

THAILAND

Gulf of Tonkin

Hainan Dao (to China)

Mekong

CAMBODIA

South China Sea

PARACEL ISLANDS
(disputed by China, Taiwan and Vietnam)

SPRATLY ISLANDS
(disputed by China, Malaysia, Philippines, Taiwan and Vietnam)

Andaman Sea

Nicobar Islands (to India)

Isthmus of Kra

Gulf of Thailand

Mouths of the Mekong

Banda Aceh
Sigli
George Town
Butterworth
Pulau Pinang
Taiping
Ipoh
Kota Bharu
Kuala Terengganu
Dungun
Cukai
Kuantan
Kepulauan Natuna

Gunung Kina
Kota Kinabalu
BANDAR SERI BEGAWAN
BRUNEI
Miri

Strait of Malacca

Langsa
Meulaboh
Medan
Tebingtinggi
Klang
KUALA LUMPUR
MALAYSIA
Bintulu
Pulau Simeulue
Pematangsiantar
Danau Toba
Sibolga
PUTRAJAYA
Melaka
Keluang
Johor Bahru
Kuching
Singkawang
Sidas
Sibu
Batang Rajang
Sri Aman
Sarawak
Sungai Kat
Kepulauan Banyak
Muar
Batu Pahat
SINGAPORE
Selat Serasan
Borneo
Pulau Nias
Pekanbaru
Kepulauan Lingga
Pontianak
Sungai Kapuas
Pegunungan Muller
Sungai Maha

Equator

Solok
Rengat
Kualatungkal
Selat Karimata
Kalimantan
Samarinda
Balikpapan
Padang
Pulau Siberut
Batang Hari
Jambi
Bangka
Sampit
Amunta
Kandar
Kepulauan Mentawai
Sungaipenuh
Pangkalpinang
Palembang
INDONESIA
Pulau Belitung
Banjarmasin
Pulau Laut
Bengkulu
Lahat
Sumatera (Sumatra)
Kotabumi
Java Sea
Mat

Cirebon
Tegal
Pekalongan
Semarang
Kudus
Pulau Madura
Bandar Lampung
JAKARTA
Serang
Bogor
Sukabumi
Bandung
Tasikmalaya
Surabaya
Probolinggo
Jember
Mat

INDIAN OCEAN

Selat Sunda

Cilacap
Magelang
Yogyakarta
Surakarta
Jawa (Java)
Kediri
Madiun
Malang
Denpa
Bali
Pulau Lombok

MALAYSIA'S TWO CAPITALS

KUALA LUMPUR – Capital
PUTRAJAYA – Administrative capital

0 km 200
0 miles 200

Population

● National capital

○ below 50,000
◎ 50,000 to 100,000
◉ 100,000 to 500,000
◼ above 500,000

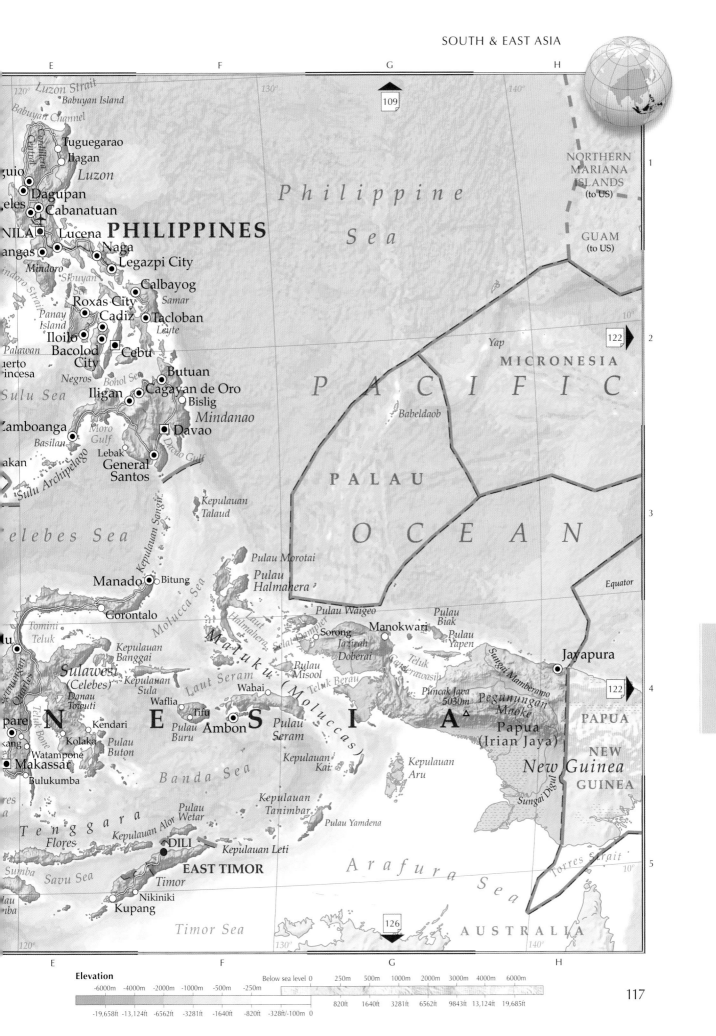

Luzon Strait
120°
Babuyan Island
Babuyan Channel
guio
Tuguegarao
Ilagan
Cordillera
Central
Luzon
Dagupan
eles
Cabanatuan
NILA
Lucena
PHILIPPINES
angas
Naga
Mindoro
Legazpi City
Sibuyan
Sea
Calbayog
Mindoro Strait
Samar
Roxas City
Cadiz
Tacloban
Panay
Island
Leyte
Iloilo
Palawan
Bacolod
City
Cebu
uerto
ncesa
Negros
Bohol Sea
Butuan
Sulu Sea
Iligan
Cagayan de Oro
Bislig
Zamboanga
Mindanao
Basilan
Moro
Gulf
Davao
akan
Lebak
Davao Gulf
General
Santos
Sulu Archipelago

Philippine
Sea

130°

109

NORTHERN
MARIANA
ISLANDS
(to US)

GUAM
(to US)

140°

1

MICRONESIA

Yap

10°

122

2

PACIFIC

Babeldaob

PALAU

OCEAN

3

Kepulauan
Talaud

elebes Sea

Kepulauan Sangir

Pulau Morotai

Pulau
Halmahera

Equator

Manado
Bitung

Molucca Sea

Gorontalo

Tomini
Teluk

Pulau Waigeo

Laut Halmahera

Selat Dampier
Sorong
Manokwari

Pulau
Biak

Pulau
Yapen

Jazirah
Doberai

Teluk
Cenderawasih

Sungai Mamberamo

Jayapura

122

4

u

Sulawesi
(Celebes)

Kepulauan
Banggai

Kepulauan
Sula

Maluku (Moluccas)

Laut Seram

Pulau
Misool

Teluk Berau

Puncak Jaya
5030m

Pegunungan
Maoke

PAPUA

Danau
Towuti

Waflia

Wahai

Papua
(Irian Jaya)

NEW

pare
N
Kendari
E
Tifu
Pulau
Buru
Ambon
S
Pulau
Seram
I
A
New Guinea

GUINEA

ang

Kolaka

Pulau
Buton

Kepulauan
Kai

Kepulauan
Aru

Watampone

Makassar

Banda Sea

Sungai Digul

Bulukumba

Kepulauan
Tanimbar

res

Tenggara

Pulau
Wetar

Kepulauan Alor

Kepulauan Leti

Pulau Yamdena

Torres Strait

10°

5

Flores

DILI

Arafura Sea

Sumba

EAST TIMOR

Timor

Nikiniki

Savu Sea

Kupang

mba

Timor Sea

126

AUSTRALIA

120°
130°
140°

E
F
G
H

Elevation

-6000m -4000m -2000m -1000m -500m -250m

-19,658ft -13,124ft -6562ft -3281ft -1640ft -820ft -328ft/-100m 0

Below sea level 0   250m   500m   1000m   2000m   3000m   4000m   6000m

820ft   1640ft   3281ft   6562ft   9843ft 13,124ft 19,685ft

The Indian Ocean

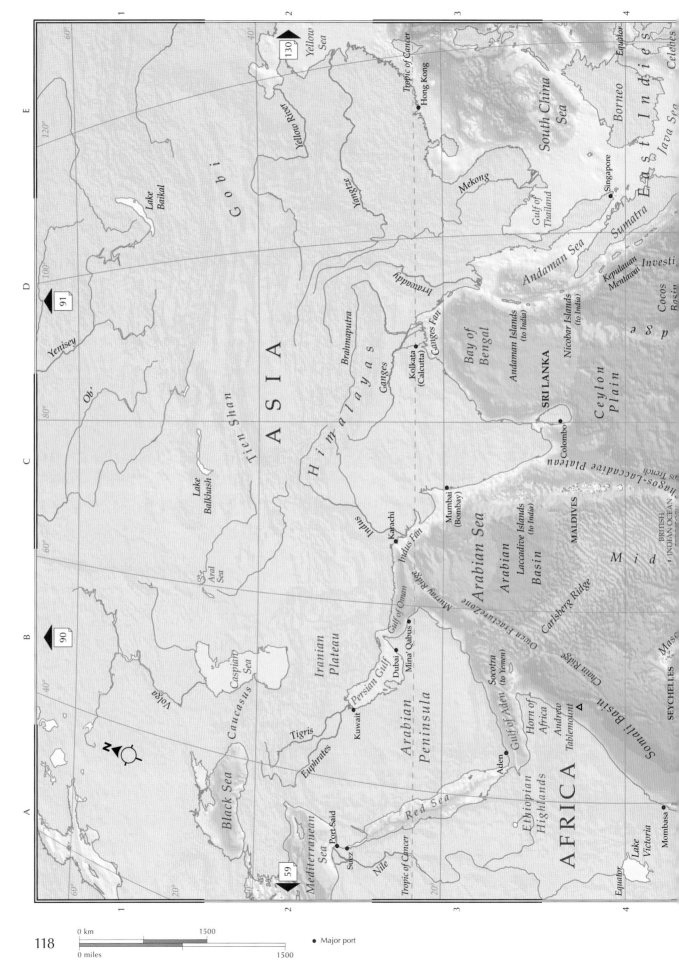

0 km — 1500
0 miles — 1500

● Major port

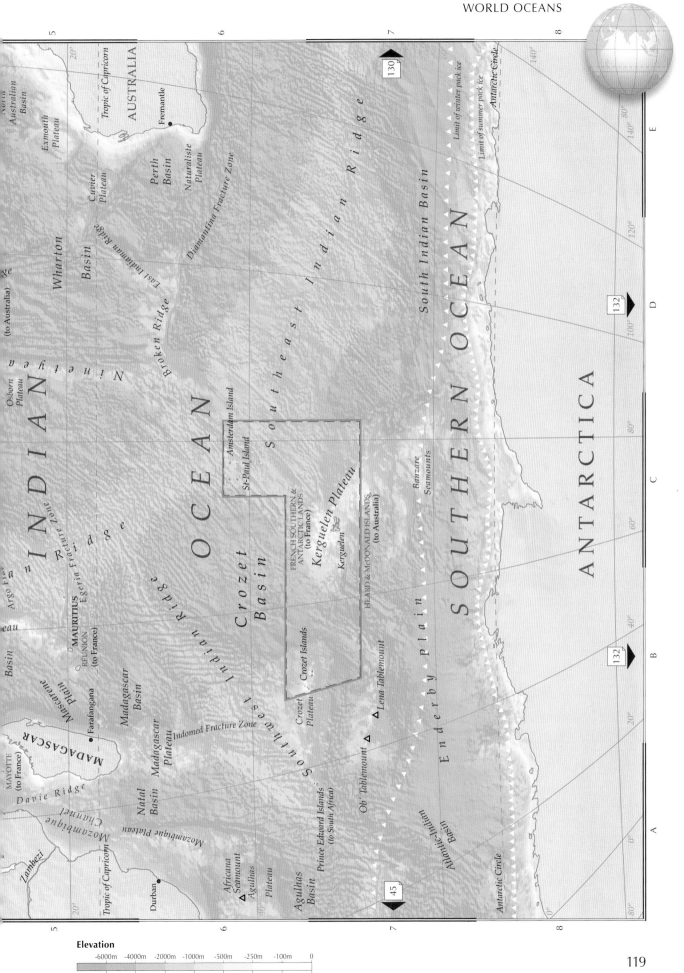

AUSTRALIA

North Australian Basin

Exmouth Plateau

Fremantle

Cuvier Plateau

Perth Basin

Naturaliste Plateau

Tropic of Capricorn

Wharton Basin

Osborn Plateau

East Indiaman Ridge

Diamantina Fracture Zone

Broken Ridge

Nineteyah

I N D I A N

(to Australia)

Egeria Fracture Zone

Argo Plain

MAURITIUS
RÉUNION
(to France)

Ridge

Southwest Indian Ridge

Madagascar Basin

Madagascar Plateau

Mascarene Plain

Faratangana

MADAGASCAR

MAYOTTE
(to France)

Davie Ridge

Zambezi

Mozambique Channel

Durban

Tropic of Capricorn

Mozambique Plateau

Natal Basin

Agulhas Plateau

Agulhas Basin

Africana Seamount

Prince Edward Islands
(to South Africa)

Indomed Fracture Zone

Crozet Islands

Crozet Plateau

O C E A N

Crozet Basin

Amsterdam Island

St-Paul Island

FRENCH SOUTHERN &
ANTARCTIC ISLANDS
(to France)

Kerguelen Plateau

Kerguelen

HEARD & McDONALD ISLANDS
(to Australia)

Ob´ Tablemount

Lena Tablemount

South east Indian Ridge

South Indian Basin

Banzare Seamounts

S O U T H E R N   O C E A N

Enderby Plain

Atlantic-Indian Basin

A N T A R C T I C A

Antarctic Circle

Limit of winter pack ice
Limit of summer pack ice

Antarctic Circle

Antarctic Circle

Elevation

-6000m	-4000m	-2000m	-1000m	-500m	-250m	-100m	0
-19,658ft	-13,124ft	-6562ft	-3281ft	-1640ft	-820ft	-328ft/-100m	0

Australasia & Oceania

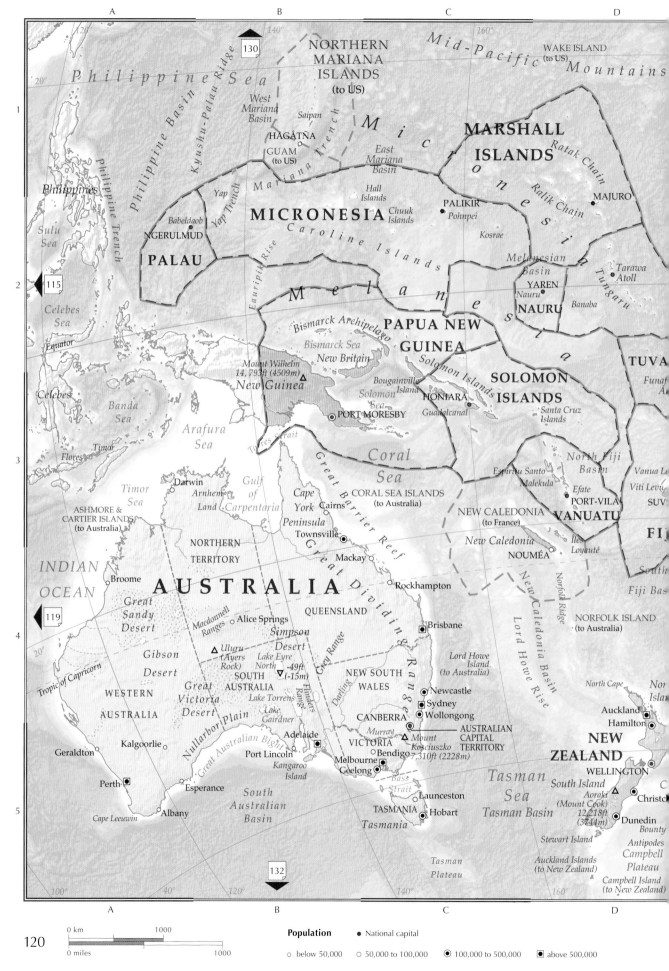

Philippine Sea

Mid-Pacific Mountains

WAKE ISLAND (to US)

NORTHERN MARIANA ISLANDS (to US)

West Mariana Basin

Saipan

HAGÅTÑA
GUAM (to US)

East Mariana Basin

Hall Islands

MARSHALL ISLANDS

Ratak Chain

Ratik Chain

MAJURO

Philippines

Philippine Trench

Kyushu-Palau Ridge

Yap Trench

Yap

Babeldaob

NGERULMUD

PALAU

Sulu Sea

Celebes Sea

MICRONESIA

Caroline Islands

Chuuk Islands

PALIKIR
• Pohnpei

Kosrae

Melanesian Basin

Tarawa Atoll

YAREN
Nauru •

NAURU

Banaba

Tungaru

TUVA

Equator

Celebes

Banda Sea

Eauripik Rise

Melanesia

Bismarck Archipelago

Bismarck Sea

New Britain

PAPUA NEW GUINEA

Solomon Islands

SOLOMON ISLANDS

Mount Wilhelm 14,793ft (4509m) △

New Guinea

Bougainville Island

Solomon Sea

HONIARA
Guadalcanal

Santa Cruz Islands

North Fiji Basin

Vanua Le

Viti Levu

SUV

FIJ

Timor

Flores

Arafura Sea

Torres Strait

Coral Sea

Espíritu Santo
Malekula

Efate

PORT-VILA

VANUATU

Timor Sea

ASHMORE & CARTIER ISLANDS (to Australia)

Darwin

Arnhem Land

Gulf of Carpentaria

Cape York Peninsula

Cairns

CORAL SEA ISLANDS (to Australia)

NEW CALEDONIA (to France)

New Caledonia

NOUMÉA

Îles Loyauté

South

Fiji Bas

INDIAN OCEAN

Broome

NORTHERN TERRITORY

AUSTRALIA

Townsville

Mackay

Rockhampton

QUEENSLAND

Great Barrier Reef

Great Dividing Range

Lord Howe Island (to Australia)

New Caledonia Ridge

Lord Howe Rise

Norfolk Ridge

NORFOLK ISLAND (to Australia)

North Cape

Nor
Islan

Great Sandy Desert

Macdonnell Ranges

Alice Springs

Simpson Desert

Brisbane

Gibson Desert

△ Uluru (Ayers Rock)

Lake Eyre North

-49ft (-15m) ▽

SOUTH AUSTRALIA

Grey Range

NEW SOUTH WALES

Darling

Lord Howe Basin

Auckland

Hamilton

Tropic of Capricorn

WESTERN AUSTRALIA

Great Victoria Desert

Lake Torrens

Flinders Range

Newcastle

Sydney

Wollongong

NEW ZEALAND

Kalgoorlie

Lake Gairdner

Adelaide

CANBERRA

AUSTRALIAN CAPITAL TERRITORY

WELLINGTON

Geraldton

Nullarbor Plain

Port Lincoln

VICTORIA

Murray

△ Mount Kosciuszko 7,310ft (2228m)

Tasman Sea

South Island

Aoraki (Mount Cook) 12,218ft (3744m) △

Christ

Perth

Esperance

Great Australian Bight

South Australian Basin

Kangaroo Island

Bendigo

Melbourne

Geelong

Bass Strait

Dunedin

Bounty

Cape Leeuwin

Albany

Launceston

TASMANIA

Hobart

Tasman Basin

Stewart Island

Antipodes

Campbell Plateau

Tasmania

Tasman Plateau

Auckland Islands (to New Zealand)

Campbell Island (to New Zealand)

of the map

Population
- National capital
- ○ below 50,000
- ○ 50,000 to 100,000
- ◉ 100,000 to 500,000
- ▣ above 500,000

0 km 1000
0 miles 1000

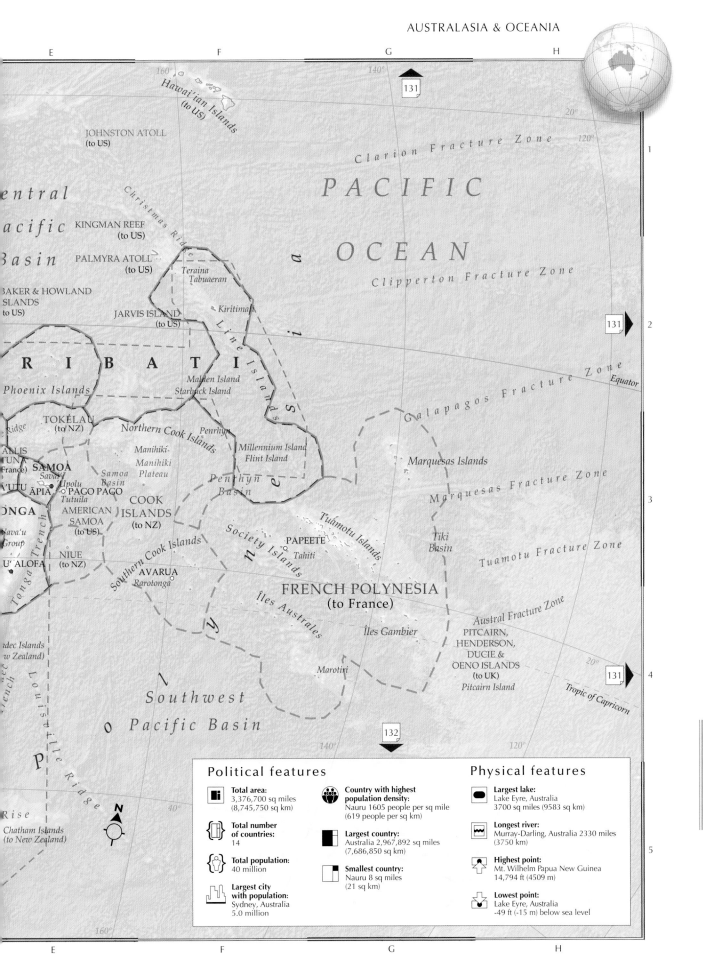

E   F   G   H

160°   140°   20°   120°

131

Hawai'ian Islands (to US)

JOHNSTON ATOLL (to US)

Clarion Fracture Zone

P A C I F I C

entral

acific

KINGMAN REEF (to US)

Christmas Ridge

Basin

PALMYRA ATOLL (to US)

O C E A N

Teraina
Tabuaeran

Clipperton Fracture Zone

BAKER & HOWLAND ISLANDS (to US)

Kiritimati

JARVIS ISLAND (to US)

131

Equator

Line Islands

Galapagos Fracture Zone

R I B A T I

Malden Island
Starbuck Island

Phoenix Islands

Marquesas Islands

Ridge

TOKELAU (to NZ)

Northern Cook Islands

Penrhyn

Marquesas Fracture Zone

ALLIS UNA France)

Manihiki
Manihiki Plateau

Millennium Island
Flint Island

Line Islands

SAMOA

Savai'i
Upolu

Samoa Basin

Penrhyn Basin

Tiki Basin

'VUTU
APIA PAGO PAGO
Tutuila

COOK
ISLANDS
(to NZ)

Tuamotu Islands

Tuamotu Fracture Zone

ONGA

AMERICAN
SAMOA
(to US)

Society Islands

PAPEETE
Tahiti

*Nava'u
Group*

Southern Cook Islands

FRENCH POLYNESIA
(to France)

Austral Fracture Zone

U' ALOFA

NIUE
(to NZ)

AVARUA
Rarotonga

Îles Australes

Îles Gambier

PITCAIRN,
HENDERSON,
DUCIE &
OENO ISLANDS
(to UK)
Pitcairn Island

Tonga Trench

20°

131

Marotiri

Tropic of Capricorn

*dec Islands
w Zealand)*

Louisville Ridge

Southwest

Pacific Basin

40°

132

Rise

Chatham Islands
(to New Zealand)

160°

N

Political features

Total area:
3,376,700 sq miles
(8,745,750 sq km)

**Total number
of countries:**
14

Total population:
40 million

**Largest city
with population:**
Sydney, Australia
5.0 million

**Country with highest
population density:**
Nauru 1605 people per sq mile
(619 people per sq km)

Largest country:
Australia 2,967,892 sq miles
(7,686,850 sq km)

Smallest country:
Nauru 8 sq miles
(21 sq km)

Physical features

Largest lake:
Lake Eyre, Australia
3700 sq miles (9583 sq km)

Longest river:
Murray-Darling, Australia 2330 miles
(3750 km)

Highest point:
Mt. Wilhelm Papua New Guinea
14,794 ft (4509 m)

Lowest point:
Lake Eyre, Australia
-49 ft (-15 m) below sea level

E   F   G   H

The Southwest Pacific

A | B | C | D

140° | 150° | 160° | 170°

130

Saipan
Tinian
Rota
NORTHERN
MARIANA
ISLANDS
(to US)

GUAM
(to US)
HAGATÑA

MARSHALL
ISLANDS

1

10°

Enewetak
Atoll
Bikini Atoll
Rongelap
Atoll

Ailuk At

Ratak Chain
Ralik Chain

Ujelang Atoll
Wotje A

Kwajalein
Atoll
Malo
Atoll

Yap

MICRONESIA

Namu Atoll
Ailinglaplap Atoll
Jaluit Atoll

Majuro A

Babeldaob

NGERULMUD

Chuuk
Islands

PALIKIR
Pohnpei

Mili A

Caroline Islands

Kosrae

Ebon Atoll

M
i
c
r
o
n
e
s
i
a

PALAU

2

117

Equator

Ma

Tare
Ato

YAREN
NAURU

Abem

Banaba

Nor

Admiralty
Islands
St.Matthias Group

Bismarck Archipelago

Bismarck Sea
New Ireland

New Guinea

PAPUA NEW GUINEA

M
e
l
a
n
e
s
i
a

INDONESIA

Madang

3

△ Mount Wilhelm
4509m
Lae

Central Range

New
Britain

Bougainville
Island

Solomon Sea

Choiseul

Santa Isabel

SOLOMON

Owen Stanley Range

Gulf of
Papua

New Georgia
Islands

Solomon Islands

Malaita

ISLANDS

PORT MORESBY

D'Entrecasteaux
Islands

HONIARA

Guadalcanal

10°

Torres Strait

San Cristobal

Santa Cruz
Islands

Rennell

Arafura Sea

Louisiade
Archipelago

Coral Sea

Banks Islands

Arnhem
Land

Groote
Eylandt

Gulf of
Carpentaria

Cape
York
Peninsula

CORAL SEA ISLANDS
(to Australia)

Espiritu Santo

Maéwo
Pentecost
Ambrym
Epi

VANUATU

4

124

Barkly Tableland

Great Barrier Reef

Malekula

Efate
PORT-VILA

NEW
CALEDONIA
(to France)

Erromango
Tanna
Aneityum

20°

NORTHERN

Ouvéa
Lifou

New
Caledonia

Iles Loyauté

Maré

TERRITORY

Tropic of Capricorn

QUEENSLAND

Great Dividing Range

NOUMÉA

Macdonnell

5

Ranges

AUSTRALIA

127

140° | 150° | 160° | 170°

A | B | C | D

122

0 km   750
0 miles   750

Population ● National capital

○ below 50,000   ○ 50,000 to 100,000   ◉ 100,000 to 500,000   ■ above 500,000

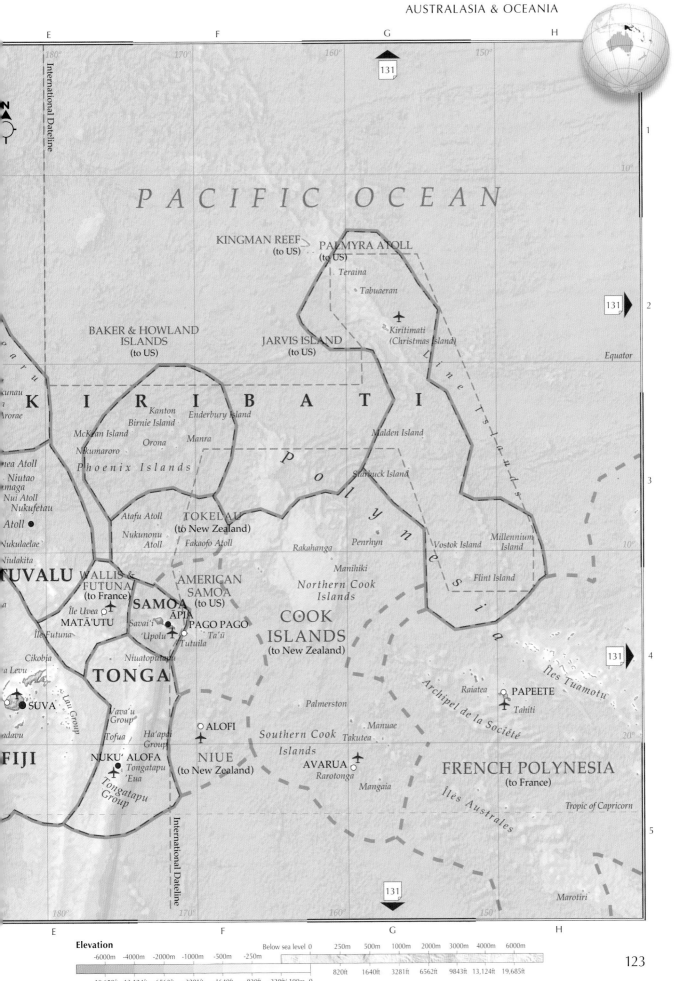

PACIFIC OCEAN

KINGMAN REEF
(to US)
PALMYRA ATOLL
(to US)

Teraina

Tabuaeran

BAKER & HOWLAND
ISLANDS
(to US)

JARVIS ISLAND
(to US)

Kiritimati
(Christmas Island)

Equator

K I R I B A T I

Line Islands

Kanton
Birnie Island
Enderbury Island

McKean Island
Orona
Manra

Malden Island

Nikumaroro

Phoenix Islands

Nea Atoll
Niutao
maga
Nui Atoll
Nukufetau

Starbuck Island

P
o
l
y
n
e
s
i
a

Atoll

Atafu Atoll

TOKELAU
(to New Zealand)

Vostok Island

Millennium
Island

10°

Nukulaelae

Nukunonu
Atoll

Fakaofo Atoll

Penrhyn

Niulakita

Rakahanga

Flint Island

TUVALU
WALLIS &
FUTUNA
(to France)

AMERICAN
SAMOA
(to US)

Manihiki

Northern Cook
Islands

Île Uvea
SAMOA
ĀPIA
MATĀ'UTU
Savai'i
PAGO PAGO

COOK
ISLANDS
(to New Zealand)

Île Futuna
Upolu
Ta'ū
Tutuila

Cikobia
a Levu
Niuatoputapu

Lau Group
TONGA

Îles Tuamotu

131

SUVA

Vava'u
Group

Palmerston

Raiatea
PAPEETE

Archipel de la Société

adavu
Tofua
Ha'apai
Group
ALOFI

Southern Cook
Islands
Manuae
Takutea

Tahiti

20°

FIJI
NUKU'ALOFA
Tongatapu
'Eua
NIUE
(to New Zealand)

AVARUA
Rarotonga

FRENCH POLYNESIA
(to France)

Tongatapu
Group

Mangaia

Îles Australes

Tropic of Capricorn

Marotiri

Elevation

-6000m	-4000m	-2000m	-1000m	-500m	-250m		

Below sea level 0   250m 500m 1000m 2000m 3000m 4000m 6000m

-19,658ft -13,124ft -6562ft -3281ft -1640ft -820ft -328ft/-100m 0

820ft 1640ft 3281ft 6562ft 9843ft 13,124ft 19,685ft

International Dateline

Western Australia

E

1

2

3

4

10°

*Arafura
Sea*

Croker Island
South Goulburn
Island

126

*Arnhem
Land*

Daly Waters

20°

Katherine

Tanimbar Kepulauan

I N D O N E S I A

130°

Van Diemen
Gulf

Melville Island

Pine Creek

Top Springs
Roadhouse

Tennant Creek

N O R T H E R N

T E R R I T O R Y

Bathurst Island

Darwin

Victoria River

*Tanami
Desert*

Kununurra

125°

Timor

EAST TIMOR

*T i m o r

S e a*

Joseph Bonaparte
Gulf

Wyndham

Kununurra

Halls Creek

117

Great Sandy Desert

Lake Mackay

Cape Londonderry

*Kimberley
Plateau*

Fitzroy
Crossing

Bonaparte
Archipelago

Bigge Island

Heywood
Islands

Fitzroy River

Percival
Lakes

Flores

King Sound

120°

Pulau Wetar

Pulau Sumba

Broome

Eighty Mile Beach

Pulau Lombok

116

I N D I A N

O C E A N

Port Hedland

Marble Bar

Newman

W E S T E R N

115°

Bali

Jawa

Dampier

Onslow

Hamersley Range

Fortescue River

Ashburton River

Barrow Island

Exmouth

Exmouth Gulf

119

10°

15°

20°

A

B

C

D

0 km   300

0 miles   300

Population

○ below 50,000   ○ 50,000 to 100,000   ◉ 100,000 to 500,000   ■ above 500,000

● Internal administrative capital

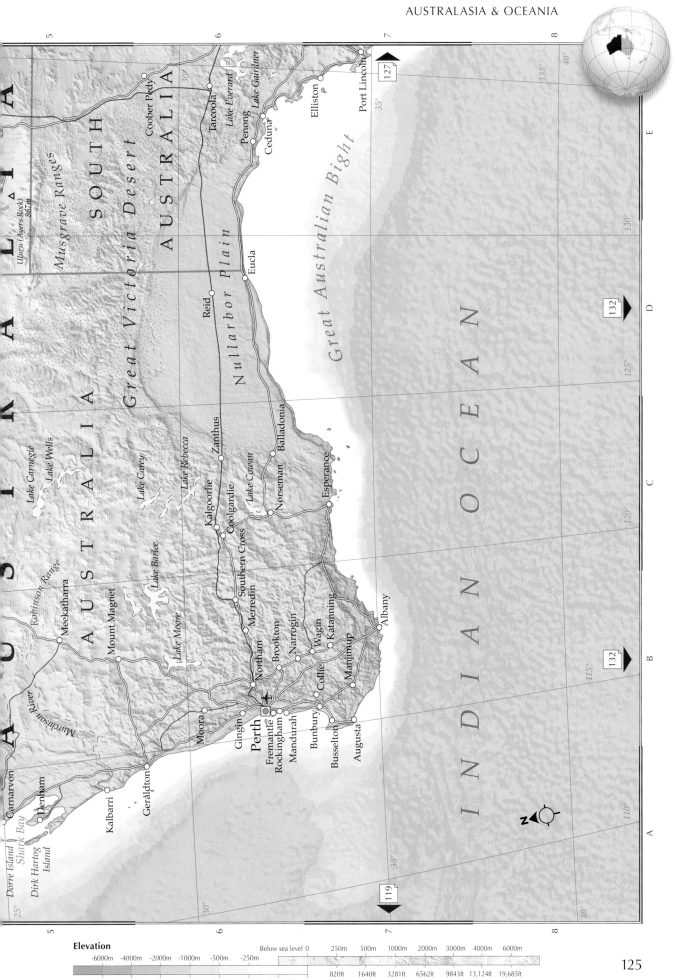

AUSTRALASIA & OCEANIA

SOUTH AUSTRALIA

AUSTRALIA

Great Victoria Desert

Musgrave Ranges

Uluru (Ayers Rock) △ 867m

Coober Pedy

Tarcoola

Lake Everard

Lake Gairdner

Penong

Ceduna

Elliston

Port Lincoln

Great Australian Bight

Nullarbor Plain

Reid

Eucla

Lake Wells

Lake Carnegie

Lake Carey

Lake Rebecca

Zanthus

Balladonia

Lake Cowan

Kalgoorlie

Coolgardie

Norseman

Esperance

Robinson Range

Meekatharra

Mount Magnet

Lake Barlee

Lake Moore

Southern Cross

Merredin

Albany

Northam

Brookton

Narrogin

Wagin

Katanning

Manjimup

Collie

Murchison River

Mingenew

Gingin

Perth

Fremantle

Rockingham

Mandurah

Bunbury

Busselton

Augusta

Carnarvon

Shark Bay

Denham

Dorre Island

Dirk Hartog Island

Kalbarri

Geraldton

INDIAN OCEAN

N

127

132

132

119

Elevation

-6000m -4000m -2000m -1000m -500m -250m

-19,658ft -13,124ft -6562ft -3281ft -1640ft -820ft -328ft/-100m 0

Below sea level 0 250m 500m 1000m 2000m 3000m 4000m 6000m

820ft 1640ft 3281ft 6562ft 9843ft 13,124ft 19,685ft

Eastern Australia

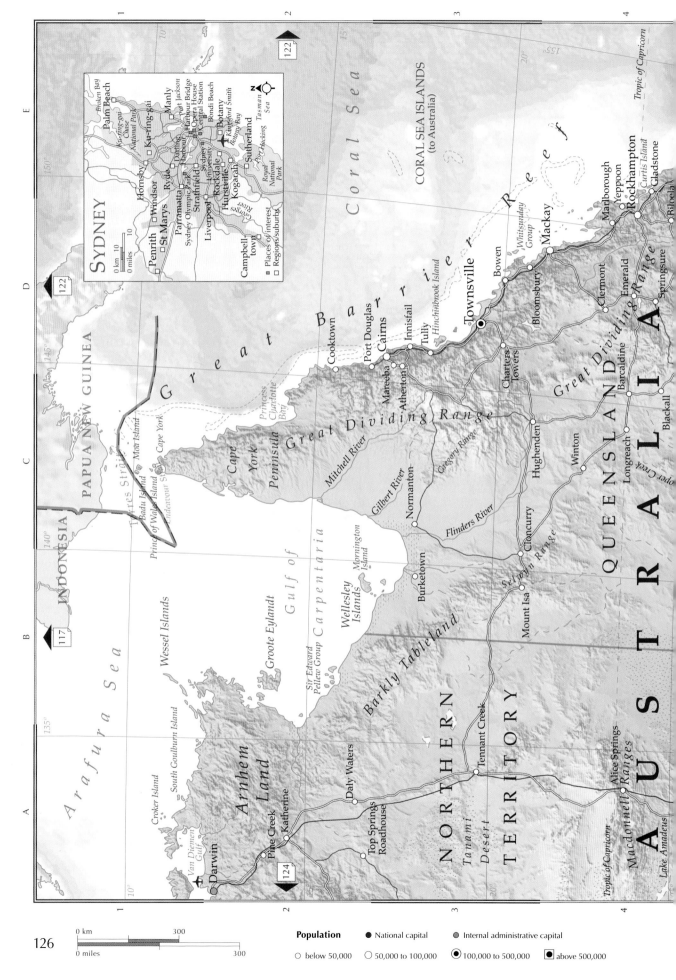

SYDNEY

Broken Bay
Palm Beach
Ku-ring-gai Chase National Park
Ku-ring-gai
Manly
Port Jackson
Harbour Bridge
Opera House
Bondi Beach
Hornsby
Darling Harbour
Botany
Windsor
Ryde
Central Station
Sydney
Sutherland
Botany Bay
Parramatta
Sydney Olympic Park
Strathfield
University
St Marys
Rockdale
Port Hacking River
Penrith
Liverpool
Hurstville
Kogarah
Royal National Park
Campbell-town
Georges River

■ Places of interest
□ Regions/suburbs

0 km   10
0 miles   10

Coral Sea

CORAL SEA ISLANDS
(to Australia)

Great Barrier Reef

PAPUA NEW GUINEA

INDONESIA

Arafura Sea

Tasman Sea

Torres Strait
Mua Island
Badu Island
Prince of Wales Island
Endeavour Strait
Cape York
Cape York Peninsula

Cooktown
Port Douglas
Cairns
Mareeba
Atherton
Innisfail
Tully
Hinchinbrook Island
Townsville
Bowen
Bloomsbury
Whitsunday Group
Mackay

Marlborough
Yeppoon
Rockhampton
Curtis Island
Gladstone
Biloela

Charters Towers

Great Dividing Range

Princess Charlotte Bay

Great Dividing Range

Mitchell River

Gilbert River

Normanton

Gregory Range

Hughenden

Clermont
Emerald
Springsure
Barcaldine
Blackall

Flinders River

Winton

Longreach

Cooper Creek

Cloncurry

Selwyn Range

Mount Isa

Burketown

Wellesley Islands
Mornington Island

Sir Edward Pellew Group

Gulf of Carpentaria

Groote Eylandt

Wessel Islands

Croker Island

South Goulburn Island

Van Diemen Gulf

Darwin

Pine Creek
Katherine

Daly Waters

Top Springs Roadhouse

Arnhem Land

Barkly Tableland

Tennant Creek

Tanami Desert

NORTHERN

TERRITORY

Alice Springs
Macdonnell Ranges
Lake Amadeus

Tropic of Capricorn

QUEENSLAND

AUSTRALIA

122

117

122

124

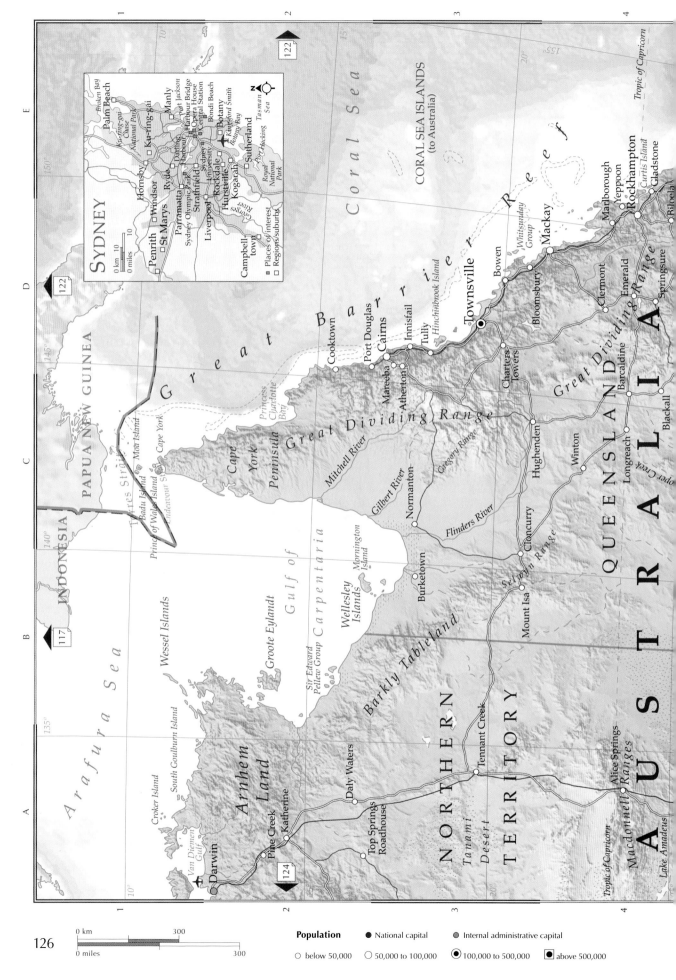

0 km   300
0 miles   300

Population   ● National capital   ● Internal administrative capital
○ below 50,000   ○ 50,000 to 100,000   ◉ 100,000 to 500,000   ◼ above 500,000

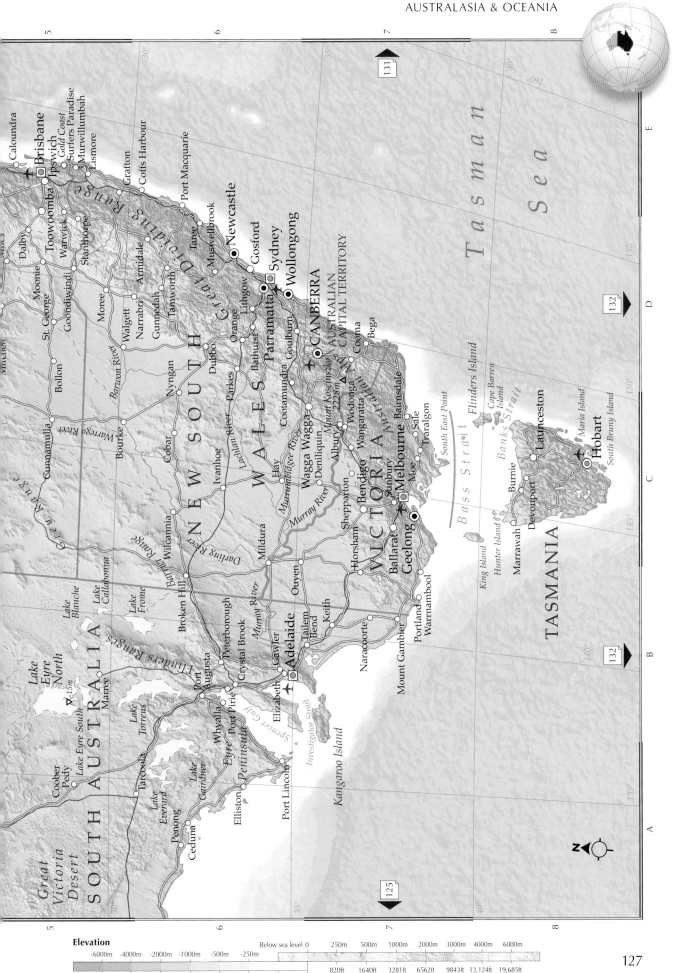

Elevation

-6000m	-4000m	-2000m	-1000m	-500m	-250m	Below sea level 0

-19,658ft -13,124ft -6562ft -3281ft -1640ft -820ft -328ft/-100m 0

250m	500m	1000m	2000m	3000m	4000m	6000m
820ft	1640ft	3281ft	6562ft	9843ft	13,124ft	19,685ft

New Zealand

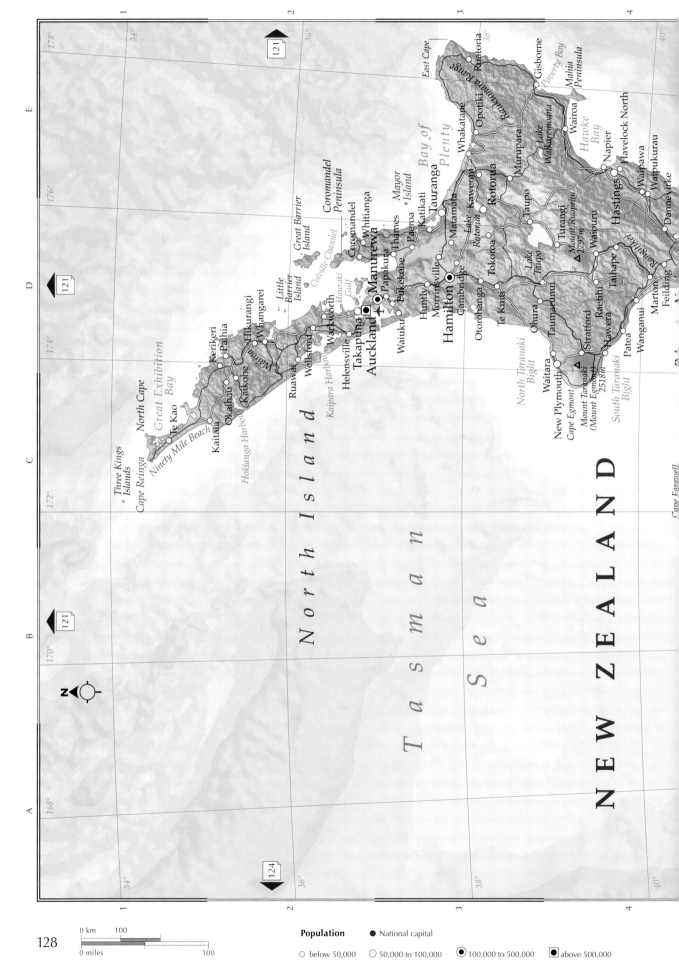

0 km   100

0 miles   100

Population   ● National capital

○ below 50,000   ○ 50,000 to 100,000   ◉ 100,000 to 500,000   ◼ above 500,000

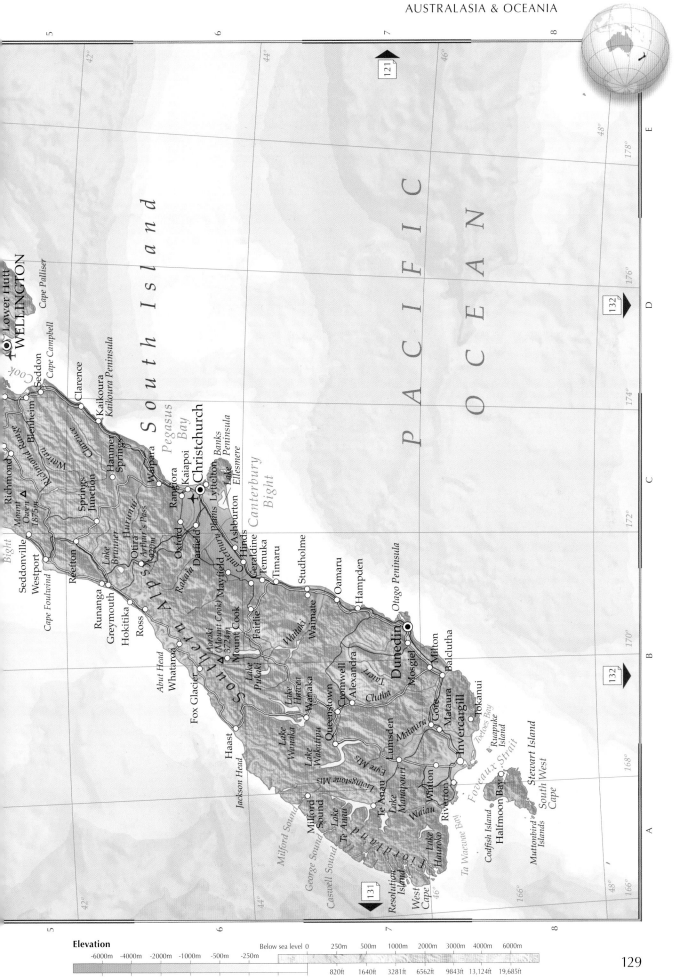

PACIFIC

OCEAN

South Island

WELLINGTON
Lower Hutt
Seddon
Cape Campbell
Cape Palliser

Cook

Clarence
Blenheim
Richmond
Kaikoura
Kaikoura Peninsula

Mount
Oven
1875m
Hanmer
Springs
Waipara
Rangiora
Kaiapoi
Christchurch
Lyttelton
Banks
Peninsula

Springs
Junction
Lake Ellesmere

Pegasus
Bay

Richmond Range
Wairau
Nelson

Reefton
Lake
Brunner
Otira
Arthur's Pass
920m
Oxford
Darfield

Hurunui

Rakaia
Methven
Ashburton
Hinds

Canterbury
Bight

Canterbury Plains

Runanga
Greymouth
Hokitika
Ross

Seddonville
Westport
Cape Foulwind

Southern Alps

Geraldine
Temuka
Timaru

Studholme

Oamaru

Hampden

Abut Head
Whataroa

Aoraki
(Mount Cook)
3724m
Mount Cook
Fairlie

Waitaki

Waimate

Fox Glacier

Lake
Pukaki

Otago Peninsula

Haast

Lake
Hawea
Wanaka
Cromwell
Alexandra

Dunedin
Mosgiel
Milton
Balclutha

Jackson Head

Lake
Wanaka

Lake
Wakatipu

Queenstown

Taieri

Clutha

Lumsden

Gore
Mataura

Tokanui

Livingstone Mts

Eyre Mts

Milford
Sound

Te Anau
Lake
Te Anau

Lake
Manapouri

Waiau

Winton
Mataura
Invercargill
Riverton

Bluff

Toetoes Bay
Ruapuke
Island

Fiordland

Stewart Island

South West
Cape

George Sound

Caswell Sound

Lake
Hauroko

Codfish Island
Halfmoon Bay

Muttonbird
Islands

Foveaux Strait

Resolution
Island

West
Cape

Ta Waewae Bay

S E A   B I G H T

S e d d o n v i l l e

Elevation

							Below sea level 0	250m	500m	1000m	2000m	3000m	4000m	6000m

-6000m   -4000m   -2000m   -1000m   -500m   -250m

-19,658ft -13,124ft -6562ft   -3281ft   -1640ft   -820ft -328ft/-100m 0

820ft   1640ft   3281ft   6562ft   9843ft 13,124ft 19,685ft

The Pacific Ocean

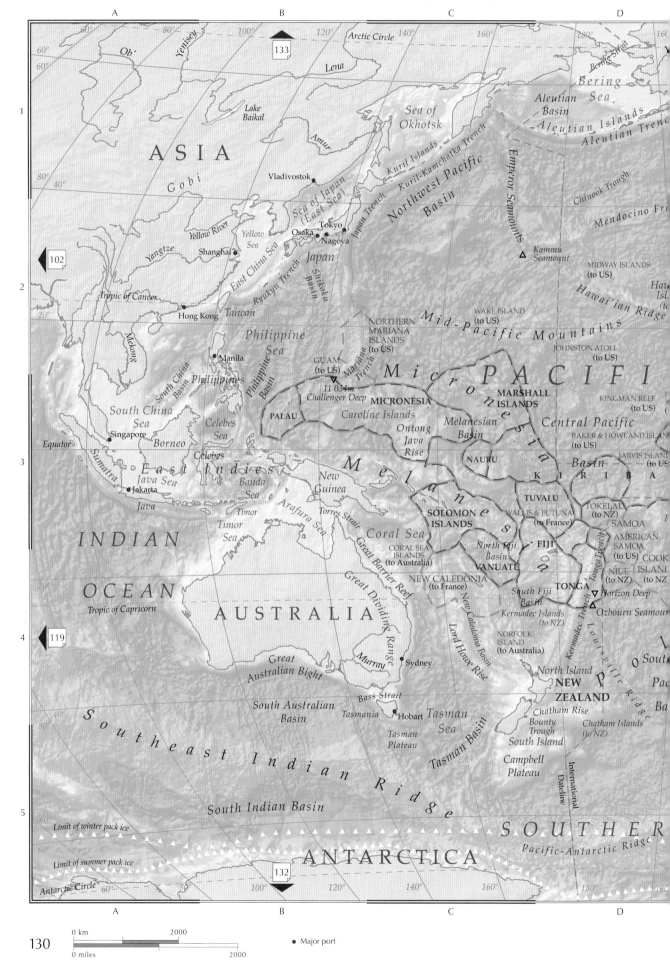

0 km   2000
0 miles   2000

● Major port

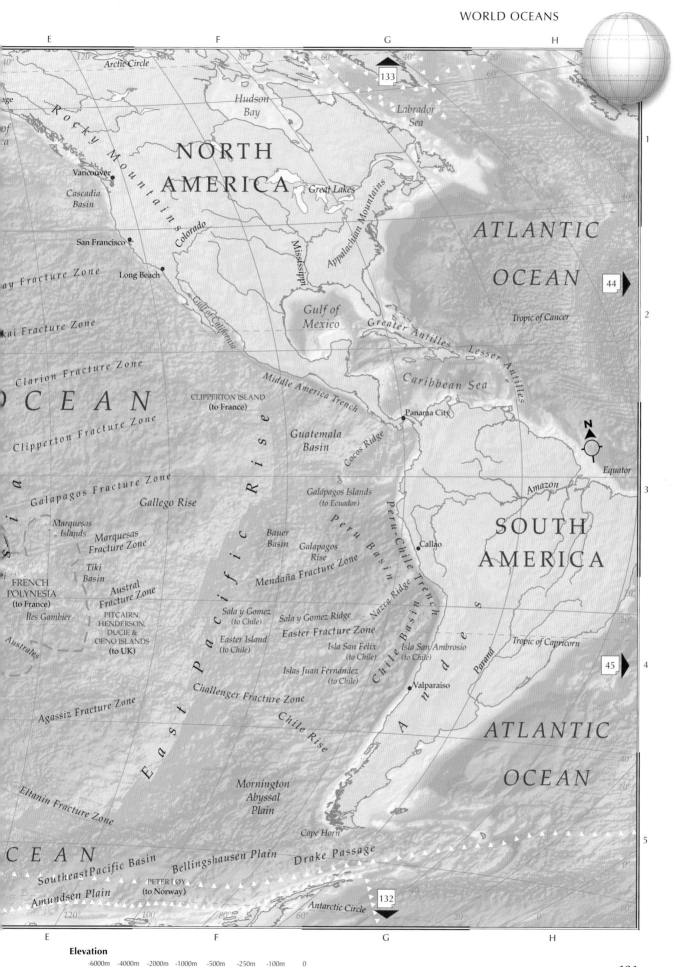

NORTH AMERICA

SOUTH AMERICA

ATLANTIC OCEAN

ATLANTIC

OCEAN

OCEAN

Arctic Circle

Rocky Mountains

Vancouver

Cascadia Basin

San Francisco

Long Beach

Colorado

Hudson Bay

Great Lakes

Appalachian Mountains

Mississippi

Gulf of California

Gulf of Mexico

Greater Antilles

Lesser Antilles

Caribbean Sea

Tropic of Cancer

Labrador Sea

...ay Fracture Zone

...kai Fracture Zone

Clarion Fracture Zone

Clipperton Fracture Zone

CLIPPERTON ISLAND (to France)

Middle America Trench

Guatemala Basin

Cocos Ridge

Panama City

Galapagos Fracture Zone

Gallego Rise

Marquesas Islands

Marquesas Fracture Zone

Tiki Basin

FRENCH POLYNESIA (to France)

Îles Gambier

Australes

Austral Fracture Zone

PITCAIRN, HENDERSON, DUCIE & OENO ISLANDS (to UK)

Sala y Gomez (to Chile)

Easter Island (to Chile)

Islas Juan Fernández (to Chile)

Challenger Fracture Zone

Agassiz Fracture Zone

Eltanin Fracture Zone

Bauer Basin

Galapagos Rise

Mendaña Fracture Zone

Sala y Gomez Ridge

Easter Fracture Zone

Isla San Félix (to Chile)

Isla San Ambrosio (to Chile)

Galápagos Islands (to Ecuador)

Peru Basin

Peru-Chile Trench

Nazca Ridge

Chile Basin

Callao

Andes

Paraná

Amazon

Equator

Tropic of Capricorn

Valparaiso

East Pacific Rise

Chile Rise

Mornington Abyssal Plain

Cape Horn

Southeast Pacific Basin

Bellingshausen Plain

Drake Passage

Amundsen Plain

PETER I ØY (to Norway)

Antarctic Circle

N

Elevation

-6000m	-4000m	-2000m	-1000m	-500m	-250m	-100m	0
-19,658ft	-13,124ft	-6562ft	-3281ft	-1640ft	-820ft	-328ft/-100m	0

Antarctica

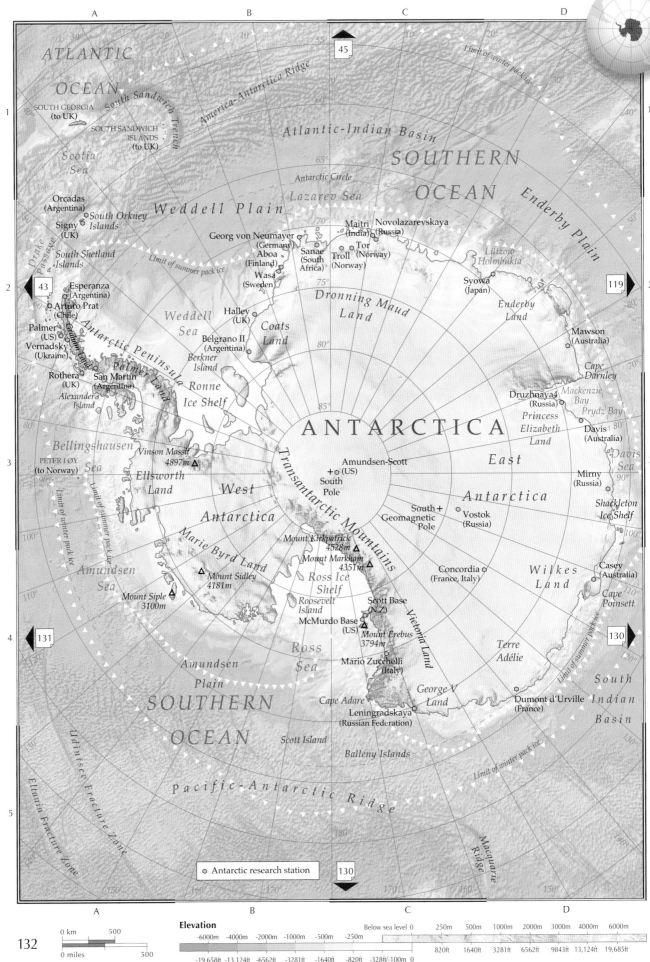

ATLANTIC

OCEAN

SOUTH GEORGIA
(to UK)

SOUTH SANDWICH
ISLANDS
(to UK)

South Sandwich Trench

America-Antarctica Ridge

Atlantic-Indian Basin

SOUTHERN

OCEAN

Limit of winter pack ice

Enderby Plain

*Scotia
Sea*

Antarctic Circle

Lazarev Sea

Weddell Plain

Orcadas
(Argentina)

South Orkney
Islands

Signy
(UK)

Drake Passage

South Shetland
Islands

Maitri
(India)

Novolazarevskaya
(Russia)

Georg von Neumayer
(Germany)

Aboa
(Finland)

Sanae
(South
Africa)

Tor
(Norway)

Troll
(Norway)

Wasa
(Sweden)

*Lützow
Holmbukta*

Syowa
(Japan)

Limit of summer pack ice

Esperanza
(Argentina)

Arturo Prat
(Chile)

Palmer
(US)

Vernadsky
(Ukraine)

Graham Land

Antarctic Peninsula

Halley
(UK)

*Weddell
Sea*

*Coats
Land*

*Dronning Maud
Land*

*Enderby
Land*

Mawson
(Australia)

Belgrano II
(Argentina)

Rothera
(UK)

San Martin
(Argentina)

Palmer Land

*Berkner
Island*

*Ronne
Ice Shelf*

Druzhnaya4
(Russia)

*Cape
Darnley*

*Mackenzie
Bay*

*Alexander
Island*

Prydz Bay

*Princess
Elizabeth
Land*

Davis
(Australia)

*Bellingshausen
Sea*

Vinson Massif
4897m

ANTARCTICA

East

*Davis
Sea*

PETER I ØY
(to Norway)

*Ellsworth
Land*

West

Amundsen-Scott
(US)
South
Pole

South
Geomagnetic
Pole

Vostok
(Russia)

Antarctica

Mirny
(Russia)

*Shackleton
Ice Shelf*

Limit of winter pack ice

Antarctica

Transantarctic Mountains

Mount Kirkpatrick
4528m

Mount Markham
4351m

Concordia
(France, Italy)

*Wilkes
Land*

Casey
(Australia)

*Amundsen
Sea*

Marie Byrd Land

Mount Sidley
4181m

*Ross Ice
Shelf*

*Cape
Poinsett*

Mount Siple
3100m

*Roosevelt
Island*

Scott Base
(N.Z.)

McMurdo Base
(US)

Mount Erebus
3794m

Victoria Land

*Terre
Adélie*

130

Ross

Sea

Mario Zucchelli
(Italy)

*George V
Land*

Dumont d'Urville
(France)

South

Indian

Amundsen

Plain

SOUTHERN

OCEAN

Cape Adare

Leningradskaya
(Russian Federation)

Basin

Udintsev Fracture Zone

Eltanin Fracture Zone

Scott Island

Balleny Islands

*Macquarie
Ridge*

Pacific-Antarctic Ridge

○ Antarctic research station

0 km   500

0 miles   500

132

Elevation

Below sea level 0   250m   500m   1000m   2000m   3000m   4000m   6000m

-6000m -4000m -2000m -1000m -500m   -250m

-19,658ft -13,124ft -6562ft -3281ft -1640ft -820ft -328ft/-100m 0

820ft   1640ft 3281ft 6562ft 9843ft 13,124ft 19,685ft

Arctic Ocean

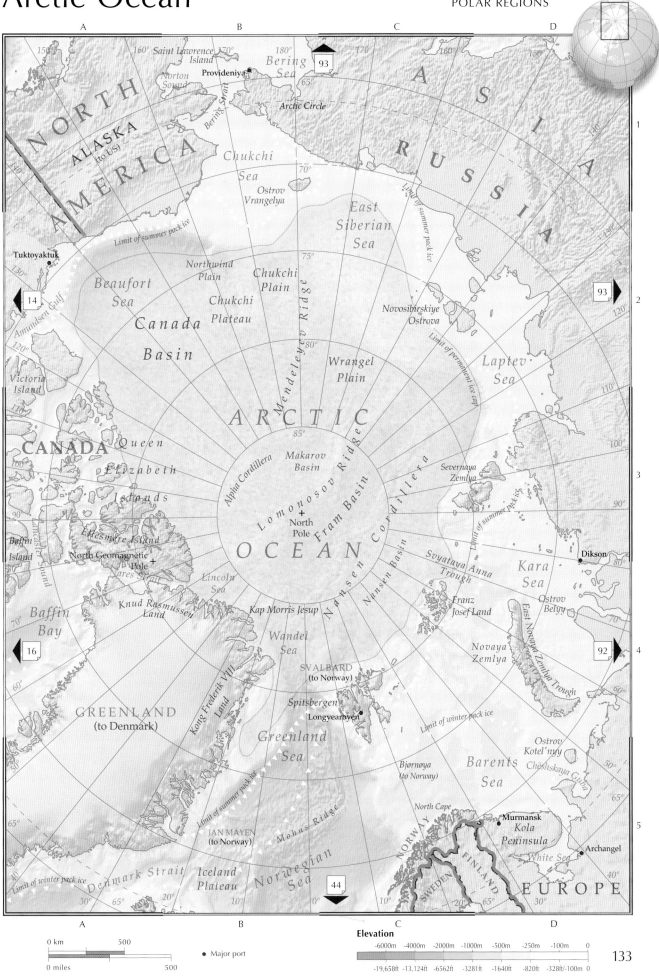

NORTH AMERICA

ALASKA (to US)

Saint Lawrence Island

Provideniya

Norton Sound

Bering Sea

93

65°

Arctic Circle

Bering Strait

Chukchi Sea

70°

ASIA

RUSSIA

Ostrov Vrangelya

East Siberian Sea

Limit of summer pack ice

Tuktoyaktuk

Limit of summer pack ice

14

Beaufort Sea

Northwind Plain

75°

Chukchi Plain

Mendeleyev Ridge

Novosibirskiye Ostrova

93

Amundsen Gulf

Chukchi Plateau

Limit of permanent ice cap

130°

Canada

80°

Wrangel Plain

Laptev Sea

120°

Basin

ARCTIC

110°

Victoria Island

Alpha Cordillera

Makarov Basin

Severnaya Zemlya

100°

CANADA

Queen

85°

Lomonosov Ridge

Fram Basin

Nansen Cordillera

90°

Elizabeth

North Pole

80°

Islands

OCEAN

Nansen Basin

Svyataya Anna Trough

Kara Sea

Dikson

Baffin Island

Ellesmere Island

North Geomagnetic Pole

Nares Strait

Lincoln Sea

Kap Morris Jesup

Franz Josef Land

Ostrov Belyy

70°

Lancaster Sound

Baffin Bay

Knud Rasmussen Land

Wandel Sea

Novaya Zemlya

East Novaya Zemlya Trough

16

Svalbard (to Norway)

92

60°

GREENLAND (to Denmark)

Kong Frederik VIII Land

Spitsbergen

Longyearbyen

Limit of winter pack ice

Ostrov Kotel'nyy

Chëshskaya Guba

50°

Greenland Sea

Bjørnøya (to Norway)

Barents Sea

65°

Limit of summer pack ice

JAN MAYEN (to Norway)

Mohns Ridge

North Cape

NORWAY

Murmansk

Kola Peninsula

Archangel

White Sea

Limit of winter pack ice

Denmark Strait

Iceland Plateau

Norwegian Sea

44

SWEDEN

FINLAND

EUROPE

40°

Elevation

-6000m	-4000m	-2000m	-1000m	-500m	-250m	-100m	0
-19,658ft	-13,124ft	-6562ft	-3281ft	-1640ft	-820ft	-328ft/-100m	0

0 km   500

0 miles   500

● Major port

Country Profiles

This Factfile is intended as a guide to a world that is continually changing as political fashions and personalities come and go. Nevertheless, all the material in these factfiles has been researched from the most up-to-date and authoritative sources to give an incisive portrait of the geographical, political, and social characteristics that make each country so unique.

There are currently 196 independent countries in the world - more than at any previous time - and over 50 dependencies. Antarctica is the only land area on Earth that is not officially part of, and does not belong to, any single country.

Country profile key

Formation Date of formation denotes the date of political origin or independence of a state, i.e. its emergence as a recognizable entity in the modern political world / date current borders were established

Population Total population / population density – based on total *land* area

Languages An asterisk (*) denotes the official language(s)

Calorie consumption Average number of kilocalories consumed daily per person

---

AFGHANISTAN
Central Asia

Page 100 D4

Landlocked in Central Asia, Afghanistan has suffered decades of conflict. The Islamist taliban, ousted by a US-led offensive in 2001, continue to resist subsequent elected governments.

Official name Islamic Republic of Afghanistan
Formation 1919 / 1919
Capital Kabul
Population 35.5 million / 141 people per sq mile (54 people per sq km)
Total area 250,000 sq miles (647,500 sq km)
Languages Pashtu*, Tajik, Dari*, Farsi, Uzbek, Turkmen
Religions Sunni Muslim 80%, Shi'a Muslim 19%, Other 1%
Ethnic mix Pashtun 38%, Tajik 25%, Hazara 19%, Uzbek and Turkmen 15%, Other 3%
Government Nonparty system
Currency Afghani = 100 puls
Literacy rate 32%
Calorie consumption 2090 kilocalories

ALBANIA
Southeast Europe

Page 79 C6

Lying at the southeastern end of the Adriatic Sea, Albania – or the "land of the eagles" – underwent upheavals after 1991 to emerge from its communist-period isolation.

Official name Republic of Albania
Formation 1912 / 1921
Capital Tirana
Population 2.9 million / 274 people per sq mile (106 people per sq km)
Total area 11,100 sq miles (28,748 sq km)
Languages Albanian*, Greek
Religions Muslim (mainly Sunni) 68%, Roman Catholic 12%, Albanian Orthodox 8%, Nonreligious 6%, Other 6%
Ethnic mix Albanian 98%, Greek 1%, Other 1%
Government Parliamentary system
Currency Lek = 100 qindarka (qintars)
Literacy rate 97%
Calorie consumption 3193 kilocalories

ALGERIA
North Africa

Page 48 C3

Lying mostly in the Sahara, this former French colony was riven by civil war after Islamists were denied electoral victory in 1992. Fighting has subsided but Islamic extremists remain a threat.

Official name People's Democratic Republic of Algeria
Formation 1962 / 1962
Capital Algiers
Population 41.3 million / 45 people per sq mile (17 people per sq km)
Total area 919,590 sq miles (2,381,740 sq km)
Languages Arabic*, Tamazight* (Kabyle, Shawia, Tamashek), French
Religions Sunni Muslim 99%, Christian & Jewish 1%
Ethnic mix Arab 75%, Berber 24%, European & Jewish 1%
Government Presidential system
Currency Algerian dinar = 100 centimes
Literacy rate 75%
Calorie consumption 3296 kilocalories

ANDORRA
Southwest Europe

Page 69 B6

A tiny landlocked principality, Andorra lies between France and Spain, high in the eastern Pyrenees. Its economy, based on tourism, also features low tax and duty-free shopping.

Official name Principality of Andorra
Formation 1278 / 1278
Capital Andorra la Vella
Population 77,000 / 428 people per sq mile (166 people per sq km)
Total area 181 sq miles (468 sq km)
Languages Spanish, Catalan*, French, Portuguese
Religions Roman Catholic 94%, Other 6%
Ethnic mix Spanish 46%, Andorran 28%, Other 18%, French 8%
Government Parliamentary system
Currency Euro = 100 cents
Literacy rate 99%
Calorie consumption Not available

ANGOLA
Southern Africa

Page 56 B2

An oil- and diamond-rich former Portuguese colony in Southwest Africa, Angola is badly scarred from the 1975–2002 civil war. The removal of thousands of land mines continues.

Official name Republic of Angola
Formation 1975 / 1975
Capital Luanda
Population 29.8 million / 62 people per sq mile (24 people per sq km)
Total area 481,351 sq miles (1,246,700 sq km)
Languages Portuguese*, Umbundu, Kimbundu, Kikongo
Religions Roman Catholic 40%, Protestant 38%, Nonreligious 12%, Other (including animist) 10%
Ethnic mix Ovimbundu 37%, Ambundu 25%, Other 25%, Bakongo 13%
Government Presidential system
Currency Readjusted kwanza = 100 lwei
Literacy rate 66%
Calorie consumption 2473 kilocalories

ANTIGUA & BARBUDA
West Indies

Page 33 H3

Lying on the Atlantic edge of the Leeward Islands, Antigua was in turn a Spanish, French, and British colony. Tourism is key, but Barbuda's beaches have suffered hurricane damage.

Official name Antigua and Barbuda
Formation 1981 / 1981
Capital St. John's
Population 100,000 / 588 people per sq mile (227 people per sq km)
Total area 170 sq miles (442 sq km)
Languages English*, English patois
Religions Other Christian 49%, Other 19%, Anglican 19%, Seventh-day Adventist 13%
Ethnic mix Black African 87%, Mixed race 5%, Hispanic 3%, Other 3%, White 2%
Government Parliamentary system
Currency East Caribbean dollar = 100 cents
Literacy rate 99%
Calorie consumption 2417 kilocalories

ARGENTINA
South America

Page 43 B5

From semiarid lowlands, through fertile grasslands, to the glacial southern tip of South America, Argentina has enjoyed democratic rule since 1983 but struggled with high foreign debts.

Official name Argentine Republic
Formation 1816 / 1816
Capital Buenos Aires
Population 44.3 million / 42 people per sq mile (16 people per sq km)
Total area 1,068,296 sq miles (2,766,890 sq km)
Languages Spanish*, Italian, Amerindian languages
Religions Roman Catholic 71%, Protestant 15%, Nonreligious 11%, Other 3%
Ethnic mix Indo-European 97%, Mestizo 2%, Amerindian 1%
Government Presidential system
Currency Argentine peso = 100 centavos
Literacy rate 98%
Calorie consumption 3229 kilocalories

ARMENIA
Southwest Asia

Page 95 F3

The smallest of the ex-Soviet republics, landlocked Armenia lies in the Lesser Caucasus mountains. It was the first country to adopt Christianity as the state religion, in the 4th century AD.

Official name Republic of Armenia
Formation 1991 / 1991
Capital Yerevan
Population 2.9 million / 252 people per sq mile (97 people per sq km)
Total area 11,506 sq miles (29,800 sq km)
Languages Armenian*, Azeri, Russian
Religions Orthodox Christian 89%, Other 8%, Nonreligious 2%, Armenian Catholic Church 1%
Ethnic mix Armenian 98%, Other 1%, Yezidi 1%
Government Parliamentary system
Currency Dram = 100 luma
Literacy rate 99%
Calorie consumption 2928 kilocalories

AUSTRALIA
Australasia & Oceania

Page 125 B5

An island continent between the Indian and Pacific oceans, Australia was settled by Europeans from 1788, but recent immigrants are mostly Asian. Minerals underpin the economy.

Official name Commonwealth of Australia
Formation 1901 / 1901
Capital Canberra
Population 24.5 million / 8 people per sq mile (3 people per sq km)
Total area 2,967,893 sq miles (7,686,850 sq km)
Languages English*, Italian, Cantonese, Greek, Arabic, Vietnamese, Aboriginal languages
Religions Roman Catholic 28%, Nonreligious 24%, Other Christian 20%, Anglican 19%, Other 9%,
Ethnic mix British 34%, Australian 27%, Other 18%, Irish 8%, Italian 4%, German 3%, Chinese 3%
Government Parliamentary system
Currency Australian dollar = 100 cents
Literacy rate 99%
Calorie consumption 3276 kilocalories

AUSTRIA
Central Europe

Page 73 D7

Nestled in Central Europe, Austria was created after the Austro-Hungarian Empire was defeated in World War I. Absorbed into Hitler's Germany in 1938, it re-emerged in 1955.

Official name Republic of Austria
Formation 1918 / 1919
Capital Vienna
Population 8.7 million / 272 people per sq mile (105 people per sq km)
Total area 32,378 sq miles (83,858 sq km)
Languages German*, Croatian, Slovenian, Hungarian (Magyar)
Religions Roman Catholic 75%, Nonreligious 12%, Other Christian 8%, Muslim 4%, Other 1%
Ethnic mix Austrian 93%, Croat, Slovene, and Hungarian 6%, Other 1%
Government Parliamentary system
Currency Euro = 100 cents
Literacy rate 99%
Calorie consumption 3768 kilocalories

AZERBAIJAN
Southwest Asia

Page 95 G2

On the west coast of the Caspian Sea, oil-rich Azerbaijan regained its independence from the USSR in 1991. A territorial dispute with Armenia remains unresolved.

Official name Republic of Azerbaijan
Formation 1991 / 1991
Capital Baku
Population 9.8 million / 293 people per sq mile (113 people per sq km)
Total area 33,436 sq miles (86,600 sq km)
Languages Azeri*, Russian
Religions Shi'a Muslim 68%, Sunni Muslim 26%, Russian Orthodox 3%, Armenian Apostolic Church (Orthodox) 2%, Other 1%
Ethnic mix Azeri 91%, Other 3%, Armenian 2%, Russian 2%, Lazs 2%
Government Presidential system
Currency New manat = 100 gopik
Literacy rate 99%
Calorie consumption 3118 kilocalories

BAHAMAS, THE
West Indies

Page 32 C1

Located in the western Atlantic, off the Florida coast, the Bahamas comprise some 700 islands and 2400 cays; only 30 are inhabited. Financial services and shipping support the economy.

Official name Commonwealth of The Bahamas
Formation 1973 / 1973
Capital Nassau
Population 400,000 / 103 people per sq mile (40 people per sq km)
Total area 5382 sq miles (13,940 sq km)
Languages English*, English Creole, French Creole
Religions Baptist 36%, Other 20%, Anglican 14%, Roman Catholic 12%, Pentecostal 9%, Seventh-day Adventist 5%, Methodist 4%
Ethnic mix Black African 85%, European 12%, Asian and Hispanic 3%
Government Parliamentary system
Currency Bahamian dollar = 100 cents
Literacy rate 96%
Calorie consumption 2670 kilocalories

BAHRAIN
Southwest Asia

Page 98 C4

Only three of Bahrain's 33 islands lying between the Qatar peninsula and Saudi Arabian are inhabited. The first Gulf emirate to export oil, reserves are expected to last another 10 to 15 years.

Official name Kingdom of Bahrain
Formation 1971 / 1971
Capital Manama
Population 1.5 million / 5495 people per sq mile (2125 people per sq km)
Total area 239 sq miles (620 sq km)
Languages Arabic*
Religions Muslim (mainly Shi'a) 70%, Other 30%
Ethnic mix Bahraini 46%, Asian 46%, Other Arab 5%, Other 3%
Government Monarchical / parliamentary system
Currency Bahraini dinar = 1000 fils
Literacy rate 95%
Calorie consumption Not available

BANGLADESH
South Asia

Page 113 G3

Low-lying Bangladesh on the Bay of Bengal suffers annual monsoon flooding. It seceded from Pakistan in 1971. Political instability and corruption are ongoing problems.

Official name People's Republic of Bangladesh
Formation 1971 / 1971
Capital Dhaka
Population 165 million / 3186 people per sq mile (1230 people per sq km)
Total area 55,598 sq miles (144,000 sq km)
Languages Bengali*, Urdu, Chakma, Marma (Magh), Garo, Khasi, Santhali, Tripuri, Mro
Religions Muslim (mainly Sunni) 90%, Hindu 9%, Other 1%
Ethnic mix Bengali 98%, Other 2%
Government Parliamentary system
Currency Taka = 100 poisha
Literacy rate 73%
Calorie consumption 2450 kilocalories

BARBADOS
West Indies

Page 33 H4

The most easterly of the Windward Islands, Barbados was under British rule from the 1620s. A sugar exporter in the 18th century, it now relies on tourism and financial services.

Official name Barbados
Formation 1966 / 1966
Capital Bridgetown
Population 300,000 / 1807 people per sq mile (698 people per sq km)
Total area 166 sq miles (430 sq km)
Languages Bajan (Barbadian English), English*
Religions Anglican 24%, Nonreligious 21%, Other 21%, Pentecostal 20%, Seventh-day Adventist 6%, Methodist 4%, Roman Catholic 4%
Ethnic mix Black African 93%, Mixed race 3%, White 3%, Other 1%
Government Parliamentary system
Currency Barbados dollar = 100 cents
Literacy rate 99%
Calorie consumption 2937 kilocalories

BELARUS
Eastern Europe

Page 85 B6

Landlocked in eastern Europe, forested Belarus, which means "White Russia," was reluctant to become independent of the USSR in 1991, and has been slow to reform its economy since.

Official name Republic of Belarus
Formation 1991 / 1991
Capital Minsk
Population 9.5 million / 119 people per sq mile (46 people per sq km)
Total area 80,154 sq miles (207,600 sq km)
Languages Belarussian*, Russian
Religions Orthodox Christian 73%, Roman Catholic 12%, Other 12%, Nonreligious 3%
Ethnic mix Belarussian 86%, Russian 8%, Polish 3%, Other 2%, Ukrainian 1%
Government Presidential system
Currency Belarussian rouble = 100 kopeks
Literacy rate 99%
Calorie consumption 3250 kilocalories

BELGIUM
Northwest Europe

Page 65 B6 [flag]

Located in Northwest Europe, Belgium has forests in the south and canals in the flat north. Its history and politics are marked by the division between its Flemish and Walloon communities.

Official name Kingdom of Belgium
Formation 1830 / 1919
Capital Brussels
Population 11.4 million / 900 people per sq mile (347 people per sq km)
Total area 11,780 sq miles (30,510 sq km)
Languages Dutch*, French*, German*
Religions Roman Catholic 88%, Other 10%, Muslim 2%
Ethnic mix Fleming 58%, Walloon 33%, Other 6%, Italian 2%, Moroccan 1%
Government Parliamentary system
Currency Euro = 100 cents
Literacy rate 99%
Calorie consumption 3733 kilocalories

BELIZE
Central America

Page 30 B1 [flag]

The last Central American country to gain independence, this former British colony lies on the eastern shore of the Yucatan Peninsula. Offshore is the world's second-largest barrier reef.

Official name Belize
Formation 1981 / 1981
Capital Belmopan
Population 400,000 / 45 people per sq mile (18 people per sq km)
Total area 8867 sq miles (22,966 sq km)
Languages English Creole, Spanish, English*, Mayan, Garifuna (Carib)
Religions Roman Catholic 40%, Other Christian 34%, Nonreligious 16%, Other 10%
Ethnic mix Mestizo 49%, Creole 24%, Maya 10%, Other 7%, Garifuna 6%, Asian Indian 4%
Government Parliamentary system
Currency Belizean dollar = 100 cents
Literacy rate 75%
Calorie consumption 2751 kilocalories

BENIN
West Africa

Page 53 F4 [flag]

Stretching north from the West African coast, this ex-French colony suffered military rule after independence but in recent decades has been a leading example of African democratization.

Official name Republic of Benin
Formation 1960 / 1960
Capital Porto-Novo; Cotonou
Population 11.2 million / 262 people per sq mile (101 people per sq km)
Total area 43,483 sq miles (112,620 sq km)
Languages Fon, Bariba, Yoruba, Adja, Houeda, Somba, French*
Religions Indigenous beliefs and Voodoo 50%, Christian 30%, Muslim 20%
Ethnic mix Fon 41%, Other 21%, Adja 16%, Yoruba 12%, Bariba 10%
Government Presidential system
Currency CFA franc = 100 centimes
Literacy rate 33%
Calorie consumption 2619 kilocalories

BHUTAN
South Asia

Page 113 G3 [flag]

This landlocked Buddhist kingdom, perched in the eastern Himalayas between India and China, is carefully protecting its cultural identity from modernization and the outside world.

Official name Kingdom of Bhutan
Formation 1656 / 1865
Capital Thimphu
Population 800,000 / 44 people per sq mile (17 people per sq km)
Total area 18,147 sq miles (47,000 sq km)
Languages Dzongkha*, Nepali, Assamese
Religions Mahayana Buddhist 75%, Hindu 25%
Ethnic mix Drukpa 50%, Nepalese 35%, Other 15%
Government Monarchical / parliamentary system
Currency Ngultrum = 100 chetrum
Literacy rate 57%
Calorie consumption Not available

BOLIVIA
South America

Page 39 F3

Bolivia lies landlocked high in central South America. Mineral riches once made it the region's wealthiest state, but wars, coups, and poor governance have reduced it to the poorest.

Official name Plurinational State of Bolivia
Formation 1825 / 1938
Capital La Paz (administrative); Sucre (judicial)
Population 11.1 million / 27 people per sq mile (10 people per sq km)
Total area 424,162 sq miles (1,098,580 sq km)
Languages Aymara*, Quechua*, Spanish*
Religions Roman Catholic 77%, Protestant 16%, Nonreligious 4%, Other 3%
Ethnic mix Quechua 37%, Aymara 32%, Mixed race 13%, European 10%, Other 8%
Government Presidential system
Currency Boliviano = 100 centavos
Literacy rate 92%
Calorie consumption 2256 kilocalories

BRAZIL
South America

Page 40 C2

Brazil covers more than half of South America and is the site of the world's largest rain forest. It has immense natural resources and produces a third of the world's coffee.

Official name Federative Republic of Brazil
Formation 1822 / 1828
Capital Brasília
Population 209 million / 64 people per sq mile (25 people per sq km)
Total area 3,286,470 sq miles (8,511,965 sq km)
Languages Portuguese*, German, Italian, Spanish, Polish, Japanese, Amerindian languages
Religions Roman Catholic 61%, Protestant 26%, Nonreligious 8%, Other 5%
Ethnic mix White 48%, Mixed race 43%, Black 8%, Other 1%
Government Presidential system
Currency Real = 100 centavos
Literacy rate 92%
Calorie consumption 3263 kilocalories

BURKINA FASO
West Africa

Page 53 E4

Known as Upper Volta until 1984, Burkina Faso is landlocked in the semiarid Sahel of West Africa. It has been under military rule for most of its post-independence history.

Official name Burkina Faso
Formation 1960 / 1960
Capital Ouagadougou
Population 19.2 million / 182 people per sq mile (70 people per sq km)
Total area 105,869 sq miles (274,200 sq km)
Languages Mossi, Fulani, French*, Tuareg, Dyula, Songhai
Religions Muslim 61%, Roman Catholic 19%, Traditional beliefs 15%, Protestant 4%, Other 1%
Ethnic mix Mossi 48%, Other 21%, Peul 10%, Lobi 7%, Bobo 7%, Mandé 7%
Government Presidential system
Currency CFA franc = 100 centimes
Literacy rate 35%
Calorie consumption 2720 kilocalories

CAMEROON
Central Africa

Page 54 A4

A former trading hub on the central West African coast, Cameroon was effectively a one-party state for 30 years. Elections since 1992 have brought no change in leadership.

Official name Republic of Cameroon
Formation 1960 / 1961
Capital Yaoundé
Population 24.1 million / 134 people per sq mile (52 people per sq km)
Total area 183,567 sq miles (475,400 sq km)
Languages Bamileke, Fang, Fulani, French*, English*
Religions Roman Catholic 35%, Traditional beliefs 25%, Muslim 22%, Protestant 18%
Ethnic mix Cameroon highlanders 31%, Other 21%, Equatorial Bantu 19%, Kirdi 11%, Fulani 10%, Northwestern Bantu 8%
Government Presidential system
Currency CFA franc = 100 centimes
Literacy rate 71%
Calorie consumption 2671 kilocalories

BOSNIA & HERZEGOVINA
Southeast Europe

Page 78 B3

In the mountainous western Balkans this state, born out of the bitter conflicts of Yugoslavia's collapse, has two key concerns: balancing ethnic rivalries, and integrating with Europe.

Official name Bosnia and Herzegovina
Formation 1992 / 1992
Capital Sarajevo
Population 3.5 million / 177 people per sq mile (68 people per sq km)
Total area 19,741 sq miles (51,129 sq km)
Languages Bosnian*, Serbian*, Croatian*
Religions Muslim (mainly Sunni) 53%, Orthodox Christian 35%, Roman Catholic 8%, Nonreligious 3%, Other 1%
Ethnic mix Bosniak 48%, Serb 34%, Croat 16%, Other 2%
Government Parliamentary system
Currency Marka = 100 pfeninga
Literacy rate 97%
Calorie consumption 3154 kilocalories

BRUNEI
Southeast Asia

Page 116 D3

On the northwest coast of the island of Borneo, Brunei is surrounded and divided in two by the Malaysian state of Sarawak. Oil and gas revenues have brought a high standard of living.

Official name Brunei Darussalam
Formation 1984 / 1984
Capital Bandar Seri Begawan
Population 400,000 / 197 people per sq mile (76 people per sq km)
Total area 2228 sq miles (5770 sq km)
Languages Malay*, English, Chinese
Religions Muslim (mainly Sunni) 79%, Christian 9%, Buddhist 8%, Other 4%
Ethnic mix Malay 66%, Other 21%, Chinese 10%, Indigenous 3%
Government Monarchy
Currency Brunei dollar = 100 cents
Literacy rate 95%
Calorie consumption 2985 kilocalories

BURUNDI
Central Africa

Page 51 B7

Small, landlocked Burundi lies just south of the Equator, on the Nile–Congo watershed. A decade of brutal conflict between Hutu and Tutsi from 1993 led to power-sharing in governance.

Official name Republic of Burundi
Formation 1962 / 1962
Capital Bujumbura
Population 10.9 million / 1101 people per sq mile (425 people per sq km)
Total area 10,745 sq miles (27,830 sq km)
Languages Kirundi*, French*, Kiswahili
Religions Roman Catholic 65%, Protestant 23%, Other 7%, Muslim 3%, Seventh-day Adventist 2%
Ethnic mix Hutu 85%, Tutsi 14%, Twa 1%
Government Presidential system
Currency Burundian franc = 100 centimes
Literacy rate 6.2%
Calorie consumption 1604 kilocalories

CANADA
North America

Page 15 E4

The world's second-largest country spans six time zones, extends north from its US border into the Arctic, and is rich in natural resources. Separatism is strong in French-speaking Québec.

Official name Canada
Formation 1867 / 1949
Capital Ottawa
Population 36.6 million / 10 people per sq mile (4 people per sq km)
Total area 3,855,171 sq miles (9,984,670 sq km)
Languages English*, French*, Chinese, Italian, German, Ukrainian, Portuguese, Inuktitut, Cree
Religions Roman Catholic 39%, Other Christian 28%, Nonreligious 24%, Other 6%, Muslim 3%
Ethnic mix European descent 80%, Asian 15%, First Nations, Métis, and Inuit 5%
Government Parliamentary system
Currency Canadian dollar = 100 cents
Literacy rate 99%
Calorie consumption 3494 kilocalories

BOTSWANA
Southern Africa

Page 56 C3

Botswana, once the British protectorate of Bechuanaland, lies landlocked in Southern Africa. Diamonds provide it with a relatively prosperous economy, but the rate of HIV infection is high.

Official name Republic of Botswana
Formation 1966 / 1966
Capital Gaborone
Population 2.3 million / 11 people per sq mile (4 people per sq km)
Total area 231,803 sq miles (600,370 sq km)
Languages Setswana, English*, Shona, San, Khoikhoi, isiNdebele
Religions Christian (mainly Protestant) 80%, Nonreligious 15%, Traditional beliefs 4%, Other (including Muslim) 1%
Ethnic mix Tswana 79%, Kalanga 11%, Other 10%
Government Presidential system
Currency Pula = 100 thebe
Literacy rate 87%
Calorie consumption 2326 kilocalories

BULGARIA
Southeast Europe

Page 82 C2

Bulgaria is located on the western shore of the Black Sea. After the fall of its communist regime in 1990, economic and political reform were slow, but EU membership was achieved in 2007.

Official name Republic of Bulgaria
Formation 1908 / 1947
Capital Sofia
Population 7.1 million / 166 people per sq mile (64 people per sq km)
Total area 42,822 sq miles (110,910 sq km)
Languages Bulgarian*, Turkish, Romani
Religions Orthodox Christian 75%, Muslim 15%, Nonreligious 5%, Other 3%, Protestant 1%, Roman Catholic 1%
Ethnic mix Bulgarian 85%, Turkish 9%, Roma 5%, Other 1%
Government Parliamentary system
Currency Lev = 100 stotinki
Literacy rate 98%
Calorie consumption 2829 kilocalories

CAMBODIA
Southeast Asia

Page 115 D5

This ancient Southeast Asian kingdom suffered the brutal totalitarian Khmer Rouge regime in the 1970s and then a decade of Vietnamese puppet rule. Free elections were only held in 1993.

Official name Kingdom of Cambodia
Formation 1953 / 1953
Capital Phnom Penh
Population 16 million / 235 people per sq mile (91 people per sq km)
Total area 69,900 sq miles (181,040 sq km)
Languages Khmer*, French, Chinese, Vietnamese, Cham
Religions Buddhist 97%, Muslim 2%, Other (mostly Christian) 1%
Ethnic mix Khmer 90%, Vietnamese 5%, Other 4%, Chinese 1%
Government Parliamentary system
Currency Riel = 100 sen
Literacy rate 74%
Calorie consumption 2477 kilocalories

CAPE VERDE
Atlantic Ocean

Page 52 A2

The mostly volcanic islands that make up Cape Verde lie off Africa's west coast. A Portuguese colony until 1975, it has been a stable democracy since its first multiparty elections in 1991.

Official name Republic of Cabo Verde
Formation 1975 / 1975
Capital Praia
Population 500,000 / 321 people per sq mile (124 people per sq km)
Total area 1557 sq miles (4033 sq km)
Languages Portuguese Creole, Portuguese*
Religions Roman Catholic 97%, Other 2%, Protestant (Church of the Nazarene) 1%
Ethnic mix Mestiço 71%, African 28%, European 1%
Government Presidential / parliamentary system
Currency Escudo = 100 centavos
Literacy rate 87%
Calorie consumption 2609 kilocalories

CENTRAL AFRICAN REPUBLIC
Central Africa

Page 54 C4

A landlocked plateau dividing the Chad and Congo river basins, the CAR has been plagued by rebellions since military rule ended in 1993. The arid north is sparsely populated.

Official name Central African Republic
Formation 1960 / 1960
Capital Bangui
Population 4.7 million / 20 people per sq mile (8 people per sq km)
Total area 240,534 sq miles (622,984 sq km)
Languages Sango, Banda, Gbaya, French*
Religions Traditional beliefs 35%, Roman Catholic 25%, Protestant 25%, Muslim 15%
Ethnic mix Baya 33%, Banda 27%, Other 17%, Mandjia 13%, Sara 10%
Government Presidential system
Currency CFA franc = 100 centimes
Literacy rate 37%
Calorie consumption 1879 kilocalories

CHAD
Central Africa

Page 54 C3

Landlocked in north Central Africa, Chad has been torn by intermittent periods of civil war since it gained independence from France in 1960. It became a net oil exporter in 2003.

Official name Republic of Chad
Formation 1960 / 1960
Capital N'Djaména
Population 14.9 million / 31 people per sq mile (12 people per sq km)
Total area 495,752 sq miles (1,284,000 sq km)
Languages French*, Sara, Arabic*, Maba
Religions Muslim 51%, Christian 35%, Animist 7%, Traditional beliefs 7%
Ethnic mix Other 30%, Sara 28%, Mayo-Kebbi 12%, Arab 12%, Ouaddai 9%, Kanem-Bornou 9%
Government Presidential system
Currency CFA franc = 100 centimes
Literacy rate 22%
Calorie consumption 2110 kilocalories

CHILE
South America

Page 42 B3

Extending in a ribbon down the Pacific coast of South America, Chile restored democracy in 1989 after a referendum rejected its military dictator. It is the world's largest copper producer.

Official name Republic of Chile
Formation 1818 / 1883
Capital Santiago
Population 18.1 million / 63 people per sq mile (24 people per sq km)
Total area 292,258 sq miles (756,950 sq km)
Languages Spanish*, Amerindian languages
Religions Roman Catholic 64%, Protestant 17%, Nonreligious 16%, Other 3%
Ethnic mix Mestizo and European 95%, Mapuche 4%, Other Amerindian 1%
Government Presidential system
Currency Chilean peso = 100 centavos
Literacy rate 96%
Calorie consumption 2979 kilocalories

CHINA
East Asia

Page 104 C4

This vast East Asian country, home to a fifth of the global population, became a communist state in 1949. It has now emerged as one of the world's major political and economic powers.

Official name People's Republic of China
Formation 960 / 1999
Capital Beijing
Population 1.41 billion / 391 people per sq mile (151 people per sq km)
Total area 3,705,386 sq miles (9,596,960 sq km)
Languages Mandarin*, Wu, Cantonese, Hsiang, Min, Hakka, Kan
Religions Nonreligious or traditional beliefs 73%, Buddhist 16%, Other 7%, Christian 3%, Muslim 1%
Ethnic mix Han 92%, Other 4%, Zhuang 1%, Hui 1%, Manchu 1%, Uighur 1%
Government One-party state
Currency Renminbi (or yuan) = 10 jiao = 100 fen
Literacy rate 95%
Calorie consumption 3108 kilocalories

COLOMBIA
South America

Page 36 B3

Lying in northwest South America, Colombia is noted for coffee, gold, emeralds, and narcotics trafficking. A 52-year civil war that displaced seven million people ended in 2016.

Official name Republic of Colombia
Formation 1819 / 1903
Capital Bogotá
Population 49.1 million / 122 people per sq mile (47 people per sq km)
Total area 439,733 sq miles (1,138,910 sq km)
Languages Spanish*, Amerindian languages
Religions Roman Catholic 79%, Protestant 13%, Nonreligious 6%, Other 2%
Ethnic mix Mestizo 58%, White 20%, European–African 14%, African 4%, African–Amerindian 3%, Amerindian 1%
Government Presidential system
Currency Colombian peso = 100 centavos
Literacy rate 94%
Calorie consumption 2804 kilocalories

COMOROS
Indian Ocean

Page 57 F2

The Comoros islands lie between Mozambique and Madagascar. There have been many coups and secession attempts by the smaller islands since independence from France in 1975.

Official name Union of the Comoros
Formation 1975 / 1975
Capital Moroni
Population 800,000 / 929 people per sq mile (359 people per sq km)
Total area 838 sq miles (2170 sq km)
Languages Arabic*, Comoran*, French*
Religions Muslim (mainly Sunni) 98%, Other 1%, Roman Catholic 1%
Ethnic mix Comoran 97%, Other 3%
Government Presidential system
Currency Comoros franc = 100 centimes
Literacy rate 49%
Calorie consumption 2139 kilocalories

CONGO
Central Africa

Page 55 B5

Astride the Equator in Central Africa, this former French colony emerged from 26 years of Marxist-Leninist rule in 1990, though the Marxist-era dictator seized power again in 1997.

Official name Republic of the Congo
Formation 1960 / 1960
Capital Brazzaville
Population 5.3 million / 40 people per sq mile (16 people per sq km)
Total area 132,046 sq miles (342,000 sq km)
Languages Kongo, Teke, Lingala, French*
Religions Traditional beliefs 50%, Roman Catholic 35%, Protestant 13%, Muslim 2%
Ethnic mix Bakongo 51%, Teke 17%, Other 16%, Mbochi 11%, Mbédé 5%
Government Presidential system
Currency CFA franc = 100 centimes
Literacy rate 79%
Calorie consumption 2208 kilocalories

CONGO, DEM. REP.
Central Africa

Page 55 C6

Straddling the Equator in east Central Africa, mineral-rich Dem. Rep. Congo is Africa's second-largest country. The former Belgian colony has endured years of corrupt rule and conflict.

Official name Democratic Republic of the Congo
Formation 1960 / 1960
Capital Kinshasa
Population 81.3 million / 93 people per sq mile (36 people per sq km)
Total area 905,563 sq miles (2,345,410 sq km)
Languages Kiswahili, Tshiluba, Kikongo, Lingala, French*
Religions Roman Catholic 50%, Protestant 20%, Traditional beliefs and other 20%, Muslim 10%
Ethnic mix Other 55%, Mongo, Luba, Kongo, and Mangbetu-Azande 45%
Government Presidential system
Currency Congolese franc = 100 centimes
Literacy rate 77%
Calorie consumption 1585 kilocalories

COSTA RICA
Central America

Page 31 E4

Costa Rica is the most stable country in Central America. It abolished its army in 1948 and its neutrality in foreign affairs is long-standing, but it has very strong ties with the US.

Official name Republic of Costa Rica
Formation 1838 / 1838
Capital San José
Population 4.9 million / 249 people per sq mile (96 people per sq km)
Total area 19,730 sq miles (51,100 sq km)
Languages Spanish*, English Creole, Bribri, Cabecar
Religions Roman Catholic 62%, Protestant 25%, Nonreligious 9%, Other 4%
Ethnic mix Mestizo and European 96%, Amerindian 3%, Black 1%
Government Presidential system
Currency Costa Rican colón = 100 céntimos
Literacy rate 97%
Calorie consumption 2848 kilocalories

CROATIA
Southeast Europe

Page 78 B2

Post-independence fighting afflicted this former Yugoslav republic until 1995. It is now capitalizing on its location on the eastern Adriatic coast and joined the EU in 2013.

Official name Republic of Croatia
Formation 1991 / 1991
Capital Zagreb
Population 4.2 million / 192 people per sq mile (74 people per sq km)
Total area 21,831 sq miles (56,542 sq km)
Languages Croatian*
Religions Roman Catholic 84%, Nonreligious 7%, Orthodox Christian 4%, Other 3%, Muslim 2%
Ethnic mix Croat 92%, Serb 4%, Other 3%, Bosniak 1%
Government Parliamentary system
Currency Kuna = 100 lipa
Literacy rate 99%
Calorie consumption 3059 kilocalories

CUBA
West Indies

Page 32 C2

Cuba is the largest island in the Caribbean and the only communist country in the Americas. It was led by Fidel Castro for almost 40 years until he handed over to his brother in 2008.

Official name Republic of Cuba
Formation 1902 / 1902
Capital Havana
Population 11.5 million / 269 people per sq mile (104 people per sq km)
Total area 42,803 sq miles (110,860 sq km)
Languages Spanish
Religions Nonreligious 49%, Roman Catholic 40%, Atheist 6%, Other 4%, Protestant 1%
Ethnic mix White 65%, Mulatto (mixed race) 25%, Black 10%
Government One-party state
Currency Cuban peso = 100 centavos
Literacy rate 99%
Calorie consumption 3409 kilocalories

CYPRUS
Southeast Europe

Page 80 C5

Cyprus lies south of Turkey in the eastern Mediterranean. Since 1974, it has been partitioned between the Turkish-occupied north and the Greek south (which joined the EU in 2004).

Official name Republic of Cyprus
Formation 1960 / 1960
Capital Nicosia
Population 1.2 million / 336 people per sq mile (130 people per sq km)
Total area 3571 sq miles (9250 sq km)
Languages Greek*, Turkish*
Religions Orthodox Christian 78%, Muslim 18%, Other 4%
Ethnic mix Greek 81%, Turkish 11%, Other 8%
Government Presidential system
Currency Euro = 100 cents (In TRNC, Turkish lira = 100 kurus)
Literacy rate 99%
Calorie consumption 2649 kilocalories

CZECH REPUBLIC (CZECHIA)
Central Europe

Page 77 A5

Landlocked in Central Europe, and formerly part of communist Czechoslovakia, it peacefully dissolved its federal union with Slovakia in 1993, and joined the EU in 2004.

Official name Czech Republic
Formation 1993 / 1993
Capital Prague
Population 10.6 million / 348 people per sq mile (134 people per sq km)
Total area 30,450 sq miles (78,866 sq km)
Languages Czech*, Slovak, Hungarian (Magyar)
Religions Nonreligious 72%, Roman Catholic 21%, Other 6%, Orthodox Christian 1%
Ethnic mix Czech 86%, Moravian 7%, Other 5%, Slovak 2%
Government Parliamentary system
Currency Czech koruna = 100 haleru
Literacy rate 99%
Calorie consumption 3256 kilocalories

DENMARK
Northern Europe

Page 63 A7

Denmark occupies the low-lying Jutland peninsula and over 400 islands. In the 1930s it set up one of the first welfare systems. Greenland and the Faroe Islands are self-governing territories.

Official name Kingdom of Denmark
Formation 950 / 1944
Capital Copenhagen
Population 5.7 million / 348 people per sq mile (135 people per sq km)
Total area 16,639 sq miles (43,094 sq km)
Languages Danish*
Religions Evangelical Lutheran 95%, Roman Catholic 3%, Muslim 2%
Ethnic mix Danish 96%, Other (including Scandinavian and Turkish) 3%, Faroese and Inuit 1%
Government Parliamentary system
Currency Danish krone = 100 øre
Literacy rate 99%
Calorie consumption 3367 kilocalories

DJIBOUTI
East Africa

Page 50 D4

Once known as French Somaliland, this city state with a desert hinterland lies on the coast of the Horn of Africa. Its economy relies on its Red Sea port, a vital trade link for landlocked Ethiopia.

Official name Republic of Djibouti
Formation 1977 / 1977
Capital Djibouti
Population 1 million / 112 people per sq mile (43 people per sq km)
Total area 8494 sq miles (22,000 sq km)
Languages Somali, Afar, French*, Arabic*
Religions Muslim (mainly Sunni) 94%, Christian 6%
Ethnic mix Issa 60%, Afar 35%, Other 5%
Government Presidential system
Currency Djibouti franc = 100 centimes
Literacy rate 70%
Calorie consumption 2607 kilocalories

DOMINICA
West Indies

Page 33 H4

This Caribbean island, known for its lush flora and fauna, resisted European colonization until the 18th century, when it came first under French and then British rule.

Official name Commonwealth of Dominica
Formation 1978 / 1978
Capital Roseau
Population 74,000 / 254 people per sq mile (98 people per sq km)
Total area 291 sq miles (754 sq km)
Languages French Creole, English*
Religions Roman Catholic 62%, Protestant 30%, Nonreligious 6%, Other 2%
Ethnic mix Black 87%, Mixed race 9%, Carib 3%, Other 1%
Government Parliamentary system
Currency East Caribbean dollar = 100 cents
Literacy rate 88%
Calorie consumption 2931 kilocalories

DOMINICAN REPUBLIC
West Indies

Page 33 E2

Occupying the eastern two-thirds of the island of Hispaniola, the Dominican Republic is the Caribbean's top tourist destination and largest economy. Ties with the US are strong.

Official name Dominican Republic
Formation 1865 / 1865
Capital Santo Domingo
Population 10.8 million / 578 people per sq mile (223 people per sq km)
Total area 18,679 sq miles (48,380 sq km)
Languages Spanish*, French Creole
Religions Roman Catholic 57%, Protestant 23%, Nonreligious 18%, Other 2%
Ethnic mix Mixed race 73%, European 16%, Black African 11%
Government Presidential system
Currency Dominican Republic peso = 100 centavos
Literacy rate 92%
Calorie consumption 2614 kilocalories

EAST TIMOR
Southeast Asia

Page 116 F5

This former Portuguese colony on the island of Timor in the East Indies was invaded by Indonesia in 1975. In 1999 it voted for independence, achieved in 2002 after a turbulent transition.

Official name Democratic Republic of Timor-Leste
Formation 2002 / 2002
Capital Dili
Population 1.3 million / 230 people per sq mile 89 people per sq km)
Total area 5756 sq miles (14,874 sq km)
Languages Tetum* (Portuguese/Austronesian), Bahasa Indonesia, Portuguese*
Religions Roman Catholic 96%, Protestant 2%, Other 2%
Ethnic mix Papuan groups approx. 85%, Indonesian groups approx. 13%, Chinese 2%
Government Parliamentary system
Currency US dollar = 100 cents
Literacy rate 58%
Calorie consumption 2131 kilocalories

ECUADOR
South America

Page 38 A2

Once part of the Inca heartland on the northwest coast of South America, Ecuador is the world's leading banana exporter. Its territory includes the wildlife-rich Galapagos Islands.

Official name Republic of Ecuador
Formation 1830 / 1942
Capital Quito
Population 16.6 million / 155 people per sq mile (60 people per sq km)
Total area 109,483 sq miles (283,560 sq km)
Languages Spanish*, Quechua, other Amerindian languages
Religions Roman Catholic 79%, Protestant 13%, Nonreligious 5%, Other 3%
Ethnic mix Mestizo 79%, Black African 7%, Amerindian 7%, White 6%, Other 1%
Government Presidential system
Currency US dollar = 100 cents
Literacy rate 94%
Calorie consumption 2344 kilocalories

EGYPT
North Africa

Page 50 B2

Egypt lies in Africa's northeast corner; the fertile Nile valley divides desert lands. Nearly 50 years of de facto military rule was interrupted in 2011 by the "Arab Spring" popular uprising.

Official name Arab Republic of Egypt
Formation 1936 / 1982
Capital Cairo
Population 97.6 million / 254 people per sq mile (98 people per sq km)
Total area 386,660 sq miles (1,001,450 sq km)
Languages Arabic*, French, English, Berber
Religions Muslim (mainly Sunni) 90%, Coptic Christian and other 9%, Other Christian 1%
Ethnic mix Egyptian 99%, Other 1%
Government Presidential system
Currency Egyptian pound = 100 piastres
Literacy rate 75%
Calorie consumption 3522 kilocalories

EL SALVADOR
Central America

Page 30 B3

El Salvador is Central America's smallest country. Since a 12-year war between the US-backed army and left-wing guerrillas ended in 1992, crime and gang violence have been key issues.

Official name Republic of El Salvador
Formation 1841 / 1841
Capital San Salvador
Population 6.4 million / 800 people per sq mile (309 people per sq km)
Total area 8124 sq miles (21,040 sq km)
Languages Spanish*
Religions Roman Catholic 50%, Protestant 36%, Nonreligious 12%, Other 2
Ethnic mix Mestizo 86%, White 13%, Other and Amerindian 1%
Government Presidential system
Currency Salvadorean colón = 100 centavos; US dollar = 100 cents
Literacy rate 88%
Calorie consumption 2577 kilocalories

EQUATORIAL GUINEA
Central Africa

Page 55 A5

Equatorial Guinea comprises the Rio Muni mainland in west Central Africa and five islands. Free elections were first held in 1988, but the former ruling party still dominates.

Official name Republic of Equatorial Guinea
Formation 1968 / 1968
Capital Malabo
Population 1.3 million / 120 people per sq mile (46 people per sq km)
Total area 10,830 sq miles (28,051 sq km)
Languages Spanish*, Fang, Bubi, French*
Religions Roman Catholic 90%, Other 10%
Ethnic mix Fang 85%, Other 11%, Bubi 4%
Government Presidential system
Currency CFA franc = 100 centimes
Literacy rate 94%
Calorie consumption Not available

ERITREA
East Africa

Page 50 C4

Lying on the shores of the Red Sea, this former Italian colony was annexed by Ethiopia in 1952. It successfully seceded in 1993, following a 30-year war for independence.

Official name State of Eritrea
Formation 1993 / 2002
Capital Asmara
Population 5.1 million / 112 people per sq mile (43 people per sq km)
Total area 46,842 sq miles (121,320 sq km)
Languages Tigrinya*, English*, Tigre, Afar, Arabic*, Saho, Bilen, Kunama, Nara, Hadareb
Religions Christian 50%, Muslim 48%, Other 2%
Ethnic mix Tigray 50%, Tigre 31%, Other 9%, Saho 5%, Afar 5%
Government Presidential / parliamentary system
Currency Nakfa = 100 cents
Literacy rate 70%
Calorie consumption 1640 kilocalories

ESTONIA
Northeast Europe

Page 84 D2

The smallest, richest, most developed Baltic state, Estonia has emphasized advanced IT and integration with Europe since renouncing the Soviet model. It joined the EU in 2004.

Official name Republic of Estonia
Formation 1991 / 1991
Capital Tallinn
Population 1.3 million / 75 people per sq mile (29 people per sq km)
Total area 17,462 sq miles (45,226 sq km)
Languages Estonian*, Russian
Religions Nonreligious 45%, Orthodox Christian 25%, Lutheran 20%, Other 10%
Ethnic mix Estonian 70%, Russian 25%, Other 2%, Ukrainian 2%, Belarussian 1%
Government Parliamentary system
Currency Euro = 100 cents
Literacy rate 99%
Calorie consumption 3253 kilocalories

ETHIOPIA
East Africa

Page 51 C5

Ethiopia, the only African country to escape colonization, was a Marxist regime in 1974–1991. Now landlocked in the Horn of Africa, it has suffered economic, civil, and natural crises.

Official name Federal Democratic Republic of Ethiopia
Formation 1896 / 2002
Capital Addis Ababa
Population 105 million / 245 people per sq mile (95 people per sq km)
Total area 435,184 sq miles (1,127,127 sq km)
Languages Amharic*, Tigrinya, Galla, Sidamo, Somali, English, Arabic
Religions Christian 62%, Muslim 34%, Other 4%
Ethnic mix Oromo 34%, Amhara 27%, Other 23%, Somali 6%, Tigray 6%, Sidama 4%
Government Parliamentary system
Currency Birr = 100 cents
Literacy rate 39%
Calorie consumption 2131 kilocalories

FIJI
Australasia & Oceania

Page 123 E5

Fiji is a volcanic archipelago of 882 islands in the southern Pacific Ocean. Tensions between ethnic Fijians and Indo-Fijians have provoked several coups. Sugar is the main export.

Official name Republic of Fiji
Formation 1970 / 1970
Capital Suva
Population 900,000 / 128 people per sq mile (49 people per sq km)
Total area 7054 sq miles (18,270 sq km)
Languages Fijian, English*, Hindi, Urdu, Tamil, Telugu
Religions Methodist 35%, Hindu 28%, Other Christian 21%, Roman Catholic 9%, Muslim 6%, Other and nonreligious 1%
Ethnic mix Melanesian 57%, Indian 38%, Other 5%
Government Parliamentary system
Currency Fiji dollar = 100 cents
Literacy rate 94%
Calorie consumption 2943 kilocalories

FINLAND
Northern Europe

Page 62 D4

A low-lying country of forests and lakes, Finland joins Scandinavia to Russia. Its language is related to only two others in Europe. Finnish women were the first in Europe to get the vote, in 1906.

Official name Republic of Finland
Formation 1917 / 1947
Capital Helsinki
Population 5.5 million / 47 people per sq mile (18 people per sq km)
Total area 130,127 sq miles (337,030 sq km)
Languages Finnish*, Swedish*, Sámi
Religions Evangelical Lutheran 78%, Nonreligious 19%, Other 2%, Orthodox Christian 1%
Ethnic mix Finnish 93%, Other (including Sámi) 7%
Government Parliamentary system
Currency Euro = 100 cents
Literacy rate 99%
Calorie consumption 3368 kilocalories

FRANCE
Western Europe

Page 68 B4

Straddling Western Europe from the English Channel to the Mediterranean Sea, France was Europe's first modern republic. It is now one of the world's leading industrial powers.

Official name French Republic
Formation 987 / 1919
Capital Paris
Population 65 million / 306 people per sq mile (118 people per sq km)
Total area 211,208 sq miles (547,030 sq km)
Languages French*, Provençal, German, Breton, Catalan, Basque
Religions Christian 51%, Nonreligious 40%, Muslim 6%, Other 2%, Jewish 1%
Ethnic mix French 86%, North African 5%, Black 5%, German (Alsace) 2%, Breton 1%, Other 1%
Government Presidential / parliamentary system
Currency Euro = 100 cents
Literacy rate 99%
Calorie consumption 3482 kilocalories

GABON
Central Africa

Page 55 A5

A former French colony straddling the Equator on Central Africa's west coast, it returned to multiparty politics in 1990, after 22 years of one-party rule. The economy relies on oil revenue.

Official name Gabonese Republic
Formation 1960 / 1960
Capital Libreville
Population 2 million / 20 people per sq mile (8 people per sq km)
Total area 103,346 sq miles (267,667 sq km)
Languages Fang, French*, Punu, Sira, Nzebi, Mpongwe
Religions Christian (mainly Roman Catholic) 55%, Traditional beliefs 40%, Other 4%, Muslim 1%
Ethnic mix Fang 26%, Shira-punu 24%, Other 16%, Foreign residents 15%, Nzabi-duma 11%, Mbédé-Teke 8%
Government Presidential system
Currency CFA franc = 100 centimes
Literacy rate 82%
Calorie consumption 2830 kilocalories

GAMBIA, THE
West Africa

Page 52 B3

A narrow state along the Gambia River on Africa's west coast and surrounded by Senegal, Gambia was renowned for its stability until a coup in 1994; the coup leader then ruled until 2016.

Official name Republic of The Gambia
Formation 1965 / 1965
Capital Banjul
Population 2.1 million / 544 people per sq mile (210 people per sq km)
Total area 4363 sq miles (11,300 sq km)
Languages Mandinka, Fulani, Wolof, Jola, Soninke, English*
Religions Sunni Muslim 90%, Christian 8%, Traditional beliefs 2%
Ethnic mix Mandinka 42%, Fulani 18%, Wolof 16%, Jola 10%, Serahuli 9%, Other 5%
Government Presidential system
Currency Dalasi = 100 butut
Literacy rate 42%
Calorie consumption 2628 kilocalories

GEORGIA
Southwest Asia

Page 95 F2

Located in the Caucasus on the Black Sea's eastern shore, Georgia is noted for its wine. Conflict broke out after the breakup of the USSR; the northern provinces have de facto autonomy.

Official name Georgia
Formation 1991 / 1991
Capital Tbilisi
Population 3.9 million / 145 people per sq mile (56 people per sq km)
Total area 26,911 sq miles (69,700 sq km)
Languages Georgian*, Russian, Azeri, Armenian, Mingrelian, Ossetian, Abkhazian (* in Abkhazia)
Religions Orthodox Christian 89%, Muslim 9%, Roman Catholic 1%, Other 1%
Ethnic mix Georgian 87%, Azeri 6%, Armenian 4%, Other 2%, Russian 1%
Government Presidential / Parliamentary system
Currency Lari = 100 tetri
Literacy rate 99%
Calorie consumption 2905 kilocalories

GERMANY
Northern Europe

Page 72 B4

Germany is Europe's major economic power and a leading influence in the EU. Divided after World War II, its democratic west and communist east were re-unified in 1990.

Official name Federal Republic of Germany
Formation 1871 / 1990
Capital Berlin
Population 82.1 million / 608 people per sq mile (235 people per sq km)
Total area 137,846 sq miles (357,021 sq km)
Languages German*, Turkish
Religions Nonreligious 36%, Roman Catholic 29%, Protestant 26%, Muslim 5%, Other 4%
Ethnic mix German 81%, Other European 10%, Other 4%, Turkish 3%, Polish 2%
Government Parliamentary system
Currency Euro = 100 cents
Literacy rate 99%
Calorie consumption 3499 kilocalories

GHANA
West Africa

Page 53 E5

Once known as the Gold Coast, Ghana was the first colony in West Africa to gain independence. In recent decades multiparty democracy has been consolidated despite economic issues.

Official name Republic of Ghana
Formation 1957 / 1957
Capital Accra
Population 28.8 million / 324 people per sq mile (125 people per sq km)
Total area 92,100 sq miles (238,540 sq km)
Languages Twi, Fanti, Ewe, Ga, Adangbe, Gurma, Dagomba (Dagbani), English*
Religions Christian 71%, Muslim 18%, Traditional beliefs 5%, Nonreligious 5%, Other 1%
Ethnic mix Akan 47%, Gurma 17%, Ga-Dangme 14%, Other 9%, Ewe 7%, Guan 6%
Government Presidential system
Currency Cedi = 100 pesewas
Literacy rate 72%
Calorie consumption 3016 kilocalories

GREECE
Southeast Europe

Page 83 A5

The southernmost Balkan nation has a mountainous mainland and over 2000 islands, engendering its seafaring tradition. High state debt has led to recent unpopular austerity measures.

Official name Hellenic Republic
Formation 1829 / 1947
Capital Athens
Population 11.2 million / 222 people per sq mile (86 people per sq km)
Total area 50,942 sq miles (131,940 sq km)
Languages Greek*, Turkish, Macedonian, Albanian
Religions Orthodox Christian 90%, Nonreligious 4%, Other 4%, Muslim 2%
Ethnic mix Greek 98%, Other 2%
Government Parliamentary system
Currency Euro = 100 cents
Literacy rate 97%
Calorie consumption 3400 kilocalories

GRENADA
West Indies

Page 33 G5

The most southerly Windward gained worldwide notoriety in 1983, when the US invaded to sever its growing links with Cuba. It is the world's second-biggest nutmeg producer.

Official name Grenada
Formation 1974 / 1974
Capital St. George's
Population 100,000 / 763 people per sq mile (294 people per sq km)
Total area 131 sq miles (340 sq km)
Languages English*, English Creole
Religions Roman Catholic 68%, Anglican 17%, Other 15%
Ethnic mix Black African 82%, Mulatto (mixed race) 13%, East Indian 3%, Other 2%
Government Parliamentary system
Currency East Caribbean dollar = 100 cents
Literacy rate 96%
Calorie consumption 2447 kilocalories

GUATEMALA
Central America

Page 30 A2

Once the heart of the Mayan civilization, the largest and most populous state on the Central American isthmus suffered years of civil war and military rule, but democracy is now flourishing.

Official name Republic of Guatemala
Formation 1838 / 1838
Capital Guatemala City
Population 16.9 million / 404 people per sq mile (156 people per sq km)
Total area 42,042 sq miles (108,890 sq km)
Languages Quiché, Mam, Cakchiquel, Kekchí, Spanish*
Religions Roman Catholic 50%, Protestant 41%, Nonreligious 6%, Other 3%
Ethnic mix Amerindian 60%, Mestizo 30%, Other 10%
Government Presidential system
Currency Quetzal = 100 centavos
Literacy rate 81%
Calorie consumption 2419 kilocalories

GUINEA
West Africa

Page 52 C4

A former French colony on Africa's west coast, Guinea chose a Marxist path, then came under army rule. Polls in 2010 brought fresh hope, though the recent Ebola epidemic was a setback.

Official name Republic of Guinea
Formation 1958 / 1958
Capital Conakry
Population 12.7 million / 134 people per sq mile (52 people per sq km)
Total area 94,925 sq miles (245,857 sq km)
Languages Pulaar, Malinké, Soussou, French*
Religions Muslim 89%, Christian 7%, Nonreligious 2%, Traditional beliefs and other 2%
Ethnic mix Peul 40%, Malinké 30%, Soussou 20%, Other 10%
Government Presidential system
Currency Guinea franc = 100 centimes
Literacy rate 32%
Calorie consumption 2566 kilocalories

GUINEA-BISSAU
West Africa

Page 52 B4

Known as Portuguese Guinea in colonial times, Guinea-Bissau is situated on Africa's west coast. One of the world's poorest countries, it has now become a transit point for cocaine trafficking.

Official name Republic of Guinea-Bissau
Formation 1974 / 1974
Capital Bissau
Population 1.9 million / 175 people per sq mile (68 people per sq km)
Total area 13,946 sq miles (36,120 sq km)
Languages Portuguese Creole, Balante, Fulani, Malinké, Portuguese*
Religions Muslim 54%, Christian 26%, Traditional beliefs 18%, Nonreligious 2%
Ethnic mix Balante 30%, Fulani 20%, Other 16%, Mandyako 14%, Mandinka 13%, Papel 7%
Government Presidential system
Currency CFA franc = 100 centimes
Literacy rate 46%
Calorie consumption 2292 kilocalories

GUYANA
South America

Page 37 F3

A land of rain forest, mountains, coastal plains, and savanna, Guyana is South America's only English-speaking state. It became a republic in 1970, four years after independence from Britain.

Official name Cooperative Republic of Guyana
Formation 1966 / 1966
Capital Georgetown
Population 800,000 / 11 people per sq mile (4 people per sq km)
Total area 83,000 sq miles (214,970 sq km)
Languages English Creole, Hindi, Tamil, Amerindian languages, English*
Religions Christian 57%, Hindu 28%, Muslim 10%, Other 5%
Ethnic mix East Indian 43%, Black African 30%, Mixed race 17%, Amerindian 9%, Other 1%
Government Presidential system
Currency Guyanese dollar = 100 cents
Literacy rate 86%
Calorie consumption 2764 kilocalories

HAITI
West Indies

Page 32 D3

The western third of the Caribbean island of Hispaniola, Haiti became the world's first black republic in 1804. Natural disasters and periodic anarchy perpetuate its endemic poverty.

Official name Republic of Haiti
Formation 1804 / 1884
Capital Port-au-Prince
Population 11 million / 1034 people per sq mile (399 people per sq km)
Total area 10,714 sq miles (27,750 sq km)
Languages French Creole*, French*
Religions Roman Catholic 55%, Protestant 28%, Other (including Voodoo) 16%, Nonreligious 1%
Ethnic mix Black African 95%, Mulatto (mixed race) and European 5%
Government Presidential system
Currency Gourde = 100 centimes
Literacy rate 49%
Calorie consumption 2091 kilocalories

HONDURAS
Central America

Page 30 C2

Straddling the Central American isthmus, Honduras returned to civilian rule in 1984, after a succession of military regimes. Crime is high and it has one of the world's worst murder rates.

Official name Republic of Honduras
Formation 1838 / 1838
Capital Tegucigalpa
Population 9.3 million / 215 people per sq mile (83 people per sq km)
Total area 43,278 sq miles (112,090 sq km)
Languages Spanish*, Garifuna (Carib), English Creole
Religions Roman Catholic 46%, Protestant 41%, Nonreligious 10%, Other 3%
Ethnic mix Mestizo 90%, Black African 5%, Amerindian 4%, White 1%
Government Presidential system
Currency Lempira = 100 centavos
Literacy rate 89%
Calorie consumption 2641 kilocalories

HUNGARY
Central Europe

Page 77 C6

Hungary is bordered by seven states in Central Europe. After the fall of communism in 1989, it introduced political and economic reforms and joined the EU in 2004.

Official name Hungary
Formation 1918 / 1947
Capital Budapest
Population 9.7 million / 272 people per sq mile (105 people per sq km)
Total area 35,919 sq miles (93,030 sq km)
Languages Hungarian* (Magyar)
Religions Roman Catholic 56%, Nonreligious 21%, Presbyterian 13%, Other (mostly Protestant) 10%
Ethnic mix Magyar 92%, Roma 3%, Other 3%, German 2%
Government Parliamentary system
Currency Forint = 100 fillér
Literacy rate 99%
Calorie consumption 3037 kilocalories

ICELAND
Northwest Europe

Page 61 E4

This northerly island outpost of Europe, sitting on the mid-Atlantic ridge, has stunning, sparsely inhabited volcanic terrain. Its economy crashed heavily in the 2008 global credit crunch.

Official name Republic of Iceland
Formation 1944 / 1944
Capital Reykjavik
Population 300,000 / 8 people per sq mile (3 people per sq km)
Total area 39,768 sq miles (103,000 sq km)
Languages Icelandic*
Religions Evangelical Lutheran 84%, Other (mostly Christian) 10%, Nonreligious 3%, Roman Catholic 3%
Ethnic mix Icelandic 94%, Other 5%, Danish 1%
Government Parliamentary system
Currency Icelandic króna = 100 aurar
Literacy rate 99%
Calorie consumption 3380 kilocalories

INDIA
South Asia

Page 112 D4

The Indian subcontinent, divided from the rest of Asia by the Himalayas, was once the jewel of the British empire. India is the world's largest democracy and second most populous country.

Official name Republic of India
Formation 1947 / 1947
Capital New Delhi
Population 1.34 billion / 1167 people per sq mile (450 people per sq km)
Total area 1,269,338 sq miles (3,287,590 sq km)
Languages Hindi*, English*, Urdu, Bengali, Marathi, Telugu, Tamil, Bihari, Gujarati, Kanarese
Religions Hindu 81%, Muslim 13%, Sikh 2%, Christian 2%, Buddhist 1%, Other 1%
Ethnic mix Indo-Aryan 72%, Dravidian 25%, Mongoloid and other 3%
Government Parliamentary system
Currency Indian rupee = 100 paise
Literacy rate 69%
Calorie consumption 2459 kilocalories

INDONESIA
Southeast Asia

Page 116 C4

The world's largest archipelago spans over 3100 miles (5000 km), from the Indian to the Pacific Ocean. Formerly the Dutch East Indies, it produces palm oil, rubber, spices, and natural gas.

Official name Republic of Indonesia
Formation 1949 / 1999
Capital Jakarta
Population 264 million / 381 people per sq mile (147 people per sq km)
Total area 741,096 sq miles (1,919,440 sq km)
Languages Javanese, Sundanese, Madurese, Bahasa Indonesia*, Dutch
Religions Sunni Muslim 87%, Protestant 7%, Roman Catholic 3%, Hindu 2%, Buddhist 1%
Ethnic mix Javanese 40%, Other 27%, Sundanese 16%, Coastal Malays 14%, Madurese 3%
Government Presidential system
Currency Rupiah = 100 sen
Literacy rate 95%
Calorie consumption 2777 kilocalories

IRAN
Southwest Asia

Page 98 C3

After the 1979 Islamist revolution led by Ayatollah Khomeini deposed the shah, this Middle Eastern country became the world's largest theocracy. It has large oil and natural gas reserves.

Official name Islamic Republic of Iran
Formation 1502 / 1990
Capital Tehran
Population 81.2 million / 129 people per sq mile (50 people per sq km)
Total area 636,293 sq miles (1,648,000 sq km)
Languages Farsi*, Azeri, Luri, Gilaki, Mazanderani, Kurdish, Turkmen, Arabic, Baluchi
Religions Shi'a Muslim 90%, Sunni Muslim 9%, Other 1%
Ethnic mix Persian 51%, Azari 24%, Other 10%, Lur and Bakhtiari 8%, Kurdish 7%
Government Islamic theocracy
Currency Iranian rial = 100 dinars
Literacy rate 85%
Calorie consumption 3094 kilocalories

IRAQ
Southwest Asia

Page 98 B3

Oil-rich Iraq is situated in the central Middle East. A US-led invasion in 2003 toppled Saddam Hussein's regime, but sectarian violence since then has caused political and social turmoil.

Official name Republic of Iraq
Formation 1932 / 1990
Capital Baghdad
Population 38.3 million / 227 people per sq mile (88 people per sq km)
Total area 168,753 sq miles (437,072 sq km)
Languages Arabic*, Kurdish*, Turkic languages, Armenian, Assyrian
Religions Shi'a Muslim 60%, Sunni Muslim 35%, Other (including Christian) 5%
Ethnic mix Arab 80%, Kurdish 15%, Turkmen 3%, Other 2%
Government Parliamentary system
Currency New Iraqi dinar = 1000 fils
Literacy rate 44%
Calorie consumption 2545 kilocalories

IRELAND
Northwest Europe

Page 67 A6

British rule ended in 1922 for 80% of the island of Ireland, which became the Irish Republic in 1949. The economy is now recovering after suffering heavily in the 2008 global financial crisis.

Official name Ireland
Formation 1922 / 1922
Capital Dublin
Population 4.8 million / 180 people per sq mile (70 people per sq km)
Total area 27,135 sq miles (70,280 sq km)
Languages English*, Irish*
Religions Roman Catholic 86%, Other Christian 6%, Nonreligious 6%, Muslim 1%, Other 1%
Ethnic mix Irish 86%, Other White 9%, Asian 2%, Other 2%, Black 1%
Government Parliamentary system
Currency Euro = 100 cents
Literacy rate 99%
Calorie consumption 3600 kilocalories

ISRAEL
Southwest Asia

Page 97 A7

In 1948 this Jewish state was carved out of Palestine on the east coast of the Mediterranean. It has gained land from its Arab neighbors, and the status of the Palestinians remains unresolved.

Official name State of Israel
Formation 1948 / 1994
Capital Jerusalem (not internationally recognized)
Population 8.3 million / 1057 people per sq mile (408 people per sq km)
Total area 8019 sq miles (20,770 sq km)
Languages Hebrew*, Arabic*, Yiddish, German, Russian, Polish, Romanian, Persian
Religions Jewish 81%, Muslim (mainly Sunni) 14%, Druze 2%, Christian 2%, Other and nonreligious 1%
Ethnic mix Jewish 81%, Arab 18%, Other 1%
Government Parliamentary system
Currency Shekel = 100 agorot
Literacy rate 98%
Calorie consumption 3610 kilocalories

ITALY
Southern Europe

Page 74 B3

A boot-shaped peninsula jutting into the Mediterranean, Italy is a world leader in product design, fashion, and textiles. Divisions exist between the industrial north and poorer south.

Official name Italian Republic
Formation 1861 / 1947
Capital Rome
Population 59.4 million / 523 people per sq mile (202 people per sq km)
Total area 116,305 sq miles (301,230 sq km)
Languages Italian*, German, French, Rhaeto-Romanic, Sardinian
Religions Roman Catholic 90%, Nonreligious 6%, Muslim 2%, Other Christian 2%
Ethnic mix Italian 92%, Other European 5%, Other 2%, North African (mainly Moroccan) 1%
Government Parliamentary system
Currency Euro = 100 cents
Literacy rate 99%
Calorie consumption 3579 kilocalories

IVORY COAST (CÔTE D'IVOIRE)
West Africa

Page 52 D4

One of the larger countries on the West African coast, this ex-French colony is the world's biggest cocoa producer. Coups and recent conflicts have destroyed its reputation for stability.

Official name Republic of Côte d'Ivoire
Formation 1960 / 1960
Capital Yamoussoukro
Population 24.3 million / 198 people per sq mile (76 people per sq km)
Total area 124,502 sq miles (322,460 sq km)
Languages Akan, French*, Krou, Voltaique
Religions Muslim 43%, Nonreligious or traditional beliefs 23%, Roman Catholic 17%, Evangelical 12%, Other Christian 4%, Other 1%
Ethnic mix Akan 42%, Voltaique 18%, Mandé du Nord 17%, Krou 11%, Mandé du Sud 10%, Other 2%
Government Presidential system
Currency CFA franc = 100 centimes
Literacy rate 44%
Calorie consumption 2799 kilocalories

JAMAICA
West Indies

Page 32 C3

Colonized by Spain and then Britain, Jamaica was the first Caribbean island to gain independence in the postwar era. Jamaican popular music culture developed reggae, ska, and dancehall.

Official name Jamaica
Formation 1962 / 1962
Capital Kingston
Population 2.9 million / 694 people per sq mile (268 people per sq km)
Total area 4243 sq miles (10,990 sq km)
Languages English Creole, English*
Religions Church of God 26%, Nonreligious 22%, Other Christian 21%, Seventh-day Adventist 12%, Pentecostal 11%, Other 8%
Ethnic mix Black African 92%, Mulatto (mixed race) 6%, East Indian 1%, Other 1%
Government Parliamentary system
Currency Jamaican dollar = 100 cents
Literacy rate 88%
Calorie consumption 2746 kilocalories

JAPAN
East Asia

Page 108 C4

Japan has four main islands and over 3000 smaller ones. It rebuilt after defeat in World War II to become one of the world's biggest economies. It retains its emperor as head of state.

Official name Japan
Formation 1861 / 1972
Capital Tokyo
Population 128 million / 877 people per sq mile (339 people per sq km)
Total area 145,882 sq miles (377,835 sq km)
Languages Japanese*, Korean, Chinese
Religions Buddhist 50%, Nonreligious 23%, Shinto 16%, Christian 10%, Muslim 1%
Ethnic mix Japanese 99%, Other (mainly Korean) 1%
Government Parliamentary system
Currency Yen = 100 sen
Literacy rate 99%
Calorie consumption 2726 kilocalories

JORDAN
Southwest Asia

Page 97 B6

This Middle Eastern kingdom stretches from the east bank of the Jordan River into largely uninhabited desert. Calls for greater democratization have engendered some reforms.

Official name Hashemite Kingdom of Jordan
Formation 1946 / 1967
Capital Amman
Population 9.7 million / 283 people per sq mile (109 people per sq km)
Total area 35,637 sq miles (92,300 sq km)
Languages Arabic*
Religions Sunni Muslim 92%, Christian 6%, Other 2%
Ethnic mix Arab 98%, Circassian 1%, Armenian 1%
Government Monarchy
Currency Jordanian dinar = 1000 fils
Literacy rate 98%
Calorie consumption 3100 kilocalories

KAZAKHSTAN
Central Asia

Page 92 B4

Second-largest of the former Soviet republics, mineral-rich Kazakhstan is Central Asia's major economic power. The former communist leader remains in charge, facing little opposition.

Official name Republic of Kazakhstan
Formation 1991 / 1991
Capital Astana
Population 18.2 million / 17 people per sq mile (7 people per sq km)
Total area 1,049,150 sq miles (2,717,300 sq km)
Languages Kazakh*, Russian, Ukrainian, German, Uzbek, Uighur
Religions Muslim (mainly Sunni) 71%, Christian (mainly Orthodox) 26%, Nonreligious 3%
Ethnic mix Kazakh 63%, Russian 24%, Other 6%, Uzbek 3%, Ukrainian 2%, Uighur 1%, Tatar 1%
Government Presidential system
Currency Tenge = 100 tiyn
Literacy rate 99%
Calorie consumption 3264 kilocalories

KENYA
East Africa

Page 51 C6

Straddling the Equator on Africa's east coast, Kenya has known both stable periods and internal strife since independence in 1963. Corruption is now a key political issue.

Official name Republic of Kenya
Formation 1963 / 1963
Capital Nairobi
Population 49.7 million / 227 people per sq mile (88 people per sq km)
Total area 224,961 sq miles (582,650 sq km)
Languages Kiswahili*, English*, Kikuyu, Luo, Kalenjin, Kamba
Religions Other Christian 60%, Roman Catholic 23%, Muslim 11%, Other 4%, Nonreligious 2%
Ethnic mix Other 35%, Kikuyu 17%, Luhya 14%, Kalenjin 13%, Luo 11%, Kamba 10%
Government Presidential system
Currency Kenya shilling = 100 cents
Literacy rate 79%
Calorie consumption 2206 kilocalories

KIRIBATI
Australasia & Oceania

Page 123 F3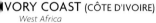

Part of the British colony of the Gilbert and Ellice Islands until independence in 1979, Kiribati comprises 33 islands in the mid-Pacific Ocean. Phosphate deposits on Banaba ran out in 1980.

Official name Republic of Kiribati
Formation 1979 / 1979
Capital Tarawa Atoll
Population 100,000 / 365 people per sq mile (141 people per sq km)
Total area 277 sq miles (717 sq km)
Languages English*, Kiribati
Religions Roman Catholic 56%, Kiribati Protestant Church 34%, Mormon 3%, Baha'i 2%, Seventh-day Adventist 2%, Other 1%
Ethnic mix Micronesian 99%, Other 1%
Government Presidential system
Currency Australian dollar = 100 cents
Literacy rate 99%
Calorie consumption 3040 kilocalories

KOSOVO (not fully recognized)
Southeast Europe

Page 79 D5

NATO intervention in 1999 ended ethnic cleansing by the Serbs of Kosovo's majority Albanian population, and nine years later the region unilaterally declared independence from Serbia.

Official name Republic of Kosovo
Formation 2008 / 2008
Capital Prishtinë/Priština
Population 1.9 million / 451 people per sq mile (174 people per sq km)
Total area 4212 sq miles (10,908 sq km)
Languages Albanian*, Serbian*, Bosniak, Gorani, Roma, Turkish
Religions Muslim 92%, Orthodox Christian 4%, Roman Catholic 4%
Ethnic mix Albanian 92%, Serb 4%, Bosniak and Gorani 2%, Roma 1%, Turkish 1%
Government Parliamentary system
Currency Euro = 100 cents
Literacy rate 92%
Calorie consumption Not available

KUWAIT
Southwest Asia

Page 98 C4

Kuwait, on the Persian Gulf, was a British protectorate from 1914 to 1961. Oil-rich since the 1950s, it was annexed briefly in 1990 by Iraq but US-led intervention restored the ruling amir.

Official name State of Kuwait
Formation 1961 / 1961
Capital Kuwait City
Population 4.1 million / 596 people per sq mile (230 people per sq km)
Total area 6880 sq miles (17,820 sq km)
Languages Arabic*, English
Religions Sunni Muslim 45%, Shi'a Muslim 40%, Christian, Hindu, and other 15%
Ethnic mix Asian 39%, Kuwaiti 37%, Other Arab 21%, African 2%, Other 1%
Government Monarchy
Currency Kuwaiti dinar = 1000 fils
Literacy rate 96%
Calorie consumption 3501 kilocalories

KYRGYZSTAN
Central Asia

Page 101 F2

This mountainous, landlocked state in Central Asia is the most rural of the ex-Soviet republics. Popular protests ousted the long-term president in 2005 and his successor in 2010.

Official name Kyrgyz Republic
Formation 1991 / 1991
Capital Bishkek
Population 6 million / 78 people per sq mile (30 people per sq km)
Total area 76,641 sq miles (198,500 sq km)
Languages Kyrgyz*, Russian*, Uzbek, Tatar, Ukrainian
Religions Muslim (mainly Sunni) 70%, Orthodox Christian 30%
Ethnic mix Kyrgyz 71%, Uzbek 14%, Russian 8%, Other 4%, Dungan 1%, Uighur 1%, Tajik 1%
Government Presidential / parliamentary system
Currency Som = 100 tyiyn
Literacy rate 99%
Calorie consumption 2817 kilocalories

LAOS
Southeast Asia

Page 114 D4

Landlocked Laos suffered a long civil war after French rule ended, and was badly bombed by US forces engaged in Vietnam. It has been under communist rule since 1975.

Official name Lao People's Democratic Republic
Formation 1953 / 1953
Capital Viangchan (Vientiane)
Population 6.9 million / 77 people per sq mile (30 people per sq km)
Total area 91,428 sq miles (236,800 sq km)
Languages Lao*, Mon-Khmer, Yao, Vietnamese, Chinese, French
Religions Buddhist 67%, Other 31%, Christian 2%
Ethnic mix Lao Loum 66%, Lao Theung 30%, Other 2%, Lao Soung 2%
Government One-party state
Currency Kip = 100 at
Literacy rate 58%
Calorie consumption 2451 kilocalories

LATVIA
Northeast Europe

Page 84 C3

Situated on the low-lying eastern shores of the Baltic Sea, Latvia, like its Baltic neighbors, regained its independence at the collapse of the USSR in 1991. It retains a large Russian population.

Official name Republic of Latvia
Formation 1991 / 1991
Capital Riga
Population 1.9 million / 76 people per sq mile (29 people per sq km)
Total area 24,938 sq miles (64,589 sq km)
Languages Latvian*, Russian
Religions Orthodox Christian 31%, Roman Catholic 23%, Nonreligious 21%, Lutheran 19%, Other 6%
Ethnic mix Latvian 62%, Russian 27%, Belarussian 3%, Other 3%, Polish 2%, Ukrainian 2%, Lithuanian 1%
Government Parliamentary system
Currency Euro = 100 cents
Literacy rate 99%
Calorie consumption 3174 kilocalories

LEBANON
Southwest Asia

Page 96 A4

Lebanon is dwarfed by its two powerful neighbors, Syria and Israel. Muslims and Christians fought a 14-year civil war until agreeing to share power in 1989, however, instability continues.

Official name Lebanese Republic
Formation 1941 / 1941
Capital Beirut
Population 6.1 million / 1544 people per sq mile (596 people per sq km)
Total area 4015 sq miles (10,400 sq km)
Languages Arabic*, French, Armenian, Assyrian
Religions Muslim 60%, Christian 39%, Other 1%
Ethnic mix Arab 95%, Armenian 4%, Other 1%
Government Parliamentary system
Currency Lebanese pound = 100 piastres
Literacy rate 91%
Calorie consumption 3066 kilocalories

LESOTHO
Southern Africa

Page 56 D5

Lesotho lies within South Africa, on whom it is economically dependent. Elections in 1993 ended military rule, but South Africa has had to intervene in politics since. AIDS is a problem.

Official name Kingdom of Lesotho
Formation 1966 / 1966
Capital Maseru
Population 2.2 million / 188 people per sq mile (72 people per sq km)
Total area 11,720 sq miles (30,355 sq km)
Languages English*, Sesotho*, isiZulu
Religions Christian 90%, Traditional beliefs 10%
Ethnic mix Sotho 99%, European and Asian 1%
Government Parliamentary system
Currency Loti = 100 lisente; South African rand = 100 cents
Literacy rate 77%
Calorie consumption 2529 kilocalories

LIBERIA
West Africa

Page 52 C5

Facing the Atlantic Ocean, Liberia is Africa's oldest republic, founded in 1847 by freed US slaves. Recovery from the 1990s' civil war has been set back by the recent Ebola epidemic.

Official name Republic of Liberia
Formation 1847 / 1847
Capital Monrovia
Population 4.7 million / 126 people per sq mile (49 people per sq km)
Total area 43,000 sq miles (111,370 sq km)
Languages Kpelle, Vai, Bassa, Kru, Grebo, Kissi, Gola, Loma, English*
Religions Christian 86%, Muslim 12%, Nonreligious 1%, Traditional beliefs and other 1%
Ethnic mix Indigenous tribes (12 groups) 50%, Kpellé 20%, Bassa 14%, Gio 8%, Krou 6%, Other 2%
Government Presidential system
Currency Liberian dollar = 100 cents
Literacy rate 43%
Calorie consumption 2204 kilocalories

LIBYA
North Africa

Page 49 F3

On the Mediterranean coast, Libya was ruled from 1969 by the idiosyncratic Col. Gaddafi. The 2011 "Arab Spring" turned to civil war, toppling his regime, but leaving Libya in anarchy.

Official name Libya
Formation 1951 / 1951
Capital Tripoli
Population 6.4 million / 9 people per sq mile (4 people per sq km)
Total area 679,358 sq miles (1,759,540 sq km)
Languages Arabic*, Tuareg
Religions Muslim (mainly Sunni) 97%, Other 3%
Ethnic mix Arab and Berber 97%, Other 3%
Government Transitional regime
Currency Libyan dinar = 1000 dirhams
Literacy rate 90%
Calorie consumption 3211 kilocalories

LIECHTENSTEIN
Central Europe

Page 73 B7

Tucked in the Alps between Switzerland and Austria, Liechtenstein became an independent principality of the Holy Roman Empire in 1719. Switzerland handles its foreign affairs and defense.

Official name Principality of Liechtenstein
Formation 1719 / 1719
Capital Vaduz
Population 38,000 / 613 people per sq mile (238 people per sq km)
Total area 62 sq miles (160 sq km)
Languages German*, Alemannish dialect, Italian
Religions Roman Catholic 78%, Protestant 9%, Muslim 6%, Nonreligious 5%, Orthodox Christian 1%, Other 1%
Ethnic mix Liechtensteiner 66%, Other 12%, Swiss 10%, Austrian 6%, German 3%, Italian 3%
Government Parliamentary system
Currency Swiss franc = 100 rappen/centimes
Literacy rate 99%
Calorie consumption Not available

LITHUANIA
Northeast Europe

Page 84 B4

A flat land of lakes, moors, and bogs, Lithuania is the largest of the three Baltic states. It has historical ties to Poland and was the first former Soviet republic to declare independence.

Official name Republic of Lithuania
Formation 1991 / 1991
Capital Vilnius
Population 2.9 million / 115 people per sq mile (44 people per sq km)
Total area 25,174 sq miles (65,200 sq km)
Languages Lithuanian*, Russian
Religions Roman Catholic 75%, Christian 14%, Nonreligious 6%, Orthodox Christian 3%, Other 2%
Ethnic mix Lithuanian 85%, Polish 7%, Russian 6%, Belarussian 1%, Other 1%
Government Parliamentary system
Currency Euro = 100 cents
Literacy rate 99%
Calorie consumption 3417 kilocalories

LUXEMBOURG
Northwest Europe

Page 65 D8

Part of the forested Ardennes plateau in Northwest Europe, Luxembourg is Europe's last independent duchy and one of its richest states. It is a banking center and hosts EU institutions.

Official name Grand Duchy of Luxembourg
Formation 1867 / 1867
Capital Luxembourg
Population 600,000 / 601 people per sq mile (232 people per sq km)
Total area 998 sq miles (2586 sq km)
Languages Luxembourgish*, German*, French*
Religions Roman Catholic 97%, Protestant, Orthodox Christian, and Jewish 3%
Ethnic mix Luxembourger 62%, Foreign residents 38%
Government Parliamentary system
Currency Euro = 100 cents
Literacy rate 99%
Calorie consumption 3539 kilocalories

MACEDONIA
Southeast Europe

Page 79 D6

This ex-Yugoslav state is landlocked in the southern Balkans. Its EU candidacy is held back over Greek fears that its name implies a claim to its own northern province of Macedonia.

Official name Republic of Macedonia
Formation 1991 / 1991
Capital Skopje
Population 2.1 million / 212 people per sq mile (82 people per sq km)
Total area 9781 sq miles (25,333 sq km)
Languages Macedonian*, Albanian*, Turkish, Romani, Serbian
Religions Orthodox Christian 65%, Muslim 33%, Other 2%
Ethnic mix Macedonian 64%, Albanian 25%, Turkish 4%, Roma 3%, Other 2%, Serb 2%
Government Presidential / parliamentary system
Currency Macedonian denar = 100 deni
Literacy rate 98%
Calorie consumption 2949 kilocalories

MADAGASCAR
Indian Ocean

Page 57 F4

Off Africa's southeast coast, this former French colony is the world's fourth-largest island. Free elections in 1993 ended 18 years of socialism, but power struggles have blighted politics since.

Official name Republic of Madagascar
Formation 1960 / 1960
Capital Antananarivo
Population 25.6 million / 114 people per sq mile (44 people per sq km)
Total area 226,656 sq miles (587,040 sq km)
Languages Malagasy*, French*, English*
Religions Traditional beliefs 52%, Christian (mainly Roman Catholic) 41%, Muslim 7%
Ethnic mix Other Malay 46%, Merina 26%, Betsimisaraka 15%, Betsileo 12%, Other 1%
Government Presidential / parliamentary system
Currency Ariary = 5 iraimbilanja
Literacy rate 72%
Calorie consumption 2052 kilocalories

MALAWI
Southern Africa

Page 57 E1

This landlocked former British colony lies along the Great Rift Valley and Lake Nyasa, Africa's third-largest lake. Multiparty elections in 1994 ended three decades of single-party rule.

Official name Republic of Malawi
Formation 1964 / 1964
Capital Lilongwe
Population 18.6 million / 512 people per sq mile (198 people per sq km)
Total area 45,745 sq miles (118,480 sq km)
Languages Chewa, Lomwe, Yao, Ngoni, English*
Religions Christian (mainly Protestant) 83%, Muslim 13%, Nonreligious 2%, Other 2%
Ethnic mix Bantu 99%, Other 1%
Government Presidential system
Currency Malawi kwacha = 100 tambala
Literacy rate 62%
Calorie consumption 2367 kilocalories

MALAYSIA
Southeast Asia

Page 116 B3

Three separate territories, Peninsular Malaysia, and Sarawak and Sabah on Borneo, make up Malaysia. Relations between indigenous Malays and the Chinese minority dominate politics.

Official name Malaysia
Formation 1963 / 1965
Capital Kuala Lumpur; Putrajaya (administrative)
Population 31.6 million / 249 people per sq mile (96 people per sq km)
Total area 127,316 sq miles (329,750 sq km)
Languages Bahasa Malaysia*, Malay, Chinese, Tamil, English
Religions Muslim (mainly Sunni) 62%, Buddhist 20%, Christian 9%, Hindu 6%, Other 3%
Ethnic mix Malay 50%, Chinese 22%, Indigenous tribes 12%, Other 9%, Indian 7%
Government Parliamentary system
Currency Ringgit = 100 sen
Literacy rate 93%
Calorie consumption 2916 kilocalories

MALDIVES
Indian Ocean

Page 110 A4

Of this group of over 1000 small low-lying coral islands in the Indian Ocean, only 200 are inhabited. A few families dominate politics and have reversed the electoral upsets of 2008 and 2009.

Official name Republic of Maldives
Formation 1965 / 1965
Capital Maale (Male')
Population 400,000 / 3448 people per sq mile (1333 people per sq km)
Total area 116 sq miles (300 sq km)
Languages Dhivehi* (Maldivian), Sinhala, Tamil, Arabic
Religions Sunni Muslim 94%, Hindu 3%, Christian 2%, Buddhist 1%
Ethnic mix Arab–Sinhalese–Malay 100%
Government Presidential system
Currency Rufiyaa = 100 laari
Literacy rate 99%
Calorie consumption 2732 kilocalories

MALI
West Africa

Page 53 E2

Mali's power as a trans-Saharan trading empire peaked 700 years ago. Modern Mali, a one-party state until 1992, called on former colonial power France to suppress Islamist rebels since 2013.

Official name Republic of Mali
Formation 1960 / 1960
Capital Bamako
Population 18.5 million / 39 people per sq mile (15 people per sq km)
Total area 478,764 sq miles (1,240,000 sq km)
Languages Bambara, Fulani, Senufo, Soninke, French*
Religions Muslim (mainly Sunni) 90%, Traditional beliefs 6%, Christian 4%
Ethnic mix Bambara 52%, Other 14%, Fulani 11%, Saracolé 7%, Soninka 7%, Tuareg 5%, Mianka 4%
Government Presidential system
Currency CFA franc = 100 centimes
Literacy rate 33%
Calorie consumption 2890 kilocalories

MALTA
Southern Europe

Page 80 A5

The Maltese archipelago lies off Sicily. Only Malta, Kemmuna, and Gozo are inhabited. Its mid-Mediterranean location has made it a gateway for illegal migration from Africa to Europe.

Official name Republic of Malta
Formation 1964 / 1964
Capital Valletta
Population 400,000 / 3226 people per sq mile (1250 people per sq km)
Total area 122 sq miles (316 sq km)
Languages Maltese*, English*
Religions Roman Catholic 98%, Other and nonreligious 2%
Ethnic mix Maltese 96%, Other 4%
Government Parliamentary system
Currency Euro = 100 cents
Literacy rate 93%
Calorie consumption 3378 kilocalories

MARSHALL ISLANDS
Australasia & Oceania

Page 122 D1

This group of 34 atolls was under US rule as part of the UN Trust Territory of the Pacific Islands until 1986. The economy depends on US aid and rent for the US missile base on Kwajalein.

Official name Republic of the Marshall Islands
Formation 1986 / 1986
Capital Majuro Atoll
Population 53,000 / 757 people per sq mile (292 people per sq km)
Total area 70 sq miles (181 sq km)
Languages Marshallese*, English*, Japanese, German
Religions Protestant 81%, Other 11%, Roman Catholic 8%
Ethnic mix Micronesian 90%, Other 10%
Government Presidential system
Currency US dollar = 100 cents
Literacy rate 98%
Calorie consumption Not available

MAURITANIA
West Africa

Page 52 C2

Two-thirds of this former French colony is desert. The Maures oppress the black minority. Despite multiparty polls since 1991, military leaders have held power until 2005 and since 2008.

Official name Islamic Republic of Mauritania
Formation 1960 / 1960
Capital Nouakchott
Population 4.4 million / 11 people per sq mile (4 people per sq km)
Total area 397,953 sq miles (1,030,700 sq km)
Languages Arabic*, Hassaniyah Arabic, Wolof, French
Religions Sunni Muslim 100%
Ethnic mix Maure 81%, Wolof 7%, Tukolor 5%, Other 4%, Soninka 3%
Government Presidential system
Currency Ouguiya = 5 khoums
Literacy rate 46%
Calorie consumption 2876 kilocalories

MAURITIUS
Indian Ocean

Page 57 H3

East of Madagascar in the Indian Ocean, Mauritius became a republic 24 years after independence from Britain. Its diversified economy includes tourism, financial services, and outsourcing.

Official name Republic of Mauritius
Formation 1968 / 1968
Capital Port Louis
Population 1.3 million / 1811 people per sq mile (699 people per sq km)
Total area 718 sq miles (1860 sq km)
Languages French Creole, Hindi, Urdu, Tamil, Chinese, English*, French
Religions Hindu 48%, Roman Catholic 26%, Muslim 17%, Other Christian 7%, Other 2%
Ethnic mix Indo-Mauritian 68%, Creole 27%, Sino-Mauritian 3%, Franco-Mauritian 2%
Government Parliamentary system
Currency Mauritian rupee = 100 cents
Literacy rate 93%
Calorie consumption 3065 kilocalories

MEXICO
North America

Page 28 D3

Located between the US and the Central American states, Mexico was a Spanish colony for 300 years. Sprawling Mexico City is built on the site of the Aztec capital, Tenochtitlán.

Official name United Mexican States
Formation 1836 / 1848
Capital Mexico City
Population 129 million / 175 people per sq mile (68 people per sq km)
Total area 761,602 sq miles (1,972,550 sq km)
Languages Spanish*, Nahuatl, Mayan, Zapotec, Mixtec, Otomi, Totonac, Tzotzil, Tzeltal
Religions Roman Catholic 81%, Protestant 9%, Nonreligious 7%, Other 3%
Ethnic mix Mestizo 60%, Amerindian 30%, European 9%, Other 1%
Government Presidential system
Currency Mexican peso = 100 centavos
Literacy rate 94%
Calorie consumption 3072 kilocalories

MICRONESIA
Australasia & Oceania

Page 122 B1

The Federated States of Micronesia, situated in the western Pacific, comprises 607 islands and atolls grouped into four main island states. The economy relies on US aid.

Official name Federated States of Micronesia
Formation 1986 / 1986
Capital Palikir (Pohnpei Island)
Population 100,000 / 369 people per sq mile (142 people per sq km)
Total area 271 sq miles (702 sq km)
Languages Trukese, Pohnpeian, Kosraean, Yapese, English*
Religions Roman Catholic 53%, Protestant 43%, Other 3%, Nonreligious 1%
Ethnic mix Chuukese 49%, Pohnpeian 24%, Other 14%, Kosraean 6%, Yapese 5%, Asian 2%
Government Nonparty system
Currency US dollar = 100 cents
Literacy rate 81%
Calorie consumption Not available

MOLDOVA
Southeast Europe

Page 86 D3

The smallest and most densely populated of the ex-Soviet republics, Moldova has strong linguistic and cultural ties with Romania to the west. It exports tobacco, wine, and fruit.

Official name Republic of Moldova
Formation 1991 / 1991
Capital Chisinau
Population 4.1 million / 315 people per sq mile (122 people per sq km)
Total area 13,067 sq miles (33,843 sq km)
Languages Moldovan*, Ukrainian, Russian
Religions Orthodox Christian 92%, Other 6%, Nonreligious 2%
Ethnic mix Moldovan 76%, Ukrainian 9%, Russian 6%, Gagauz 4%, Romanian 2%, Bulgarian 2%, Other 1%
Government Parliamentary system
Currency Moldovan leu = 100 bani
Literacy rate 99%
Calorie consumption 2714 kilocalories

MONACO
Southern Europe

Page 69 E6

The destiny of this tiny enclave on France's Côte d'Azur was changed in 1863 when its prince opened a casino. A jet-set image and thriving service sector define its modern identity.

Official name Principality of Monaco
Formation 1861 / 1861
Capital Monaco
Population 38,000 / 50,667 people per sq mile (19,487 people per sq km)
Total area 0.75 sq miles (1.95 sq km)
Languages French*, Italian, Monégasque, English
Religions Roman Catholic 89%, Protestant 6%, Other 5%
Ethnic mix French 47%, Other 21%, Italian 16%, Monégasque 16%
Government Monarchical / parliamentary system
Currency Euro = 100 cents
Literacy rate 99%
Calorie consumption Not available

MONGOLIA
East Asia

Page 104 D2

Vast Mongolia is sparsely populated and mostly desert. Under the sway of its giant neighbors, Russia and China, it was communist from independence from China in 1924 until 1990.

Official name Mongolia
Formation 1924 / 1924
Capital Ulaanbaatar
Population 3.1 million / 5 people per sq mile (2 people per sq km)
Total area 604,247 sq miles (1,565,000 sq km)
Languages Khalkha Mongolian*, Kazakh, Chinese, Russian
Religions Tibetan Buddhist 53%, Nonreligious 38%, Muslim 3%, Shamanist 3%, Christian 2%, Other 1%
Ethnic mix Khalkh 82%, Other 9%, Kazakh 4%, Dorvod 3%, Bayad 2%
Government Presidential / parliamentary system
Currency Tugrik (tögrög) = 100 möngö
Literacy rate 98%
Calorie consumption 2510 kilocalories

MONTENEGRO
Southeast Europe

Page 79 C5

Part of the former Yugoslavia, the tiny republic of Montenegro broke away from Serbia in 2006. Its attractive coast and mountains are a big tourist draw. It hopes to join the EU soon.

Official name Montenegro
Formation 2006 / 2006
Capital Podgorica
Population 600,000 / 113 people per sq mile (43 people per sq km)
Total area 5332 sq miles (13,812 sq km)
Languages Montenegrin*, Serbian, Albanian, Bosniak, Croatian
Religions Orthodox Christian 74%, Muslim 20%, Roman Catholic 4%, Nonreligious 1%, Other 1%
Ethnic mix Montenegrin 43%, Serb 32%, Other 12%, Bosniak 8%, Albanian 5%
Government Parliamentary system
Currency Euro = 100 cents
Literacy rate 98%
Calorie consumption 3491 kilocalories

MOROCCO
North Africa

Page 48 C2

A former French colony in northwest Africa, Morocco has occupied the disputed territory of Western Sahara since 1975. The king has handed more power to parliament since 2011.

Official name Kingdom of Morocco
Formation 1956 / 1969
Capital Rabat
Population 35.7 million / 207 people per sq mile (80 people per sq km)
Total area 172,316 sq miles (446,300 sq km)
Languages Arabic*, Tamazight* (Berber), French, Spanish
Religions Muslim (mainly Sunni) 99%, Other (mostly Christian) 1%
Ethnic mix Arab 70%, Berber 29%, European 1%
Government Monarchical / parliamentary system
Currency Moroccan dirham = 100 centimes
Literacy rate 69%
Calorie consumption 3403 kilocalories

MOZAMBIQUE
Southern Africa

Page 57 E3

Mozambique, on the southeast African coast, frequently suffers both floods and droughts. It was torn by civil war from 1977 to 1992 as the Marxist state fought South African-backed rebels.

Official name Republic of Mozambique
Formation 1975 / 1975
Capital Maputo
Population 29.7 million / 98 people per sq mile (38 people per sq km)
Total area 309,494 sq miles (801,590 sq km)
Languages Makua, Xitsonga, Sena, Lomwe, Portuguese*
Religions Roman Catholic 28%, Nonreligious 19%, Muslim 18%, Traditional beliefs 16%, Other 19%
Ethnic mix Makua Lomwe 47%, Tsonga 23%, Malawi 12%, Shona 11%, Yao 4%, Other 3%
Government Presidential system
Currency New metical = 100 centavos
Literacy rate 51%
Calorie consumption 2283 kilocalories

MYANMAR (BURMA)
Southeast Asia

Page 114 A3

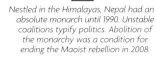

Myanmar, on the eastern shores of the Bay of Bengal and the Andaman Sea, suffered years of ethnic conflict and repressive military rule following independence from Britain in 1948.

Official name Republic of the Union of Myanmar
Formation 1948 / 1948
Capital Nay Pyi Taw
Population 53.4 million / 210 people per sq mile (81 people per sq km)
Total area 261,969 sq miles (678,500 sq km)
Languages Burmese* (Myanmar), Shan, Karen, Rakhine, Chin, Yangbye, Kachin, Mon
Religions Buddhist 90%, Christian 6%, Muslim 2%, Animist 1%, Other 1%
Ethnic mix Burman (Bamah) 68%, Other 12%, Shan 9%, Karen 7%, Rakhine 4%
Government Presidential system
Currency Kyat = 100 pyas
Literacy rate 76%
Calorie consumption 2571 kilocalories

NAMIBIA
Southern Africa

Page 56 B3

On Africa's southwest coast, this mineral-rich ex-German colony was governed by South Africa from 1915 to 1990. The white minority controls the economy, a legacy of apartheid.

Official name Republic of Namibia
Formation 1990 / 1994
Capital Windhoek
Population 2.5 million / 8 people per sq mile (3 people per sq km)
Total area 318,694 sq miles (825,418 sq km)
Languages Ovambo, Kavango, English*, Bergdama, German, Afrikaans
Religions Christian 90%, Traditional beliefs 10%
Ethnic mix Ovambo 50%, Other tribes 22%, Kavango 9%, Herero 7%, Damara 7%, Other 5%
Government Presidential system
Currency Namibian dollar = 100 cents; South African rand = 100 cents
Literacy rate 88%
Calorie consumption 2171 kilocalories

NAURU
Australasia & Oceania

Page 122 D3

The world's smallest republic, 2480 miles (4000 km) northeast of Australia, grew rich from its phosphate deposits, but these have almost run out and poor investment has caused financial crisis.

Official name Republic of Nauru
Formation 1968 / 1968
Capital None (Yaren de facto capital)
Population 13,000 / 1605 people per sq mile (619 people per sq km)
Total area 8.1 sq miles (21 sq km)
Languages Nauruan*, Kiribati, Chinese, Tuvaluan, English
Religions Nauruan Congregational Church 60%, Roman Catholic 35%, Other 5%
Ethnic mix Nauruan 93%, Chinese 5%, Other Pacific islanders 1%, European 1%
Government Nonparty system
Currency Australian dollar = 100 cents
Literacy rate 95%
Calorie consumption Not available

NEPAL
South Asia

Page 113 E3

Nestled in the Himalayas, Nepal had an absolute monarch until 1990. Unstable coalitions typify politics. Abolition of the monarchy was a condition for ending the Maoist rebellion in 2008.

Official name Federal Democratic Republic of Nepal
Formation 1769 / 1769
Capital Kathmandu
Population 29.3 million / 555 people per sq mile (214 people per sq km)
Total area 54,363 sq miles (140,800 sq km)
Languages Nepali*, Maithili, Bhojpuri
Religions Hindu 82%, Buddhist 9%, Other (including Christian) 5%, Muslim 4%
Ethnic mix Other 52%, Chhetri 17%, Hill Brahman 12%, Magar 7%, Tharu 7%, Tamang 6%
Government Parliamentary system
Currency Nepalese rupee = 100 paisa
Literacy rate 60%
Calorie consumption 2673 kilocalories

NETHERLANDS
Northwest Europe

Page 64 C3

Astride the delta of four major rivers in northwest Europe, the Netherlands was ruled by Spain until 1648. It has a long trading tradition, and Rotterdam remains Europe's largest port.

Official name Kingdom of the Netherlands
Formation 1648 / 1839
Capital Amsterdam; The Hague (administrative)
Population 17 million / 1298 people per sq mile (501 people per sq km)
Total area 16,033 sq miles (41,526 sq km)
Languages Dutch*, Frisian
Religions Roman Catholic 36%, Other 34%, Protestant 27%, Muslim 3%
Ethnic mix Dutch 82%, Other 12%, Surinamese 2%, Turkish 2%, Moroccan 2%
Government Parliamentary system
Currency Euro = 100 cents
Literacy rate 99%
Calorie consumption 3228 kilocalories

NEW ZEALAND
Australasia & Oceania

Page 128 A4

This former British colony, on the Pacific Rim, has a volcanic, more populous North Island and a mountainous South Island. It was the first country to give women the vote, in 1893.

Official name New Zealand
Formation 1947 / 1947
Capital Wellington
Population 4.7 million / 45 people per sq mile (17 people per sq km)
Total area 103,737 sq miles (268,680 sq km)
Languages English*, Maori*
Religions Nonreligious 36%, Other Christian 16%, Anglican 15%, Roman Catholic 14%, Presbyterian 11%, Other 8%
Ethnic mix European 60%, Other 19%, Maori 14%, Chinese 4%, Samoan 3%
Government Parliamentary system
Currency New Zealand dollar = 100 cents
Literacy rate 99%
Calorie consumption 3137 kilocalories

NICARAGUA
Central America

Page 30 D3

Nicaragua, at the heart of Central America, plans to build a canal to rival Panama. Left-wing Sandinistas threw out a brutal dictator in 1978, then faced conflict with US-backed Contras.

Official name Republic of Nicaragua
Formation 1838 / 1838
Capital Managua
Population 6.2 million / 135 people per sq mile (52 people per sq km)
Total area 49,998 sq miles (129,494 sq km)
Languages Spanish*, English Creole, Miskito
Religions Roman Catholic 50%, Protestant 40%, Nonreligious 7%, Other 3%
Ethnic mix Mestizo 69%, White 17%, Black 9%, Amerindian 5%
Government Presidential system
Currency Córdoba oro = 100 centavos
Literacy rate 78%
Calorie consumption 2638 kilocalories

NIGER
West Africa

Page 53 G3

Landlocked Niger is linked to the sea by the River Niger. This ex-French colony has suffered coups, military rule, civil unrest, and severe droughts. It is one of the poorest countries in the world.

Official name Republic of Niger
Formation 1960 / 1960
Capital Niamey
Population 21.5 million / 44 people per sq mile (17 people per sq km)
Total area 489,188 sq miles (1,267,000 sq km)
Languages Hausa, Djerma, Fulani, Tuareg, Teda, French*
Religions Muslim 99%, Other (including Christian) 1%
Ethnic mix Hausa 55%, Djerma and Songhai 21%, Tuareg 9%, Peul 9%, Kanuri 5%, Other 1%
Government Presidential system
Currency CFA franc = 100 centimes
Literacy rate 16%
Calorie consumption 2547 kilocalories

NIGERIA
West Africa

Page 53 G4

Nigeria has Africa's largest population, whose religious and ethnic rivalries have brought down both civilian and military regimes in the past. Islamic extremists are one current challenge.

Official name Federal Republic of Nigeria
Formation 1960 / 1961
Capital Abuja
Population 191 million / 543 people per sq mile (210 people per sq km)
Total area 356,667 sq miles (923,768 sq km)
Languages Hausa, English*, Yoruba, Ibo
Religions Muslim 50%, Christian 40%, Traditional beliefs 10%
Ethnic mix Other 29%, Hausa 21%, Yoruba 21%, Ibo 18%, Fulani 11%
Government Presidential system
Currency Naira = 100 kobo
Literacy rate 51%
Calorie consumption 2700 kilocalories

NORTH KOREA
East Asia

Page 106 E3

The maverick communist state in Korea's northern half has been isolated from the outside world since 1948. Its shattered state-run economy leaves people short of food and power.

Official name Democratic People's Republic of Korea
Formation 1948 / 1953
Capital Pyongyang
Population 25.5 million / 549 people per sq mile (212 people per sq km)
Total area 46,540 sq miles (120,540 sq km)
Languages Korean*
Religions Atheist 100%
Ethnic mix Korean 100%
Government One-party state
Currency North Korean won = 100 chon
Literacy rate 99%
Calorie consumption 2094 kilocalories

NORWAY
Northern Europe

Page 63 A5

Lying on the rugged western coast of Scandinavia, most people live in southern, coastal areas. Oil and gas wealth has brought one of the world's best standards of living.

Official name Kingdom of Norway
Formation 1905 / 1905
Capital Oslo
Population 5.3 million / 45 people per sq mile (17 people per sq km)
Total area 125,181 sq miles (324,220 sq km)
Languages Norwegian* (Bokmål "book language" and Nynorsk "new Norsk"), Sámi
Religions Evangelical Lutheran 88%, Other and nonreligious 8%, Muslim 2%, Roman Catholic 1%, Pentecostal 1%
Ethnic mix Norwegian 93%, Other 6%, Sámi 1%
Government Parliamentary system
Currency Norwegian krone = 100 øre
Literacy rate 99%
Calorie consumption 3485 kilocalories

OMAN
Southwest Asia

Page 99 D6

Situated on the eastern corner of the Arabian Peninsula, Oman is the least developed of the Gulf states, despite modest oil exports. The current sultan has been in power since 1970.

Official name Sultanate of Oman
Formation 1951 / 1951
Capital Muscat
Population 4.6 million / 56 people per sq mile (22 people per sq km)
Total area 82,031 sq miles (212,460 sq km)
Languages Arabic*, Baluchi, Farsi, Hindi, Punjabi
Religions Ibadi Muslim 75%, Other Muslim and Hindu 25%
Ethnic mix Arab 88%, Baluchi 4%, Indian and Pakistani 3%, Persian 3%, African 2%
Government Monarchy
Currency Omani rial = 1000 baisa
Literacy rate 93%
Calorie consumption 3143 kilocalories

PAKISTAN
South Asia

Page 112 B2

Once part of British India, Pakistan was created in 1947 as a Muslim state. Today, this nuclear-armed country is struggling to deal with complex domestic and international tensions.

Official name Islamic Republic of Pakistan
Formation 1947 / 1971
Capital Islamabad
Population 197 million / 662 people per sq mile (256 people per sq km)
Total area 310,401 sq miles (803,940 sq km)
Languages Punjabi, Sindhi, Pashtu, Urdu*, Baluchi, Brahui
Religions Sunni Muslim 77%, Shi'a Muslim 20%, Hindu 2%, Christian 1%
Ethnic mix Punjabi 56%, Pathan (Pashtun) 15%, Sindhi 14%, Mohajir 7%, Other 4%, Baluchi 4%
Government Parliamentary system
Currency Pakistani rupee = 100 paisa
Literacy rate 57%
Calorie consumption 2440 kilocalories

PALAU
Australasia & Oceania

Page 122 A2

This archipelago of over 200 islands, only ten of which are inhabited, lies in the western Pacific Ocean. Until 1994 it was under US administration. The economy relies on US aid and tourism.

Official name Republic of Palau
Formation 1994 / 1994
Capital Ngerulmud
Population 21,000 / 107 people per sq mile (41 people per sq km)
Total area 177 sq miles (458 sq km)
Languages Palauan*, English*, Japanese, Angaur, Tobi, Sonsorolese
Religions Roman Catholic 49%, Protestant 33%, Modekngei 9%, Other 8%, Nonreligious 1%
Ethnic mix Palauan 73%, Filipino 16%, Other Asian 7%, Other Micronesian 3%, Other 1%
Government Nonparty system
Currency US dollar = 100 cents
Literacy rate 97%
Calorie consumption Not available

PANAMA
Central America

Page 31 F5

The US invaded Central America's southernmost country in 1989 to oust its dictator. The Panama Canal is a vital shortcut for shipping between the Atlantic and Pacific oceans.

Official name Republic of Panama
Formation 1903 / 1903
Capital Panama City
Population 4.1 million / 140 people per sq mile (54 people per sq km)
Total area 30,193 sq miles (78,200 sq km)
Languages English Creole, Spanish*, Amerindian languages, Chibchan languages
Religions Roman Catholic 70%, Protestant 19%, Nonreligious 7%, Other 4%
Ethnic mix Mestizo 70%, Black 14%, White 10%, Amerindian 6%
Government Presidential system
Currency Balboa = 100 centésimos; US dollar
Literacy rate 94%
Calorie consumption 2733 kilocalories

PAPUA NEW GUINEA
Australasia & Oceania

Page 122 B3

The world's most linguistically diverse country, mineral-rich PNG occupies the east of the island of New Guinea and several other island groups. It was administered by Australia before 1975.

Official name Independent State of Papua New Guinea
Formation 1975 / 1975
Capital Port Moresby
Population 8.3 million / 47 people per sq mile (18 people per sq km)
Total area 178,703 sq miles (462,840 sq km)
Languages Pidgin English, Papuan, English*, Motu, 800 (est.) native languages
Religions Protestant 60%, Roman Catholic 37%, Other 3%
Ethnic mix Melanesian and mixed race 100%
Government Parliamentary system
Currency Kina = 100 toea
Literacy rate 63%
Calorie consumption 2193 kilocalories

PARAGUAY
South America

Page 42 D2

South America's longest dictatorship held power in landlocked Paraguay from 1954 to 1989. Now under democratic rule, the country's economy is still largely agricultural.

Official name Republic of Paraguay
Formation 1811 / 1938
Capital Asunción
Population 6.8 million / 44 people per sq mile (17 people per sq km)
Total area 157,046 sq miles (406,750 sq km)
Languages Guaraní*, Spanish*, German
Religions Roman Catholic 89%, Protestant (including Mennonite) 7%, Other 3%, Nonreligious 1%
Ethnic mix Mestizo 91%, Other 7%, Amerindian 2%
Government Presidential system
Currency Guaraní = 100 céntimos
Literacy rate 95%
Calorie consumption 2589 kilocalories

PERU
South America

Page 38 C3

On the Pacific coast of South America, Peru was once the heart of the Inca empire, before the Spanish conquest in the 16th century. It elected its first Amerindian president in 2001.

Official name Republic of Peru
Formation 1824 / 1941
Capital Lima
Population 32.2 million / 65 people per sq mile (25 people per sq km)
Total area 496,223 sq miles (1,285,200 sq km)
Languages Spanish*, Quechua*, Aymara
Religions Roman Catholic 76%, Protestant 17%, Nonreligious 4%, Other 3%
Ethnic mix Amerindian 45%, Mestizo 37%, White 15%, Other 3%
Government Presidential system
Currency New sol = 100 céntimos
Literacy rate 94%
Calorie consumption 2700 kilocalories

PHILIPPINES
Southeast Asia

Page 117 E1

This 7107-island archipelago between the South China Sea and the Pacific is prone to earthquakes and volcanic activity. A 21-year dictatorship ended in 1986; politics since has been volatile.

Official name Republic of the Philippines
Formation 1946 / 1946
Capital Manila
Population 105 million / 911 people per sq mile (352 people per sq km)
Total area 115,830 sq miles (300,000 sq km)
Languages Filipino*, English*, Tagalog, Cebuano, Ilocano, Hiligaynon, many other local languages
Religions Roman Catholic 81%, Other Christian 11%, Muslim 5%, Other 3%
Ethnic mix Other 34%, Tagalog 28%, Cebuano 13%, Ilocano 9%, Hiligaynon 8%, Bisaya 8%
Government Presidential system
Currency Philippine peso = 100 centavos
Literacy rate 96%
Calorie consumption 2570 kilocalories

POLAND
Northern Europe

Page 76 B3

Poland's low-lying plains extend from the Baltic Sea into the heart of Europe. It has undergone massive political and economic change since the fall of communism. It joined the EU in 2004.

Official name Republic of Poland
Formation 1918 / 1945
Capital Warsaw
Population 38.2 million / 325 people per sq mile (125 people per sq km)
Total area 120,728 sq miles (312,685 sq km)
Languages Polish*
Religions Roman Catholic 87%, Nonreligious 7%, Other 5%, Orthodox Christian 1%
Ethnic mix Polish 97%, Silesian 2%, Other 1%
Government Parliamentary system
Currency Zloty = 100 groszy
Literacy rate 99%
Calorie consumption 3451 kilocalories

PORTUGAL
Southwest Europe

Page 70 B3

Portugal, on the Iberian Peninsula, is the westernmost country in mainland Europe. Isolated under 44 years of dictatorship until 1974, it modernized fast after joining the EU in 1986.

Official name Portuguese Republic
Formation 1139 / 1640
Capital Lisbon
Population 10.3 million / 290 people per sq mile (112 people per sq km)
Total area 35,672 sq miles (92,391 sq km)
Languages Portuguese*
Religions Roman Catholic 88%, Nonreligious 7%, Other Christian 4%, Other 1%
Ethnic mix Portuguese 98%, African and other 2%
Government Parliamentary system
Currency Euro = 100 cents
Literacy rate 94%
Calorie consumption 3477 kilocalories

145

QATAR
Southwest Asia

Page 98 C4

Projecting north from the Arabian Peninsula into the Persian Gulf, Qatar is mostly flat, semiarid desert. Massive reserves of oil and gas have made it one of the world's wealthiest states.

Official name State of Qatar
Formation 1971 / 1971
Capital Doha
Population 2.6 million / 612 people per sq mile (236 people per sq km)
Total area 4416 sq miles (11,437 sq km)
Languages Arabic*
Religions Muslim (mainly Sunni) 78%, Other 14%, Christian 8%
Ethnic mix Qatari 20%, Other Arab 20%, Indian 20%, Nepalese 13%, Filipino 10%, Other 10%, Pakistani 7%
Government Monarchy
Currency Qatar riyal = 100 dirhams
Literacy rate 98%
Calorie consumption Not available

ROMANIA
Southeast Europe

Page 86 B4

Romania lies on the western shores of the Black Sea. Its communist regime was overthrown in 1989 and, despite a slow transition to a free-market economy, it joined the EU in 2007.

Official name Romania
Formation 1878 / 1947
Capital Bucharest
Population 19.7 million / 222 people per sq mile (86 people per sq km)
Total area 91,699 sq miles (237,500 sq km)
Languages Romanian*, Hungarian (Magyar), Romani, German
Religions Orthodox Christian 86%, Other 8%, Roman Catholic 5%, Nonreligious 1%
Ethnic mix Romanian 89%, Magyar 7%, Roma 3%, Other 1%
Government Presidential / Parliamentary system
Currency New Romanian leu = 100 bani
Literacy rate 99%
Calorie consumption 3358 kilocalories

RUSSIA
Europe / Asia

Page 92 D4

The world's largest country, with vast mineral and energy reserves, Russia dominated the former USSR and is still a major power. It has over 150 ethnic groups, many with their own territory.

Official name Russian Federation
Formation 1480 / 1991
Capital Moscow
Population 144 million / 22 people per sq mile (8 people per sq km)
Total area 6,592,735 sq miles (17,075,200 sq km)
Languages Russian*, Tatar, Ukrainian, Chavash, various other national languages
Religions Orthodox Christian 71%, Nonreligious 15%, Muslim 11%, Other Christian 2%, Other 1%
Ethnic mix Russian 81%, Other 11%, Tatar 4%, Ukrainian 1%, Bashkir 1%, Chavash 1%, Chechen 1%
Government Presidential / parliamentary system
Currency Russian rouble = 100 kopeks
Literacy rate 99%
Calorie consumption 3361 kilocalories

RWANDA
Central Africa

Page 51 B6

Rwanda lies just south of the Equator in Central Africa. Ethnic violence flared into genocide in 1994, when almost a million died. The main victims, the Tutsi, dominate government now.

Official name Republic of Rwanda
Formation 1962 / 1962
Capital Kigali
Population 12.2 million / 1266 people per sq mile (489 people per sq km)
Total area 10,169 sq miles (26,338 sq km)
Languages Kinyarwanda*, French*, Kiswahili, English*
Religions Roman Catholic 44%, Protestant 38%, Seventh-day Adventist 12%, Other 4%, Muslim 2%
Ethnic mix Hutu 85%, Tutsi 14%, Other (including Twa) 1%
Government Presidential system
Currency Rwanda franc = 100 centimes
Literacy rate 68%
Calorie consumption 2228 kilocalories

ST KITTS AND NEVIS
West Indies

Page 33 G3

Saint Kitts and Nevis are part of the Caribbean Leeward Islands. A former British colony, the country is a popular tourist destination. Less-developed Nevis is famed for its hot springs.

Official name Federation of Saint Christopher and Nevis
Formation 1983 / 1983
Capital Basseterre
Population 55,000 / 545 people per sq mile (211 people per sq km)
Total area 101 sq miles (261 sq km)
Languages English*, English Creole
Religions Anglican 33%, Methodist 29%, Other 22%, Moravian 9%, Roman Catholic 7%
Ethnic mix Black 95%, Mixed race 3%, White 1%, Other and Amerindian 1%
Government Parliamentary system
Currency East Caribbean dollar = 100 cents
Literacy rate 98%
Calorie consumption 2492 kilocalories

ST LUCIA
West Indies

Page 33 G4

One of the most beautiful Caribbean Windward Islands, Saint Lucia retains both French and British influences from its colonial history. Tourism and fruit production dominate the economy.

Official name Saint Lucia
Formation 1979 / 1979
Capital Castries
Population 178,000 / 745 people per sq mile (287 people per sq km)
Total area 239 sq miles (620 sq km)
Languages English*, French Creole
Religions Roman Catholic 68%, Seventh-day Adventist 9%, Other Christian 9%, Pentecostal 6%, Nonreligious 5%, Rastafarian 2%, Other 1%
Ethnic mix Black 84%, Mulatto (mixed race) 12%, Asian 3%, Other 1%
Government Parliamentary system
Currency East Caribbean dollar = 100 cents
Literacy rate 95%
Calorie consumption 2595 kilocalories

ST VINCENT & THE GRENADINES
West Indies

Page 33 G4

Formerly ruled by Britain, these volcanic islands form part of the Caribbean Windward Islands. The economy relies on tourism and bananas, and it is the world's largest arrowroot producer.

Official name Saint Vincent and the Grenadines
Formation 1979 / 1979
Capital Kingstown
Population 110,000 / 733 people per sq mile (282 people per sq km)
Total area 150 sq miles (389 sq km)
Languages English*, English Creole
Religions Other Christian 48%, Anglican 18%, Pentecostal 18%, Nonreligious 9%, Other 7%
Ethnic mix Black 73%, Mulatto (mixed race) 20%, Carib 4%, Asian 2%, Other 1%
Government Parliamentary system
Currency East Caribbean dollar = 100 cents
Literacy rate 88%
Calorie consumption 2968 kilocalories

SAMOA
Australasia & Oceania

Page 123 F4

Samoa, in the southern Pacific, was ruled by New Zealand before 1962. Four of the nine islands are inhabited. The traditional Samoan way of life is communal and conservative.

Official name Independent State of Samoa
Formation 1962 / 1962
Capital Apia
Population 200,000 / 183 people per sq mile (71 people per sq km)
Total area 1104 sq miles (2860 sq km)
Languages Samoan*, English*
Religions Other Christian 78%, Roman Catholic 20%, Other 2%
Ethnic mix Polynesian 91%, Euronesian 7%, Other 2%
Government Parliamentary system
Currency Tala = 100 sene
Literacy rate 99%
Calorie consumption 2960 kilocalories

SAN MARINO
Southern Europe

Page 74 C3

Perched on the slopes of Monte Titano in the Italian Appennino, San Marino has been a city-state since the 4th century AD, and was recognized as independent by the pope in 1631.

Official name Republic of San Marino
Formation 1631 / 1631
Capital San Marino
Population 34,000 / 1417 people per sq mile (557 people per sq km)
Total area 23.6 sq miles (61 sq km)
Languages Italian*
Religions Roman Catholic 93%, Other and nonreligious 7%
Ethnic mix Sammarinese 88%, Italian 10%, Other 2%
Government Parliamentary system
Currency Euro = 100 cents
Literacy rate 99%
Calorie consumption Not available

SAO TOME & PRINCIPE
West Africa

Page 55 A5

This ex-Portuguese colony off Africa's west coast has two main islands and smaller islets. Multiparty democracy, adopted in 1990, ended 15 years of Marxism. Cocoa is the main export.

Official name Democratic Republic of São Tomé and Principe
Formation 1975 / 1975
Capital São Tomé
Population 200,000 / 539 people per sq mile (208 people per sq km)
Total area 386 sq miles (1001 sq km)
Languages Portuguese Creole, Portuguese*
Religions Roman Catholic 56%, Nonreligious 21%, Other Christian 15%, Other 8%
Ethnic mix Black 90%, Portuguese and Creole 10%
Government Presidential / Parliamentary system
Currency Dobra = 100 cêntimos
Literacy rate 90%
Calorie consumption 2400 kilocalories

SAUDI ARABIA
Southwest Asia

Page 99 B5

The desert kingdom of Saudi Arabia, rich in oil and gas, covers an area the size of Western Europe. It includes Islam's holiest cities, Medina and Mecca. Women's rights are restricted.

Official name Kingdom of Saudi Arabia
Formation 1932 / 1932
Capital Riyadh
Population 32.9 million / 40 people per sq mile (16 people per sq km)
Total area 756,981 sq miles (1,960,582 sq km)
Languages Arabic*
Religions Sunni Muslim 85%, Shi'a Muslim 15%
Ethnic mix Arab 72%, Foreign residents (mostly south and southeast Asian) 20%, Afro-Asian 8%
Government Monarchy
Currency Saudi riyal = 100 halalat
Literacy rate 94%
Calorie consumption 3255 kilocalories

SENEGAL
West Africa

Page 52 B3

This ex-French colony was ruled by one party for 40 years after independence, despite the adoption of multipartyism in 1981. Its capital, Dakar, stands on the westernmost cape of Africa.

Official name Republic of Senegal
Formation 1960 / 1960
Capital Dakar
Population 15.9 million / 214 people per sq mile (83 people per sq km)
Total area 75,749 sq miles (196,190 sq km)
Languages Wolof, Pulaar, Serer, Diola, Mandinka, Malinké, Soninké, French*
Religions Sunni Muslim 95%, Christian (mainly Roman Catholic) 4%, Traditional beliefs 1%
Ethnic mix Wolof 43%, Serer 15%, Peul 14%, Other 14%, Toucouleur 9%, Diola 5%
Government Presidential system
Currency CFA franc = 100 centimes
Literacy rate 43%
Calorie consumption 2456 kilocalories

SERBIA
Southeast Europe

Page 78 D4

The former Yugoslavia began breaking up in 1991, and Serbia has found itself the sole successor republic. It refuses to acknowledge the 2008 secession of Albanian-dominated Kosovo.

Official name Republic of Serbia
Formation 2006 / 2008
Capital Belgrade
Population 8.8 million / 294 people per sq mile (114 people per sq km)
Total area 29,905 sq miles (77,453 sq km)
Languages Serbian*, Hungarian (Magyar)
Religions Orthodox Christian 88%, Roman Catholic 4%, Nonreligious 4%, Muslim 2%, Other 2%
Ethnic mix Serb 87%, Magyar 4%, Other 3%, Roma 2%, Bosniak 2%, Croat 1%, Slovak 1%
Government Parliamentary system
Currency Serbian dinar = 100 para
Literacy rate 99%
Calorie consumption 2728 kilocalories

SEYCHELLES
Indian Ocean

Page 57 G1

This ex-British colony spans 115 islands in the Indian Ocean. Multiparty polls in 1993 ended 14 years of one-party rule. Unique flora includes the world's largest seed, the coco-de-mer.

Official name Republic of Seychelles
Formation 1976 / 1976
Capital Victoria
Population 100,000 / 962 people per sq mile (370 people per sq km)
Total area 176 sq miles (455 sq km)
Languages French Creole*, English*, French*
Religions Roman Catholic 84%, Anglican 6%, Other Christian 5%, Hindu 2%, Other and nonreligious 2%, Muslim 1%
Ethnic mix Creole 89%, Indian 5%, Other 4%, Chinese 2%
Government Presidential system
Currency Seychelles rupee = 100 cents
Literacy rate 92%
Calorie consumption 2426 kilocalories

SIERRA LEONE
West Africa

Page 52 C4

Founded in 1787 as a British colony for freed slaves, Sierra Leone gained independence in 1961. Recovery from civil war in the 1990s was set back by West Africa's recent Ebola epidemic.

Official name Republic of Sierra Leone
Formation 1961 / 1961
Capital Freetown
Population 7.6 million / 275 people per sq mile (106 people per sq km)
Total area 27,698 sq miles (71,740 sq km)
Languages Mende, Temne, Krio, English*
Religions Muslim 60%, Christian 30%, Traditional beliefs 10%
Ethnic mix Mende 35%, Temne 32%, Other 21%, Limba 8%, Kuranko 4%
Government Presidential system
Currency Leone = 100 cents
Literacy rate 32%
Calorie consumption 2404 kilocalories

SINGAPORE
Southeast Asia

Page 116 A1

A city state linked to the southern tip of the Malay Peninsula by a causeway, Singapore is one of Asia's major commercial centers. Politics has been dominated for decades by one party.

Official name Republic of Singapore
Formation 1965 / 1965
Capital Singapore
Population 5.7 million / 24,153 people per sq mile (9344 people per sq km)
Total area 250 sq miles (648 sq km)
Languages Mandarin*, Malay*, Tamil*, English*
Religions Christian 31%, Buddhist 28%, Nonreligious 14%, Muslim 13%, Taoist 9%, Hindu 4%, Other 1%
Ethnic mix Chinese 74%, Malay 14%, Indian 9%, Other 3%
Government Parliamentary system
Currency Singapore dollar = 100 cents
Literacy rate 97%
Calorie consumption Not available

SLOVAKIA
Central Europe

Page 77 C6

After 900 years of Hungarian control, Slovakia was the less-developed half of communist Czechoslovakia in the 20th century. It became independent in 1993 and joined the EU in 2004.

Official name Slovak Republic
Formation 1993 / 1993
Capital Bratislava
Population 5.4 million / 285 people per sq mile (110 people per sq km)
Total area 18,859 sq miles (48,845 sq km)
Languages Slovak*, Hungarian (Magyar), Czech
Religions Roman Catholic 69%, Nonreligious 15%, Other Christian 11%, Greek Catholic (Uniate) 4%, Other 1%
Ethnic mix Slovak 87%, Magyar 9%, Roma 2%, Other 1%, Czech 1%
Government Parliamentary system
Currency Euro = 100 cents
Literacy rate 99%
Calorie consumption 2944 kilocalories

SLOVENIA
Central Europe

Page 73 D8

The northernmost of the ex-Yugoslav republics was the first to break away, with little violence, in 1991. It always had the closest links with Western Europe, and joined the EU in 2004.

Official name Republic of Slovenia
Formation 1991 / 1991
Capital Ljubljana
Population 2.1 million / 269 people per sq mile (104 people per sq km)
Total area 7820 sq miles (20,253 sq km)
Languages Slovenian*
Religions Roman Catholic 75%, Nonreligious 18%, Muslim 3%, Orthodox Christian 3%, Other (mostly Protestant) 1%
Ethnic mix Slovene 92%, Other 3%, Serb 2%, Croat 2%, Bosniak 1%
Government Parliamentary system
Currency Euro = 100 cents
Literacy rate 99%
Calorie consumption 3168 kilocalories

SOLOMON ISLANDS
Australasia & Oceania

Page 122 C3

This archipelago of around 1000 islands scattered in the southwest Pacific was formerly ruled by Britain. Most people live on six main islands. Ethnic conflict from 1998 led to devolved governance.

Official name Solomon Islands
Formation 1978 / 1978
Capital Honiara
Population 600,000 / 56 people per sq mile (21 people per sq km)
Total area 10,985 sq miles (28,450 sq km)
Languages English*, Pidgin English, Melanesian Pidgin, around 120 native languages
Religions Church of Melanesia (Anglican) 34%, Roman Catholic 19%, South Seas Evangelical Church 17%, Methodist 11%, Other 19%
Ethnic mix Melanesian 93%, Polynesian 4%, Other 3%
Government Parliamentary system
Currency Solomon Islands dollar = 100 cents
Literacy rate 77%
Calorie consumption 2391 kilocalories

SOMALIA
East Africa

Page 51 E5

Italian and British Somaliland were united to create this semiarid state on the Horn of Africa. Anarchy since 1991 has caused mass hunger, a refugee crisis, and ineffective central authority.

Official name Federal Republic of Somalia
Formation 1960 / 1960
Capital Mogadishu
Population 14.7 million / 61 people per sq mile (23 people per sq km)
Total area 246,199 sq miles (637,657 sq km)
Languages Somali*, Arabic*, English, Italian
Religions Sunni Muslim 99%, Christian 1%
Ethnic mix Somali 85%, Other 15%
Government Nonparty system
Currency Somali shilin = 100 senti
Literacy rate 24%
Calorie consumption 1696 kilocalories

SOUTH AFRICA
Southern Africa

Page 56 C4

Mineral-rich South Africa was settled by the Dutch and the British. Multiracial polls in 1994 ended decades of white minority rule and apartheid. AIDS, poverty, and crime are problems.

Official name Republic of South Africa
Formation 1934 / 1994
Capital Pretoria (Tshwane); Cape Town; Bloemfontein
Population 56.7 million / 120 people per sq mile (46 people per sq km)
Total area 471,008 sq miles (1,219,912 sq km)
Languages English*, isiZulu*, isiXhosa*, Afrikaans*, Sepedi*, Setswana*, Sesotho*, Xitsonga*, siSwati*, Tshivenda*, isiNdebele*
Religions Christian 81%, Nonreligious 15%, Muslim 2%, Hindu 1%, Other 1%
Ethnic mix Black 80%, White 9%, Colored 9%, Asian 2%
Government Presidential system
Currency Rand = 100 cents
Literacy rate 94%
Calorie consumption 3022 kilocalories

SOUTH KOREA
East Asia

Page 106 E4

Allied with the US, the southern half of the Korean peninsula was separated from the communist North in 1948. It is the world's leading shipbuilder and a major force in high-tech industries.

Official name Republic of Korea
Formation 1948 / 1953
Capital Seoul; Sejong City (administrative)
Population 51 million / 1338 people per sq mile (517 people per sq km)
Total area 38,023 sq miles (98,480 sq km)
Languages Korean*
Religions Nonreligious 47%, Mahayana Buddhist 23%, Other Christian 18%, Roman Catholic 11%, Other 1%
Ethnic mix Korean 100%
Government Presidential system
Currency South Korean won = 100 chon
Literacy rate 99%
Calorie consumption 3334 kilocalories

SOUTH SUDAN
East Africa

Page 51 B5

Landlocked and little developed, this mostly Christian region seceded from the mainly Muslim north of Sudan in 2011 after years of civil war. Oil production is the economic mainstay.

Official name Republic of South Sudan
Formation 2011 / 2011
Capital Juba
Population 12.6 million / 51 people per sq mile (20 people per sq km)
Total area 248,777 sq miles (644,329 sq km)
Languages Arabic, Dinka, Nuer, Zande, Bari, Shilluk, Lotuko, English*
Religions Over 50% Christian/traditional beliefs
Ethnic mix Dinka 40%, Nuer 15%, Shilluk/Anwak 10%, Azande 10%, Arab 10%, Bari 10%, Other 5%
Government Transitional regime
Currency South Sudan Pound = 100 piastres
Literacy rate 37%
Calorie consumption Not available

SPAIN
Southwest Europe

Page 70 D2

At the gateway to the Mediterranean, Spain became a world power once united in 1492. A vigorous regionalism now exists, with separatist movements in the Basque Country and Catalonia.

Official name Kingdom of Spain
Formation 1492 / 1713
Capital Madrid
Population 46.4 million / 241 people per sq mile (93 people per sq km)
Total area 194,896 sq miles (504,782 sq km)
Languages Spanish*, Catalan*, Galician*, Basque*
Religions Roman Catholic 71%, Nonreligious 26%, Other 3%
Ethnic mix Castilian Spanish 72%, Catalan 17%, Galician 6%, Basque 2%, Other 2%, Roma 1%
Government Parliamentary system
Currency Euro = 100 cents
Literacy rate 98%
Calorie consumption 3174 kilocalories

SRI LANKA
South Asia

Page 110 D3

A former British colony, the island republic of Sri Lanka is separated from India by the narrow Palk Strait. A brutal 26-year civil war between the Sinhalese and Tamils ended in 2009.

Official name Democratic Socialist Republic of Sri Lanka
Formation 1948 / 1948
Capital Colombo; Sri Jayewardenapura Kotte
Population 20.9 million / 836 people per sq mile (323 people per sq km)
Total area 25,332 sq miles (65,610 sq km)
Languages Sinhala*, Tamil*, Sinhala-Tamil, English
Religions Buddhist 70%, Hindu 13%, Muslim 10%, Christian (mainly Roman Catholic) 7%
Ethnic mix Sinhalese 75%, Tamil 15%, Moor 9%, Other 1%
Government Presidential / parliamentary system
Currency Sri Lanka rupee = 100 cents
Literacy rate 91%
Calorie consumption 2539 kilocalories

SUDAN
East Africa

Page 50 B4

On the west coast of the Red Sea, Sudan has been ruled by a military Islamic regime since a coup in 1989. In 2011, it lost its southern third (and most of its oil reserves) after years of civil war.

Official name Republic of the Sudan
Formation 1956 / 2011
Capital Khartoum
Population 40.5 million / 56 people per sq mile (22 people per sq km)
Total area 718,722 sq miles (1,861,481 sq km)
Languages Arabic*, Nubian, Beja, Fur
Religions Almost 100% Muslim (mainly Sunni)
Ethnic mix Arab 60%, Other 18%, Nubian 10%, Beja 8%, Fur 3%, Zaghawa 1%
Government Presidential system
Currency New Sudanese pound = 100 piastres
Literacy rate 73%
Calorie consumption 2336 kilocalories

SURINAME
South America

Page 37 G3

This former Dutch colony on the north coast of South America has some of the world's richest bauxite reserves. The military ruler in the 1980s was elected president from 2010.

Official name Republic of Suriname
Formation 1975 / 1975
Capital Paramaribo
Population 600,000 / 10 people per sq mile (4 people per sq km)
Total area 63,039 sq miles (163,270 sq km)
Languages Sranan (creole), Dutch*, Javanese, Sarnami Hindi, Saramaccan, Chinese, Carib
Religions Christian 50%, Hindu 23%, Muslim 14%, Other 13%
Ethnic mix East Indian 27%, Creole 18%, Black 15%, Javanese 15%, Mixed race 13%, Other 12%
Government Presidential / parliamentary system
Currency Surinamese dollar = 100 cents
Literacy rate 93%
Calorie consumption 2753 kilocalories

SWAZILAND
Southern Africa

Page 56 D4

This tiny kingdom, ruled by Britain until 1968, depends economically on its neighbor South Africa. Its absolute monarch has banned political parties. It has the world's highest rate of HIV.

Official name Kingdom of Swaziland
Formation 1968 / 1968
Capital Mbabane; Lobamba
Population 1.4 million / 211 people per sq mile (81 people per sq km)
Total area 6704 sq miles (17,363 sq km)
Languages English*, siSwati*, isiZulu, Xitsonga
Religions Traditional beliefs 40%, Other 30%, Roman Catholic 20%, Muslim 10%
Ethnic mix Swazi 97%, Other 3%
Government Monarchy
Currency Lilangeni = 100 cents
Literacy rate 83%
Calorie consumption 2329 kilocalories

SWEDEN
Northern Europe

Page 62 B4

Densely forested Sweden is the largest and most populous Scandinavian country and stretches into the Arctic Circle. Its strong industrial base helps to fund its extensive welfare system.

Official name Kingdom of Sweden
Formation 1523 / 1921
Capital Stockholm
Population 9.9 million / 62 people per sq mile (24 people per sq km)
Total area 173,731 sq miles (449,964 sq km)
Languages Swedish*, Finnish, Sámi
Religions Evangelical Lutheran 75%, Other 13%, Muslim 5%, Other Protestant 5%, Roman Catholic 2%
Ethnic mix Swedish 86%, Foreign-born or first-generation immigrant 12%, Finnish & Sámi 2%
Government Parliamentary system
Currency Swedish krona = 100 öre
Literacy rate 99%
Calorie consumption 3179 kilocalories

SWITZERLAND
Central Europe

Page 73 A7

One of the world's richest countries, with a long tradition of neutrality, this mountainous nation lies at the center of Europe geographically, but outside it politically, having not joined the EU.

Official name Swiss Confederation
Formation 1291 / 1857
Capital Bern
Population 8.5 million / 554 people per sq mile (214 people per sq km)
Total area 15,942 sq miles (41,290 sq km)
Languages German*, Swiss-German, French*, Italian*, Romansch*
Religions Roman Catholic 39%, Other Christian 34%, Nonreligious 21%, Muslim 5%, Other 1%
Ethnic mix German 64%, French 20%, Other 9.5%, Italian 6%, Romansch 0.5%
Government Parliamentary system
Currency Swiss franc = 100 rappen/centimes
Literacy rate 99%
Calorie consumption 3391 kilocalories

SYRIA
Southwest Asia

Page 96 B3

Syria's borders were drawn in 1941 at the end of French rule. Suppression of pro-democracy protests in 2011 erupted into civil war; Islamists controlled the Euphrates Valley in 2014-2017.

Official name Syrian Arab Republic
Formation 1941 / 1967
Capital Damascus
Population 18.3 million / 258 people per sq mile (99 people per sq km)
Total area 71,498 sq miles (184,180 sq km)
Languages Arabic*, French, Kurdish, Armenian, Circassian, Turkic languages, Assyrian, Aramaic
Religions Sunni Muslim 74%, Alawi 12%, Christian 10%, Druze 3%, Other 1%
Ethnic mix Arab 90%, Kurdish 9%, Armenian, Turkmen, and Circassian 1%
Government Presidential system
Currency Syrian pound = 100 piastres
Literacy rate 85%
Calorie consumption 3106 kilocalories

TAIWAN
East Asia

Page 107 D6

China's nationalist government fled to Taiwan in 1949 when ousted by the communists. China regards the island, 80 miles (130 km) southeast of the mainland, as a renegade province.

Official name Republic of China (ROC)
Formation 1949 / 1949
Capital Taipei
Population 23.5 million / 1887 people per sq mile (728 people per sq km)
Total area 13,892 sq miles (35,980 sq km)
Languages Amoy Chinese, Mandarin Chinese*, Hakka Chinese
Religions Buddhist, Confucianist, and Taoist 93%, Christian 5%, Other 2%
Ethnic mix Han (pre-20th-century migration) 84%, Han (20th-century migration) 14%, Aboriginal 2%
Government Presidential system
Currency Taiwan dollar = 100 cents
Literacy rate 98%
Calorie consumption 2997 kilocalories

TAJIKISTAN
Central Asia

Page 101 F3

This resource-poor ex-Soviet republic lies landlocked on the western slopes of the Pamirs. Tajiks are of Persian (Iranian) origin rather than Turkic like their Central Asian neighbors.

Official name Republic of Tajikistan
Formation 1991 / 1991
Capital Dushanbe
Population 8.9 million / 161 people per sq mile (62 people per sq km)
Total area 55,251 sq miles (143,100 sq km)
Languages Tajik*, Uzbek, Russian
Religions Sunni Muslim 95%, Shi'a Muslim 3%, Other 2%
Ethnic mix Tajik 84%, Uzbek 12%, Other 2%, Kyrgyz 1%, Russian 1%
Government Presidential system
Currency Somoni = 100 diram
Literacy rate 99%
Calorie consumption 2201 kilocalories

TANZANIA
East Africa

Page 51 B7

This East African state was formed in 1964 by the union of Tanganyika and Zanzibar. A third of its area is game reserve or national park, including Africa's highest peak, Mt. Kilimanjaro.

Official name United Republic of Tanzania
Formation 1964 / 1964
Capital Dodoma
Population 57.3 million / 167 people per sq mile (65 people per sq km)
Total area 364,898 sq miles (945,087 sq km)
Languages Kiswahili*, Sukuma, Chagga, Nyamwezi, Hehe, Makonde, Yao, Sandawe, English*
Religions Christian 63%, Muslim 35%, Other 2%
Ethnic mix Native African (over 120 tribes) 99%, European, Asian, and Arab 1%
Government Presidential system
Currency Tanzanian shilling = 100 cents
Literacy rate 78%
Calorie consumption 2208 kilocalories

THAILAND
Southeast Asia

Page 115 C5

Thailand lies at the heart of the Indochinese Peninsula. Formerly Siam, it has been an independent kingdom for most of its history. The military has frequently intervened in politics.

Official name Kingdom of Thailand
Formation 1238 / 1907
Capital Bangkok
Population 69 million / 350 people per sq mile (135 people per sq km)
Total area 198,455 sq miles (514,000 sq km)
Languages Thai*, Chinese, Malay, Khmer, Mon, Karen, Miao
Religions Buddhist 94%, Muslim 5%, Other (including Christian) 1%
Ethnic mix Thai 83%, Chinese 12%, Malay 3%, Khmer and Other 2%
Government Transitional regime
Currency Baht = 100 satang
Literacy rate 93%
Calorie consumption 2784 kilocalories

TOGO
West Africa

Page 53 F4

Togo lies sandwiched between Ghana and Benin in West Africa. Its long-term military leader, and then his son and successor, have won every election held there since 1993.

Official name Togolese Republic
Formation 1960 / 1960
Capital Lomé
Population 7.8 million / 371 people per sq mile (143 people per sq km)
Total area 21,924 sq miles (56,785 sq km)
Languages Ewe, Kabye, Gurma, French*
Religions Christian 47%, Traditional beliefs 33%, Muslim 14%, Other 6%
Ethnic mix Ewe 46%, Other African 41%, Kabye 12%, European 1%
Government Presidential system
Currency CFA franc = 100 centimes
Literacy rate 64%
Calorie consumption 2454 kilocalories

TONGA
Australasia & Oceania

Page 123 E4

Northeast of New Zealand, Tonga is a 170-island archipelago, 45 of which are inhabited. Politics is effectively controlled by the king, though limited democratic reforms are taking place.

Official name Kingdom of Tonga
Formation 1970 / 1970
Capital Nuku'alofa
Population 100,000/ 360 people per sq mile (139 people per sq km)
Total area 289 sq miles (748 sq km)
Languages English*, Tongan*
Religions Free Wesleyan 38%, Church of Jesus Christ of Latter-day Saints 17%, Roman Catholic 16%, Other Christian 16%, Free Church of Tonga 12%, Other 1%
Ethnic mix Tongan 98%, Other 2%
Government Monarchy
Currency Pa'anga (Tongan dollar) = 100 seniti
Literacy rate 99%
Calorie consumption Not available

TRINIDAD AND TOBAGO
West Indies

Page 33 H5

This former British colony is the most southerly of the Windward Islands, just 9 miles (15 km) off the coast of Venezuela. Politics is mainly polarized by race. Oil and gas are exported.

Official name Republic of Trinidad and Tobago
Formation 1962 / 1962
Capital Port of Spain
Population 1.4 million / 707 people per sq mile (273 people per sq km)
Total area 1980 sq miles (5128 sq km)
Languages English Creole, English*, Hindi, French, Spanish
Religions Protestant 38%, Roman Catholic 24%, Hindu 20%, Other 12%, Muslim 6%
Ethnic mix East Indian 38%, Black 36%, Mixed race 24%, White and Chinese 1%, Other 1%
Government Parliamentary system
Currency Trinidad and Tobago dollar = 100 cents
Literacy rate 99%
Calorie consumption 3052 kilocalories

TUNISIA
North Africa

Page 49 E2

North Africa's smallest country, one of the more liberal yet stable Arab states, had only two post-independence rulers until the "Arab Spring" of 2011. Moderate Islamists then won power.

Official name Republic of Tunisia
Formation 1956 / 1956
Capital Tunis
Population 11.5 million / 192 people per sq mile (74 people per sq km)
Total area 63,169 sq miles (163,610 sq km)
Languages Arabic*, French
Religions Muslim (mainly Sunni) 98%, Christian 1%, Jewish 1%
Ethnic mix Arab and Berber 98%, European 1%, Jewish 1%
Government Presidential / parliamentary system
Currency Tunisian dinar = 1000 millimes
Literacy rate 79%
Calorie consumption 3349 kilocalories

TURKEY
Asia / Europe

Page 94 B3

With land in Europe and Asia, Turkey guards the entrance to the Black Sea. The secular/Islamic divide is key to its politics. It is the only Muslim member of NATO, and hopes to join the EU.

Official name Republic of Turkey
Formation 1923 / 1939
Capital Ankara
Population 80.7 million / 272 people per sq mile (105 people per sq km)
Total area 301,382 sq miles (780,580 sq km)
Languages Turkish*, Kurdish, Arabic, Circassian, Armenian, Greek, Georgian, Ladino
Religions Muslim (mainly Sunni) 99%, Other 1%
Ethnic mix Turkish 70%, Kurdish 20%, Other 8%, Arab 2%
Government Parliamentary system
Currency Turkish lira = 100 kurus
Literacy rate 96%
Calorie consumption 3706 kilocalories

TURKMENISTAN
Central Asia

Page 100 B2

Stretching from the Caspian Sea into Central Asia's deserts, this ex-Soviet state exploits vast gas reserves. The pre-independence president built a personality cult and ruled until 2007.

Official name Turkmenistan
Formation 1991 / 1991
Capital Ashgabat
Population 5.8 million / 31 people per sq mile (12 people per sq km)
Total area 188,455 sq miles (488,100 sq km)
Languages Turkmen*, Uzbek, Russian, Kazakh, Tatar
Religions Sunni Muslim 89%, Orthodox Christian 9%, Other 2%
Ethnic mix Turkmen 85%, Other 6%, Uzbek 5%, Russian 4%
Government Presidential system
Currency New manat = 100 tenge
Literacy rate 99%
Calorie consumption 2840 kilocalories

TUVALU
Australasia & Oceania

Page 123 E3

Known as the Ellice Islands under British rule, Tuvalu is a chain of nine atolls in the Central Pacific. It has the world's smallest GNI, but made substantial earnings leasing its ".tv" internet suffix.

Official name Tuvalu
Formation 1978 / 1978
Capital Funafuti Atoll
Population 11,000 / 1100 people per sq mile (423 people per sq km)
Total area 10 sq miles (26 sq km)
Languages Tuvaluan, Kiribati, English*
Religions Church of Tuvalu 91%, Other (mostly Protestant) 5%, Seventh-day Adventist 2%, Baha'i 2%
Ethnic mix Polynesian 96%, Micronesian 4%
Government Nonparty system
Currency Australian dollar = 100 cents; Tuvaluan dollar = 100 cents
Literacy rate 95%
Calorie consumption Not available

UGANDA
East Africa

Page 51 B6

Landlocked Uganda faced ethnic strife under 1970s' dictator Idi Amin. From 1986, reconciliation was aided by two decades of "no-party" democracy, but insurgency continued in the north.

Official name Republic of Uganda
Formation 1962 / 1962
Capital Kampala
Population 42.9 million / 557 people per sq mile (215 people per sq km)
Total area 91,135 sq miles (236,040 sq km)
Languages Luganda, Nkole, Chiga, Lango, Acholi, Teso, Lugbara, English*
Religions Roman Catholic 42%, Protestant 42%, Muslim (mainly Sunni) 12%, Other 4%
Ethnic mix Other 50%, Baganda 17%, Banyakole 10%, Basoga 9%, Bakiga 7%, Iteso 7%
Government Presidential system
Currency Uganda shilling = 100 cents
Literacy rate 70%
Calorie consumption 2130 kilocalories

UKRAINE
Eastern Europe

Page 86 C2

Bordered by seven states, fertile Ukraine was the "breadbasket" of the USSR. Its political divide between pro-Russian sentiment and assertive nationalism exploded into civil war in 2014.

Official name Ukraine
Formation 1991 / 1991
Capital Kiev
Population 44.2 million / 190 people per sq mile (73 people per sq km)
Total area 223,089 sq miles (603,700 sq km)
Languages Ukrainian*, Russian, Tatar
Religions Orthodox Christian 78%, Roman Catholic 10%, Nonreligious 7%, Other 5%
Ethnic mix Ukrainian 78%, Russian 17%, Other 4%, Belarussian 1%
Government Presidential / Parliamentary system
Currency Hryvna = 100 kopiykas
Literacy rate 99%
Calorie consumption 3138 kilocalories

UNITED ARAB EMIRATES
Southwest Asia

Page 99 D5

Bordering the Persian Gulf on the north of the Arabian Peninsula, the United Arab Emirates is a federation of seven states. Wealth once relied on pearls, but oil and gas are now exported.

Official name United Arab Emirates
Formation 1971 / 1972
Capital Abu Dhabi
Population 9.4 million / 291 people per sq mile (112 people per sq km)
Total area 32,000 sq miles (82,880 sq km)
Languages Arabic*, Farsi, Indian and Pakistani languages, English
Religions Muslim (mainly Sunni) 96%, Christian, Hindu, and other 4%
Ethnic mix Asian 60%, Emirian 25%, Other Arab 12%, European 3%
Government Monarchy
Currency UAE dirham = 100 fils
Literacy rate 90%
Calorie consumption 3280 kilocalories

UNITED KINGDOM
Northwest Europe

Page 67 C5

Lying across the English Channel from France, the UK comprises England, Wales, Scotland, and Northern Ireland. Its prominent role in world affairs is a legacy of its once-vast empire.

Official name United Kingdom of Great Britain and Northern Ireland
Formation 1707 / 1922
Capital London
Population 66.2 million / 710 people per sq mile (274 people per sq km)
Total area 94,525 sq miles (244,820 sq km)
Languages English*, Welsh (* in Wales), Gaelic, Irish
Religions Christian 64%, Nonreligious 28%, Muslim 5%, Other 2%, Hindu 1%
Ethnic mix White 87%, Indian and Pakistani 4%, Other 3%, Black 3%, Other Asian 2%, Bengali 1%
Government Parliamentary system
Currency Pound sterling = 100 pence
Literacy rate 99%
Calorie consumption 3424 kilocalories

UNITED STATES
North America

Page 13 B5

Stretching across the most temperate part of North America, and with many natural resources, the USA is the sole truly global superpower and has the world's largest economy.

Official name United States of America
Formation 1776 / 1959
Capital Washington D.C.
Population 324 million / 92 people per sq mile (35 people per sq km)
Total area 3,717,792 sq miles (9,626,091 sq km)
Languages English*, Spanish, Chinese, French, Polish, German, Tagalog, Vietnamese, Italian, Korean, Russian
Religions Protestant 47%, Nonreligious 23%, Roman Catholic 21%, Other 6%, Jewish 2%, Muslim 1%
Ethnic mix White 60%, Hispanic 17%, Black American/African 14%, Asian 6%, Other 3%
Government Presidential system
Currency US dollar = 100 cents
Literacy rate 99%
Calorie consumption 3682 kilocalories

URUGUAY
South America

Page 42 D4

Uruguay, in southeastern South America, has much rich low-lying pasture land and is a major wool exporter. Military rule from 1973 to 1985 has given way to democracy.

Official name Oriental Republic of Uruguay
Formation 1828 / 1828
Capital Montevideo
Population 3.5 million / 52 people per sq mile (20 people per sq km)
Total area 68,039 sq miles (176,220 sq km)
Languages Spanish*
Religions Roman Catholic 42%, Nonreligious 37%, Protestant 15%, Other 6%
Ethnic mix White 87%, Black 7%, Mestizo 5%, Other 1%
Government Presidential system
Currency Uruguayan peso = 100 centésimos
Literacy rate 98%
Calorie consumption 3050 kilocalories

UZBEKISTAN
Central Asia

Page 100 D2

The most populous of the Central Asian republics lies on the ancient Silk Road. Today, its main exports are cotton, oil, gas, and gold. Its pre-independence ruler held power until he died in 2016.

Official name Republic of Uzbekistan
Formation 1991 / 1991
Capital Tashkent
Population 31.9 million / 185 people per sq mile (71 people per sq km)
Total area 172,741 sq miles (447,400 sq km)
Languages Uzbek*, Russian, Tajik, Kazakh
Religions Sunni Muslim 88%, Orthodox Christian 9%, Other 3%
Ethnic mix Uzbek 80%, Other 6%, Russian 6%, Tajik 5%, Kazakh 3%
Government Presidential system
Currency Som = 100 tiyin
Literacy rate 99%
Calorie consumption 2760 kilocalories

VANUATU
Australasia & Oceania

Page 122 D4

This South Pacific archipelago of 82 islands and islets boasts the world's highest per capita density of languages. Until independence, it was under joint Anglo-French rule.

Official name Republic of Vanuatu
Formation 1980 / 1980
Capital Port Vila
Population 300,000 / 64 people per sq mile (25 people per sq km)
Total area 4710 sq miles (12,200 sq km)
Languages Bislama* (Melanesian pidgin), English*, French*, other indigenous languages
Religions Other 33%, Presbyterian 28%, Anglican 15%, Seventh-day Adventist 12%, Roman Catholic 12%
Ethnic mix ni-Vanuatu 99%, Other 1%
Government Parliamentary system
Currency Vatu = 100 centimes
Literacy rate 83%
Calorie consumption 2836 kilocalories

VATICAN CITY
Southern Europe

Page 75 A8

The Vatican City, seat of the Roman Catholic Church, is a walled enclave in Rome. It is the world's smallest country. Its head, the pope, is elected for life by a college of cardinals.

Official name Vatican City
Formation 1929 / 1929
Capital Vatican City
Population 1000 / 5882 people per sq mile (2273 people per sq km)
Total area 0.17 sq miles (0.44 sq km)
Languages Italian*, Latin*
Religions Roman Catholic 100%
Ethnic mix Most resident lay persons are Italian
Government Papal state
Currency Euro = 100 cents
Literacy rate 99%
Calorie consumption Not available

VENEZUELA
South America

Page 36 D2

Located on the Caribbean coast of South America, Venezuela has the continent's most urbanized society, and some of the largest known oil deposits outside the Middle East.

Official name Bolivarian Republic of Venezuela
Formation 1830 / 1830
Capital Caracas
Population 32 million / 94 people per sq mile (36 people per sq km)
Total area 352,143 sq miles (912,050 sq km)
Languages Spanish*, Amerindian languages
Religions Roman Catholic 73%, Protestant 17%, Nonreligious 7%, Other 3%
Ethnic mix Mestizo 69%, White 20%, Black 9%, Amerindian 2%
Government Presidential system
Currency Bolívar fuerte = 100 céntimos
Literacy rate 97%
Calorie consumption 2631 kilocalories

VIETNAM
Southeast Asia

Page 114 D4

The eastern strip of the Indochinese Peninsula, Vietnam was partitioned in 1954, and only reunited after the communist north's victory in the devastating 1962–75 Vietnam War.

Official name Socialist Republic of Vietnam
Formation 1976 / 1976
Capital Hanoi
Population 95.5 million / 760 people per sq mile (294 people per sq km)
Total area 127,243 sq miles (329,560 sq km)
Languages Vietnamese*, Chinese, Thai, Khmer, Muong, Nung, Miao, Yao, Jarai
Religions Nonreligious 81%, Buddhist 9%, Roman Catholic 7%, Hoa Hao 1%, Cao Dai 1%, Other 1%
Ethnic mix Vietnamese 86%, Other 8%, Tay 2%, Thai 2%, Muong 1%, Khome 1%
Government One-party state
Currency Dông = 10 hao = 100 xu
Literacy rate 94%
Calorie consumption 2745 kilocalories

YEMEN
Southwest Asia

Page 99 C7

The Arab world's only Marxist regime and a military-run republic united in 1990 to form Yemen, stretching across southern Arabia. Tribal insurgency and militant Islamism have caused conflict.

Official name Republic of Yemen
Formation 1990 / 1990
Capital Saana
Population 28.3 million / 130 people per sq mile (50 people per sq km)
Total area 203,849 sq miles (527,970 sq km)
Languages Arabic*
Religions Sunni Muslim 55%, Shi'a Muslim 42%, Christian, Hindu, and Jewish 3%
Ethnic mix Arab 99%, Afro-Arab, Indian, Somali, and European 1%
Government Transitional regime
Currency Yemeni rial = 100 fils
Literacy rate 66%
Calorie consumption 2223 kilocalories

ZAMBIA
Southern Africa

Page 56 C2

Landlocked in southern Africa, copper-rich Zambia (once known as Northern Rhodesia) has seen its politics dogged by corruption both before and after the end of single-party rule in 1991.

Official name Republic of Zambia
Formation 1964 / 1964
Capital Lusaka
Population 17.1 million / 60 people per sq mile (23 people per sq km)
Total area 290,584 sq miles (752,614 sq km)
Languages Bemba, Tonga, Nyanja, Lozi, Lala-Bisa, Nsenga, English*
Religions Protestant 75%, Roman Catholic 20%, Other (including Muslim) 3%, Nonreligious 2%
Ethnic mix Bemba 34%, Other African 26%, Tonga 16%, Nyanja 14%, Lozi 9%, European 1%
Government Presidential system
Currency New Zambian kwacha = 100 ngwee
Literacy rate 83%
Calorie consumption 1930 kilocalories

ZIMBABWE
Southern Africa

Page 56 D3

Full independence from Britain in 1980 ended 15 years of troubled white-minority rule. Poor governance, violent land redistribution, and severe drought have destroyed the economy.

Official name Republic of Zimbabwe
Formation 1980 / 1980
Capital Harare
Population 16.5 million / 111 people per sq mile (43 people per sq km)
Total area 150,803 sq miles (390,580 sq km)
Languages Shona, isiNdebele, English*
Religions Syncretic 50%, Christian 25%, Traditional beliefs 24%, Other (including Muslim) 1%
Ethnic mix Shona 71%, Ndebele 16%, Other African 11%, White 1%, Asian 1%,
Government Presidential system
Currency Zimbabwe dollar suspended in 2009; US dollar, SA rand, & 7 other currencies legal tender
Literacy rate 89%
Calorie consumption 2110 kilocalories

Overseas Territories and Dependencies

Despite the rapid process of decolonization since the end of the Second World War, around 10 million people in more than 50 territories around the world continue to live under the protection of a parent state.

AUSTRALIA

ASHMORE & CARTIER ISLANDS
Indian Ocean
Claimed 1931
Capital not applicable
Area 2 sq miles (5 sq km)
Population None

CHRISTMAS ISLAND
Indian Ocean
Claimed 1958
Capital The Settlement
Area 52 sq miles (135 sq km)
Population 2205

COCOS ISLANDS
Indian Ocean
Claimed 1955
Capital West Island
Area 5.5 sq miles (14 sq km)
Population 596

CORAL SEA ISLANDS
Southwest Pacific
Claimed 1969
Capital None
Area Less than 1.2 sq miles (3 sq km)
Population below 10 (scientists)

HEARD & McDONALD ISLANDS
Indian Ocean
Claimed 1947
Capital not applicable
Area 161 sq miles (417 sq km)
Population None

NORFOLK ISLAND
Southwest Pacific
Claimed 1774
Capital Kingston
Area 13.3 sq miles (34 sq km)
Population 1748

DENMARK

FAROE ISLANDS
North Atlantic
Claimed 1380
Capital Tórshavn
Area 540 sq miles (1399 sq km)
Population 49,120

GREENLAND
North Atlantic
Claimed 1380
Capital Nuuk
Area 840,000 sq miles (2,175,516 sq km)
Population 56,190

FRANCE

CLIPPERTON ISLAND
East Pacific
Claimed 1935
Capital not applicable
Area 2.7 sq miles (7 sq km)
Population None

FRENCH GUIANA
South America
Claimed 1817
Capital Cayenne
Area 35,135 sq miles (90,996 sq km)
Population 276,000

FRENCH POLYNESIA
South Pacific
Claimed 1843
Capital Papeete
Area 1,608 sq miles (4165 sq km)
Population 280,210

GUADELOUPE
West Indies
Claimed 1635
Capital Basse-Terre
Area 629 sq miles (1628 sq km)
Population 400,187

MARTINIQUE
West Indies
Claimed 1635
Capital Fort-de-France
Area 425 sq miles (1100 sq km)
Population 396,000

MAYOTTE
Indian Ocean
Claimed 1843
Capital Mamoudzou
Area 144 sq miles (374 sq km)
Population 212,645

NEW CALEDONIA
Southwest Pacific
Claimed 1853
Capital Nouméa
Area 7,374 sq miles (19,103 sq km)
Population 278,000

RÉUNION
Indian Ocean
Claimed 1638
Capital Saint-Denis
Area 970 sq miles (2512 sq km)
Population 867,000

ST. PIERRE & MIQUELON
North America
Claimed 1604
Capital Saint-Pierre
Area 93 sq miles (242 sq km)
Population 5533

WALLIS & FUTUNA
South Pacific
Claimed 1842
Capital Matá'Utu
Area 106 sq miles (274 sq km)
Population 15,714

NETHERLANDS

ARUBA
West Indies
Claimed 1643
Capital Oranjestad
Area 75 sq miles (194 sq km)
Population 104,810

BONAIRE
West Indies
Claimed 1816
Capital Kralendijk
Area 113 sq miles (294 sq km)
Population 19,400

CURAÇAO
West Indies
Claimed 1815
Capital Willemstad
Area 171 sq miles (444 sq km)
Population 160,000

SABA
West Indies
Claimed 1816
Capital The Bottom
Area 5 sq miles (13 sq km)
Population 1846

SINT-EUSTATIUS
West Indies
Claimed 1784
Capital Oranjestad
Area 8 sq miles (21 sq km)
Population 3200

SINT-MAARTEN
West Indies
Claimed 1648
Capital Phillipsburg
Area 13 sq miles (34 sq km)
Population 40,010

NEW ZEALAND

COOK ISLANDS
South Pacific
Claimed 1901
Capital Avarua
Area 91 sq miles (235 sq km)
Population 11,700

NIUE
South Pacific
Claimed 1901
Capital Alofi
Area 102 sq miles (264 sq km)
Population 1618

TOKELAU
South Pacific
Claimed 1926
Capital not applicable
Area 4 sq miles (10 sq km)
Population 1285

NORWAY

BOUVET ISLAND
South Atlantic
Claimed 1928
Capital not applicable
Area 22 sq miles (58 sq km)
Population None

JAN MAYEN
North Atlantic
Claimed 1929
Capital not applicable
Area 147 sq miles (381 sq km)
Population 18 (scientists)

PETER I ISLAND
Antarctica
Claimed 1931
Capital not applicable
Area 69 sq miles (180 sq km)
Population None

SVALBARD
Arctic Ocean
Claimed 1920
Capital Longyearbyen
Area 24,289 sq miles (62,906 sq km)
Population 2752

UNITED KINGDOM

ANGUILLA
West Indies
Claimed 1650
Capital The Valley
Area 37 sq miles (96 sq km)
Population 17,087

ASCENSION ISLAND
South Atlantic
Claimed 1673
Capital Georgetown
Area 34 sq miles (88 sq km)
Population 806

BERMUDA
North Atlantic
Claimed 1612
Capital Hamilton
Area 20 sq miles (53 sq km)
Population 65,330

BRITISH INDIAN OCEAN TERRITORY
Indian Ocean
Claimed 1814
Capital Diego Garcia
Area 23 sq miles (60 sq km)
Population 4200

BRITISH VIRGIN ISLANDS
West Indies
Claimed 1672
Capital Road Town
Area 59 sq miles (153 sq km)
Population 30,660

CAYMAN ISLANDS
West Indies
Claimed 1670
Capital George Town
Area 100 sq miles (259 sq km)
Population 60,770

FALKLAND ISLANDS
South Atlantic
Claimed 1832
Capital Stanley
Area 4699 sq miles (12,173 sq km)
Population 3198

GIBRALTAR
Southwest Europe
Claimed 1713
Capital Gibraltar
Area 2.5 sq miles (6.5 sq km)
Population 34,410

GUERNSEY
Northwest Europe
Claimed 1066
Capital St Peter Port
Area 25 sq miles (65 sq km)
Population 66,502

ISLE OF MAN
Northwest Europe
Claimed 1765
Capital Douglas
Area 221 sq miles (572 sq km)
Population 83,740

JERSEY
Northwest Europe
Claimed 1066
Capital St. Helier
Area 45 sq miles (116 sq km)
Population 98,840

MONTSERRAT
West Indies
Claimed 1632
Capital Brades *(de facto)*; Plymouth *(de jure)*
Area 40 sq miles (102 sq km)
Population 5215

PITCAIRN GROUP OF ISLANDS
South Pacific
Claimed 1887
Capital Adamstown
Area 18 sq miles (47 sq km)
Population 54

ST. HELENA
South Atlantic
Claimed 1673
Capital Jamestown
Area 47 sq miles (122 sq km)
Population 4800

SOUTH GEORGIA &
   THE SOUTH SANDWICH ISLANDS
South Atlantic
Capital not applicable
Claimed 1775
Area 1387 sq miles (3592 sq km)
Population None

TRISTAN DA CUNHA
South Atlantic
Claimed 1612
Capital Edinburgh
Area 38 sq miles (98 sq km)
Population 293

TURKS & CAICOS ISLANDS
West Indies
Claimed 1766
Capital Cockburn Town
Area 166 sq miles (430 sq km)
Population 34,900

UNITED STATES OF AMERICA

AMERICAN SAMOA
South Pacific
Claimed 1900
Capital Pago Pago
Area 75 sq miles (195 sq km)
Population 55,600

BAKER & HOWLAND ISLANDS
Central Pacific
Claimed 1856
Capital not applicable
Area 0.54 sq miles (1.4 sq km)
Population None

GUAM
West Pacific
Claimed 1898
Capital Hagåtña
Area 212 sq miles (549 sq km)
Population 162,900

JARVIS ISLAND
Central Pacific
Claimed 1856
Capital not applicable
Area 1.7 sq miles (4.5 sq km)
Population None

NORTHERN MARIANA ISLANDS
West Pacific
Claimed 1947
Capital Saipan
Area 177 sq miles (457 sq km)
Population 55,020

PALMYRA ATOLL
Central Pacific
Claimed 1898
Capital not applicable
Area 5 sq miles (12 sq km)
Population None

PUERTO RICO
West Indies
Claimed 1898
Capital San Juan
Area 3515 sq miles (9104 sq km)
Population 3.7 million

VIRGIN ISLANDS
West Indies
Claimed 1917
Capital Charlotte Amalie
Area 137 sq miles (355 sq km)
Population 102,950

WAKE ISLAND
Central Pacific
Claimed 1898
Capital not applicable
Area 2.5 sq miles (6.5 sq km)
Population 150 (US air base)

Geographical comparisons

Largest countries

Russia	6,592,735 sq miles	(17,075,200 sq km)
Canada	3,855,171 sq miles	(9,984,670 sq km)
USA	3,717,792 sq miles	(9,626,091 sq km)
China	3,705,386 sq miles	(9,596,960 sq km)
Brazil	3,286,470 sq miles	(8,511,965 sq km)
Australia	2,967,893 sq miles	(7,686,850 sq km)
India	1,269,338 sq miles	(3,287,590 sq km)
Argentina	1,068,296 sq miles	(2,766,890 sq km)
Kazakhstan	1,049,150 sq miles	(2,717,300 sq km)
Algeria	919,590 sq miles	(2,381,740 sq km)

Smallest countries

Vatican City	0.17 sq miles	(0.44 sq km)
Monaco	0.75 sq miles	(1.95 sq km)
Nauru	8 sq miles	(21 sq km)
Tuvalu	10 sq miles	(26 sq km)
San Marino	24 sq miles	(61 sq km)
Liechtenstein	62 sq miles	(160 sq km)
Marshall Islands	70 sq miles	(181 sq km)
St. Kitts & Nevis	101 sq miles	(261 sq km)
Maldives	116 sq miles	(300 sq km)
Malta	122 sq miles	(316 sq km)

Largest islands

Greenland	840,000 sq miles (2,175,600 sq km)
New Guinea	312,000 sq miles (808,000 sq km)
Borneo	292,222 sq miles (757,050 sq km)
Madagascar	226,656 sq miles (587,040 sq km)
Sumatra	202,300 sq miles (524,000 sq km)
Baffin Island	183,800 sq miles (476,000 sq km)
Honshu	88,800 sq miles (230,000 sq km)
Britain	88,700 sq miles (229,800 sq km)
Victoria Island	81,900 sq miles (212,000 sq km)
Ellesmere Island	75,700 sq miles (196,000 sq km)

Richest countries

(GNI per capita, in US$)

Monaco	186,950
Liechtenstein	115,530
Norway	93,530
Switzerland	84,550
Qatar	83,990
Luxembourg	77,480
Denmark	60,270
Australia	60,050
Sweden	57,900
USA	55,990

Poorest countries

(GNI per capita, in US$)

Burundi	260
Central African Republic	320
Malawi	350
Liberia	380
Niger	390
Congo, Dem. Republic	410
Madagascar	420
The Gambia	470
Guinea	470
North Korea	500

Most populous countries

China	1.41 billion
India	1.34 billion
USA	324 million
Indonesia	264 million
Brazil	209 million
Pakistan	197 million

Most populous countries *continued*

Nigeria	174 million
Bangladesh	157 million
Russia	143 million
Japan	127 million

Least populous countries

Vatican City	1000
Tuvalu	11,000
Nauru	13,000
Palau	21,000
San Marino	34,000
Monaco	38,000
Liechtenstein	38,000
Marshall Islands	53,000
St. Kitts & Nevis	55,000
Dominica	74,000

Most densely populated countries

Monaco	50,667 people per sq mile (19,487 per sq km)
Singapore	24,153 people per sq mile (9344 per sq km)
Vatican City	5882 people per sq mile (2273 per sq km)
Bahrain	5495 people per sq mile (2125 per sq km)
Maldives	3448 people per sq mile (1333 per sq km)
Malta	3226 people per sq mile (1250 per sq km)
Bangladesh	3186 people per sq mile (1230 per sq km)
Taiwan	1887 people per sq mile (728 per sq km)
Mauritius	1811 people per sq mile (699 per sq km)
Barbados	1807 people per sq mile (698 per sq km)

Most sparsely populated countries

Mongolia	5 people per sq mile	(2 per sq km)
Namibia	8 people per sq mile	(3 per sq km)
Australia	8 people per sq mile	(3 per sq km)
Iceland	8 people per sq mile	(3 per sq km)
Libya	9 people per sq mile	(4 per sq km)
Suriname	10 people per sq mile	(4 per sq km)
Canada	10 people per sq mile	(4 per sq km)
Mauritania	11 people per sq mile	(4 per sq km)
Botswana	11 people per sq mile	(4 per sq km)
Guyana	11 people per sq mile	(4 per sq km)

Most widely spoken languages

1. Chinese (Mandarin)	6. Portuguese
2. Spanish	7. Bengali
3. English	8. Russian
4. Hindi	9. Japanese
5. Arabic	10. Javanese

Largest conurbations

Guangzhou (China)	48,600,000
Tokyo (Japan)	39,800,000
Shanghai (China)	31,100,000
Jakarta (Indonesia)	28,900,000
Delhi (India)	27,200,000
Karachi (Pakistan)	25,100,000
Seoul (South Korea)	24,800,000
Manila (Philippines)	24,100,000
Mumbai (India)	23,600,000
Mexico City (Mexico)	22,300,000
New York (USA)	22,200,000
São Paulo (Brazil)	21,900,000
Beijing (China)	20,700,000
Dhaka (Bangladesh)	17,900,000
Osaka (Japan)	17,800,000
Los Angeles (USA)	17,700,000
Lagos (Nigeria)	17,600,000
Bangkok (Thailand)	17,400,000

Largest conurbations *continued*

Cairo (Egypt)	17,100,000
Moscow (Russia)	17,100,000
Kolkata (India)	16,200,000
Buenos Aires (Argentina)	16,000,000
Istanbul (Turkey)	14,600,000
London (UK)	14,500,000
Tehran (Iran)	14,000,000

Longest rivers

Nile (Northeast Africa)	4160 miles	(6695 km)
Amazon (South America)	4049 miles	(6516 km)
Yangtze (China)	3915 miles	(6299 km)
Mississippi/Missouri (USA)	3710 miles	(5969 km)
Ob'-Irtysh (Russia)	3461 miles	(5570 km)
Yellow River (China)	3395 miles	(5464 km)
Congo (Central Africa)	2900 miles	(4667 km)
Mekong (Southeast Asia)	2749 miles	(4425 km)
Lena (Russia)	2734 miles	(4400 km)
Mackenzie (Canada)	2640 miles	(4250 km)
Yenisey (Russia)	2541 miles	(4090 km)

Highest mountains

(Height above sea level)

Everest	29,029 ft	(8848 m)
K2	28,251 ft	(8611 m)
Kanchenjunga I	28,210 ft	(8598 m)
Makalu I	27,767 ft	(8463 m)
Cho Oyu	26,907 ft	(8201 m)
Dhaulagiri I	26,796 ft	(8167 m)
Manaslu I	26,783 ft	(8163 m)
Nanga Parbat I	26,661 ft	(8126 m)
Annapurna I	26,547 ft	(8091 m)
Gasherbrum I	26,471 ft	(8068 m)

Largest bodies of inland water

(Area & depth)

Caspian Sea	143,243 sq miles (371,000 sq km)	3215 ft (980 m)
Lake Superior	32,151 sq miles (83,270 sq km)	1289 ft (393 m)
Lake Victoria	26,560 sq miles (68,880 sq km)	328 ft (100 m)
Lake Huron	23,436 sq miles (60,700 sq km)	751 ft (229 m)
Lake Michigan	22,402 sq miles (58,020 sq km)	922 ft (281 m)
Lake Tanganyika	12,703 sq miles (32,900 sq km)	4700 ft (1435 m)
Great Bear Lake	12,274 sq miles (31,790 sq km)	1047 ft (319 m)
Lake Baikal	11,776 sq miles (30,500 sq km)	5712 ft (1741 m)
Great Slave Lake	10,981 sq miles (28,440 sq km)	459 ft (140 m)
Lake Erie	9915 sq miles (25,680 sq km)	197 ft (60 m)

Deepest ocean features

Challenger Deep, Mariana Trench (Pacific)	36,201 ft (11,034 m)
Vityaz III Depth, Tonga Trench (Pacific)	35,704 ft (10,882 m)
Vityaz Depth, Kurile-Kamchatka Trench (Pacific)	34,588 ft (10,542 m)
Cape Johnson Deep, Philippine Trench (Pacific)	34,441 ft (10,497 m)
Kermadec Trench (Pacific)	32,964 ft (10,047 m)
Ramapo Deep, Japan Trench (Pacific)	32,758 ft (9984 m)
Milwaukee Deep, Puerto Rico Trench (Atlantic)	30,185 ft (9200 m)
Argo Deep, Torres Trench (Pacific)	30,070 ft (9165 m)
Meteor Depth, South Sandwich Trench (Atlantic)	30,000 ft (9144 m)
Planet Deep, New Britain Trench (Pacific)	29,988 ft (9140 m)

Greatest waterfalls

(Mean flow of water)

Boyoma (Congo, Dem. Rep.)	600,400 cu. ft/sec (17,000 cu.m/sec)
Khone (Laos/Cambodia)	410,000 cu. ft/sec (11,600 cu.m/sec)
Niagara (USA/Canada)	195,000 cu. ft/sec (5500 cu.m/sec)
Grande (Uruguay)	160,000 cu. ft/sec (4500 cu.m/sec)
Paulo Afonso (Brazil)	100,000 cu. ft/sec (2800 cu.m/sec)
Urubupunga (Brazil)	97,000 cu. ft/sec (2750 cu.m/sec)
Iguaçu (Argentina/Brazil)	62,000 cu. ft/sec (1700 cu.m/sec)
Maribondo (Brazil)	53,000 cu. ft/sec (1500 cu.m/sec)
Victoria (Zimbabwe)	39,000 cu. ft/sec (1100 cu.m/sec)

Greatest waterfalls *continued*

Kabalega (Uganda)	42,000 cu. ft/sec (1200 cu.m/sec)
Churchill (Canada)	35,000 cu. ft/sec (1000 cu.m/sec)
Cauvery (India)	33,000 cu. ft/sec (900 cu.m/sec)

Highest waterfalls

Angel (Venezuela)	3212 ft	(979 m)
Tugela (South Africa)	3110 ft	(948 m)
Utigard (Norway)	2625 ft	(800 m)
Mongefossen (Norway)	2539 ft	(774 m)
Mtarazi (Zimbabwe)	2500 ft	(762 m)
Yosemite (USA)	2425 ft	(739 m)
Ostre Mardola Foss (Norway)	2156 ft	(657 m)
Tyssestrengane (Norway)	2119 ft	(646 m)
*Cuquenan (Venezuela)	2001 ft	(610 m)
Sutherland (New Zealand)	1903 ft	(580 m)
*Kjellfossen (Norway)	1841 ft	(561 m)

* indicates that the total height is a single leap

Largest deserts

Sahara	3,450,000 sq miles (9,065,000 sq km)
Gobi	500,000 sq miles (1,295,000 sq km)
Ar Rub al Khali	289,600 sq miles (750,000 sq km)
Great Victorian	249,800 sq miles (647,000 sq km)
Sonoran	120,000 sq miles (311,000 sq km)
Kalahari	120,000 sq miles (310,800 sq km)
Garagum	115,800 sq miles (300,000 sq km)
Takla Makan	100,400 sq miles (260,000 sq km)
Namib	52,100 sq miles (135,000 sq km)
Thar	33,670 sq miles (130,000 sq km)

NB – Most of Antarctica is a polar desert, with only 2 inches (50 mm) of precipitation annually

Hottest inhabited places

Djibouti (Djibouti)	86.0°F	(30.0°C)
Timbouctou (Mali)	84.7°F	(29.3°C)
Tirunelveli (India)	84.7°F	(29.3°C)
Tuticorin (India)	84.7°F	(29.3°C)
Nellore (India)	84.5°F	(29.2°C)
Santa Marta (Colombia)	84.5°F	(29.2°C)
Aden (Yemen)	84.0°F	(29.0°C)
Madurai (India)	84.0°F	(29.0°C)
Niamey (Niger)	84.0°F	(29.0°C)

Driest inhabited places

Aswān (Egypt)	0.02 in	(0.5 mm)
Luxor (Egypt)	0.03 in	(0.7 mm)
Arica (Chile)	0.04 in	(1.1 mm)
Ica (Peru)	0.10 in	(2.3 mm)
Antofagasta (Chile)	0.20 in	(4.9 mm)
El Minya (Egypt)	0.20 in	(5.1 mm)
Asyūt (Egypt)	0.20 in	(5.2 mm)
Callao (Peru)	0.50 in	(12.0 mm)
Trujillo (Peru)	0.55 in	(14.0 mm)
El Faiyūm (Egypt)	0.80 in	(19.0 mm)

Wettest inhabited places

Buenaventura (Colombia)	265 in	(6743 mm)
Monrovia (Liberia)	202 in	(5131 mm)
Pago Pago (American Samoa)	196 in	(4990 mm)
Moulmein (Myanmar)	191 in	(4852 mm)
Lae (Papua New Guinea)	183 in	(4645 mm)
Baguio (Luzon I., Philippines)	180 in	(4573 mm)
Sylhet (Bangladesh)	176 in	(4457 mm)
Padang (Sumatra, Indonesia)	166 in	(4225 mm)
Bogor (Java, Indonesia)	166 in	(4225 mm)
Conakry (Guinea)	171 in	(4341 mm)

Aa – Almere

A

Aa see Gauja
Aachen 72 A4 Dut. Aken, Fr. Aix-la-Chapelle; anc. Aquae Grani, Aquisgranum. Nordrhein-Westfalen, W Germany
Aaiún see Laâyoune
Aalborg 63 B7 var. Ålborg, Ålborg-Nørresundby; anc. Alburgum. Nordjylland, N Denmark
Aalen 73 B6 Baden-Württemberg, S Germany
Aalsmeer 64 C3 Noord-Holland, C Netherlands
Aalst 65 B6 Oost-Vlaanderen, C Belgium
Aalten 64 E4 Gelderland, E Netherlands
Aalter 65 B5 Oost-Vlaanderen, NW Belgium
Aanaarjävri see Inarijärvi
Äänekoski 63 D5 Keski-Suomi, W Finland
Aar see Aare
Aare 73 A7 var. Aar. river W Switzerland
Aarhus 63 B7 var. Århus, C Denmark
Aarlen see Arlon
Aat see Ath
Aba 55 E5 Orientale, NE Dem. Rep. Congo
Aba 53 G5 Abia, S Nigeria
Abā as Su'ūd see Najrān
Abaco Island see Great Abaco, N The Bahamas
Ābādān 98 C4 Khūzestān, SW Iran
Abadan 100 C3 prev. Bezmein, Büzmeýin, Rus. Byuzmeyin. Ahal Welaýaty, C Turkmenistan
Abai see Blue Nile
Abakan 92 D4 Respublika Khakasiya, S Russia
Abancay 38 D4 Apurímac, SE Peru
Abariringa see Kanton
Abashiri 108 D2 var. Abasiri. Hokkaidō, NE Japan
Abasiri see Abashiri
Ābay Wenz see Blue Nile
Abbaia see Ābaya Hāyk'
Abbatis Villa see Abbeville
Abbazia see Opatija
Abbeville 68 C2 anc. Abbatis Villa. Somme, N France
'Abd al 'Azīz, Jabal 96 D2 mountain range NE Syria
Abéché 54 C3 var. Abécher, Abeshr. Ouaddaï, SE Chad
Abécher see Abéché
Abela see Ávila
Abellinum see Avellino
Abemama 122 D2 var. Apamama; prev. Roger Simpson Island. atoll Tungaru, W Kiribati
Abengourou 53 E5 E Ivory Coast
Aberbrothock see Arbroath
Abercorn see Mbala
Aberdeen 66 D3 anc. Devana. NE Scotland, United Kingdom
Aberdeen 23 E2 South Dakota, N USA
Aberdeen 24 B2 Washington, NW USA
Abergwaun see Fishguard
Abertawe see Swansea
Aberystwyth 67 C6 W Wales, United Kingdom
Abeshr see Abéché
Abhā 99 B6 'Asīr, SW Saudi Arabia
Abidavichy 85 D7 Rus. Obidovichi. Mahilyowskaya Voblasts', E Belarus
Abidjan 53 E5 S Ivory Coast
Abilene 27 F3 Texas, SW USA
Abingdon see Pinta, Isla
Abkhazia see Ap'khazet'i
Åbo see Turku
Aboa 132 B2 Finnish research station Antarctica
Aboisso 53 E5 SE Ivory Coast
Abo, Massif d' 54 B1 mountain range NW Chad
Abomey 53 F5 S Benin
Abou-Déïa 54 C3 Salamat, SE Chad
Aboudouhour see Abū aḍ Ḍuhūr
Abou Kémal see Abū Kamāl
Abrantes 70 B3 var. Abrántes. Santarém, C Portugal
Abrashlare see Brezovo
Abrolhos Bank 34 E4 undersea bank W Atlantic Ocean
Abrova 85 B6 Rus. Obrovo. Brestskaya Voblasts', SW Belarus
Abrud 86 B4 Ger. Gross-Schlatten, Hung. Abrudbánya. Alba, SW Romania
Abrudbánya see Abrud
Abruzzese, Appennino 74 C4 mountain range C Italy
Absaroka Range 22 B2 mountain range Montana/Wyoming, NW USA
Abū aḍ Ḍuhūr 96 B3 Fr. Aboudouhour. Idlib, NW Syria
Abu Dhabi see Abū Ẓabī
Abu Hamed 50 C3 River Nile, N Sudan
Abū Ḥardān 96 E3 var. Hajîne. Dayr az Zawr, E Syria
Abuja 53 G4 country capital (Nigeria) Federal Capital District, C Nigeria
Abū Kamāl 96 E3 Fr. Abou Kémal. Dayr az Zawr, E Syria
Abula see Ávila
Abunã, Rio 40 C2 var. Río Abuná. river Bolivia/Brazil
Abut Head 129 B6 headland South Island, New Zealand
Abuye Meda 50 D4 mountain C Ethiopia
Abū Ẓabī 99 C5 var. Abū Ẓabī, Eng. Abu Dhabi. country capital (United Arab Emirates) Abū Ẓaby, C United Arab Emirates
Abū Ẓabī see Abū Ẓabī
Abyaḍ, Al Baḥr al see White Nile
Abyla see Ávila
Abyssinia see Ethiopia
Acalayong 55 A5 SW Equatorial Guinea
Acaponeta 28 D4 Nayarit, C Mexico
Acapulco 29 E5 var. Acapulco de Juárez. Guerrero, S Mexico
Acapulco de Juárez see Acapulco
Acarai Mountains 37 F4 Sp. Serra Acaraí. mountain range Brazil/Guyana
Acaraí, Serra see Acarai Mountains
Acarigua 36 D2 Portuguesa, N Venezuela
Accra 53 E5 country capital (Ghana) SE Ghana
Achacachi 39 E4 La Paz, W Bolivia
Ach'ara 95 F2 var. Ajaria. autonomous republic SW Georgia
Acklins Island 32 C2 island SE The Bahamas
Aconcagua, Cerro 42 B4 mountain W Argentina
Açores/Açores, Arquipélago dos/Açores, Ilhas dos see Azores
A Coruña 70 B1 Cast. La Coruña, Eng. Corunna; anc. Caronium. Galicia, NW Spain
Acre 40 C2 off. Estado do Acre. state/region W Brazil
Açu see Assu
Acunum Acusio see Montélimar
Ada 78 D3 Vojvodina, N Serbia
Ada 27 G2 Oklahoma, C USA
Ada Bazar see Adapazarı
Adalia see Antalya
Adalia, Gulf of see Antalya Körfezi
Adama see Nazrēt
'Adan 99 B7 Eng. Aden. SW Yemen
Adana 94 D4 var. Seyhan. Adana, S Turkey
Adâncata see Horlivka
Adapazarı 94 B2 prev. Ada Bazar. Sakarya, NW Turkey
Adare, Cape 132 B4 cape Antarctica
Ad Dahna 98 C4 desert E Saudi Arabia
Ad Dakhla 48 A4 var. Dakhla. SW Western Sahara
Ad Dalanj see Dilling
Ad Damar see Ed Damer
Ad Damazīn see Ed Damazin
Ad Dāmir see Ed Damer
Ad Dammān 98 C4 var. Dammām. Ash Sharqīyah, NE Saudi Arabia
Ad Dāmūr see Damoûr
Ad Dawḥah 98 C4 Eng. Doha. country capital (Qatar) C Qatar
Aḍ Ḍiffah see Libyan Plateau
Addis Ababa see Ādīs Ābeba
Addoo Atoll see Addu Atoll
Addu Atoll 110 A5 var. Addoo Atoll, Seenu Atoll. atoll S Maldives
Adelaide 127 B6 state capital South Australia
Adelsberg see Postojna
Aden see 'Adan
Aden, Gulf of 99 C7 gulf SW Arabian Sea
Adige 74 C2 Ger. Etsch. river N Italy
Adirondack Mountains 19 F2 mountain range New York, NE USA
Ādīs Ābeba 51 C5 Eng. Addis Ababa. country capital (Ethiopia) Ādīs Ābeba, C Ethiopia
Adıyaman 95 E4 Adıyaman, SE Turkey
Adjud 86 C4 Vrancea, E Romania
Admiralty Islands 122 B3 island group N Papua New Guinea
Adra 71 E5 Andalucía, S Spain
Adrar 48 D3 C Algeria
Adrian 18 C3 Michigan, N USA
Adrianople/Adrianopolis see Edirne
Adriatico, Mare see Adriatic Sea
Adriatic Sea 81 E2 Alb. Deti Adriatik, It. Mare Adriatico, SCr. Jadransko More, Slvn. Jadransko Morje. sea N Mediterranean Sea
Adriatik, Deti see Adriatic Sea
Adycha 93 F2 river NE Russia
Aegean Sea 83 C5 Gk. Aigaíon Pelagos, Aigaío Pélagos, Turk. Ege Denizi. sea NE Mediterranean Sea
Aegviidu 84 D2 Ger. Charlottenhof. Harjumaa, NW Estonia
Aegyptus see Egypt
Aelana see Al 'Aqabah
Aelok see Ailuk Atoll
Aelōnlaplap see Ailinglaplap Atoll
Aemona see Ljubljana
Aeolian Islands 75 C6 var. Isole Lipari, Eng. Aeolian Islands, Lipari Islands. island group S Italy
Aeolian Islands see Eolie, Isole
Æsernia see Isernia
Afar Depression see Danakil Desert
Afars et des Issas, Territoire Français des see Djibouti
Afghānestān, Dowlat-e Eslāmī-ye see Afghanistan
Afghanistan 100 C4 off. Islamic Republic of Afghanistan, Per. Dowlat-e Eslāmī-ye Afghānestān; prev. Republic of Afghanistan. country C Asia
Afmadow 51 D6 Jubbada Hoose, S Somalia
Africa 46 continent
Africa, Horn of 46 E4 physical region Ethiopia/Somalia
Africana Seamount 119 A6 seamount SW Indian Ocean
'Afrīn 96 B2 Ḥalab, N Syria
Afyon 94 B3 prev. Afyonkarahisar. Afyon, W Turkey
Agadès see Agadez
Agadez 53 G3 prev. Agadès. Agadez, C Niger
Agadir 48 B3 SW Morocco
Agana/Agaña see Hagåtña
Āgaro 51 C5 Oromīya, C Ethiopia
Agassiz Fracture Zone 121 G5 fracture zone S Pacific Ocean
Agatha see Agde
Agathónisi 83 D6 island Dodekánisa, Greece, Aegean Sea
Agde 69 C6 anc. Agatha. Hérault, S France
Agedabia see Ajdābiyā
Agen 69 B5 anc. Aginnum. Lot-et-Garonne, SW France
Agendicum see Sens
Aghri Dagh see Büyükağrı Dağı
Agiá 82 B4 var. Ayiá. Thessalía, C Greece
Agialoúsa see Yenierenköy
Agía Marína 83 E6 Léros, Dodekánisa, Greece, Aegean Sea
Aginnum see Agen
Ágios Efstrátios 82 D4 var. Áyios Evstrátios, Hagios Evstrátios. island E Greece
Ágios Nikólaos 83 D8 var. Áyios Nikólaos. Kríti, Greece, E Mediterranean Sea
Āgra 112 D3 Uttar Pradesh, N India
Agra and Oudh, United Provinces of see Uttar Pradesh
Agram see Zagreb
Ağrı 95 F3 var. Karaköse; prev. Karakılısse. Ağrı, NE Turkey
Agri Dagi see Büyükağrı Dağı
Agrigento 75 C7 Gk. Akragas; prev. Girgenti. Sicilia, Italy, C Mediterranean Sea
Agriovótano 83 C5 Évvoia, C Greece
Agropoli 75 D5 Campania, S Italy
Aguachica 36 B2 Cesar, N Colombia
Aguadulce 31 F5 Coclé, S Panama
Agua Prieta 28 B1 Sonora, NW Mexico
Aguascalientes 28 D4 Aguascalientes, C Mexico
Aguaytía 38 C3 Ucayali, C Peru
Aguilas 71 E4 Murcia, SE Spain
Aguililla 28 D4 Michoacán, SW Mexico
Agulhas Basin 47 D8 undersea basin SW Indian Ocean
Agulhas, Cape 56 C5 headland SW South Africa
Agulhas Plateau 45 D6 undersea plateau SW Indian Ocean
Ahaggar 53 F2 high plateau region SE Algeria
Ahlen 72 B4 Nordrhein-Westfalen, W Germany
Ahmadābād 112 C4 var. Ahmedabad. Gujarāt, W India
Ahmadnagar 112 C5 var. Ahmednagar. Mahārāshtra, W India
Ahmedabad see Ahmadābād
Ahmednagar see Ahmadnagar
Ahuachapán 30 B3 Ahuachapán, W El Salvador
Ahvāz 98 C3 var. Ahwāz; prev. Nāsiri. Khūzestān, SW Iran
Ahvenanmaa see Åland
Ahwāz see Ahvāz
Aigaíon Pelagos/Aigaío Pélagos see Aegean Sea
Aígina 83 C6 var. Aíyina, Egina. Aígina, C Greece
Aígio 83 B5 var. Egio; prev. Aíyion. Dytikí Ellás, S Greece
Aiken 21 E2 South Carolina, SE USA
Ailinglaplap Atoll 122 D2 var. Aelōnlaplap. atoll Ralik Chain, S Marshall Islands
Ailuk Atoll 122 D1 var. Aelok. atoll Ratak Chain, NE Marshall Islands
Ainaži 84 D3 Est. Heinaste, Ger. Hainasch. Limbaži, N Latvia
'Aïn Ben Tili 52 D1 Tiris Zemmour, N Mauritania
Aintab see Gaziantep
Aïoun el Atrous/Aïoun el Atroûss see 'Ayoûn el 'Atroûs
Aiquile 39 F4 Cochabamba, C Bolivia
Aïr see Aïr, Massif de l'
Air du Azbine see Aïr, Massif de l'
Aïr, Massif de l' 53 G2 var. Aïr, Air du Azbine, Asben. mountain range NC Niger
Aiud 86 B4 Ger. Strassburg, Hung. Nagyenyed; prev. Engeten. Alba, SW Romania
Aix see Aix-en-Provence
Aix-en-Provence 69 D6 var. Aix; anc. Aquae Sextiae. Bouches-du-Rhône, SE France
Aix-la-Chapelle see Aachen
Aíyina see Aígina
Aíyion see Aígio
Aizkraukle 84 C4 Aizkraukle, S Latvia
Ajaccio 69 E7 Corse, France, C Mediterranean Sea
Ajaria see Ach'ara
Ajastan see Armenia
Aj Bogd Uul 104 D2 mountain SW Mongolia
Ajdābiyā 49 G2 var. Agedabia, Ajdābiyah. NE Libya
Ajdābiyah see Ajdābiyā
Ajjinena see El Geneina
Ajmer 112 D3 var. Ajmere. Rājasthān, N India
Ajo 26 A3 Arizona, SW USA
Akaba see Al 'Aqabah
Akamagaseki see Shimonoseki
Akasha 50 B3 Northern, N Sudan
Akchâr 52 C2 desert W Mauritania
Aken see Aachen
Akermanceaster see Bath
Akhaltsikhe 95 F2 prev. Akhalts'ikhe. SW Georgia
Akhalts'ikhe see Akhaltsikhe
Akhisar 94 A3 Manisa, W Turkey
Akhmīm 50 B2 var. Akhmîm; anc. Panopolis. C Egypt
Akhtubinsk 89 C7 Astrakhanskaya Oblast', SW Russia
Akhtyrka see Okhtyrka
Akimiski Island 16 C3 island Nunavut, C Canada
Akinovka 87 F4 Zaporiz'ka Oblast', S Ukraine
Akita 108 D4 Akita, Honshū, C Japan
Akjoujt 52 C2 prev. Fort-Repoux. Inchiri, W Mauritania
Akkeshi 108 E2 Hokkaidō, NE Japan
Aklavik 14 D3 Northwest Territories, NW Canada
Akmola see Astana
Akmolinsk see Astana
Aknavásár see Târgu Ocna
Akpatok Island 17 E1 island Nunavut, E Canada
Akragas see Agrigento
Akron 18 D4 Ohio, N USA
Akrotiri 80 C5 var. Akrotírion. UK air base S Cyprus
Akrotírion see Akrotiri
Akrotírion Aksai Chin 102 B2 Chin. Aksayqin. disputed region China/India
Aksaray 94 C4 Aksaray, C Turkey
Aksayqin see Aksai Chin
Akşehir 94 B4 Konya, W Turkey
Aktash see Oqtosh
Aktau 92 A4 Kaz. Aqtaū; prev. Shevchenko. Mangistau, W Kazakhstan
Aktjubinsk/Aktyubinsk see Aktobe
Aktobe 92 B4 Kaz. Aqtöbe; prev. Aktjubinsk, Aktyubinsk. Aktyubinsk, NW Kazakhstan
Aktsyabrski 85 C7 Rus. Oktyabr'skiy; prev. Karpilovka. Homyel'skaya Voblasts', SE Belarus
Aktyubinsk see Aktobe
Akula 55 C5 Equateur, NW Dem. Rep. Congo
Akureyri 61 E4 Nordhurland Eystra, N Iceland
Akyab see Sittwe
Alabama 20 C2 off. State of Alabama, also known as Camellia State, Heart of Dixie, The Cotton State, Yellowhammer State. state S USA
Alabama River 20 C3 river Alabama, S USA
Alaca 94 C3 Çorum, N Turkey
Alacant see Alicante
Alagoas 41 G2 off. Estado de Alagoas. state/region E Brazil
Alais see Alès
Alajuela 31 E4 Alajuela, C Costa Rica
Alakanuk 14 C2 Alaska, USA
Al 'Alamayn 50 B1 var. El 'Alamein. N Egypt
Al 'Amārah 98 C3 var. Amara. Maysān, E Iraq
Alamo 25 D6 Nevada, W USA
Alamogordo 26 D3 New Mexico, SW USA
Alamosa 22 C5 Colorado, C USA
Åland 63 C6 var. Aland Islands, Fin. Ahvenanmaa. island group SW Finland
Aland Islands see Åland
Aland Sea see Ålands Hav
Ålands Hav 63 C6 var. Aland Sea. strait Baltic Sea/Gulf of Bothnia
Alanya 94 C4 Antalya, S Turkey
Alappuzha 110 C3 var. Alleppey. Kerala, SW India
Al 'Aqabah 97 B8 var. Akaba, Aqaba, 'Aqaba; anc. Aelana, Elath. Al 'Aqabah, SW Jordan
Al 'Arabīyah as Su'ūdīyah see Saudi Arabia
Alasca, Golfo de see Alaska, Gulf of
Alaşehir 94 A4 Manisa, W Turkey
Al 'Ashārah 96 E3 var. Ashara. Dayr az Zawr, E Syria
Alaska 14 C3 off. State of Alaska, also known as Land of the Midnight Sun, The Last Frontier, Seward's Folly; prev. Russian America. state NW USA
Alaska, Gulf of 14 C4 var. Golfo de Alasca. gulf Canada/USA
Alaska Peninsula 14 C3 peninsula Alaska, USA
Alaska Range 12 B2 mountain range Alaska, USA
Al-Asnam see Chlef
Alattio see Alta
Al Awaynāt see Al 'Uwaynāt
Alaykel'/Alay-Kuu see Këk-Art
Al 'Aynā 97 B7 Al Karak, W Jordan
Alazeya 93 G2 river NE Russia
Al Bāb 96 B2 Ḥalab, N Syria
Albacete 71 E3 Castilla-La Mancha, C Spain
Al Baghdādī 98 B3 var. Khān al Baghdādī. Al Anbār, SW Iraq
Al Bāha see Al Bāḥah
Al Bāḥah 99 B5 var. Al Bāha. Al BāḤah, SW Saudi Arabia
Al Baḥrayn see Bahrain
Alba Iulia 86 B4 Ger. Weissenburg, Hung. Gyulafehérvár; prev. Bălgrad, Karlsburg, Károly-Fehérvár. Alba, W Romania
Albania 79 C7 off. Republic of Albania, Alb. Republika e Shqipqërisë, Shqipëria; prev. People's Socialist Republic of Albania. country SE Europe
Albania see Aubagne
Albany 125 B7 Western Australia
Albany 20 D3 Georgia, SE USA
Albany 19 F3 state capital New York, NE USA
Albany 24 B3 Oregon, NW USA
Albany 16 C3 river Ontario, S Canada
Alba Regia see Székesfehérvár
Al Bāridah 96 C4 var. Bāridah. Ḥimṣ, C Syria
Al Baṣrah 98 C3 Eng. Basra, hist. Busra, Bussora. Al Baṣrah, SE Iraq
Al Batrūn see Batroûn
Al Bawītī 50 B2 var. Bawîṭi C Egypt
Al Bayḍā' 49 G2 var. Beida. NE Libya
Albemarle Island see Isabela, Isla
Albemarle Sound 21 G1 inlet W Atlantic Ocean
Albergaria-a-Velha 70 B2 Aveiro, N Portugal
Albert 68 C3 Somme, N France
Alberta 15 E4 province SW Canada
Albert Edward Nyanza see Edward, Lake
Albert, Lake 51 B6 var. Albert Nyanza, Lac Mobutu Sese Seko. lake Uganda/Dem. Rep. Congo
Albert Lea 23 F3 Minnesota, N USA
Albert Nyanza see Albert, Lake
Albertville see Kalemie
Albi 69 C6 anc. Albiga. Tarn, S France
Albiga see Albi
Ålborg see Aalborg
Ålborg-Nørresundby see Aalborg
Alborz, Reshteh-ye Kūhhā-ye 98 C2 Eng. Elburz Mountains. mountain range N Iran
Albuquerque 26 D2 New Mexico, SW USA
Al Burayqah see Marsá al Burayqah
Alburgum see Aalborg
Albury 127 C7 New South Wales, SE Australia
Alcácer do Sal 70 B4 Setúbal, W Portugal
Alcalá de Henares 71 E3 Ar. Alkal'a; anc. Complutum. Madrid, C Spain
Alcamo 75 C7 Sicilia, Italy, C Mediterranean Sea
Alcañiz 71 F2 Aragón, NE Spain
Alcántara, Embalse de 70 C3 reservoir W Spain
Alcaudete 70 D4 Andalucía, S Spain
Alcázar see Ksar-el-Kebir
Alcazarquivir see Ksar-el-Kebir
Alcoi see Alcoy
Alcoy 71 F4 Cat. Alcoi. País Valenciano, E Spain
Aldabra Group 57 G2 island group SW Seychelles
Aldan 93 F3 river NE Russia
al Dar al Baida see Rabat
Alderney 68 A2 island Channel Islands
Aleg 52 C3 Brakna, SW Mauritania
Aleksandriya see Oleksandriya
Aleksandropol' see Gyumri
Aleksandrovka see Oleksandrivka
Aleksandrovsk see Zaporizhzhya
Aleksin 89 B5 Tul'skaya Oblast', W Russia
Aleksinac 78 E4 Serbia, SE Serbia
Alençon 68 B3 Orne, N France
Alenquer 41 E2 Pará, NE Brazil
Alep/Aleppo see Ḥalab
Alert 15 F1 Ellesmere Island, Nunavut, N Canada
Alès 69 C6 prev. Alais. Gard, S France
Aleşd 86 B3 Hung. Élesd. Bihor, SW Romania
Alessandria 74 B2 Fr. Alexandrie. Piemonte, N Italy
Ålesund 63 A5 Møre og Romsdal, S Norway
Aleutian Basin 91 G3 undersea basin Bering Sea
Aleutian Islands 14 A3 island group Alaska, USA
Aleutian Range 12 A2 mountain range Alaska, USA
Aleutian Trench 91 H3 trench S Bering Sea
Alexander Archipelago 14 D4 island group Alaska, USA
Alexander City 20 D2 Alabama, S USA
Alexander Island 132 A3 island Antarctica
Alexander Range see Kirghiz Range
Alexandra 129 B7 Otago, South Island, New Zealand
Alexándreia 82 B4 var. Alexándria. Kentrikí Makedonía, N Greece
Alexandretta see İskenderun
Alexandretta, Gulf of see İskenderun Körfezi
Alexandria 50 B1 Ar. Al Iskandarīyah. N Egypt
Alexandria 86 C5 Teleorman, S Romania
Alexandria 20 B3 Louisiana, S USA
Alexandria 23 F2 Minnesota, N USA
Alexándria see Alexándreia
Alexandrie see Alessandria
Alexandroúpoli 82 D3 var. Alexandroúpolis, Turk. Dedeagaç, Dedeagach. Anatolikí Makedonía kai Thráki, NE Greece
Alexandroúpolis see Alexandroúpoli
Al Fāshir see El Fasher
Alfatar 82 E1 Silistra, NE Bulgaria
Alfeiós 83 B6 prev. Alfiós; anc. Alpheius, Alpheus. river S Greece
Alfiós see Alfeiós
Alföld see Great Hungarian Plain
Al-Furāt see Euphrates
Alga 92 B4 Kaz. Algha. Aktyubinsk, NW Kazakhstan
Algarve 70 B4 cultural region S Portugal
Algeciras 70 C5 Andalucía, SW Spain
Algemesí 71 F3 País Valenciano, E Spain
Al-Genain see El Geneina
Alger 49 E1 var. Algiers, El Djazaïr, Al Jazair. country capital (Algeria) N Algeria
Algeria 48 C3 off. Democratic and Popular Republic of Algeria. country N Africa
Algeria, Democratic and Popular Republic of see Algeria
Algerian Basin 58 C5 var. Balearic Plain. undersea basin W Mediterranean Sea
Algha see Alga
Al Ghābah 99 E5 var. Ghaba. C Oman
Alghero 75 A5 Sardegna, Italy, C Mediterranean Sea
Al Ghurdaqah 50 C2 var. Hurghada, Ghurdaqah. E Egypt
Algiers see Alger
Al Golea see El Goléa
Algona 23 F3 Iowa, C USA
Al Hajar al Gharbi 99 D5 mountain range N Oman
Al Hamad see Syrian Desert
Al Ḥasakah 96 D2 var. Al Hasijah, El Haseke, Fr. Hassetché. Al Ḥasakah, NE Syria
Al Hasijah see Al Ḥasakah
Al Ḥillah 98 B3 var. Hilla. Bābil, C Iraq
Al Ḥisā 97 B7 Aṭ Ṭafīlah, W Jordan
Al Ḥudaydah 99 B6 Eng. Hodeida. W Yemen
Al Hufūf 98 C4 var. Hofuf. Ash Sharqīyah, NE Saudi Arabia
Aliákmon see Aliákmonas
Aliákmonas 82 B4 prev. Aliákmon; anc. Haliacmon. river N Greece
Alíartos 83 C5 Stereá Ellás, C Greece
Alicante 71 F4 Cat. Alacant, Lat. Lucentum. País Valenciano, SE Spain
Alice 27 G5 Texas, SW USA
Alice Springs 126 A4 Northern Territory, C Australia
Alifu Atoll see Ari Atoll
Aligandí 31 G4 Kuna Yala, NE Panama
Aliki see Alykí
Alima 55 B6 river C Congo
Al Imārāt al 'Arabīyahal Muttaḥidah see United Arab Emirates
Alindao 54 C4 Basse-Kotto, S Central African Republic
Aliquippa 18 D4 Pennsylvania, NE USA
Al Iskandarīyah see Alexandria
Al Ismā'īlīya 50 B1 var. Ismailia, Ismâ'ilîya. N Egypt
Alistráti 82 C3 Kentrikí Makedonía, NE Greece
Alivéri 83 C5 var. Alivérion. Évvoia, C Greece
Alivérion see Alivéri
Al Jabal al Akhḍar 49 G2 mountain range NE Libya
Al Jafr 97 B7 Ma'ān, S Jordan
Al Jaghbūb 49 H3 NE Libya
Al Jahrā' 98 C4 var. Al Jahrah, Jahra. C Kuwait
Al Jahrah see Al Jahrā'
Al Jamāhīrīyah al 'Arabīyah al Lībīyah ash Sha'bīyah al Ishtirākīy see Libya
Al Jawf 98 B4 off. Jauf. Al Jawf, NW Saudi Arabia
Al Jawlān see Golan Heights
Al Jazair see Alger
Al Jazirah 96 E2 physical region Iraq/Syria
Al Jīzah see Gîza
Al Junaynah see El Geneina
Alkal'a see Alcalá de Henares
Al Karak 97 B7 var. El Kerak, Karak, Kerak; anc. Kir Moab, Kir of Moab. Al Karak, W Jordan
Al-Kasr al-Kebir see Ksar-el-Kebir
Al Khalīl see Hebron
Al Khārijah 50 B2 var. El Khârga. C Egypt
Al Khums 49 F2 var. Homs, Khoms, Khums. NW Libya
Alkmaar 64 C2 Noord-Holland, NW Netherlands
Al Kufrah 49 H4 SE Libya
Al Kūt 98 C3 var. Kūt al 'Amārah, Kut al Imara. Wāsiṭ, E Iraq
Al-Kuwait see Al Kuwayt
Al Kuwayt 98 C4 var. Al-Kuwait, Eng. Kuwait, Kuwait City; prev. Qurein. country capital (Kuwait) E Kuwait
Al Lādhiqīyah 96 A3 Eng. Latakia, Fr. Lattaquié; anc. Laodicea, Laodicea ad Mare. Al Lādhiqīyah, W Syria
Allahābād 113 E3 Uttar Pradesh, N India
Allanmyo see Aunglan
Allegheny Plateau 19 E3 mountain range New York/Pennsylvania, NE USA
Allenstein see Olsztyn
Allentown 19 F4 Pennsylvania, NE USA
Alleppey see Alappuzha
Alliance 22 D3 Nebraska, C USA
Al Līth 99 B5 Makkah, SW Saudi Arabia
Al Lubnān see Lebanon
Alma-Ata see Almaty
Almada 70 B4 Setúbal, W Portugal
Al Madīnah 99 A5 Eng. Medina. Al Madīnah, W Saudi Arabia
Al Mafraq 97 B6 var. Mafraq. Al Mafraq, N Jordan
Al Mahdīyah see Mahdia
Al Mahrah 99 C6 mountain range E Yemen
Al Majma'ah 98 B4 Ar Riyāḍ, C Saudi Arabia
Al Mālikīyah 96 E1 var. Malkiye. Al Ḥasakah, N Syria
Almalyk see Olmaliq
Al Mamlakah see Morocco
Al Mamlaka al Urdunīya al Hashemīyah see Jordan
Al Manāmah 98 C4 Eng. Manama. country capital (Bahrain) N Bahrain
Al Manāṣif 96 E3 mountain range E Syria
Almansa 71 F4 Castilla-La Mancha, C Spain
Al-Mariyya see Almería
Al Marj 49 G2 var. Barka, It. Barce. NE Libya
Almaty 92 C5 var. Alma-Ata. Almaty, SE Kazakhstan
Al Mawṣil 98 B2 Eng. Mosul. Nīnawá, N Iraq
Al Mayādīn 96 D3 var. Mayadin, Fr. Meyadine. Dayr az Zawr, E Syria
Al Mazra' see Al Mazra'ah
Al Mazra'ah 97 B6 var. Al Mazra', Mazra'a. Al Karak, W Jordan
Almelo 64 E3 Overijssel, E Netherlands
Almendra, Embalse de 70 C2 reservoir Castilla-León, NW Spain
Almendralejo 70 C4 Extremadura, W Spain
Almere 64 C3 var. Almere-stad. Flevoland, C Netherlands

Almere-stad – Arinsal

Almere-stad *see* Almere
Almería *71 E5 Ar.* Al-Mariyya; *anc.* Unci, *Lat.* Portus Magnus. Andalucía, S Spain
Al'met'yevsk *89 D5* Respublika Tatarstan, W Russia
Al Mīnā' *see* El Mina
Al Minyā *50 B2 var.* El Minya, Minya. C Egypt
Almirante *31 E4* Bocas del Toro, NW Panama
Al Mudawwarah *97 B8* Ma'ān, SW Jordan
Al Mukallā *99 C6 var.* Mukalla. SE Yemen
Al Obayyid *see* El Obeid
Alofi *123 F4 dependent territory capital* (Niue) W Niue
Aloha State *see* Hawaii
Aloja *84 D3* Limbaži, N Latvia
Alónnisos *83 C5 island* Vóreies Sporádes, Greece, Aegean Sea
Álora *70 D5* Andalucía, S Spain
Alor, Kepulauan *117 E5 island group* E Indonesia
Al Oued *see* El Oued
Alpen *see* Alps
Alpena *18 D2* Michigan, N USA
Alpes *see* Alps
Alpha Cordillera *133 B3 var.* Alpha Ridge. *seamount range* Arctic Ocean
Alpha Ridge *see* Alpha Cordillera
Alpheius *see* Alfeiós
Alphen *see* Alphen aan den Rijn
Alphen aan den Rijn *64 C3 var.* Alphen. Zuid-Holland, C Netherlands
Alpheus *see* Alfeiós
Alpi *see* Alps
Alpine *27 E4* Texas, SW USA
Alps *80 C1 Fr.* Alpes, *Ger.* Alpen, *It.* Alpi. *mountain range* C Europe
Al Qaḍārif *see* Gedaref
Al Qāhirah *see* Cairo
Al Qāmishlī *96 E1 var.* Kamishli, Qamishly. Al Ḥasakah, NE Syria
Al Qaşrayn *see* Kasserine
Al Qayrawān *see* Kairouan
Al-Qsar al-Kbir *see* Ksar-el-Kebir
Al Qubayyāt *see* Qoubaïyât
Al Quds/Al Quds ash Sharīf *see* Jerusalem
Alqueva, Barragem do *70 C4 reservoir* Portugal/Spain
Al Qunayṭirah *97 B5 var.* El Kuneitra, El Quneitra, Kuneitra, Qunaytra. Al Qunayṭirah, SW Syria
Al Quşayr *96 B4 var.* El Quseir, Quşayr, *Fr.* Kousseir. Ḥimṣ, W Syria
Al Quwayrah *97 B8 var.* El Quweira. Al 'Aqabah, SW Jordan
Alsace *68 E3 Ger.* Elsass; *anc.* Alsatia. *cultural region* NE France
Alsatia *see* Alsace
Alsdorf *72 A4* Nordrhein-Westfalen, W Germany
Alt *see* Olt
Alta *62 D2 Fin.* Alattio. Finnmark, N Norway
Altai *see* Altai Mountains
Altai Mountains *104 C2 var.* Altai, *Chin.* Altay Shan, *Rus.* Altay. *mountain range* Asia/Europe
Altamaha River *21 E3 river* Georgia, SE USA
Altamira *41 E2* Pará, NE Brazil
Altamura *75 E5 anc.* Lupatia. Puglia, SE Italy
Altar, Desierto de *28 A1 var.* Sonoran Desert. *desert* Mexico/USA
Altar, Desierto de *see* Sonoran Desert
Altay *104 C2* Xinjiang Uygur Zizhiqu, NW China
Altay *104 D2 prev.* Yösönbulag. Govĭ-Altay, W Mongolia
Altay Altai Mountains, Asia/Europe
Altay Shan *see* Altai Mountains
Altbetsche *see* Bečej
Altenburg *see* București, Romania
Altin Köprü *98 B3 var.* Altun Kupri. At Ta'mīn, N Iraq
Altiplano *39 F4 physical region* W South America
Altkanischa *see* Kanjiža
Alton *18 B4* Illinois, C USA
Altoona *19 E4* Pennsylvania, NE USA
Alto Paraná *see* Paraná
Altpasua *see* Stara Pazova
Alt-Schwanenburg *see* Gulbene
Altsohl *see* Zvolen
Altun Kupri *see* Altin Köprü
Altun Shan *104 C3 var.* Altyn Tagh. *mountain range* NW China
Altus *27 F2* Oklahoma, C USA
Altyn Tagh *see* Altun Shan
Al Ubayyiḍ *see* El Obeid
Alūksne *84 D3 Ger.* Marienburg. Alūksne, NE Latvia
Al 'Ulā *98 A4* Al Madīnah, NW Saudi Arabia
Al 'Umarī *97 C6* 'Ammān, E Jordan
Alupka *87 F5* Respublika Krym, S Ukraine
Al Uqşur *see* Luxor
Al Urdunn *see* Jordan
Alushta *87 F5* Respublika Krym, S Ukraine
Al 'Uwaynāt *49 F4 var.* Al Awaynāt. SW Libya
Alva *27 F1* Oklahoma, C USA
Alvarado *29 F4* Veracruz-Llave, E Mexico
Alvin *27 H4* Texas, SW USA
Al Wajh *98 A4* Tabūk, NW Saudi Arabia
Alwar *112 D3* Rājasthān, N India
Al Wari'ah *98 C4* Ash Sharqīyah, N Saudi Arabia
Al Yaman *see* Yemen
Alykí *82 C4 var.* Aliki. Thásos, N Greece
Alytus *85 B5 Pol.* Olita. Alytus, S Lithuania
Alzette *65 D8 river* S Luxembourg
Amadeus, Lake *125 D5 seasonal lake* Northern Territory, C Australia
Amadi *51 B5* W Equatoria, S South Sudan
Amadjuak Lake *15 G3 lake* Baffin Island, Nunavut, N Canada
Amakusa-nada *109 A7 gulf* SW Japan
Åmål *63 B6* Västra Götaland, S Sweden
Amami-gunto *108 A3 island group* SW Japan
Amami-o-shim *108 A3 island* S Japan
Amantea *75 D6* Calabria, SW Italy
Amapá *41 E1 off.* Estado de Amapá; *prev.* Território de Amapá. *state/region* NE Brazil
Amapá, Estado de *see* Amapá
Amapá, Território de *see* Amapá
Amara *see* Al 'Amārah
Amarapura *114 B3* Mandalay, C Myanmar (Burma)
Amarillo *27 E2* Texas, SW USA
Amay *65 C6* Liège, E Belgium
Amazon *41 E1 Sp.* Amazonas. *river* Brazil/Peru
Amazonas *see* Amazon
Amazon Basin *40 D2 basin* N South America
Amazonía *36 B4 region* S Colombia
Amazon, Mouths of the *41 F1 delta* NE Brazil
Ambam *55 B5* Sud, S Cameroon
Ambanja *57 G2* Antsirañana, N Madagascar
Ambarchik *93 G2* Respublika Sakha (Yakutiya), NE Russia
Ambato *38 B1* Tungurahua, C Ecuador
Ambérieu-en-Bugey *69 D5* Ain, E France
Ambianum *see* Amiens
Amboasary *57 F4* Toliara, S Madagascar
Amboina *see* Ambon
Ambon *117 F4 prev.* Amboina, Amboyna. Pulau Ambon, E Indonesia
Ambositra *57 G3* Fianarantsoa, SE Madagascar
Amboyna *see* Ambon
Ambracia *see* Árta
Ambre, Cap d' *see* Bobaomby, Tanjona
Ambrim *see* Ambrym
Ambriz *56 A1* Bengo, NW Angola
Ambrym *122 D4 var.* Ambrim. *island* C Vanuatu
Amchitka Island *14 A2 island* Aleutian Islands, Alaska, USA
Amdo *104 C5* Xizang Zizhiqu, W China
Ameland *64 D1 Fris.* It Amelân. *island* Waddeneilanden, N Netherlands
Amelân, It *see* Ameland
America *see* United States of America
America-Antarctica Ridge *45 C7 undersea ridge* S Atlantic Ocean
America in Miniature *see* Maryland
American Falls Reservoir *24 E4 reservoir* Idaho, NW USA
American Samoa *123 E4 US unincorporated territory* W Polynesia
Amersfoort *64 D3* Utrecht, C Netherlands
Ames *23 F3* Iowa, C USA
Amfilochía *83 A5 var.* Amfilokhía. Dytikí Ellás, C Greece
Amfilokhía *see* Amfilochía
Amga *93 F3 river* NE Russia
Amherst *17 F4* Nova Scotia, SE Canada
Amherst *see* Kyaikkami
Amida *see* Diyarbakır
Amiens *68 C3 anc.* Ambianum, Samarobriva. Somme, N France
Amíndaion/Amindeo *see* Amýntaio
Amindivi Islands *110 A2 island group* Lakshadweep, India, N Indian Ocean
Amirante Islands *57 G1 var.* Amirantes Group. *island group* C Seychelles
Amirantes Group *see* Amirante Islands
Amistad, Presa de la *see* Amistad Reservoir
Amistad Reservoir *27 F4 var.* Presa de la Amistad. *reservoir* Mexico/USA
Amisus *see* Samsun
Ammaia *see* Portalegre
'Ammān *97 B6 anc.* Philadelphia, *Bibl.* Rabbah Ammon, Rabbath Ammon. *country capital* (Jordan) 'Ammān, NW Jordan
Ammassalik *60 D4 var.* Angmagssalik. Tunu, S Greenland
Ammóchostos *see* Gazimağusa
Ammóchostos, Kólpos *see* Gazimağusa Körfezi
Amnok-kang *see* Yalu
Amoea *see* Portalegre
Amoentai *see* Amuntai
Āmol *98 D2 var.* Amul. Māzandarān, N Iran
Amorgós *83 D6* Amorgós, Kykládes, Greece, Aegean Sea
Amorgós *83 D6 island* Kykládes, Greece, Aegean Sea
Amos *16 D4* Québec, SE Canada
Amourj *52 D3* Hodh ech Chargui, SE Mauritania
Amoy *see* Xiamen
Ampato, Nevado *39 E4 mountain* S Peru
Amposta *71 F2* Cataluña, NE Spain
Amraoti *see* Amrāvati
Amrāvati *112 D4 prev.* Amraoti. Mahārāshtra, C India
Amritsar *112 D2* Punjab, N India
Amstelveen *64 C3* Noord-Holland, C Netherlands
Amsterdam *64 C3 country capital* (Netherlands) Noord-Holland, C Netherlands
Amsterdam Island *119 C6 island* NE French Southern and Antarctic Lands
Am Timan *54 C3* Salamat, SE Chad
Amu Darya *100 D2 Rus.* Amudar'ya, *Taj.* Dar"yoi Amu, *Turkm.* Amyderya, *Uzb.* Amudaryo; *anc.* Oxus. *river* C Asia
Amu-Dar'ya *see* Amyderya
Amudar'ya/Amudaryo/Amu, Dar"yoi *see* Amu Darya
Amul *see* Āmol
Amund Ringnes Island *15 F2 Island* Nunavut, N Canada
Amundsen Basin *see* Fram Basin
Amundsen Plain *132 A4 abyssal plain* S Pacific Ocean
Amundsen-Scott *132 B3 US research station* Antarctica
Amundsen Sea *132 A4 sea* S Pacific Ocean
Amuntai *116 D4 prev.* Amoentai. Borneo, C Indonesia
Amur *93 G4 Chin.* Heilong Jiang. *river* China/Russia
Amvrosiyivka *see* Amvrosiyivka
Amvrosiyivka *87 H3 Rus.* Amvrosiyevka. Donets'ka Oblast', SE Ukraine
Amyderya *101 E3 Rus.* Amu-Dar'ya. Lebap Welaýaty, NE Turkmenistan
Amyderya *see* Amu Darya
Amýntaio *82 B4 var.* Amindeo; *prev.* Amíndaion. Dytikí Makedonía, N Greece
Anabar *93 E2 river* NE Russia
An Abhainn Mhór *see* Blackwater
Anaco *37 E2* Anzoátegui, NE Venezuela
Anaconda *22 B2* Montana, NW USA
Anacortes *24 B1* Washington, NW USA
Anadolu Dağları *see* Doğu Karadeniz Dağları
Anadyr' *93 H1* Chukotskiy Avtonomnyy Okrug, NE Russia
Anadyr' *93 G1 river* NE Russia
Anadyr, Gulf of *see* Anadyrskiy Zaliv
Anadyrskiy Zaliv *93 H1 Eng.* Gulf of Anadyr. *gulf* NE Russia
Anáfi *83 D7 anc.* Anaphe. *island* Kykládes, Greece, Aegean Sea
'Anah *see* 'Annah
Anaheim *24 E2* California, W USA
Anaiza *see* 'Unayzah
Analalava *57 G2* Mahajanga, NW Madagascar
Anamur *94 C5* İçel, S Turkey
Anantapur *110 C2* Andhra Pradesh, S India
Anaphe *see* Anáfi
Anápolis *41 F3* Goiás, C Brazil
Anār *98 D3* Kermān, C Iran
Anatolia *94 C4 plateau* C Turkey
Anatom *see* Aneityum
Añatuya *42 C3* Santiago del Estero, N Argentina
An Bhearú *see* Barrow
Anchorage *14 C3* Alaska, USA
Ancona *74 C3* Marche, C Italy
Ancud *43 B6 prev.* San Carlos de Ancud. Los Lagos, S Chile
Ancyra *see* Ankara
Åndalsnes *63 A5* Møre og Romsdal, S Norway
Andalucía *70 D4 cultural region* S Spain
Andalusia *20 C3* Alabama, S USA
Andaman Islands *102 B4 island group* India, NE Indian Ocean
Andaman Sea *102 C4 sea* NE Indian Ocean
Andenne *65 C6* Namur, SE Belgium
Anderlues *65 B7* Hainaut, S Belgium
Anderson *18 C4* Indiana, N USA
Andes *42 B3 mountain range* W South America
Andhra Pradesh *113 E5 cultural region* E India
Andijon *101 F2 Rus.* Andizhan. Andijon Viloyati, E Uzbekistan
Andikíthira *see* Antikýthira
Andipaxi *see* Antípaxoi
Andípsara *see* Antípsara
Ándissa *see* Ántissa
Andizhan *see* Andijon
Andkhvoy *100 D3* Fāryāb, N Afghanistan
Andorra *69 A7 off.* Principality of Andorra, *Cat.* Valls d'Andorra, *Fr.* Vallée d'Andorre. *country* SW Europe
Andorra *see* Andorra la Vella
Andorra la Vella *69 A8 var.* Andorra, *Fr.* Andorre la Vielle, *Sp.* Andorra la Vieja. *country capital* (Andorra) C Andorra
Andorra la Vieja *see* Andorra la Vella
Andorra, Principality of *see* Andorra
Andorra, Valls d'/Andorra, Vallée d' *see* Andorra
Andorre la Vielle *see* Andorra la Vella
Andover *67 D7* S England, United Kingdom
Andøya *62 C2 island* C Norway
Andreanof Islands *14 A3 island group* Aleutian Islands, Alaska, USA
Andrews *27 E3* Texas, SW USA
Andrew Tablemount *118 A4 var.* Gora Andryu. *seamount* W Indian Ocean
Andria *75 D5* Puglia, SE Italy
An Droichead Nua *see* Newbridge
Andropov *see* Rybinsk
Ándros *83 D6* Ándros, Kykládes, Greece, Aegean Sea
Ándros *83 C6 island* Kykládes, Greece, Aegean Sea
Andros Island *32 B2 island* NW The Bahamas
Andros Town *32 C1* Andros Island, NW The Bahamas
Andryu, Gora *see* Andrew Tablemount
Aneityum *122 D5 var.* Anatom; *prev.* Kéamu. *island* S Vanuatu
Ānewetak *see* Enewetak Atoll
Angara *93 E4 river* C Russia
Angarsk *93 E4* Irkutskaya Oblast', S Russia
Ange *63 C5* Västernorrland, C Sweden
Ángel de la Guarda, Isla *28 B2 island* NW Mexico
Angeles *117 E1 off.* Angeles City. Luzon, N Philippines
Angeles City *see* Angeles
Angel Falls *37 E3 Eng.* Angel Falls. *waterfall* E Venezuela
Angel Falls *see* Ángel, Salto
Angerburg *see* Węgorzewo
Ångermanälven *62 C4 river* N Sweden
Angermünde *72 D3* Brandenburg, NE Germany
Angers *68 B4 anc.* Juliomagus. Maine-et-Loire, NW France
Anglesey *67 C5 island* NW Wales, United Kingdom
Anglet *69 A6* Pyrénées-Atlantiques, SW France
Angleton *27 H4* Texas, SW USA
Anglia *see* England
Anglo-Egyptian Sudan *see* Sudan
Angmagssalik *see* Ammassalik
Ang Nam Ngum *114 C4 lake* C Laos
Angola *56 B2 off.* Republic of Angola; *prev.* People's Republic of Angola, Portuguese West Africa. *country* SW Africa
Angola Basin *47 B5 undersea basin* E Atlantic Ocean
Angola, People's Republic of *see* Angola
Angola, Republic of *see* Angola
Angora *see* Ankara
Angostura *see* Ciudad Bolívar
Angostura, Presa de la *29 G5 reservoir* SE Mexico
Angoulême *69 B5 anc.* Iculisma. Charente, W France
Angoumois *69 B5 cultural region* W France
Angra Pequena *see* Lüderitz
Angren *101 F2* Toshkent Viloyati, E Uzbekistan
Anguilla *33 G3 UK dependent territory* E West Indies
Anguilla Cays *32 B2 islets* SW The Bahamas
Anhui *106 C5 var.* Anhui Sheng, Anhwei, Wan. *province* E China
AnhuiSheng/Anhwei Wan *see* Anhui
Anicium *see* le Puy
Anina *86 A4 Ger.* Steierdorf, *Hung.* Stájerlakanina; *prev.* Ştaierdorf-Anina, Steierdorf-Anina, Steyerlak-Anina. Caraş-Severin, SW Romania
Anjou *68 B4 cultural region* NW France
Anjouan *57 F2 var.* Ndzouani, Nzwani. *island* SE Comoros
Ankara *94 C3 prev.* Angora; *anc.* Ancyra. *country capital* (Turkey) Ankara, C Turkey
Ankeny *23 F3* Iowa, C USA
Anklam *72 D2* Mecklenburg-Vorpommern, NE Germany
An Mhuir Cheilteach *see* Celtic Sea
Annaba *49 E1 prev.* Bône. NE Algeria
An Nafud *98 B4 desert* NW Saudi Arabia
'Annah *98 B3 var.* 'Ānah. Al Anbār, NW Iraq
An Najaf *98 B3 var.* Najaf. An Najaf, S Iraq
Annamite Mountains *114 D4 Fr.* Annamitique, Chaîne. *mountain range* C Laos
Annamitique, Chaîne *see* Annamite Mountains
Annapolis *19 F4 state capital* Maryland, NE USA
Annapurṇa *113 E3 mountain* C Nepal
An Nāqūrah *see* En Nâqoûra
Ann Arbor *18 C3* Michigan, N USA
An Nāṣirīyah *98 C3 var.* Nasiriya. Dhī Qār, SE Iraq
Anneciacum *see* Annecy
Annecy *69 D5 anc.* Anneciacum. Haute-Savoie, E France
An Nīl al Abyaḍ *see* White Nile
An Nīl al Azraq *see* Blue Nile
Anniston *20 D2* Alabama, S USA
Annotto Bay *32 B4* C Jamaica
An Ómaigh *see* Omagh
Anqing *106 D5* Anhui, E China
Anse La Raye *33 F1* NW Saint Lucia
Anshun *106 B6* Guizhou, S China
Ansongo *53 E3* Gao, E Mali
An Srath Bán *see* Strabane
Antakya *94 D4 anc.* Antioch, Antiochia. Hatay, S Turkey
Antalaha *57 G2* Antsirañana, NE Madagascar
Antalya *94 B4 prev.* Adalia; *anc.* Attaleia, *Bibl.* Attalia. Antalya, SW Turkey
Antalya, Gulf of *94 B4 var.* Gulf of Adalia, *Eng.* Gulf of Antalya. *gulf* SW Turkey
Antalya, Gulf of *see* Antalya Körfezi
Antananarivo *57 G3 prev.* Tananarive. *country capital* (Madagascar) Antananarivo, C Madagascar
Antarctica *132 B3 continent*
Antarctic Peninsula *132 A2 peninsula* Antarctica
Antep *see* Gaziantep
Antequera *70 D5 anc.* Anticaria, Antiquaria. Andalucía, S Spain
Antequera *see* Oaxaca
Antibes *69 D6 anc.* Antipolis. Alpes-Maritimes, SE France
Anticaria *see* Antequera
Anticosti, Île d' *17 F3 Eng.* Anticosti Island. *island* Québec, E Canada
Anticosti Island *see* Anticosti, Île d'
Antigua *33 G3 island* S Antigua and Barbuda, Leeward Islands
Antigua and Barbuda *33 G3 country* E West Indies
Antikythira *83 B7 var.* Andikíthira. *island* S Greece
Anti-Lebanon *96 B4 var.* Jebel esh Sharqi, *Ar.* Al Jabal ash Sharqī, *Fr.* Anti-Liban. *mountain range* Lebanon/Syria
Anti-Liban *see* Anti-Lebanon
Antioch *see* Antakya
Antiochia *see* Antakya
Antípaxoi *83 A5 var.* Andipaxi. *island* Iónia Nisiá, Greece, C Mediterranean Sea
Antipodes Islands *120 D5 island group* S New Zealand
Antipolis *see* Antibes
Antípsara *83 D5 var.* Andípsara. *island* E Greece
Antiquaria *see* Antequera
Ántissa *83 D5 var.* Ándissa. Lésvos, E Greece
An tIúr *see* Newry
Antivari *see* Bar
Antofagasta *42 B2* Antofagasta, N Chile
Antony *68 E2* Hauts-de-Seine, N France
An tSionainn *see* Shannon
Antsirañana *57 G2 province* N Madagascar
Antsohihy *57 G2* Mahajanga, NW Madagascar
An-tung *see* Dandong
Antwerpen *65 C5 Eng.* Antwerp, *Fr.* Anvers. Antwerpen, N Belgium
Anuradhapura *110 D3* North Central Province, C Sri Lanka
Anvers *see* Antwerpen
Anyang *106 C4* Henan, C China
A'nyêmaqên Shan *104 D4 mountain range* C China
Anykščiai *84 C4* Utena, E Lithuania
Anzio *75 C5* Lazio, C Italy
Ao Krung Thep *115 C5 var.* Krung Thep Mahanakhon, *Eng.* Bangkok. *country capital* (Thailand) Bangkok, C Thailand
Aomori *108 D3* Aomori, Honshū, C Japan
Aóos *see* Vjosës, Lumi i
Aoraki *129 B6 prev.* Aorangi, Mount Cook. *mountain* South Island, New Zealand
Aorangi *see* Aoraki
Aosta *74 A1 anc.* Augusta Praetoria. Valle d'Aosta, NW Italy
Aoukâr *52 D3 var.* Aouker. *plateau* C Mauritania
Aouk, Bahr *54 C4 river* Central African Republic/Chad
Aouker *see* Aoukâr
Aozou *54 C1* Borkou-Ennedi-Tibesti, N Chad
Apalachee Bay *20 D3 bay* Florida, SE USA
Apalachicola River *20 D3 river* Florida, SE USA
Apamama *see* Abemama
Apaporis, Río *36 C4 river* Brazil/Colombia
Apatity *88 C2* Murmanskaya Oblast', NW Russia
Ape *84 D3* Alūksne, NE Latvia
Apeldoorn *64 D3* Gelderland, E Netherlands
Apennines *74 E2 Eng.* Apennines. *mountain range* Italy/San Marino
Apennines *see* Appennino
Āpia *123 F4 country capital* (Samoa) Upolu, SE Samoa
Ap'khazet'i *95 E1 var.* Abkhazia. *autonomous republic* NW Georgia
Apoera *37 G3* Sipaliwini, NW Suriname
Apostle Islands *18 B1 island group* Wisconsin, N USA
Appalachian Mountains *13 D5 mountain range* E USA
Appingedam *64 E1* Groningen, NE Netherlands
Appleton *18 B2* Wisconsin, N USA
Apulia *see* Puglia
Apure, Río *36 C2 river* W Venezuela
Apurímac, Río *38 D3 river* S Peru
Apuseni, Munţii *86 A4 mountain range* W Romania
Aqaba/'Aqaba *see* Al 'Aqabah
Aqaba, Gulf of *98 A4 var.* Gulf of Elat, *Ar.* Khalīj al 'Aqabah; *anc.* Sinus Aelaniticus. *gulf* NE Red Sea
'Aqabah, Khalīj al *see* Aqaba, Gulf of
Āqchah *101 E3 var.* Āqcheh. Jowzjān, N Afghanistan
Āqcheh *see* Āqchah
Aqmola *see* Astana
Aqtöbe *see* Aktobe
Aquae Augustae *see* Dax
Aquae Calidae *see* Bath
Aquae Flaviae *see* Chaves
Aquae Grani *see* Aachen
Aquae Sextiae *see* Aix-en-Provence
Aquae Solis *see* Bath
Aquae Tarbelicae *see* Dax
Aquidauana *41 E4* Mato Grosso do Sul, S Brazil
Aquila/Aquila degli Abruzzi *see* L'Aquila
Aquisgranum *see* Aachen
Aquitaine *69 B6 cultural region* SW France
'Arabah, Wadi al *97 B7 Heb.* Ha'Arava. *dry watercourse* Israel/Jordan
Arabian Basin *102 A4 undersea basin* N Arabian Sea
Arabian Desert *see* Sahara el Sharqîya
Arabian Peninsula *99 B5 peninsula* SW Asia
Arabian Sea *102 A3 sea* NW Indian Ocean
Arabicus, Sinus *see* Red Sea
'Arabī, Khalīj al *see* Persian Gulf
'Arabīyah as Su'ūdīyah, Al Mamlakah al *see* Saudi Arabia
'Arabīyah Jumhūrīyah, Mişr al *see* Egypt
Arab Republic of Egypt *see* Egypt
Aracaju *41 G3 state capital* Sergipe, E Brazil
Araçuai *41 F3* Minas Gerais, SE Brazil
Arad *97 B7* Southern, S Israel
Arad *86 A4* Arad, W Romania
Arafura Sea *120 A3 Ind.* Laut Arafuru. *sea* W Pacific Ocean
Arafuru, Laut *see* Arafura Sea
Aragón *71 E2 autonomous community* E Spain
Araguaia, Río *41 E3 var.* Araguaya. *river* C Brazil
Araguari *41 F3* Minas Gerais, SE Brazil
Araguaya *see* Araguaia, Río
Ara Jovis *see* Aranjuez
Arāk *98 C3 prev.* Sultānābād. Markazī, W Iran
Arakan Yoma *114 A3 mountain range* W Myanmar (Burma)
Araks/Arak's *see* Aras
Aral *see* Aralsk, Kazakhstan
Aral Sea *100 C1 Kaz.* Aral Tengizi, *Rus.* Aral'skoye More, *Uzb.* Orol Dengizi. *inland sea* Kazakhstan/Uzbekistan
Aral'sk *92 B4 Kaz.* Aral. Kzylorda, SW Kazakhstan
Aral'skoye More/Aral Tengizi *see* Aral Sea
Aranda de Duero *70 D2* Castilla-León, N Spain
Arandelovac *78 D4 prev.* Arandjelovac. Serbia, C Serbia
Arandjelovac *see* Arandelovac
Aranjuez *70 D3 anc.* Ara Jovis. Madrid, C Spain
Araouane *53 E2* Tombouctou, N Mali
'Ar'ar *98 B3* Al Ḥudūd ash Shamālīyah, NW Saudi Arabia
Ararat, Mount *see* Büyükağrı Dağı
Aras *95 G3 Arm.* Arak's, *Az.* Araz Nehri, *Per.* Rūd-e Aras, *Rus.* Araks; *prev.* Araxes. *river* SW Asia
Aras, Rūd-e *see* Aras
Arauca *36 C2* Arauca, NE Colombia
Arauca, Río *36 C2 river* Colombia/Venezuela
Arausio *see* Orange
Araxes *see* Aras
Araz Nehri *see* Aras
Arbela *see* Arbīl
Arbīl *98 B2 var.* Erbil, Irbīl, *Kurd.* Hawlēr; *anc.* Arbela. Arbīl, N Iraq
Arbroath *66 D3 anc.* Aberbrothock. E Scotland, United Kingdom
Arbuzinka *see* Arbuzynka
Arbuzynka *87 E3 Rus.* Arbuzinka. Mykolayivs'ka Oblast', S Ukraine
Arcachon *69 B5* Gironde, SW France
Arcae Remorum *see* Châlons-en-Champagne
Arcata *24 A4* California, W USA
Archangel *see* Arkhangel'sk
Archangel Bay *see* Chēshskaya Guba
Archidona *70 D5* Andalucía, S Spain
Arco *74 C2* Trentino-Alto Adige, N Italy
Arctic Mid Oceanic Ridge *see* Nansen Cordillera
Arctic Ocean *133 B3 ocean*
Arda *82 C3 var.* Ardhas, *Gk.* Ardas. *river* Bulgaria/Greece
Ardabīl *98 C2 var.* Ardebil. Ardabīl, NW Iran
Ardakān *98 D3* Yazd, C Iran
Ardas *82 D3 var.* Ardhas, *Bul.* Arda. *river* Bulgaria/Greece
Arḍ aş Şawwān *97 C7 var.* Ardh es Suwwān. *plain* S Jordan
Ardeal *see* Transylvania
Ardebil *see* Ardabīl
Ardèche *69 C5 cultural region* E France
Ardennes *65 C8 physical region* Belgium/France
Ardhas *see* Arda/Ardas
Ardh es Suwwān *see* Arḍ aş Şawwān
Ardino *82 D3* Kŭrdzhali, S Bulgaria
Ard Mhacha *see* Armagh
Ardmore *27 G2* Oklahoma, C USA
Arel *see* Arlon
Arelas/Arelate *see* Arles
Arendal *63 A6* Aust-Agder, S Norway
Arensburg *see* Kuressaare
Arenys de Mar *71 G2* Cataluña, NE Spain
Areópoli *83 B7 prev.* Areópolis. Pelopónnisos, S Greece
Areópolis *see* Areópoli
Arequipa *39 E4* Arequipa, SE Peru
Arezzo *74 C3 anc.* Arretium. Toscana, C Italy
Argalastí *83 C5* Thessalía, C Greece
Argenteuil *68 D1* Val-d'Oise, N France
Argentina *43 B5 off.* Argentine Republic. *country* S South America
Argentina Basin *see* Argentine Basin
Argentine Basin *35 C7 var.* Argentina Basin. *undersea basin* SW Atlantic Ocean
Argentine Republic *see* Argentina
Argentine Rise *see* Falkland Plateau
Argentoratum *see* Strasbourg
Arghandab, Darya-ye *101 E5 river* SE Afghanistan
Argirocastro *see* Gjirokastër
Argo *50 B3* Northern, N Sudan
Argo Fracture Zone *119 C5 tectonic Feature* C Indian Ocean
Árgos *83 B6* Pelopónnisos, S Greece
Argostóli *83 A5 var.* Argostólion. Kefallonía, Iónia Nisiá, Greece, C Mediterranean Sea
Argostólion *see* Argostóli
Argun *103 E1 Chin.* Ergun He, *Rus.* Argun'. *river* China/Russia
Argyrokastron *see* Gjirokastër
Århus *see* Aarhus
Aria *see* Herāt
Ari Atoll *110 A4 var.* Alifu Atoll. *atoll* C Maldives
Arica *42 B1 hist.* San Marcos de Arica. Tarapacá, N Chile
Aridaía *82 B3 var.* Aridea, Aridhaía. Dytikí Makedonía, N Greece
Aridea *see* Aridaía
Aridhaía *see* Aridaía
Arīḥā *96 B3 Eng.* Jericho. Al Karak, W Jordan
Ariminum *see* Rimini
Arinsal *69 A7* NW Andorra Europe

Brampton – Caprivizipfel

Brampton *16 D5* Ontario, S Canada
Branco, Rio *34 C3 river* N Brazil
Brandberg *56 A3 mountain* NW Namibia
Brandenburg *72 C3 var.* Brandenburg an der Havel. Brandenburg, NE Germany
Brandenburg an der Havel *see* Brandenburg
Brandon *15 F5* Manitoba, S Canada
Braniewo *76 D2 Ger.* Braunsberg. Warmińsko-mazurskie, N Poland
Brasil *see* Brazil
Brasília *41 F3 country capital* (Brazil) Distrito Federal, C Brazil
Brasil, Repùblica Federativa do *see* Brazil
Braşov *86 C4 Ger.* Kronstadt, *Hung.* Brassó; *prev.* Oraşul Stalin. Braşov, C Romania
Brassó *see* Braşov
Bratislava *77 C6 Ger.* Pressburg, *Hung.* Pozsony. *country capital* (Slovakia) Bratislavský Kraj, W Slovakia
Bratsk *93 E4* Irkutskaya Oblast', C Russia
Brattia *see* Brač
Braunsberg *see* Braniewo
Braunschweig *72 C4 Eng./Fr.* Brunswick. Niedersachsen, N Germany
Brava *see* Baraawe
Brava, Costa *71 H2 coastal region* NE Spain
Bravo del Norte, Río/Bravo, Río *see* Grande, Rio
Bravo, Río *28 C1 river* Mexico/USA North America
Brawley *25 D8* California, W USA
Brazil *40 C2 off.* Federative Republic of Brazil, *Port.* Repùblica Federativa do Brasil, *Sp.* Brasil; *prev.* United States of Brazil. *country* South America
Brazil Basin *45 C5 var.* Brazilian Basin, Brazil'skaya Kotlovina. *undersea basin* W Atlantic Ocean
Brazil, Federative Republic of *see* Brazil
Brazilian Basin *see* Brazil Basin
Brazilian Highlands *see* Central, Planalto
Brazil'skaya Kotlovina *see* Brazil Basin
Brazil, United States of *see* Brazil
Brazos River *27 G3 river* Texas, SW USA
Brazza *see* Brač
Brazzaville *55 B6 country capital* (Congo) Capital District, S Congo
Brčko *78 C3* Republika Srpska, NE Bosnia and Herzegovina
Brecht *65 C5* Antwerpen, N Belgium
Brecon Beacons *67 C6 mountain range* S Wales, United Kingdom
Breda *64 C4* Noord-Brabant, S Netherlands
Bree *65 D5* Limburg, NE Belgium
Bregalnica *79 E6 river* E Macedonia
Bregenz *35 B7 anc.* Brigantium. Vorarlberg, W Austria
Bregovo *82 B1* Vidin, NW Bulgaria
Bremen *72 B3 Fr.* Brême. Bremen, NW Germany
Bremerhaven *72 B3* Bremen, NW Germany
Bremerton *24 B2* Washington, NW USA
Brenham *27 G3* Texas, SW USA
Brenner, Col du/Brennero, Passo del *see* Brenner Pass
Brenner Pass *74 C1 var.* Brenner Sattel, *Fr.* Col du Brenner, *Ger.* Brennerpass, *It.* Passo del Brennero. *pass* Austria/Italy
Brennerpass *see* Brenner Pass
Brenner Sattel *see* Brenner Pass
Brescia *74 B2 anc.* Brixia. Lombardia, N Italy
Breslau *see* Wrocław
Bressanone *74 C1 Ger.* Brixen. Trentino-Alto Adige, N Italy
Brest *85 A6 Pol.* Brześć nad Bugiem, *Rus.* Brest-Litovsk; *prev.* Brześć Litewski. Brestskaya Voblasts', SW Belarus
Brest *68 A3* Finistère, NW France
Brest-Litovsk *see* Brest
Bretagne *68 A3 Eng.* Brittany, *Lat.* Britannia Minor. *cultural region* NW France
Brewster, Kap *see* Kangikajik
Brewton *20 C3* Alabama, S USA
Brezhnev *see* Naberezhnyye Chelny
Brezovo *82 D2 prev.* Abrashlare. Plovdiv, C Bulgaria
Bria *54 D4* Haute-Kotto, C Central African Republic
Briançon *69 D5 anc.* Brigantio. Hautes-Alpes, SE France
Bricgstow *see* Bristol
Bridgeport *19 F3* Connecticut, NE USA
Bridgetown *33 G2 country capital* (Barbados) SW Barbados
Bridlington *67 D5* E England, United Kingdom
Bridport *67 D7* S England, United Kingdom
Brieg *see* Brzeg
Brig *73 A7 Fr.* Brigue, *It.* Briga. Valais, SW Switzerland
Briga *see* Brig
Brigantio *see* Briançon
Brigantium *see* Bregenz
Brigham City *22 B3* Utah, W USA
Brighton *67 E7* SE England, United Kingdom
Brighton *22 D4* Colorado, C USA
Brigue *see* Brig
Brindisi *75 E5 anc.* Brundisium, Brundusium. Puglia, SE Italy
Briovera *see* St-Lô
Brisbane *127 E5 state capital* Queensland, E Australia
Bristol *67 D7 anc.* Bricgstow. SW England, United Kingdom
Bristol *19 F3* Connecticut, NE USA
Bristol *18 D5* Tennessee, S USA
Bristol Bay *14 B3 bay* Alaska, USA
Bristol Channel *67 C7 inlet* England/Wales, United Kingdom
Britain *58 C3 var.* Great Britain. *island* United Kingdom
Britannia Minor *see* Bretagne
British Columbia *14 D4 Fr.* Colombie-Britannique. *province* SW Canada
British Guiana *see* Guyana
British Honduras *see* Belize
British Indian Ocean Territory *119 B5 UK dependent territory* C Indian Ocean
British Isles *67 island group* NW Europe
British North Borneo *see* Sabah
British Solomon Islands Protectorate *see* Solomon Islands
British Virgin Islands *33 F3 var.* Virgin Islands. *UK dependent territory* E West Indies
Brittany *see* Bretagne
Briva Curretia *see* Brive-la-Gaillarde
Briva Isarae *see* Pontoise
Brive *see* Brive-la-Gaillarde
Brive-la-Gaillarde *69 C5 prev.* Brive; *anc.* Briva Curretia. Corrèze, C France
Brixen *see* Bressanone
Brixia *see* Brescia
Brno *77 B5 Ger.* Brünn. Jihomoravský Kraj, SE Czech Republic (Czechia)
Brocēni *84 B3* Saldus, SW Latvia
Brod/Bród *see* Slavonski Brod
Brodeur Peninsula *15 F2 peninsula* Baffin Island, Nunavut, NE Canada
Brod na Savi *see* Slavonski Brod
Brodnica *76 C3 Ger.* Buddenbrock. Kujawski-pomorskie, C Poland
Broek-in-Waterland *64 C3* Noord-Holland, C Netherlands
Broken Arrow *27 G1* Oklahoma, C USA
Broken Bay *126 E1 bay* New South Wales, SE Australia
Broken Hill *127 B6* New South Wales, SE Australia
Broken Ridge *119 D6 undersea plateau* S Indian Ocean
Bromberg *see* Bydgoszcz
Bromley *67 B8* United Kingdom
Brookhaven *20 B3* Mississippi, S USA
Brookings *23 F3* South Dakota, N USA
Brooks Range *14 D2 mountain range* Alaska, USA
Brookton *125 B6* Western Australia
Broome *124 B3* Western Australia
Broomfield *22 D4* Colorado, C USA
Broucsella *see* Brussel/Bruxelles
Brovary *87 E2* Kyyivs'ka Oblast', N Ukraine
Brownfield *27 E2* Texas, SW USA
Brownsville *27 G5* Texas, SW USA
Brownwood *27 F3* Texas, SW USA
Brozha *85 D7* Mahilyowskaya Voblasts', E Belarus
Bruges *see* Brugge
Brugge *65 A5 Fr.* Bruges. West-Vlaanderen, NW Belgium
Brummen *64 D3* Gelderland, E Netherlands
Brundisium/Brundusium *see* Brindisi
Brunei *116 D3 off.* Brunei Darussalam, *Mal.* Negara Brunei Darussalam. *country* SE Asia
Brunei Darussalam *see* Brunei
Brunei Town *see* Bandar Seri Begawan
Brünn *see* Brno
Brunner, Lake *129 C5 lake* South Island, New Zealand
Brunswick *21 E3* Georgia, SE USA
Brunswick *see* Braunschweig
Brusa *see* Bursa
Brus Laguna *30 D2* Gracias a Dios, E Honduras
Brussa *see* Bursa
Brussel *65 C6 var.* Brussels, *Fr.* Bruxelles, *Ger.* Brüssel; *anc.* Broucsella. *country capital* (Belgium) Brussels, C Belgium
Brüssel/Brussels *see* Brussel/Bruxelles
Brüx *see* Most
Bruxelles *see* Brussel
Bryan *27 G3* Texas, SW USA
Bryansk *89 A5* Bryanskaya Oblast', W Russia
Brzeg *76 C4 Ger.* Brieg; *anc.* Civitas Altae Ripae. Opolskie, S Poland
Brześć Litewski/Brześć nad Bugiem *see* Brest
Brzeżany *see* Berezhany
Bucaramanga *36 B2* Santander, N Colombia
Buchanan *52 C5 prev.* Grand Bassa. SW Liberia
Buchanan, Lake *27 F3 reservoir* Texas, SW USA
Bucharest *see* Bucureşti
Buckeye State *see* Ohio
Bu Craa *see* Bou Craa
Bucureşti *86 C5 Eng.* Bucharest, *Ger.* Bukarest, *prev.* Altenburg; *anc.* Cetatea Damboviţei. *country capital* (Romania) Bucureşti, S Romania
Buda-Kashalyova *85 D7 Rus.* Buda-Koshelëvo. Homyel'skaya Voblasts', SE Belarus
Buda-Koshelëvo *see* Buda-Kashalyova
Budapest *77 C6 off.* Budapest Főváros, *SCr.* Budimpešta. *country capital* (Hungary) Pest, N Hungary
Budapest Főváros *see* Budapest
Budaun *112 D3* Uttar Pradesh, N India
Buddenbrock *see* Brodnica
Budimpešta *see* Budapest
Budweis *see* České Budějovice
Budyšin *see* Bautzen
Buena Park *24 E2* California, W USA North America
Buenaventura *36 A3* Valle del Cauca, W Colombia
Buena Vista *39 G4* Santa Cruz, C Bolivia
Buena Vista *71 H5* S Gibraltar Europe
Buenos Aires *42 D4 hist.* Santa Maria del Buen Aire. *country capital* (Argentina) Buenos Aires, E Argentina
Buenos Aires *31 E5* Puntarenas, SE Costa Rica
Buenos Aires, Lago *43 B6 var.* Lago General Carrera. *lake* Argentina/Chile
Buffalo *19 E3* New York, NE USA
Buffalo Narrows *15 F4* Saskatchewan, C Canada
Buff Bay *32 B5* E Jamaica
Buftea *86 C5* Ilfov, S Romania
Bug *59 E3 Bel.* Zakhodni Buh, *Eng.* Western Bug, *Rus.* Zapadnyy Bug, *Ukr.* Zakhidnyy Buh. *river* E Europe
Buga *36 B3* Valle del Cauca, W Colombia
Bughotu *see* Santa Isabel
Buguruslan *89 D6* Orenburgskaya Oblast', W Russia
Buitenzorg *see* Bogor
Bujalance *70 D4* Andalucía, S Spain
Bujanovac *79 E5* SE Serbia
Bujnurd *see* Bojnūrd
Bujumbura *51 B7 prev.* Usumbura. *country capital* (Burundi) W Burundi
Bukarest *see* Bucureşti
Bukavu *55 E6 prev.* Costermansville. Sud-Kivu, E Dem. Rep. Congo
Bukhara *see* Buxoro
Bukoba *51 B6* Kagera, NW Tanzania
Bülach *73 B7* Zürich, NW Switzerland
Bulawayo *56 D3* Matabeleland North, SW Zimbabwe
Bulgan *105 E2* Bulgan, N Mongolia
Bulgaria *82 C2 off.* Republic of Bulgaria, *Bul.* Bŭlgariya; *prev.* People's Republic of Bulgaria. *country* SE Europe
Bulgaria, People's Republic of *see* Bulgaria
Bulgaria, Republic of *see* Bulgaria
Bŭlgariya *see* Bulgaria
Bullion State *see* Missouri
Bull Shoals Lake *20 B1 reservoir* Arkansas/Missouri, C USA
Bulukumba *117 E4 prev.* Boeloekoemba. Sulawesi, C Indonesia
Bumba *55 D5* Equateur, N Dem. Rep. Congo
Bunbury *125 A7* Western Australia
Bundaberg *126 E4* Queensland, E Australia
Bungo-suidō *109 B7 strait* SW Japan
Bunia *55 E5* Orientale, NE Dem. Rep. Congo
Bünyan *94 D3* Kayseri, C Turkey
Buraida *see* Buraydah
Buraydah *98 B4 var.* Buraida. Al Qaşīm, N Saudi Arabia
Burdigala *see* Bordeaux
Burdur *94 B4 var.* Buldur. Burdur, SW Turkey
Burdur Gölü *94 B4 salt lake* SW Turkey
Burē *50 C4* Āmara, N Ethiopia
Burgas *82 E2 var.* Bourgas. Burgas, E Bulgaria
Burgaski Zaliv *82 E2 gulf* E Bulgaria
Burgos *70 D2* Castilla-León, N Spain
Burgundy *see* Bourgogne
Burhan Budai Shan *104 D4 mountain range* C China
Buriram *115 D5 var.* Buri Ram, Puriramya. Buri Ram, E Thailand
Buri Ram *see* Buriram
Burjassot *71 F3* País Valenciano, E Spain
Burkburnett *27 F2* Texas, SW USA
Burketown *126 B3* Queensland, NE Australia
Burkina *see* Burkina Faso
Burkina Faso *53 E4 off.* Burkina Faso; *var.* Burkina; *prev.* Upper Volta. *country* W Africa
Burley *24 D4* Idaho, NW USA
Burlington *23 G4* Iowa, C USA
Burlington *19 F2* Vermont, NE USA
Burma *see* Myanmar
Burnie *127 C8* Tasmania, SE Australia
Burns *24 C3* Oregon, NW USA
Burnside *15 F3 river* Nunavut, NW Canada
Burnsville *23 F2* Minnesota, N USA
Burrel *79 D6 var.* Burreli. Dibër, C Albania
Burreli *see* Burrel
Burriana *see* Borriana
Bursa *94 B3 var.* Brussa, *prev.* Brusa; *anc.* Prusa. Bursa, NW Turkey
Būr Sa'īd *50 B1 var.* Port Said. N Egypt
Burtnieks *84 C3 var.* Burtnieks Ezers. *lake* N Latvia
Burtnieks Ezers *see* Burtnieks
Burundi *51 B7 off.* Republic of Burundi; *prev.* Kingdom of Burundi, Urundi. *country* C Africa
Burundi, Kingdom of *see* Burundi
Burundi, Republic of *see* Burundi
Buru, Pulau *117 F4 prev.* Boeroe. *island* E Indonesia
Busan *107 E4 off.* Busan Gwang-yeoksi, *prev.* Pusan, *Jap.* Fusan. SE South Korea
Busan Gwang-yeoksi *see* Busan
Buşayrah *96 D3* Dayr az Zawr, E Syria
Büshehr/Bushire *see* Bandar-e Büshehr
Busra *see* Al Başrah, Iraq
Busselton *125 A7* Western Australia
Bussora *see* Al Başrah
Buta *55 D5* Orientale, N Dem. Rep. Congo
Butembo *55 E5* Nord-Kivu, NE Dem. Rep. Congo
Butler *19 E4* Pennsylvania, NE USA
Buton, Pulau *117 E4 var.* Pulau Butung; *prev.* Boetoeng. *island* C Indonesia
Bütow *see* Bytów
Butte *22 B2* Montana, NW USA
Butterworth *116 B3* Pinang, Peninsular Malaysia
Button Islands *17 E1 island group* Nunavut, NE Canada
Butuan *117 F2 off.* Butuan City. Mindanao, S Philippines
Butuan City *see* Butuan
Butung, Pulau *see* Buton, Pulau
Butuntum *see* Bitonto
Buulobarde *51 D5 var.* Buulo Berde. Hiiraan, C Somalia
Buulo Berde *see* Buulobarde
Buur Gaabo *51 D6* Jubbada Hoose, S Somalia
Buxoro *100 D2 var.* Bokhara, *Rus.* Bukhara. Buxoro Viloyati, C Uzbekistan
Buynaksk *89 B8* Respublika Dagestan, SW Russia
Büyükağrı Dağı *95 F3 var.* Aghri Dagh, Agri Dagi, Koh I Noh, Masis, *Eng.* Great Ararat, Mount Ararat. *mountain* E Turkey
Büyükmenderes Nehri *94 A4 river* SW Turkey
Buzău *86 C4* Buzău, SE Romania
Būzmeyin *see* Abadan
Buzuluk *89 D6* Orenburgskaya Oblast', W Russia
Byahoml' *85 D5 Rus.* Begoml'. Vitsyebskaya Voblasts', N Belarus
Byalynichy *85 D6 Rus.* Belynichi. Mahilyowskaya Voblasts', E Belarus
Byan Tumen *see* Choybalsan
Byarezina *85 D6 prev.* Byerezino, *Rus.* Berezina. *river* C Belarus
Bydgoszcz *76 C3 Ger.* Bromberg. Kujawski-pomorskie, C Poland
Byelaruskaya Hrada *85 B6 Rus.* Belorusskaya Gryada. *ridge* N Belarus
Byerezino *see* Byarezina
Byron Island *see* Nikunau
Bystrovka *see* Kemin
Bytča *77 C5* Žilinský Kraj, N Slovakia
Bytom *77 C5 Ger.* Beuthen. Śląskie, S Poland
Bytów *76 C2 Ger.* Bütow. Pomorskie, N Poland
Byuzmeyin *see* Abadan
Byval'ki *85 D8* Homyel'skaya Voblasts', SE Belarus
Byzantium *see* İstanbul

C

Caála *56 B2 var.* Kaala, Robert Williams, *Port.* Vila Robert Williams. Huambo, C Angola
Caazapá *42 D3* Caazapá, S Paraguay
Caballo Reservoir *26 C3 reservoir* New Mexico, SW USA
Cabanaquinta *70 D1 var.* Cabañaquinta. Asturias, N Spain
Cabañaquinta *see* Cabanaquinta
Cabanatuan *117 E1 off.* Cabanatuan City. Luzon, N Philippines
Cabanatuan City *see* Cabanatuan
Cabillonum *see* Chalon-sur-Saône
Cabimas *36 C1* Zulia, NW Venezuela
Cabinda *56 A1 var.* Kabinda. Cabinda, NW Angola
Cabinda *56 A1 var.* Kabinda. *province* NW Angola
Cabo Verde, Republic of *see* Cape Verde
Cahora Bassa, Albufeira de *56 D2 var.* Lake Cabora Bassa. *reservoir* NW Mozambique
Cabora Bassa, Lake *see* Cahora Bassa, Albufeira de
Caborca *28 B1* Sonora, NW Mexico
Cabot Strait *17 G4 strait* E Canada
Cabo Verde, Ilhas do *see* Cape Verde
Cabras, Ilha das *54 E2 island* S Sao Tome and Principe, Africa, E Atlantic Ocean
Cabrera, Illa de *71 G3 island* E Spain
Cáceres *70 C3 Ar.* Qazris. Extremadura, W Spain
Cachimbo, Serra do *41 E2 mountain range* C Brazil
Caconda *56 B2* Huíla, C Angola
Čadca *77 C5 Hung.* Csaca. Žilinský Kraj, N Slovakia
Cadillac *18 C2* Michigan, N USA
Cadiz *117 E2 off.* Cadiz City. Negros, C Philippines
Cádiz *70 C5 anc.* Gades, Gadier, Gadir, Gadire. Andalucía, SW Spain
Cadiz City *see* Cadiz
Cádiz, Golfo de *70 B5 Eng.* Gulf of Cadiz. *gulf* Portugal/Spain
Cadiz, Gulf of *see* Cádiz, Golfo de
Cadurcum *see* Cahors
Caen *68 B3* Calvados, N France
Caene/Caenepolis *see* Qinā
Caerdydd *see* Cardiff
Caer Glou *see* Gloucester
Caer Gybi *see* Holyhead
Caerleon *see* Chester
Caer Luel *see* Carlisle
Caesaraugusta *see* Zaragoza
Caesarea Mazaca *see* Kayseri
Caesarobriga *see* Talavera de la Reina
Caesarodunum *see* Tours
Caesaromagus *see* Beauvais
Caesena *see* Cesena
Cafayate *42 C2* Salta, N Argentina
Cagayan de Oro *117 E2 off.* Cagayan de Oro City. Mindanao, S Philippines
Cagayan de Oro City *see* Cagayan de Oro
Cagliari *75 A6 anc.* Caralis. Sardegna, Italy, C Mediterranean Sea
Caguas *33 F3* E Puerto Rico
Cahors *69 C5 anc.* Cadurcum. Lot, S France
Cahul *86 D4 Rus.* Kagul. S Moldova
Caicos Passage *32 D2 strait* The Bahamas/Turks and Caicos Islands
Caiffa *see* Hefa
Cailungo *74 E1* N San Marino
Caiphas *see* Hefa
Cairns *126 D3* Queensland, NE Australia
Cairo *50 B2 var.* El Qâhira, *Ar.* Al Qāhirah. *country capital* (Egypt) N Egypt
Cairo *18 B5* Illinois, N USA
Caisleán an Bharraigh *see* Castlebar
Cajamarca *38 B3 prev.* Caxamarca. Cajamarca, NW Peru
Čakovec *78 B2 Ger.* Csakathurn, *Hung.* Csáktornya; *prev. Ger.* Tschakathurn. Medimurje, N Croatia
Calabar *53 G5* Cross River, S Nigeria
Calabozo *36 D2* Guárico, C Venezuela
Calafat *86 B5* Dolj, SW Romania
Calafate *see* El Calafate
Calahorra *71 E2* La Rioja, N Spain
Calais *68 C2* Pas-de-Calais, N France
Calais *19 H2* Maine, NE USA
Calais, Pas de *see* Dover, Strait of
Calama *42 B2* Antofagasta, N Chile
Călăraşi *see* Călăraşi
Călăraşi *86 D3 var.* Călăras, *Rus.* Kalarash. C Moldova
Călăraşi *86 C5* Călăraşi, SE Romania
Calatayud *71 E2* Aragón, NE Spain
Calbayog *117 E2 off.* Calbayog City. Samar, C Philippines
Calbayog City *see* Calbayog
Calcutta *see* Kolkata
Caldas da Rainha *70 B3* Leiria, W Portugal
Caldera *42 B3* Atacama, N Chile
Caldwell *24 C3* Idaho, NW USA
Caledonia *30 C1* Corozal, N Belize
Caleta Olivia *43 B6* Santa Cruz, SE Argentina
Calgary *15 E5* Alberta, SW Canada
Cali *36 B3* Valle del Cauca, W Colombia
Calicut *see* Kozhikode
California *25 B7 off.* State of California, *also known as* El Dorado, The Golden State. *state* W USA
California, Golfo de *28 B2 Eng.* Gulf of California; *prev.* Sea of Cortez. *gulf* W Mexico
California, Gulf of *see* California, Golfo de
Călimăneşti *86 B4* Vâlcea, SW Romania
Calisia *see* Kalisz
Callabonna, Lake *127 B5 lake* South Australia
Callao *38 C4* Callao, W Peru
Callatis *see* Mangalia
Callosa de Segura *71 F4* País Valenciano, E Spain
Calmar *see* Kalmar
Caloundra *127 E5* Queensland, E Australia
Caltanissetta *75 C7* Sicilia, Italy, C Mediterranean Sea
Caluula *50 E4* Bari, NE Somalia
Calvinia *56 C5* W South Africa
Camabatela *56 B1* Cuanza Norte, NW Angola
Camacupa *56 B2 var.* General Machado, *Port.* Vila General Machado. Bié, C Angola
Camagüey *32 C2 prev.* Puerto Príncipe. Camagüey, C Cuba
Camagüey, Archipiélago de *32 C2 island group* C Cuba
Camana *39 E4 var.* Camaná. Arequipa, SW Peru
Camargue *69 D6 physical region* SE France
Ca Mau *115 D6 var.* Quan Long. Minh Hai, S Vietnam
Cambay, Gulf of *see* Khambhāt, Gulf of
Camberia *see* Chambéry
Cambodia *115 D5 off.* Kingdom of Cambodia, *var.* Democratic Kampuchea, Roat Kampuchea, *Cam.* Kampuchea; *prev.* People's Democratic Republic of Kampuchea. *country* SE Asia
Cambodia, Kingdom of *see* Cambodia
Cambrai *68 C2 Flem.* Kambryk, *prev.* Cambray; *anc.* Cameracum. Nord, N France
Cambray *see* Cambrai
Cambrian Mountains *67 C6 mountain range* C Wales, United Kingdom
Cambridge *32 A4* W Jamaica
Cambridge *128 D3* Waikato, North Island, New Zealand
Cambridge *67 E6 Lat.* Cantabrigia. E England, United Kingdom
Cambridge *19 F4* Maryland, NE USA
Cambridge *18 D4* Ohio, NE USA
Cambridge Bay *15 F3 var.* Ikaluktutiak. Victoria Island, Nunavut, NW Canada
Camden *20 B2* Arkansas, C USA
Camellia State *see* Alabama
Cameracum *see* Cambrai
Cameroon *54 A4 off.* Republic of Cameroon, *Fr.* Cameroun. *country* W Africa
Cameroon, Republic of *see* Cameroon
Cameroun *see* Cameroon
Camocim *41 F2* Ceará, E Brazil
Camopi *37 H3* E French Guiana
Campamento *30 C2* Olancho, C Honduras
Campania *75 D5 Eng.* Champagne. *region* S Italy
Campbell, Cape *129 D5 headland* South Island, New Zealand
Campbell Island *120 D5 island* S New Zealand
Campbell Plateau *120 D5 undersea plateau* SW Pacific Ocean
Campbell River *14 D5* Vancouver Island, British Columbia, SW Canada
Campeche *29 G4* Campeche, SE Mexico
Campeche, Bahía de *29 F4 Eng.* Bay of Campeche. *bay* E Mexico
Campeche, Bay of *see* Campeche, Bahía de
Câm Pha *114 E3* Quang Ninh, N Vietnam
Câmpina *86 C4 prev.* Cîmpina. Prahova, SE Romania
Campina Grande *41 G2* Paraíba, E Brazil
Campinas *41 F4* São Paulo, S Brazil
Campobasso *75 D5* Molise, C Italy
Campo Criptana *see* Campo de Criptana
Campo de Criptana *71 E3 var.* Campo Criptana. Castilla-La Mancha, C Spain
Campo dos Goytacazes *41 F4 var.* Campos. Rio de Janeiro, SE Brazil
Campo Grande *41 E4 state capital* Mato Grosso do Sul, SW Brazil
Campos *see* Campo dos Goytacazes
Câmpulung *86 B4 prev.* Câmpulung-Muşcel, Cîmpulung. Argeş, S Romania
Câmpulung-Muşcel *see* Câmpulung
Campus Stellae *see* Santiago de Compostela
Cam Ranh *115 E6* Khanh Hoa, S Vietnam
Canada *12 B4 country* N North America
Canada Basin *12 C2 undersea basin* Arctic Ocean
Canadian River *27 E2 river* SW USA
Çanakkale *94 A3 var.* Dardanelli; *prev.* Chanak, Kale Sultanie. Çanakkale, W Turkey
Cananea *28 B1* Sonora, NW Mexico
Canarreos, Archipiélago de los *32 B2 island group* W Cuba
Canary Islands *48 A2 Eng.* Canary Islands. *island group* Spain, NE Atlantic Ocean
Canary Islands *see* Canarias, Islas
Cañas *30 D4* Guanacaste, NW Costa Rica
Canaveral, Cape *21 E4 headland* Florida, SE USA
Canavieiras *41 G3* Bahia, E Brazil
Canberra *120 C4 country capital* (Australia) Australian Capital Territory, SE Australia
Cancún *29 H3* Quintana Roo, SE Mexico
Candia *see* Irákleio
Canea *see* Chaniá
Cangzhou *106 D4* Hebei, E China
Caniapiscau *17 E2 river* Québec, E Canada
Caniapiscau, Réservoir de *16 D3 reservoir* Québec, C Canada
Canik Dağları *94 D2 mountain range* N Turkey
Canillo *69 A7* Canillo, C Andorra Europe
Çankırı *94 C3 var.* Chankiri; *anc.* Gangra, Germanicopolis. Çankın, N Turkey
Cannanore *see* Kannur
Cannes *69 D6* Alpes-Maritimes, SE France
Canoas *41 E5* Rio Grande do Sul, S Brazil
Canon City *22 C5* Colorado, C USA
Cantabria *70 D1 autonomous community* N Spain
Cantábrica, Cordillera *70 C1 mountain range* N Spain
Cantabrigia *see* Cambridge
Cantaura *37 E2* Anzoátegui, NE Venezuela
Canterbury *67 E7 hist.* Cantwaraburh; *anc.* Durovernum, *Lat.* Cantuaria. SE England, United Kingdom
Canterbury Bight *129 C6 bight* South Island, New Zealand
Canterbury Plains *129 C6 plain* South Island, New Zealand
Cân Thơ *115 E6* Cân Thơ, S Vietnam
Canton *20 B2* Mississippi, S USA
Canton *18 D4* Ohio, N USA
Canton *see* Guangzhou
Canton Island *see* Kanton
Cantuaria/Cantwaraburh *see* Canterbury
Canyon *27 E2* Texas, SW USA
Cao Bāng *114 D3 var.* Caobang. Cao Bāng, N Vietnam
Caobang *see* Cao Bāng
Cap-Breton, Île du *see* Cape Breton Island
Cape Barren Island *127 C8 island* Furneaux Group, Tasmania, SE Australia
Cape Basin *47 B7 undersea basin* S Atlantic Ocean
Cape Breton Island *17 G4 Fr.* Île du Cap-Breton. *island* Nova Scotia, SE Canada
Cape Charles *19 F5* Virginia, NE USA
Cape Coast *53 E5 prev.* Cape Coast Castle. S Ghana
Cape Coast Castle *see* Cape Coast
Cape Girardeau *23 H5* Missouri, C USA
Capelle aan den IJssel *64 C4* Zuid-Holland, SW Netherlands
Cape Palmas *see* Harper
Cape Saint Jacques *see* Vung Tau
Cape Town *56 B5 var.* Ekapa, *Afr.* Kaapstad, Kapstad. *country capital* (South Africa-legislative capital) Western Cape, SW South Africa
Cape Verde *52 A2 off.* Republic of Cabo Verde, *Port.* Cabo Verde, Ilhas do Cabo Verde. *country* E Atlantic Ocean
Cape Verde Basin *44 C4 undersea basin* E Atlantic Ocean
Cape Verde Plain *44 C4 abyssal plain* E Atlantic Ocean
Cape York Peninsula *126 C2 peninsula* Queensland, N Australia
Cap-Haïtien *32 D3 var.* Le Cap. N Haiti
Capira *31 G5* Panamá, C Panama
Capitán Pablo Lagerenza *42 D1 var.* Mayor Pablo Lagerenza. Chaco, N Paraguay
Capodistria *see* Koper
Capri *75 C5 island* S Italy
Caprivi Concession *see* Caprivi Strip
Caprivi Strip *56 C3 Ger.* Caprivizipfel; *prev.* Caprivi Concession. *cultural region* NE Namibia
Caprivizipfel *see* Caprivi Strip

Cap Saint-Jacques – Ch'ing Hai

Cap Saint-Jacques *see* Vung Tau
Caquetá, Río *36 C5 var.* Rio Japurá, Yapurá. *river* Brazil/Colombia
Caquetá, Río *see* Japurá, Rio
CAR *see* Central African Republic
Caracal *86 B5* Olt, S Romania
Caracaraí *40 D1* Rondônia, W Brazil
Caracas *36 D1 country capital* (Venezuela) Distrito Federal, N Venezuela
Caralis *see* Cagliari
Caratasca, Laguna de *31 E2 lagoon* NE Honduras
Carballiño *see* O Carballiño
Carbón, Laguna del *43 B7 physical feature* SE Argentina
Carbondale *18 B5* Illinois, N USA
Carbonia *75 A6 var.* Carbonia Centro. Sardegna, Italy, C Mediterranean Sea
Carbonia Centro *see* Carbonia
Carcaso *see* Carcassonne
Carcassonne *69 C6 anc.* Carcaso. Aude, S France
Cardamomes, Chaîne des *see* Krâvanh, Chuŏr Phnum
Cardamom Mountains *see* Krâvanh, Chuŏr Phnum
Cárdenas *32 B2* Matanzas, W Cuba
Cardiff *67 C7 Wel.* Caerdydd. *national capital* S Wales, United Kingdom
Cardigan Bay *67 C6 bay* W Wales, United Kingdom
Carei *86 B3 Ger.* Gross-Karol, Karol, *Hung.* Nagykároly; *prev.* Careii-Mari. Satu Mare, NW Romania
Careii-Mari *see* Carei
Carey, Lake *125 B6 lake* Western Australia
Cariaco *37 E1* Sucre, NE Venezuela
Caribbean Sea *32 C4 sea* W Atlantic Ocean
Caribrod *see* Dimitrovgrad
Carlisle *66 C4 anc.* Caer Luel, Luguvallium, Luguvallum. NW England, United Kingdom
Carlow *67 B6 Ir.* Ceatharlach. SE Ireland
Carlsbad *26 D3* New Mexico, SW USA
Carlsbad *see* Karlovy Vary
Carlsberg Ridge *118 B4 undersea ridge* S Arabian Sea
Carlsruhe *see* Karlsruhe
Carmana/Carmania *see* Kermān
Carmarthen *67 C6* SW Wales, United Kingdom
Carmaux *69 C6* Tarn, S France
Carmel *18 C4* Indiana, N USA
Carmelita *30 B1* Petén, N Guatemala
Carmen *29 G4 var.* Ciudad del Carmen. Campeche, SE Mexico
Carmona *70 C4* Andalucía, S Spain
Carmona *see* Uíge
Carnaro *see* Kvarner
Carnarvon *125 A5* Western Australia
Carnegie, Lake *125 B5 salt lake* Western Australia
Car Nicobar *111 F3 island* Nicobar Islands, India, NE Indian Ocean
Caroço, Ilha *54 E1 island* N Sao Tome and Principe, Africa, E Atlantic Ocean
Carolina *41 F2* Maranhão, E Brazil
Caroline Island *see* Millennium Island
Caroline Islands *122 B2 island group* C Micronesia
Carolopois *see* Châlons-en-Champagne
Caroní, Río *37 E3 river* E Venezuela
Caronium *see* A Coruña
Carora *36 C1* Lara, N Venezuela
Carpathian Mountains *59 E4 var.* Carpathians, *Cz./Pol.* Karpaty, *Ger.* Karpaten. *mountain range* E Europe
Carpathians *see* Carpathian Mountains
Carpathos/Carpathus *see* Kárpathos
Carpații Meridionalii *86 B4 var.* Alpi Transilvaniei, Carpații Sudici, *Eng.* South Carpathians, Transylvanian Alps, *Ger.* Südkarpaten, Transsylvanische Alpen, *Hung.* Déli-Kárpátok, Erdélyi-Havasok. *mountain range* C Romania
Carpații Sudici *see* Carpații Meridionalii
Carpentaria, Gulf of *126 B2 gulf* N Australia
Carpi *74 C2* Emilia-Romagna, N Italy
Carrara *74 B3* Toscana, C Italy
Carson City *25 C5 state capital* Nevada, W USA
Carson Sink *25 C5 salt flat* Nevada, W USA
Carstensz, Puntjak *see* Jaya, Puncak
Cartagena *36 B1 var.* Cartagena de los Indes. Bolívar, NW Colombia
Cartagena *71 F4 anc.* Carthago Nova. Murcia, SE Spain
Cartagena de los Indes *see* Cartagena
Cartago *31 E4* Cartago, C Costa Rica
Carthage *23 F5* Missouri, C USA
Carthago Nova *see* Cartagena
Cartwright *17 F2* Newfoundland and Labrador, E Canada
Carúpano *37 E1* Sucre, NE Venezuela
Carusbur *see* Cherbourg
Caruthersville *23 H5* Missouri, C USA
Cary *21 F1* North Carolina, SE USA
Casablanca *48 C2 Ar.* Dar-el-Beida. NW Morocco
Casa Grande *26 B2* Arizona, SW USA
Cascade Range *24 B3 mountain range* Oregon/Washington, NW USA
Cascadia Basin *12 A4 undersea basin* NE Pacific Ocean
Cascais *70 B4* Lisboa, C Portugal
Caserta *75 D5* Campania, S Italy
Casey *132 D4 Australian research station* Antarctica
Čáslav *77 B5 Ger.* Tschaslau. Střední Čechy, C Czech Republic (Czechia)
Casper *22 C3* Wyoming, C USA
Caspian Depression *89 B7 Kaz.* Kaspiy Mangy Oypaty, *Rus.* Prikaspiyskaya Nizmennost'. *depression* Kazakhstan/Russia
Caspian Sea *92 A4 Az.* Xäzär Dänizi, *Kaz.* Kaspiy Tengizi, *Per.* Baḩr-e Khazar, Daryā-ye Khazar, *Rus.* Kaspiyskoye More. *inland sea* Asia/Europe
Cassai *see* Kasai
Cassel *see* Kassel
Castamoni *see* Kastamonu
Casteggio *74 B2* Lombardia, N Italy
Castelló de la Plana *see* Castellón de la Plana
Castellón *see* Castellón de la Plana
Castellón de la Plana *71 F3 var.* Castellón, *Cat.* Castelló de la Plana. País Valenciano, E Spain
Castelnaudary *69 C6* Aude, S France
Castelo Branco *70 C3* Castelo Branco, C Portugal
Castelsarrasin *69 B6* Tarn-et-Garonne, S France
Castelvetrano *75 C7* Sicilia, Italy, C Mediterranean Sea
Castilla-La Mancha *71 E3 autonomous community* NE Spain
Castilla-León *70 C2 var.* Castillia y Leon. *autonomous community* NW Spain
Castillia y Leon *see* Castilla-León
Castlebar *67 A5 Ir.* Caisleán an Bharraigh. W Ireland
Castleford *67 D5* N England, United Kingdom
Castle Harbour *20 B5 inlet* Bermuda, NW Atlantic Ocean
Castra Regina *see* Regensburg
Castricum *64 C3* Noord-Holland, W Netherlands
Castries *33 F1 country capital* (Saint Lucia) N Saint Lucia
Castro *43 B6* Los Lagos, W Chile
Castrovillari *75 D6* Calabria, SW Italy
Castuera *70 D4* Extremadura, W Spain
Caswell Sound *129 A7 sound* South Island, New Zealand
Catacamas *30 D2* Olancho, C Honduras
Catacaos *38 B3* Piura, NW Peru
Catalan Bay *71 H4 bay* E Gibraltar, Mediterranean Sea
Cataluña *71 G2* N Spain
Catamarca *see* San Fernando del Valle de Catamarca
Catania *75 D7* Sicilia, Italy, C Mediterranean Sea
Catanzaro *75 D6* Calabria, SW Italy
Catarroja *71 F3* País Valenciano, E Spain
Cat Island *32 C1 island* C The Bahamas
Catskill Mountains *19 F3 mountain range* New York, NE USA
Cattaro *see* Kotor
Cauca, Río *36 B2 river* N Colombia
Caucasia *36 B2* Antioquia, NW Colombia
Caucasus *59 G4 Rus.* Kavkaz. *mountain range* Georgia/Russia
Caura, Río *37 E3 river* C Venezuela
Cavaia *see* Kavajë
Cavalla *52 D5 var.* Cavally, Cavally Fleuve. *river* Ivory Coast/Liberia
Cavally/Cavally Fleuve *see* Cavalla
Caviana de Fora, Ilha *41 E1 var.* Ilha Caviana. *island* N Brazil
Caviana, Ilha *see* Caviana de Fora, Ilha
Cawnpore *see* Känpur
Caxamarca *see* Cajamarca
Caxito *56 B1* Bengo, NW Angola
Cayenne *37 H3 dependent territory/ arrondissement capital* (French Guiana) NE French Guiana
Cayes *32 D3 var.* Les Cayes. SW Haiti
Cayman Brac *32 B3 island* E Cayman Islands
Cayman Islands *32 B3 UK dependent territory* W West Indies
Cayo *see* San Ignacio
Cay Sal *32 B2 islet* SW The Bahamas
Cazin *78 B3* Federacija Bosna I Hercegovina, NW Bosnia and Herzegovina
Cazorla *71 E4* Andalucía, S Spain
Ceadâr-Lunga *see* Ciadîr-Lunga
Ceará *41 F2 off.* Estado do Ceará. *state/region* C Brazil
Ceará *see* Fortaleza
Ceara Abyssal Plain *see* Ceará Plain
Ceará, Estado do *see* Ceará
Ceará Plain *34 E3 var.* Ceara Abyssal Plain. *abyssal plain* W Atlantic Ocean
Ceatharlach *see* Carlow
Cébaco, Isla *31 F5 island* SW Panama
Cebu *117 E2 off.* Cebu City. Cebu, C Philippines
Cebu City *see* Cebu
Čechy *see* Bohemia
Cecina *74 B3* Toscana, C Italy
Cedar City *22 A5* Utah, W USA
Cedar Falls *23 G3* Iowa, C USA
Cedar Lake *16 A2 lake* Manitoba, C Canada
Cedar Rapids *23 G3* Iowa, C USA
Cedros, Isla *28 A2 island* W Mexico
Ceduna *127 A6* South Australia
Cefalù *75 C7 anc.* Cephaloedium. Sicilia, Italy, C Mediterranean Sea
Celebes *see* Sulawesi
Celebes Sea *117 E3 Ind.* Laut Sulawesi. *sea* Indonesia/Philippines
Celje *73 E7 Ger.* Cilli. C Slovenia
Celldömölk *77 C6* Vas, W Hungary
Celle *72 B3 var.* Zelle. Niedersachsen, N Germany
Celovec *see* Klagenfurt
Celtic Sea *67 B7 Ir.* An Mhuir Cheilteach. *sea* SW British Isles
Celtic Shelf *58 B3 continental shelf* E Atlantic Ocean
Cenderawasih, Teluk *117 G4 var.* Teluk Irian, Teluk Sarera. *bay* W Pacific Ocean
Cenon *69 B5* Gironde, SW France
Centennial State *see* Colorado
Centrafricaine, République *see* Central African Republic
Central African Republic *54 C4 var.* République Centrafricaine, *abbrev.* CAR; *prev.* Ubangi-Shari, Oubangui-Chari, Territoire de l'Oubangui-Chari. *country* C Africa
Central, Cordillera *36 B3 mountain range* W Colombia
Cordillera Central *33 E3 mountain range* C Dominican Republic
Cordillera Central *31 F5 mountain range* C Panama
Central, Cordillera *117 E1 mountain range* Luzon, N Philippines
Central Group *see* Inner Islands
Centralia *24 B2* Washington, NW USA
Central Indian Ridge *see* Mid-Indian Ridge
Central Makran Range *112 A3 mountain range* W Pakistan
Central Pacific Basin *120 D1 undersea basin* C Pacific Ocean
Central, Planalto *41 F3 var.* Brazilian Highlands. *mountain range* E Brazil
Central Provinces and Berar *see* Madhya Pradesh
Central Range *122 B3 mountain range* NW Papua New Guinea
Central Russian Upland *see* Srednerusskaya Vozvyshennost'
Central Siberian Plateau *92 D3 var.* Central Siberian Uplands, *Eng.* Central Siberian Plateau. *mountain range* N Russia
Central Siberian Plateau/Central Siberian Uplands *see* Srednesibirskoye Ploskogor'ye
Central, Sistema *70 D3 mountain range* C Spain
Central Valley *25 B6 valley* California, W USA
Centum Cellae *see* Civitavecchia
Ceos *see* Tziá
Cephaloedium *see* Cefalù
Ceram *see* Seram, Pulau
Ceram Sea *see* Laut Seram
Cerasus *see* Giresun
Cereté *36 B2* Córdoba, NW Colombia
Cergy-Pontoise *see* Pontoise
Cerignola *75 D5* Puglia, SE Italy
Çerkeş *94 C2* Çankin, N Turkey
Cernăuți *see* Chernivtsi
Cernay *68 E4* Haut-Rhin, NE France
Cerro de Pasco *38 C3* Pasco, C Peru
Cervera *71 F2* Cataluña, NE Spain
Cervino, Monte *see* Matterhorn
Cesena *74 C3 anc.* Caesena. Emilia-Romagna, N Italy
Cēsis *84 D3 Ger.* Wenden. Cēsis, C Latvia
Česká Republika *see* Czech Republic (Czechia)
České Budějovice *77 B5 Ger.* Budweis. Jihočeský Kraj, S Czech Republic (Czechia)
Český Krumlov *77 A5 var.* Böhmisch-Krumau, *Ger.* Krummau. Jihočeský Kraj, S Czech Republic (Czechia)
Ceský Les *see* Bohemian Forest
Cetatea Damboviței *see* București
Cetinje *79 C5 It.* Cettigne. S Montenegro
Cette *see* Sète
Cettigne *see* Cetinje
Ceuta *48 C2 enclave* Spain, N Africa
Cévennes *69 C6 mountain range* S France
Ceyhan *94 D4* Adana, S Turkey
Ceylanpınar *95 E4* Şanlıurfa, SE Turkey
Ceylon *see* Sri Lanka
Ceylon Plain *102 B4 abyssal plain* N Indian Ocean
Ceyre to the Caribs *see* Marie-Galante
Chachapoyas *38 B2* Amazonas, NW Peru
Chachevichy *85 D6 Rus.* Chechevichi. Mahilyowskaya Voblasts', E Belarus
Chaco *see* Gran Chaco
Chad *54 C3 off.* Republic of Chad, *Fr.* Tchad. *country* C Africa
Chad, Lake *54 B3 Fr.* Lac Tchad. *lake* C Africa
Chad, Republic of *see* Chad
Chadron *22 D3* Nebraska, C USA
Chadyr-Lunga *see* Ciadîr-Lunga
Chagai Hills *112 A2 var.* Chāh Gay. *mountain range* Afghanistan/Pakistan
Chaghasarāy *see* Asadābād
Chagos-Laccadive Plateau *102 B4 undersea plateau* N Indian Ocean
Chagos Trench *119 C5 trench* N Indian Ocean
Chāh Gay *see* Chāgai Hills
Chaillu, Massif du *55 B6 mountain range* C Gabon
Chain Ridge *118 B4 undersea ridge* W Indian Ocean
Chajul *30 B2* Quiché, W Guatemala
Chakhānsūr *100 D5* Nīmrūz, SW Afghanistan
Chala *38 D4* Arequipa, SW Peru
Chalatenango *30 C3* Chalatenango, N El Salvador
Chalcidice *see* Chalkidikí
Chalcis *see* Chalkída
Chalki *83 E7 island* Dodekánisa, Greece, Aegean Sea
Chalkída *83 C5 var.* Halkida, *prev.* Khalkís; *anc.* Chalcis. Evvoia, E Greece
Chalkidikí *82 C4 var.* Khalkidhikí; *anc.* Chalcidice. *peninsula* NE Greece
Challans *68 B4* Vendée, NW France
Challapata *39 F4* Oruro, SW Bolivia
Challenger Deep *130 B3 trench* W Pacific Ocean
Challenger Fracture Zone *131 F4 tectonic feature* SE Pacific Ocean
Châlons-en-Champagne *68 D3 prev.* Châlons-sur-Marne, *hist.* Arcae Remorum; *anc.* Carolopois. Marne, NE France
Châlons-sur-Marne *see* Châlons-en-Champagne
Chalon-sur-Saône *68 D4 anc.* Cabillonum. Saône-et-Loire, C France
Cha Mai *see* Thung Song
Chaman *112 B2* Baluchistān, SW Pakistan
Chambéry *69 D5 anc.* Camberia. Savoie, E France
Champagne *68 D3 cultural region* N France
Champagne *see* Campania
Champaign *18 B4* Illinois, N USA
Champasak *115 D5* Champasak, S Laos
Champlain, Lake *19 F2 lake* Canada/USA
Champotón *29 G4* Campeche, SE Mexico
Chanak *see* Çanakkale
Chañaral *42 B3* Atacama, N Chile
Chan-chiang/Chanchiang *see* Zhanjiang
Chandeleur Islands *20 C3 island group* Louisiana, S USA
Chandīgarh *112 D2 state capital* Punjab, N India
Chandrapur *113 E5* Mahārāshtra, C India
Changan *see* Xi'an, Shaanxi, C China
Changane *57 E3 river* S Mozambique
Changchun *106 D3 var.* Ch'angch'un, Ch'ang-ch'un; *prev.* Hsinking. *province capital* Jilin, NE China
Ch'angch'un/Ch'ang-ch'un *see* Changchun
Chang Jiang *106 A5 Eng.* Yangtze; *var.* Yangtze Kiang. *river* SW China
Changkiakow *see* Zhangjiakou
Chang, Ko *115 C6 island* S Thailand
Changsha *106 C5 var.* Ch'angsha, Ch'ang-sha. *province capital* Hunan, S China
Ch'angsha/Ch'ang-sha *see* Changsha
Changzhi *106 C4* Shanxi, C China
Chaniá *83 C7 var.* Hania, Khaniá, *Eng.* Canea; *anc.* Cydonia. Kríti, Greece, E Mediterranean Sea
Chañi, Nevado de *42 B2 mountain* NW Argentina
Chankiri *see* Çankırı
Channel Islands *67 C8 Fr.* Iles Normandes. *island group* S English Channel
Channel Islands *25 B8 island group* California, W USA
Channel-Port aux Basques *17 G4* Newfoundland and Labrador, SE Canada
Channel, The *see* English Channel
Channel Tunnel *68 C2 tunnel* France/ United Kingdom
Chantabun/Chantaburi *see* Chanthaburi
Chantada *70 C1* Galicia, NW Spain
Chanthaburi *115 C6 var.* Chantabun, Chantaburi. Chantaburi, S Thailand
Chanute *23 F5* Kansas, C USA
Chaouèn *see* Chefchaouen
Chaoyang *106 D3* Liaoning, NE China
Chapala, Lago de *28 D4 lake* C Mexico
Chapan, Gora *100 B3 mountain* C Turkmenistan
Chapayevsk *89 C6* Samarskaya Oblast', W Russia
Chaplynka *87 F4* Khersons'ka Oblast', S Ukraine
Chapra *see* Chhapra
Charcot Seamounts *58 B3 seamount range* E Atlantic Ocean
Chardzhev *see* Türkmenabat
Chardzhou/Chardzhui *see* Türkmenabat
Charente *69 B5 cultural region* W France
Charente *69 B5 river* W France
Chari *54 B3 var.* Shari. *river* Central African Republic/Chad
Chārīkār *101 E4* Parvān, NE Afghanistan
Charity *37 F2* NW Guyana
Chärjew *see* Türkmenabat
Charkhlik/Charkhliq *see* Ruoqiang
Charleroi *65 C7* Hainaut, S Belgium
Charlesbourg *17 E4* Québec, SE Canada
Charles de Gaulle *68 E1* (Paris) Seine-et-Marne, N France
Charles Island *16 D1 island* Nunavut, NE Canada
Charles Island *see* Santa María, Isla
Charleston *21 F2* South Carolina, SE USA
Charleston *18 D5 state capital* West Virginia, NE USA
Charleville *127 D5* Queensland, E Australia
Charleville-Mézières *68 D3* Ardennes, N France
Charlie-Gibbs Fracture Zone *44 C2 tectonic feature* N Atlantic Ocean
Charlotte *21 E1* North Carolina, SE USA
Charlotte Amalie *33 F3 prev.* Saint Thomas. *dependent territory capital* (Virgin Islands (US)) Saint Thomas, N Virgin Islands (US)
Charlotte Harbor *21 E5 inlet* Florida, SE USA
Charlottenhof *see* Aegviidu
Charlottesville *19 E5* Virginia, NE USA
Charlottetown *17 F4 province capital* Prince Edward Island, Prince Edward Island, SE Canada
Charlotte Town *see* Roseau, Dominica
Charsk *see* Shar
Charters Towers *126 D3* Queensland, NE Australia
Chartres *68 C3 anc.* Autricum, Civitas Carnutum. Eure-et-Loir, C France
Chashniki *85 D5* Vitsyebskaya Voblasts', N Belarus
Châteaubriant *68 B4* Loire-Atlantique, NW France
Châteaudun *68 C3* Eure-et-Loir, C France
Châteauroux *68 C4 prev.* Indreville. Indre, C France
Château-Thierry *68 C3* Aisne, N France
Châtelet *65 C7* Hainaut, S Belgium
Châtelherault *see* Châtellerault
Châtellerault *68 B4 var.* Châtelherault. Vienne, W France
Chatham Island *see* San Cristóbal, Isla
Chatham Island Rise *see* Chatham Rise
Chatham Islands *121 E5 island group* New Zealand, SW Pacific Ocean
Chatham Rise *120 D5 var.* Chatham Island Rise. *undersea rise* S Pacific Ocean
Chatkal Range *101 F2 Rus.* Chatkal'skiy Khrebet. *mountain range* Kyrgyzstan/Uzbekistan
Chatkal'skiy Khrebet *see* Chatkal Range
Chāttagām *see* Chittagong
Chattahoochee River *20 D3 river* SE USA
Chattanooga *20 D1* Tennessee, S USA
Chatyr-Tash *101 G2* Narynskaya Oblast', C Kyrgyzstan
Châu Đôc *115 D6 var.* Chauphu, Chau Phu. An Giang, S Vietnam
Chauk *114 A3* Magway, W Myanmar (Burma)
Chaumont *68 D4 prev.* Chaumont-en-Bassigny. Haute-Marne, N France
Chaumont-en-Bassigny *see* Chaumont
Chau Phu *see* Châu Đôc
Chausy *see* Chavusy
Chaves *70 C2 anc.* Aquae Flaviae. Vila Real, N Portugal
Chávez, Isla *see* Santa Cruz, Isla
Chavusy *85 E6 Rus.* Chausy. Mahilyowskaya Voblasts', E Belarus
Chaykovskiy *89 D5* Permskaya Oblast', NW Russia
Cheb *77 A5 Ger.* Eger. Karlovarský Kraj, W Czech Republic (Czechia)
Cheboksary *89 C5* Chuvashskaya Respublika, W Russia
Cheboygan *18 C2* Michigan, N USA
Chechaouèn *see* Chefchaouen
Chech, Erg *52 D1 desert* Algeria/Mali
Chechevichi *see* Chachevichy
Che-chiang *see* Zhejiang
Cheduba Island *114 A4 island* W Myanmar (Burma)
Chefchaouen *48 C2 var.* Chaouèn, Chechaouèn, *Sp.* Xauen. N Morocco
Chefoo *see* Yantai
Cheju-do *see* Jeju-do
Cheju Strait *see* Jeju Strait
Chekiang *see* Zhejiang
Cheleken *see* Hazar
Chelkar *see* Shalkar
Chełm *76 E4 Rus.* Kholm. Lubelskie, SE Poland
Chełmno *76 C3 Ger.* Culm, Kulm. Kujawski-pomorskie, C Poland
Chełmża *76 C3 Ger.* Culmsee, Kulmsee. Kujawski-pomorskie, C Poland
Cheltenham *67 D6* C England, United Kingdom
Chelyabinsk *92 C3* Chelyabinskaya Oblast', C Russia
Chemnitz *72 D4 prev.* Karl-Marx-Stadt. Sachsen, E Germany
Chemulpo *see* Incheon
Chenāb *112 C2 river* India/Pakistan
Chengchiatun *see* Liaoyuan
Ch'eng-chou/Chengchow *see* Zhengzhou
Chengde *106 D3 var.* Jehol. Hebei, E China
Chengdu *106 B5 var.* Chengtu, Ch'eng-tu. *province capital* Sichuan, C China
Chenghsien *see* Zhengzhou
Chengtu/Ch'eng-tu *see* Chengdu
Chennai *110 D2 prev.* Madras. *state capital* Tamil Nādu, S India
Chenstokhov *see* Częstochowa
Chen Xian/Chenxian/Chen Xiang *see* Chenzhou
Chenzhou *106 C6 var.* Chenxian, Chen Xian, Chen Xiang. Hunan, S China
Chepelare *82 C3* Smolyan, S Bulgaria
Chepén *38 B3* La Libertad, C Peru
Cher *68 C4 river* C France
Cherbourg *68 B3 anc.* Carusbur. Manche, N France
Cherepovets *88 B4* Vologodskaya Oblast', NW Russia
Chergui, Chott ech *48 D2 salt lake* NW Algeria
Cherikov *see* Cherykaw
Cherkassy *see* Cherkasy
Cherkasy *87 E2 Rus.* Cherkassy. Cherkas'ka Oblast', C Ukraine
Cherkessk *89 B7* Karachayevo-Cherkesskaya Respublika, SW Russia
Chernigov *see* Chernihiv
Chernihiv *87 E1 Rus.* Chernigov. Chernihivs'ka Oblast', NE Ukraine
Chernivtsi *86 C3 Ger.* Czernowitz, *Rom.* Cernăuți, *Rus.* Chernovtsy. Chernivets'ka Oblast', W Ukraine
Cherno More *see* Black Sea
Chernomorskoye *see* Chornomors'ke
Chernovtsy *see* Chernivtsi
Chernoye More *see* Black Sea
Chernyakhovsk *84 A4 Ger.* Insterburg. Kaliningradskaya Oblast', W Russia
Cherry Hill *19 F4* New Jersey, NE USA
Cherski Range *see* Cherskogo, Khrebet
Cherskiy *93 G2* Respublika Sakha (Yakutiya), NE Russia
Cherskogo, Khrebet *93 F2 var.* Cherski Range. *mountain range* NE Russia
Cherso *see* Cres
Cherven' *see* Chervyen'
Chervonograd *see* Chervonohrad
Chervonohrad *86 C2 Rus.* Chervonograd. L'vivs'ka Oblast', NW Ukraine
Chervyen' *85 D6 Rus.* Cherven'. Minskaya Voblasts', C Belarus
Cherykaw *85 E7 Rus.* Cherikov. Mahilyowskaya Voblasts', E Belarus
Chesapeake Bay *19 F5 inlet* NE USA
Chesha Bay *see* Chëshskaya Guba
Chëshskaya Guba *133 D5 var.* Archangel Bay, Chesha Bay, Dvina Bay. *bay* NW Russia
Chester *67 C6 Wel.* Caerleon, *hist.* Legaceaster, *Lat.* Deva, Devana Castra. C England, United Kingdom
Chetumal *29 H4 var.* Payo Obispo. Quintana Roo, SE Mexico
Cheviot Hills *66 D4 hill range* England/Scotland, United Kingdom
Cheyenne *22 D4 state capital* Wyoming, C USA
Cheyenne River *22 D3 river* South Dakota/ Wyoming, N USA
Chezdi-Oșorheiu *see* Târgu Secuiesc
Chhapra *113 F3 prev.* Chapra. Bihār, N India
Chhattisgarh *113 E4 cultural region* E India
Chiai *see* Chiayi
Chia-i *see* Chiayi
Chiang-hsi *see* Jiangxi
Chiang Mai *114 B4 var.* Chiangmai, Chiengmai, Kiangmai. Chiang Mai, NW Thailand
Chiangmai *see* Chiang Mai
Chiang Rai *114 C3 var.* Chianpai, Chienrai, Muang Chiang Rai. Chiang Rai, NW Thailand
Chiang-su *see* Jiangsu
Chianning/Chian-ning *see* Nanjing
Chianpai *see* Chiang Rai
Chianti *74 C3 cultural region* C Italy
Chiapa *see* Chiapa de Corzo
Chiapa de Corzo *29 G5 var.* Chiapa. Chiapas, SE Mexico
Chiayi *106 D6 var.* Chiai, Chia-i, Kiayi, Jiayi, *Jap.* Kagi. C Taiwan
Chiba *108 B1 var.* Tiba. Chiba, Honshū, S Japan
Chibougamau *16 D3* Québec, SE Canada
Chicago *18 B3* Illinois, N USA
Ch'i-ch'i-ha-erh *see* Qiqihar
Chickasha *27 G2* Oklahoma, C USA
Chiclayo *38 B3* Lambayeque, NW Peru
Chico *25 B5* California, W USA
Chico, Río *43 B7 river* SE Argentina
Chico, Río *43 B6 river* S Argentina
Chicoutimi *17 E4* Québec, SE Canada
Chiengmai *see* Chiang Mai
Chienrai *see* Chiang Rai
Chiesanuova *74 D2* SW San Marino
Chieti *74 D4 var.* Teate. Abruzzo, C Italy
Chifeng *105 G2 var.* Ulanhad. Nei Mongol Zizhiqu, N China
Chigirin *see* Chyhyryn
Chih-fu *see* Yantai
Chihli *see* Hebei
Chihli, Gulf of *see* Bo Hai
Chihuahua *28 C2* Chihuahua, NW Mexico
Childress *27 F2* Texas, SW USA
Chile *42 B3 off.* Republic of Chile. *country* SW South America
Chile Basin *35 A5 undersea basin* E Pacific Ocean
Chile Chico *43 B6* Aisén, W Chile
Chile, Republic of *see* Chile
Chile Rise *35 A7 undersea rise* SE Pacific Ocean
Chilia-Nouă *see* Kiliya
Chililabombwe *56 D2* Copperbelt, C Zambia
Chi-lin *see* Jilin
Chillán *43 B5* Bío Bío, C Chile
Chillicothe *18 D4* Ohio, N USA
Chill Mhantáin, Sléibhte *see* Wicklow Mountains
Chiloé, Isla de *43 A6 var.* Isla Grande de Chiloé. *island* W Chile
Chilpancingo *29 E5 var.* Chilpancingo de los Bravos. Guerrero, S Mexico
Chilpancingo de los Bravos *see* Chilpancingo
Chilung *see* Keelung
Chimán *31 G5* Panamá, E Panama
Chimbay *see* Chimboy
Chimborazo *38 A1 volcano* C Ecuador
Chimbote *38 C3* Ancash, W Peru
Chimboy *100 D1 Rus.* Chimbay. Qoraqalpog'iston Respublikasi, NW Uzbekistan
Chimkent *see* Shymkent
Chimoio *57 E3* Manica, C Mozambique
China *102 C2 off.* People's Republic of China, *Chin.* Chung-hua Jen-min Kung-ho-kuo, Zhonghua Renmin Gongheguo; *prev.* Chinese Empire. *country* E Asia
Chi-nan/Chinan *see* Jinan
Chinandega *30 C3* Chinandega, NW Nicaragua
China, People's Republic of *see* China
China, Republic of *see* Taiwan
Chincha Alta *38 D4* Ica, SW Peru
Chin-chiang *see* Quanzhou
Chin-chou/Chinchow *see* Jinzhou
Chindwin *see* Chindwinn
Chindwinn *114 B2 var.* Chindwin. *river* N Myanmar (Burma)
Chinese Empire *see* China
Ch'ing Hai *see* Qinghai Hu, China

Djember – El Muglad

Djember see Jember
Djérablous see Jarābulus
Djerba 49 F2 var. Djerba, Jazīrat Jarbah. island E Tunisia
Djerba see Jerba, Île de
Djérem 54 B4 river C Cameroon
Djevdjelija see Gevgelija
Djibouti 50 D4 var. Jibuti. country capital (Djibouti) E Djibouti
Djibouti 50 D4 off. Republic of Djibouti, var. Jibuti; prev. French Somaliland, French Territory of the Afars and Issas, Fr. Côte Française des Somalis, Territoire Français des Afars et des Issas. country E Africa
Djibouti, Republic of see Djibouti
Djokjakarta see Yogyakarta
Djourab, Erg du 54 C2 desert N Chad
Djúpivogur 61 E5 Austurland, SE Iceland
Dmitriyevsk see Makiyivka
Dnepr see Dnieper
Dneprodzerzhinskoye Vodokhranilishche see Dniprodzerzhyns'ke Vodoskhovyshche
Dneprorudnoye see Dniprorudne
Dnestr see Dniester
Dnieper 59 F4 Bel. Dnyapro, Rus. Dnepr, Ukr. Dnipro. river E Europe
Dnieper Lowland 87 E2 Bel. Prydnyaprowskaya Nizina, Ukr. Prydniprovs'ka Nyzovyna. lowlands Belarus/Ukraine
Dniester 59 E4 Rom. Nistru, Rus. Dnestr, Ukr. Dnister; anc. Tyras. river Moldova/Ukraine
Dnipro see Dnieper
Dniprodzerzhyns'k see Kam"yans'ke
Dniprodzerzhyns'ke Vodoskhovyshche 87 F3 Rus. Dneprodzerzhinskoye Vodokhranilishche. reservoir C Ukraine
Dnipro 87 F3 Rus. Dnipropetrovs'k; prev. Yekaterinoslav. Dnipropetrovs'ka Oblast', E Ukraine
Dnipropetrovs'k see Dnipro
Dniprorudne 87 F3 Rus. Dneprorudnoye. Zaporiz'ka Oblast', SE Ukraine
Dnister see Dniester
Dnyapro see Dnieper
Doba 54 C4 Logone-Oriental, S Chad
Döbeln 72 D4 Sachsen, E Germany
Doberai Peninsula 117 G4 Dut. Vogelkop. peninsula Papua, E Indonesia
Doboj 78 C3 Republiks Srpska, N Bosnia and Herzegovina
Dobre Miasto 76 D2 Ger. Guttstadt. Warmińsko-mazurskie, NE Poland
Dobrich 82 E1 Rom. Bazargic; prev. Tolbukhin. Dobrich, NE Bulgaria
Dobrush 85 D7 Homyel'skaya Voblasts', SE Belarus
Dobryn' see Dabryn'
Dodecanese see Dodekánisa
Dodekánisa 83 D6 var. Nóties Sporádes, Eng. Dodecanese; prev. Dhodhekánisos, Dodekanisos. island group SE Greece
Dodekanisos see Dodekánisa
Dodge City 23 E5 Kansas, C USA
Dodoma 47 D5 country capital (Tanzania) Dodoma, C Tanzania
Dogana 74 E1 NE San Marino Europe
Dogo 109 B6 island Oki-shotō, SW Japan
Dogondoutchi 53 F3 Dosso, SW Niger
Dogrular see Pravda
Doğubayazıt 95 F3 Ağrı, E Turkey
Doğu Karadeniz Dağları 95 E3 var. Anadolu Dağları. mountain range NE Turkey
Doha see Ad Dawḩah
Doire see Londonderry
Dokdo see Liancourt Rocks
Dokkum 64 D1 Friesland, N Netherlands
Dokuchayevs'k 87 G3 var. Dokuchayevsk. Donets'ka Oblast', SE Ukraine
Dokuchayevsk see Dokuchayevs'k
Doldrums Fracture Zone 44 C4 fracture zone W Atlantic Ocean
Dôle 68 D4 Jura, E France
Dolina see Dolyna
Dolinskaya see Dolyns'ka
Dolisie 55 B6 prev. Loubomo. Niari, S Congo
Dolna Oryakhovitsa 82 D2 prev. Polikrayshte. Veliko Tŭrnovo, N Bulgaria
Dolni Chiflik 82 E2 prev. Rudnik. Varna, E Bulgaria
Dolomites 74 C1 var. Dolomiti, Eng. Dolomites. mountain range NE Italy
Dolomites/Dolomiti see Dolomitiche, Alpi
Dolores 42 D4 Buenos Aires, E Argentina
Dolores 30 B1 Petén, N Guatemala
Dolores 42 D4 Soriano, SW Uruguay
Dolores Hidalgo 29 E4 var. Ciudad de Dolores Hidalgo. Guanajuato, C Mexico
Dolyna 86 B2 Rus. Dolina. Ivano-Frankivs'ka Oblast', W Ukraine
Dolyns'ka 87 F3 Rus. Dolinskaya. Kirovohrads'ka Oblast', S Ukraine
Domachëvo/Domaczewo see Damachava
Dombås 63 B5 Oppland, S Norway
Domel Island see Letsôk-aw Kyun
Domesnes, Cape see Kolkasrags
Domeyko 42 B3 Atacama, N Chile
Dominica 33 H4 off. Commonwealth of Dominica. country E West Indies
Dominica Channel see Martinique Passage
Dominica, Commonwealth of see Dominica
Dominican Republic 33 E2 country C West Indies
Domokós 83 B5 var. Dhomokós. Stereá Ellás, C Greece
Don 89 B6 var. Duna, Tanais. river SW Russia
Donau see Danube
Donauwörth 73 C6 Bayern, S Germany
Don Benito 70 C3 Extremadura, W Spain
Doncaster 67 D5 anc. Danum. N England, United Kingdom
Dondo 56 B1 Cuanza Norte, NW Angola
Donegal 67 B5 Ir. Dún na nGall. Donegal, NW Ireland
Donegal Bay 67 A5 Ir. Bá Dhún na nGall. bay NW Ireland
Donets 87 G2 river Russia/Ukraine
Donets'k 87 G3 Rus. Donetsk; prev. Stalino. Donets'ka Oblast', E Ukraine
Dongfang 106 B7 var. Basuo. Hainan, S China
Dongguan 106 C6 Guangdong, S China
Đông Ha 114 E4 Quang Tri, C Vietnam
Dong Hai see East China Sea
Đông Hơi 114 D4 Quang Binh, C Vietnam
Dongliao see Liaoyuan
Dongola 50 B3 var. Donqola, Dunqulah. Northern, N Sudan
Dongou 55 C5 Likouala, NE Congo
Dong Rak, Phanom see Dângrêk, Chuŏr Phnum
Dongting Hu 106 C5 var. Tung-t'ing Hu. lake S China
Donostia 71 E1 País Vasco, N Spain see also San Sebastián
Donqola see Dongola
Doolow 51 D5 Sumalē, E Ethiopia
Doornik see Tournai
Door Peninsula 18 C2 peninsula Wisconsin, N USA
Dooxo Nugaaleed 51 E5 var. Nogal Valley. valley E Somalia
Dordogne 69 B5 cultural region SW France
Dordogne 69 B5 river W France
Dordrecht 64 C4 var. Dordt, Dort. Zuid-Holland, SW Netherlands
Dordt see Dordrecht
Dorohoi 86 C3 Botoşani, NE Romania
Dorotea 62 C4 Västerbotten, N Sweden
Dorpat see Tartu
Dorre Island 125 A5 island Western Australia
Dort see Dordrecht
Dortmund 72 A4 Nordrhein-Westfalen, W Germany
Dos Hermanas 70 C4 Andalucía, S Spain
Dospad Dagh see Rhodope Mountains
Dospat 82 C3 Smolyan, S Bulgaria
Dothan 20 D3 Alabama, S USA
Dotnuva 84 B4 Kaunas, C Lithuania
Douai 68 C2 prev. Douay; anc. Duacum. Nord, N France
Douala 55 A5 var. Duala. Littoral, W Cameroon
Douay see Douai
Douglas 67 C5 dependent territory capital (Isle of Man) E Isle of Man
Douglas 26 C3 Arizona, SW USA
Douglas 22 D3 Wyoming, C USA
Douma see Dūmā
Douro see Duero
Douvres see Dover
Dover 67 E7 Fr. Douvres, Lat. Dubris Portus. SE England, United Kingdom
Dover 19 F4 state capital Delaware, NE USA
Dover, Strait of 68 C2 var. Straits of Dover, Fr. Pas de Calais. strait England, United Kingdom/France
Dover, Straits of see Dover, Strait of
Dovrefjell 63 B5 plateau S Norway
Downpatrick 67 B5 Ir. Dún Pádraig. SE Northern Ireland, United Kingdom
Dozen 109 B6 island Oki-shotō, SW Japan
Drâa, Hammada du see Dra, Hamada du
Drač/Draç see Durrës
Drachten 64 D2 Friesland, N Netherlands
Drăgăşani 86 B5 Vâlcea, SW Romania
Dragoman 82 B2 Sofiya, W Bulgaria
Dra, Hamada du 48 C3 var. Hammada du Drâa, Haut Plateau du Dra. plateau W Algeria
Dra, Haut Plateau du see Dra, Hamada du
Drahichyn 85 B6 Pol. Drohiczyn Poleski, Rus. Drogichin. Brestskaya Voblasts', SW Belarus
Drakensberg 56 D5 mountain range Lesotho/South Africa
Drake Passage 35 B8 passage Atlantic Ocean/Pacific Ocean
Dralfa 82 D2 Tŭrgovishte, N Bulgaria
Dráma 82 C3 var. Dhráma. Anatolikí Makedonía kai Thráki, NE Greece
Dramburg see Drawsko Pomorskie
Drammen 63 B6 Buskerud, S Norway
Drau see Drava
Drava 78 C3 var. Drau, Eng. Drave, Hung. Dráva. river C Europe
Dráva/Drave see Drau/Drava
Drawsko Pomorskie 76 B3 Ger. Dramburg. Zachodnio-pomorskie, NW Poland
Drépano, Akrotírio 82 C4 var. Akrotírio Dhrepanon. headland N Greece
Drepanum see Trapani
Dresden 72 D4 Sachsen, E Germany
Drin see Drinit, Lumi i
Drina 78 C3 river Bosnia and Herzegovina/Serbia
Drinit, Lumi i 79 D5 var. Drin. river NW Albania
Drinit të Zi, Lumi i see Black Drin
Drissa see Drysa
Drobeta-Turnu Severin 86 B5 prev. Turnu Severin. Mehedinţi, SW Romania
Drogheda 67 B5 Ir. Droichead Átha. NE Ireland
Drogichin see Drahichyn
Drogobych see Drohobych
Drohiczyn Poleski see Drahichyn
Drohobych 86 B2 Pol. Drohobycz, Rus. Drogobych. L'vivs'ka Oblast', NW Ukraine
Drohobycz see Drohobych
Droichead Átha see Drogheda
Drôme 69 D5 cultural region E France
Dronning Maud Land 132 B2 physical region Antarctica
Drontheim see Trondheim
Drug see Durg
Druk-yul see Bhutan
Drummondville 17 E4 Québec, SE Canada
Druskienniki see Druskininkai
Druskininkai 85 B5 Pol. Druskienniki. Alytus, S Lithuania
Druzhnaya4 132 D3 Russian research station Antarctica
Dryden 16 B3 Ontario, C Canada
Drysa 85 D5 Rus. Drissa. river N Belarus
Duacum see Douai
Duala see Douala
Dubai see Dubayy
Dubăsari 86 D3 Rus. Dubossary. NE Moldova
Dubawnt 15 F4 river Nunavut, NW Canada
Dubayy 98 D4 Eng. Dubai. Dubayy, NE United Arab Emirates
Dubbo 127 D6 New South Wales, SE Australia
Dublin 67 B5 Ir. Baile Átha Cliath; anc. Eblana. country capital (Ireland) Dublin, E Ireland
Dublin 21 E2 Georgia, SE USA
Dubno 86 C2 Rivnens'ka Oblast', NW Ukraine
Dubossary see Dubăsari
Dubris Portus see Dover
Dubrovnik 79 B5 It. Ragusa. Dubrovnik-Neretva, SE Croatia
Dubuque 23 G3 Iowa, C USA
Dudelange 65 D8 var. Forge du Sud, Ger. Dudelingen. Luxembourg, S Luxembourg
Dudelingen see Dudelange
Duero 70 D2 Port. Douro. river Portugal/Spain
Duesseldorf see Düsseldorf
Duffel 65 C5 Antwerpen, C Belgium
Dugi Otok 78 A4 var. Isola Grossa, It. Isola Lunga. island W Croatia
Duinekerke see Dunkerque
Duisburg 72 A4 prev. Duisburg-Hamborn. Nordrhein-Westfalen, W Germany
Duisburg-Hamborn see Duisburg
Duiven 64 D4 Gelderland, E Netherlands
Duk Faiwil 51 B5 Jonglei, C South Sudan
Dulan 104 D4 var. Qagan Us. Qinghai, C China
Dulce, Golfo 31 E5 gulf S Costa Rica
Dulce, Golfo see Izabal, Lago de
Dülmen 72 A4 Nordrhein-Westfalen, W Germany
Dulovo 82 E1 Silistra, NE Bulgaria
Duluth 23 G2 Minnesota, N USA
Dūmā 97 B5 Fr. Douma. Dimashq, SW Syria
Dumas 27 E1 Texas, SW USA
Dumfries 66 C4 S Scotland, United Kingdom
Dumont d'Urville 132 C4 French research station Antarctica
Dumyāṭ 50 B1 Eng. Damietta. N Egypt
Duna see Danube, C Europe
Düna see Western Dvina
Duna see Don
Dünaburg see Daugavpils
Dunaj see Wien, Austria
Dunaj see Danube, C Europe
Dunapentele see Dunaújváros
Dunărea see Danube
Dunaújváros 77 C7 prev. Dunapentele, Sztálinváros. Fejér, C Hungary
Dunav see Danube
Dunavska Ravnina 82 C2 Eng. Danubian Plain. lowlands N Bulgaria
Duncan 27 G2 Oklahoma, C USA
Dundalk 67 B5 Ir. Dún Dealgan. Louth, NE Ireland
Dún Dealgan see Dundalk
Dundee 56 D4 KwaZulu/Natal, E South Africa
Dundee 66 C4 E Scotland, United Kingdom
Dunedin 129 B7 Otago, South Island, New Zealand
Dunfermline 66 C4 C Scotland, United Kingdom
Dungu 55 E5 Orientale, NE Dem. Rep. Congo
Dungun 116 B3 var. Kuala Dungun. Terengganu, Peninsular Malaysia
Dunholme see Durham
Dunkerque 68 C2 Eng. Dunkirk, Flem. Duinekerke; prev. Dunquerque. Nord, N France
Dunkirk see Dunkerque
Dún Laoghaire 67 B6 Eng. Dunleary; prev. Kingstown. E Ireland
Dunleary see Dún Laoghaire
Dún Pádraig see Downpatrick
Dunquerque see Dunkerque
Dunqulah see Dongola
Dupnitsa 82 C2 prev. Marek, Stanke Dimitrov. Kyustendil, W Bulgaria
Duqm 99 E5 var. Daqm. E Oman
Durance 69 D6 river SE France
Durango 28 D3 var. Victoria de Durango. Durango, W Mexico
Durango 22 C5 Colorado, C USA
Durankulak 82 E1 Rom. Răcari; prev. Blatnitsa, Duranulac. Dobrich, NE Bulgaria
Durant 27 G2 Oklahoma, C USA
Duranulac see Durankulak
Durazzo see Durrës
Durban 56 D4 var. Port Natal. KwaZulu/Natal, E South Africa
Durbe 84 B3 Ger. Durben. Liepāja, W Latvia
Durben see Durbe
Durg 113 E4 prev. Drug. Chhattīsgarh, C India
Durham 67 D5 hist. Dunholme. N England, United Kingdom
Durham 21 F1 North Carolina, SE USA
Durocortorum see Reims
Durostorum see Silistra
Durovernum see Canterbury
Durrës 79 C6 var. Durrësi, Dursi, It. Durazzo, SCr. Drač, Turk. Draç. Durrës, W Albania
Durrësi see Durrës
Dursi see Durrës
Durūz, Jabal ad 97 C5 mountain SW Syria
D'Urville Island 128 C4 island C New Zealand
Dusa Mareb/Dusa Marreb see Dhuusa Marreeb
Dushanbe 101 E3 var. Dyushambe; prev. Stalinabad, Taj. Stalinobod. country capital (Tajikistan) W Tajikistan
Düsseldorf 72 A4 var. Duesseldorf. Nordrhein-Westfalen, W Germany
Dŭsti 101 E3 Rus. Dusti. SW Tajikistan
Dutch East Indies see Indonesia
Dutch Guiana see Suriname
Dutch Harbor 14 B3 Unalaska Island, Alaska, USA
Dutch New Guinea see Papua
Duzdab see Zāhedān
Dvina Bay see Chëshskaya Guba
Dvinsk see Daugavpils
Dyanev see Galkynyş
Dyersburg 20 C1 Tennessee, S USA
Dyushambe see Dushanbe
Dza Chu see Mekong
Dzaudzhikau see Vladikavkaz
Dzerzhinsk 89 C5 Nizhegorodskaya Oblast', W Russia
Dzerzhinskiy see Nar'yan-Mar
Dzhalal-Abad 101 F2 Kir. Jalal-Abad. Dzhalal-Abadskaya Oblast', W Kyrgyzstan
Dzhambul see Taraz
Dzhankoy 87 F4 Respublika Krym, S Ukraine
Dzharkurgan see Jarqo'rg'on
Dzhelandy 101 F3 SE Tajikistan
Dzhergalan 101 G2 Kir. Jyrgalan. Issyk-Kul'skaya Oblast', NE Kyrgyzstan
Dzhezkazgan see Zhezkazgan
Dzhizak see Jizzax
Dzhugdzhur, Khrebet 93 G3 mountain range E Russia
Dzhusaly see Zhosaly
Działdowo 76 D3 Warmińsko-Mazurskie, C Poland
Dzuunmod 105 E2 Töv, C Mongolia
Dzüün Soyonï Nuruu see Vostochnyy Sayan
Dzvina see Western Dvina

E

Eagle Pass 27 F4 Texas, SW USA
East Açores Fracture Zone see East Azores Fracture Zone
East Antarctica 132 C3 var. Greater Antarctica. physical region Antarctica
East Australian Basin see Tasman Basin
East Azores Fracture Zone 44 C3 var. East Açores Fracture Zone. tectonic feature E Atlantic Ocean
Eastbourne 67 E7 SE England, United Kingdom
East Cape 128 E3 headland North Island, New Zealand
East China Sea 103 E2 Chin. Dong Hai. sea W Pacific Ocean
Easter Fracture Zone 131 G4 tectonic feature E Pacific Ocean
Easter Island see Pascua, Isla de
Eastern Desert see Şaḩrā' ash Sharqīyah
Eastern Ghats 102 B3 mountain range SE India
Eastern Sayans see Vostochnyy Sayan
Eastern Sierra Madre see Madre Oriental, Sierra
East Falkland 43 D8 var. Isla Soledad. island E Falkland Islands
East Frisian Islands 72 A3 Eng. East Frisian Islands. island group NW Germany
East Frisian Islands see Ostfriesische Inseln
East Grand Forks 23 E1 Minnesota, N USA
East Indiaman Ridge 119 D6 undersea ridge E Indian Ocean
East Indies 130 A3 island group SE Asia
East Kilbride 66 C4 S Scotland, United Kingdom
East Korea Bay 107 E3 bay E North Korea
Eastleigh 67 D7 S England, United Kingdom
East London 56 D5 Afr. Oos-Londen; prev. Emonti, Port Rex. Eastern Cape, S South Africa
Eastmain 16 D3 river Québec, C Canada
East Mariana Basin 120 B1 undersea basin W Pacific Ocean
East Novaya Zemlya Trough 90 C1 var. Novaya Zemlya Trough. trough W Kara Sea
East Pacific Rise 131 F4 undersea rise E Pacific Ocean
East Pakistan see Bangladesh
East Saint Louis 18 B4 Illinois, N USA
East Scotia Basin 45 C7 undersea basin SE Scotia Sea
East Sea 108 A4 var. Sea of Japan, Rus. Yapanskoye More. Sea NW Pacific Ocean
East Siberian Sea see Vostochno-Sibirskoye More
East Timor 117 F5 var. Loro Sae; prev. Portuguese Timor, Timor Timur. country S Indonesia
Eau Claire 18 A2 Wisconsin, N USA
Eau Claire, Lac à L' see St. Clair, Lake
Eauripik Rise 120 B2 undersea rise W Pacific Ocean
Ebensee 73 D6 Oberösterreich, N Austria
Eberswalde-Finow 72 D3 Brandenburg, E Germany
Ebetsu 108 D2 var. Ebetu. Hokkaidō, NE Japan
Ebetu see Ebetsu
Eblana see Dublin
Ebolowa 55 A5 Sud, S Cameroon
Ebon Atoll 122 D2 var. Epoon. atoll Ralik Chain, S Marshall Islands
Ebora see Évora
Eboracum see York
Ebro 71 E2 river NE Spain
Eburacum see York
Ebusus see Eivissa
Ecbatana see Hamadān
Ech Cheliff/Ech Chleff see Chlef
Echo Bay 15 E3 Northwest Territories, NW Canada
Echt 65 D5 Limburg, SE Netherlands
Ecija 70 D4 anc. Astigi. Andalucía, SW Spain
Eckengraf see Viesīte
Ecuador 38 B1 off. Republic of Ecuador. country NW South America
Ecuador, Republic of see Ecuador
Ed Da'ein 50 A4 Southern Darfur, W Sudan
Ed Damazin 50 C4 var. Ad Damazīn. Blue Nile, E Sudan
Ed Damer 50 C3 var. Ad Dāmir, Ad Damar. River Nile, NE Sudan
Ed Debba 50 B3 Northern, N Sudan
Ede 64 D4 Gelderland, C Netherlands
Ede 53 F5 Osun, SW Nigeria
Edéa 55 A5 Littoral, SW Cameroon
Edessa see Şanlıurfa
Edfu see Idfū
Edgeøya 61 G2 island S Svalbard
Edgware 67 A7 Harrow, SE England, United Kingdom
Edinburg 27 G5 Texas, SW USA
Edinburgh 66 C4 national capital S Scotland, United Kingdom
Edingen see Enghien
Edirne 94 A2 Eng. Adrianople; anc. Adrianopolis, Hadrianopolis. Edirne, NW Turkey
Edmonds 24 B2 Washington, NW USA
Edmonton 15 E5 province capital Alberta, SW Canada
Edmundston 17 E4 New Brunswick, SE Canada
Edna 27 G4 Texas, SW USA
Edolo 74 B1 Lombardia, N Italy
Edremit 94 A3 Balıkesir, NW Turkey
Edward, Lake 55 E5 var. Albert Edward Nyanza, Edward Nyanza, Lac Idi Amin, Lake Rutanzige. lake Uganda/Dem. Rep. Congo
Edward Nyanza see Edward, Lake
Edwards Plateau 27 F3 plain Texas, SW USA
Edzo 31 E4 prev. Rae-Edzo. Northwest Territories, NW Canada
Eeklo 65 B5 var. Eekloo. Oost-Vlaanderen, NW Belgium
Eekloo see Eeklo
Eems see Ems
Eersel 65 C5 Noord-Brabant, S Netherlands
Eesti Vabariik see Estonia
Efate 122 D4 var. Efate, Fr. Vaté; prev. Sandwich Island. island C Vanuatu
Efate see Efate
Effingham 18 B4 Illinois, N USA
Eforie-Sud 86 D5 Constanţa, E Romania
Egadi Is. 75 B7 island group S Italy
Ege Denizi see Aegean Sea
Eger 77 D6 Ger. Erlau. Heves, NE Hungary
Eger see Cheb, Czech Republic (Czechia)
Egeria Fracture Zone 119 C5 tectonic feature W Indian Ocean
Éghezèe 65 C6 Namur, C Belgium
Egina see Aígina
Egio see Aígio
Egmont see Taranaki, Mount
Egmont, Cape 128 C4 headland North Island, New Zealand
Egoli see Johannesburg
Egypt 50 B2 off. Arab Republic of Egypt, Ar. Jumhūrīyah Miṣr al 'Arabīyah, prev. United Arab Republic; anc. Aegyptus. country NE Africa
Eibar 71 E1 País Vasco, N Spain
Eibergen 64 E3 Gelderland, E Netherlands
Eidfjord 63 A5 Hordaland, S Norway
Eier-Berg see Suur Munamägi
Eifel 73 A5 plateau W Germany
Eiger 73 B7 mountain C Switzerland
Eigg 66 B3 island W Scotland, United Kingdom
Eight Degree Channel 110 B3 channel India/Maldives
Eighty Mile Beach 124 B4 beach Western Australia
Eijsden 65 D6 Limburg, SE Netherlands
Eilat see Elat
Eindhoven 65 D5 Noord-Brabant, S Netherlands
Eipel see Ipel'
Éire see Ireland
Éireann, Muir see Irish Sea
Eisenhüttenstadt 72 D4 Brandenburg, E Germany
Eisenmarkt see Hunedoara
Eisenstadt 73 E6 Burgenland, E Austria
Eisleben 72 C4 Sachsen-Anhalt, C Germany
Eivissa 71 G3 var. Iviza, Cast. Ibiza; anc. Ebusus. Ibiza, Spain, W Mediterranean Sea
Ejea de los Caballeros 71 E2 Aragón, NE Spain
Ejin Qi see Dalain Hob
Ekapa see Cape Town
Ekaterinodar see Krasnodar
Ekvyvatapskiy Khrebet 93 G1 mountain range NE Russia
El 'Alamein see Al 'Alamayn
El Asnam see Chlef
Elat 97 B8 var. Eilat, Elath. Southern, S Israel
Elat, Gulf of see Aqaba, Gulf of
Elath see Elat, Israel
Elath see Al 'Aqabah, Jordan
El'Atrun 50 B3 Northern Darfur, NW Sudan
Elâzığ 95 E3 var. Elâzig, Elâziz. Elâzığ, E Turkey
Elba 74 B4 island Archipelago Toscano, C Italy
Elbasan 79 D6 var. Elbasani. Elbasan, C Albania
Elbasani see Elbasan
Elbe 58 D3 Cz. Labe. river Czech Republic (Czechia)/Germany
Elbert, Mount 22 C4 mountain Colorado, C USA
Elbing see Elbląg
Elbląg 76 C2 var. Elblag, Ger. Elbing. Warmińsko-Mazurskie, NE Poland
El Boulaida/El Boulaïda see Blida
El'brus 89 A8 var. Gora El'brus. mountain SW Russia
El'brus, Gora see El'brus
El Burgo de Osma 71 E2 Castilla-León, C Spain
Elburz Mountains see Alborz, Reshteh-ye Kūhhā-ye
El Cajon 25 C8 California, W USA
El Calafate 43 B7 var. Calafate. Santa Cruz, S Argentina
El Callao 37 E2 Bolívar, E Venezuela
El Campo 27 G4 Texas, SW USA
El Carmen de Bolívar 36 B2 Bolívar, NW Colombia
El Cayo see San Ignacio
El Centro 25 D8 California, W USA
Elche 71 F4 Cat. Elx; anc. Ilici, Lat. Illicis. País Valenciano, E Spain
Elda 71 F4 País Valenciano, E Spain
El Djazaïr see Alger
El Djelfa see Djelfa
Eldorado 42 E3 Misiones, NE Argentina
El Dorado 28 C3 Sinaloa, C Mexico
El Dorado 20 B2 Arkansas, C USA
El Dorado 23 F5 Kansas, C USA
El Dorado 37 F2 Bolívar, E Venezuela
El Dorado see California
Eldoret 51 C6 Rift Valley, W Kenya
Elektrostal' 89 B5 Moskovskaya Oblast', W Russia
Elemi Triangle 51 B5 disputed region Kenya/Sudan
Elephant Butte Reservoir 26 C2 reservoir New Mexico, SW USA
Élesd see Aleşd
Eleuthera Island 32 C1 island N The Bahamas
El Fasher 50 A4 var. Al Fāshir. Northern Darfur, W Sudan
El Ferrol/El Ferrol del Caudillo see Ferrol
El Gedaref see Gedaref
El Geneina 50 A4 var. Ajjinena, Al-Genain, Al Junaynah. Western Darfur, W Sudan
Elgin 66 C3 NE Scotland, United Kingdom
Elgin 18 B3 Illinois, N USA
El Gîza see Gîza
El Goléa 48 D3 var. Al Golea. C Algeria
El Hank 52 D1 cliff N Mauritania
El Haseke see Al Ḩasakah
Elimberrum see Auch
Eliocroca see Lorca
Élisabethville see Lubumbashi
Elista 89 B7 Respublika Kalmykiya, SW Russia
Elizabeth 127 B6 South Australia
Elizabeth City 21 G1 North Carolina, SE USA
Elizabethtown 18 C5 Kentucky, S USA
El-Jadida 48 C2 prev. Mazagan. W Morocco
Ełk 76 E2 Ger. Lyck. Warmińsko-mazurskie, NE Poland
Elk City 27 F1 Oklahoma, C USA
El Khalil see Hebron
El Khârga see Al Khārijah
Elkhart 18 C3 Indiana, N USA
El Khartûm see Khartoum
Elk River 23 F2 Minnesota, N USA
El Kuneitra see Al Qunayṭirah
Ellás see Greece
Ellef Ringnes Island 15 E1 island Nunavut, N Canada
Ellen, Mount 22 B5 mountain Utah, W USA
Ellensburg 24 B2 Washington, NW USA
Ellesmere Island 15 F1 island Queen Elizabeth Islands, Nunavut, N Canada
Ellesmere, Lake 129 C6 lake South Island, New Zealand
Ellice Islands see Tuvalu
Elliston 127 A6 South Australia
Ellsworth Land 132 A3 physical region Antarctica
El Mahbas 48 B3 var. Mahbés. SW Western Sahara
El Mina 96 B4 var. Al Mīnā'. N Lebanon
El Minya see Al Minyā
Elmira 19 E3 New York, NE USA
El Mreyyé 52 D2 desert E Mauritania
Elmshorn 72 B3 Schleswig-Holstein, N Germany
El Muglad 50 B4 Western Kordofan, C Sudan

El Obeid – Fortaleza

El Obeid 50 B4 var. Al Obayyid, Al Ubayyiḍ. Northern Kordofan, C Sudan
El Ouâdi see El Oued
El Oued 49 E2 var. Al Oued, El Ouâdi, El Wad. NE Algeria
Eloy 26 B2 Arizona, SW USA
El Paso 26 D3 Texas, SW USA
El Porvenir 31 G4 Kuna Yala, N Panama
El Progreso 30 C2 Yoro, NW Honduras
El Puerto de Santa María 70 C5 Andalucía, S Spain
El Qâhira see Cairo
El Quneitra see Al Qunayṭirah
El Quseir see Al Quṣayr
El Quweira see Al Quwayrah
El Rama 31 E3 Región Autónoma Atlántico Sur, SE Nicaragua
El Real 31 H5 var. El Real de Santa María. Darién, SE Panama
El Real de Santa María see El Real
El Reno 27 F1 Oklahoma, C USA
El Salvador 30 B3 off. Republica de El Salvador. country Central America
El Salvador, Republica de see El Salvador
Elsass see Alsace
El Sáuz 28 C2 Chihuahua, N Mexico
El Serrat 69 A7 N Andorra Europe
Elst 64 D4 Gelderland, E Netherlands
El Sueco 28 C2 Chihuahua, N Mexico
El Suweida see As Suwaydā'
El Suweis see Suez
Eltanin Fracture Zone 131 E5 tectonic feature SE Pacific Ocean
El Tigre 37 E2 Anzoátegui, NE Venezuela
Elvas 70 C4 Portalegre, C Portugal
El Vendrell 71 G2 Cataluña, NE Spain
El Vigía 36 C2 Mérida, NW Venezuela
El Wad see El Oued
Elwell, Lake 22 B1 reservoir Montana, NW USA
Elx see Elche
Ely 25 D5 Nevada, W USA
El Yopal see Yopal
Emajõgi 84 D3 Ger. Embach. river SE Estonia
Emāmrūd see Shāhrūd
Emāmshahr see Shāhrūd
Emba 92 B4 Kaz. Embi. Aktyubinsk, W Kazakhstan
Embach see Emajõgi
Embi see Emba
Emden 72 A3 Niedersachsen, NW Germany
Emerald 126 D4 Queensland, E Australia
Emerald Isle see Montserrat
Emesa see Ḥimṣ
Emmaste 84 C2 Hiiumaa, W Estonia
Emmeloord 64 D2 Flevoland, N Netherlands
Emmen 64 E2 Drenthe, NE Netherlands
Emmendingen 73 A6 Baden-Württemberg, SW Germany
Emona see Ljubljana
Emonti see East London
Emory Peak 27 E4 mountain Texas, SW USA
Empalme 28 B2 Sonora, NW Mexico
Emperor Seamounts 91 G3 seamount range NW Pacific Ocean
Empire State of the South see Georgia
Emporia 23 F5 Kansas, C USA
Empty Quarter see Ar Rub 'al Khālī
Ems 72 A3 Dut. Eems. river NW Germany
Enareträsk see Inarijärvi
Encamp 69 A8 Encamp, C Andorra Europe
Encarnación 42 D3 Itapúa, S Paraguay
Encinitas 25 C8 California, W USA
Encs 77 D6 Borsod-Abaúj-Zemplén, NE Hungary
Endeavour Strait 126 C1 strait Queensland, NE Australia
Enderbury Island 123 F3 atoll Phoenix Islands, C Kiribati
Enderby Land 132 C2 physical region Antarctica
Enderby Plain 132 D2 abyssal plain S Indian Ocean
Endersdorf see Jędrzejów
Enewetak Atoll 122 C1 var. Ānewetak, Eniwetok. atoll Ralik Chain, W Marshall Islands
Enfield 67 A7 United Kingdom
Engeten see Aiud
Enghien 65 B6 Dut. Edingen. Hainaut, SW Belgium
England 67 D5 Lat. Anglia. cultural region England, United Kingdom
Englewood 22 D4 Colorado, C USA
English Channel 67 D8 var. The Channel, Fr. la Manche. channel NW Europe
Engure 84 C3 Tukums, W Latvia
Engures Ezers 84 B3 lake NW Latvia
Enguri 95 F1 Rus. Inguri. river NW Georgia
Enid 27 F1 Oklahoma, C USA
Enikale Strait see Kerch Strait
Eniwetok see Enewetak Atoll
En Nâqoûra 97 A5 var. An Nāqūrah. SW Lebanon
En Nazira see Natzrat
Ennedi 54 D2 plateau E Chad
Ennis 67 A6 Ir. Inis. Clare, W Ireland
Ennis 27 G3 Texas, SW USA
Enniskillen 67 B5 var. Inniskilling, Ir. Inis Ceithleann. SW Northern Ireland, United Kingdom
Enns 73 D6 river C Austria
Enschede 64 E3 Overijssel, E Netherlands
Ensenada 28 A1 Baja California Norte, NW Mexico
Entebbe 51 B6 S Uganda
Entroncamento 70 B3 Santarém, C Portugal
Enugu 53 G5 Enugu, S Nigeria
Epanomí 82 B4 Kentrikí Makedonía, N Greece
Epéna 55 B5 Likouala, NE Congo
Eperies/Eperjes see Prešov
Epi 122 D4 var. Épi. island C Vanuatu
Épi see Epi
Épinal 68 D4 Vosges, NE France
Epiphania see Ḥamāh
Epoon see Ebon Atoll
Epsom 67 A8 United Kingdom
Equality State see Wyoming
Equatorial Guinea 55 A5 off. Equatorial Guinea, Republic of. country C Africa
Equatorial Guinea, Republic of see Equatorial Guinea
Erautini see Johannesburg
Erbil see Arbil
Erciş 95 F3 Van, E Turkey
Erdély see Transylvania
Erdélyi-Havasok see Carpaţii Meridionalii
Erdenet 105 E2 Orhon, N Mongolia
Erdi 54 C2 plateau NE Chad
Erdi Ma 54 D2 desert NE Chad
Erebus, Mount 132 B4 volcano Ross Island, Antarctica
Ereğli 94 C4 Konya, S Turkey
Erenhot 105 F2 var. Erlian. Nei Mongol Zizhiqu, NE China
Erfurt 72 C4 Thüringen, C Germany
Ergene Çayı see Ergene Irmaği
Ergene Irmaği 94 A2 var. Ergene Çayı. river NW Turkey
Ergun 105 F1 var. Labudalin; prev. Ergun Youqi. Nei Mongol Zizhiqu, N China
Ergun He see Argun
Ergun Youqi see Ergun
Erie 18 D3 Pennsylvania, NE USA
Érié, Lac see Erie, Lake
Erie, Lake 18 D3 Fr. Lac Érié. lake Canada/USA
Eritrea 50 C4 off. State of Eritrea, Ērtra. country E Africa
Eritrea, State of see Eritrea
Erivan see Yerevan
Erlangen 73 C5 Bayern, S Germany
Erlau see Eger
Erlian see Erenhot
Ermelo 64 D3 Gelderland, C Netherlands
Ermióni 83 C6 Pelopónnisos, S Greece
Ermoúpoli 83 D6 var. Hermoupolis; prev. Ermoúpolis. Sýyros, Kykládes, Greece, Aegean Sea
Ermoúpolis see Ermoúpoli
Ernākulam 110 C3 Kerala, SW India
Erode 110 C2 Tamil Nādu, SE India
Erquelinnes 65 B7 Hainaut, S Belgium
Er-Rachidia 48 C2 var. Ksar al Soule. E Morocco
Er Rahad 50 B4 var. Ar Rahad. Northern Kordofan, C Sudan
Erromango 122 D4 island S Vanuatu
Ertis see Irtysh, C Asia
Ērtra see Eritrea
Erzerum see Erzurum
Erzgebirge 73 C5 Cz. Krušné Hory, Eng. Ore Mountains. mountain range Czech Republic (Czechia)/Germany
Erzincan 95 E3 var. Erzinjan. Erzincan, E Turkey
Erzinjan see Erzincan
Erzurum 95 E3 prev. Erzerum. Erzurum, NE Turkey
Esbjerg 63 A7 Ribe, W Denmark
Esbo see Espoo
Escaldes 69 A8 Escaldes Engordany, C Andorra Europe
Escanaba 18 C2 Michigan, N USA
Escaut see Scheldt
Esch-sur-Alzette 65 D8 Luxembourg, S Luxembourg
Esclaves, Grand Lac des see Great Slave Lake
Escondido 25 C8 California, W USA
Escuinapa 28 D3 var. Escuinapa de Hidalgo. Sinaloa, C Mexico
Escuinapa de Hidalgo see Escuinapa
Escuintla 30 B2 Escuintla, S Guatemala
Escuintla 29 G5 Chiapas, SE Mexico
Esenguly 100 B3 Rus. Gasan-Kuli. Balkan Welaýaty, W Turkmenistan
Eşfahān 98 C3 Eng. Isfahan; anc. Aspadana. Eşfahān, C Iran
Esh Sharā see Ash Sharāh
Esil see Ishim, Kazakhstan/Russia
Eskimo Point see Arviat
Eskişehir 94 B3 var. Eskishehr. Eskişehir, W Turkey
Eskishehr see Eskişehir
Eslāmābād 98 C3 var. Eslāmābād-e Gharb; prev. Harunabad, Shāhābād. Kermānshāhān, W Iran
Eslāmābād-e Gharb see Eslāmābād
Esmeraldas 38 A1 Esmeraldas, N Ecuador
Esna see Isnā
España see Spain
Espanola 26 D1 New Mexico, SW USA
Esperance 125 B7 Western Australia
Esperanza 28 B2 Sonora, NW Mexico
Esperanza 132 A2 Argentinian research station Antarctica
Espinal 36 B3 Tolima, C Colombia
Espinhaço, Serra do 34 D4 mountain range SE Brazil
Espírito Santo 41 F4 off. Estado do Espírito Santo. region E Brazil
Espírito Santo 41 F4 off. Estado do Espírito Santo. state E Brazil
Espírito Santo, Estado do see Espírito Santo
Espiritu Santo 122 C4 var. Santo. island W Vanuatu
Espoo 63 D6 Swe. Esbo. Uusimaa, S Finland
Esquel 43 B6 Chubut, SW Argentina
Essaouira 48 B2 prev. Mogador. W Morocco
Esseg see Osijek
Es Semara see Smara
Essen 65 C5 Antwerpen, N Belgium
Essen 72 A4 var. Essen an der Ruhr. Nordrhein-Westfalen, W Germany
Essen an der Ruhr see Essen
Essequibo River 37 F3 river C Guyana
Es Suweida see As Suwaydā'
Estacado, Llano 27 E2 plain New Mexico/Texas, SW USA
Estados, Isla de los 43 C8 prev. Eng. Staten Island. island S Argentina
Estância 41 G3 Sergipe, E Brazil
Estelí 30 D3 Estelí, NW Nicaragua
Estella 71 E1 Bas. Lizarra. Navarra, N Spain
Estepona 70 D5 Andalucía, S Spain
Estevan 15 F5 Saskatchewan, S Canada
Estland see Estonia
Estonia 84 D2 off. Republic of Estonia, Est. Eesti Vabariik, Ger. Estland, Latv. Igaunija; prev. Estonian SSR, Rus. Estonskaya SSR. country NE Europe
Estonian SSR see Estonia
Estonia, Republic of see Estonia
Estonskaya SSR see Estonia
Estrela, Serra da 70 C3 mountain range C Portugal
Estremadura see Extremadura
Estremoz 70 C4 Évora, S Portugal
Eszék see Osijek
Esztergom 77 C6 Ger. Gran; anc. Strigonium. Komárom-Esztergom, N Hungary
Étalle 65 D8 Luxembourg, SE Belgium
Etāwah 112 D3 Uttar Pradesh, N India
Ethiopia 51 C5 off. Federal Democratic Republic of Ethiopia; prev. Abyssinia, People's Democratic Republic of Ethiopia. country E Africa
Ethiopia, Federal Democratic Republic of see Ethiopia
Ethiopian Highlands 51 C5 var. Ethiopian Plateau. plateau N Ethiopia
Ethiopian Plateau see Ethiopian Highlands
Ethiopia, People's Democratic Republic of see Ethiopia
Etna, Monte 75 C7 Eng. Mount Etna. volcano Sicilia, Italy, C Mediterranean Sea
Etna, Mount see Etna, Monte
Etosha Pan 56 B3 salt lake N Namibia
Etoumbi 55 B5 Cuvette Ouest, NW Congo
Etsch see Adige
Et Tafila see Aṭ Ṭafīlah
Ettelbrück 65 D8 Diekirch, C Luxembourg
'Eua 123 E5 prev. Middleburg Island. island Tongatapu Group, SE Tonga
Euboea 83 C5 Lat. Euboea. island C Greece
Euboea see Évvoia
Eucla 125 D6 Western Australia
Euclid 18 D3 Ohio, N USA
Eufaula Lake 27 G1 var. Eufaula Reservoir. reservoir Oklahoma, C USA
Eufaula Reservoir see Eufaula Lake
Eugene 24 B3 Oregon, NW USA
Eumolpias see Plovdiv
Eupen 65 D6 Liège, E Belgium
Euphrates 90 B4 Ar. Al-Furāt, Turk. Firat Nehri. river SW Asia
Eureka 25 A5 California, W USA
Eureka 22 A1 Montana, NW USA
Europa Point 71 H5 headland S Gibraltar
Europe 58 continent
Eutin 72 C2 Schleswig-Holstein, N Germany
Euxine Sea see Black Sea
Evansdale 23 G3 Iowa, C USA
Evanston 18 B3 Illinois, N USA
Evanston 22 B4 Wyoming, C USA
Evansville 18 B5 Indiana, N USA
Eveleth 23 G1 Minnesota, N USA
Everard, Lake 127 A6 salt lake South Australia
Everest, Mount 104 B5 Chin. Qomolangma Feng, Nep. Sagarmāthā. mountain China/Nepal
Everett 24 B2 Washington, NW USA
Everglades, The 21 F5 wetland Florida, SE USA
Evje 63 A6 Aust-Agder, S Norway
Evmolpia see Plovdiv
Évora 70 B4 anc. Ebora, Lat. Liberalitas Julia. Évora, C Portugal
Évreux 68 C3 anc. Civitas Eburovicum. Eure, N France
Évros see Maritsa
Évry 68 E2 Essonne, N France
Ewarton 32 B5 C Jamaica
Excelsior Springs 23 F4 Missouri, C USA
Exe 67 C7 river SW England, United Kingdom
Exeter 67 C7 anc. Isca Damnoniorum. SW England, United Kingdom
Exmoor 67 C7 moorland SW England, United Kingdom
Exmouth 124 A4 Western Australia
Exmouth 67 C7 SW England, United Kingdom
Exmouth Gulf 124 A4 gulf Western Australia
Exmouth Plateau 119 E5 undersea plateau E Indian Ocean
Extremadura 70 C3 var. Estremadura. autonomous community W Spain
Exuma Cays 32 C1 islets C The Bahamas
Exuma Sound 32 C1 sound C The Bahamas
Eyre Mountains 129 A7 mountain range South Island, New Zealand
Eyre North, Lake 127 A5 salt lake South Australia
Eyre Peninsula 127 A6 peninsula South Australia
Eyre South, Lake 127 A5 salt lake South Australia
Ezo see Hokkaidō

F

Faadhippolhu Atoll 110 B4 var. Fadiffolu, Lhaviyani Atoll. atoll N Maldives
Fabens 26 D3 Texas, SW USA
Fada 54 C2 Borkou-Ennedi-Tibesti, E Chad
Fada-Ngourma 53 E4 E Burkina
Fadiffolu see Faadhippolhu Atoll
Faenza 74 C3 anc. Faventia. Emilia-Romagna, N Italy
Faeroe Islands see Faero Islands
Færøerne see Faroe Islands
Faetano 74 E2 E San Marino
Făgăraş 86 C4 Ger. Fogarasch, Hung. Fogaras. Braşov, C Romania
Fagibina, Lake see Faguibine, Lac
Fagne 65 C7 hill range S Belgium
Faguibine, Lac 53 E3 var. Lake Fagibina. lake NW Mali
Fahlun see Falun
Fahraj 98 E4 Kermān, SE Iran
Faial 70 A5 var. Ilha do Faial. island Azores, Portugal, NE Atlantic Ocean
Faial, Ilha do see Faial
Faifo see Hôi An
Fairbanks 14 D3 Alaska, USA
Fairfield 25 B6 California, W USA
Fair Isle 66 D2 island NE Scotland, United Kingdom
Fairlie 129 B6 Canterbury, South Island, New Zealand
Fairmont 23 F3 Minnesota, N USA
Faisalābād 112 C2 prev. Lyallpur. Punjab, NE Pakistan
Faizābād 113 E3 Uttar Pradesh, N India
Faizabad/Faizābād see Feyẕābād
Fakaofo Atoll 123 F3 island SE Tokelau
Falam 114 A3 Chin State, W Myanmar (Burma)
Falconara Marittima 74 C3 Marche, C Italy
Falkenau an der Eger see Sokolov
Falkland Islands 43 D7 var. Falklands, Islas Malvinas. UK dependent territory SW Atlantic Ocean
Falkland Plateau 35 D7 var. Argentine Rise. undersea feature SW Atlantic Ocean
Falklands see Falkland Islands
Falknov nad Ohří see Sokolov
Fallbrook 25 C8 California, W USA
Falmouth 32 A4 W Jamaica
Falmouth 67 C7 SW England, United Kingdom
Falster 63 B8 island SE Denmark
Fălticeni 86 C3 Hung. Falticsén. Suceava, NE Romania
Falticsén see Fălticeni
Falun 63 C6 var. Fahlun. Kopparberg, C Sweden
Famagusta see Gazimağusa
Famagusta Bay see Gazimağusa Körfezi
Famenne 65 C7 physical region SE Belgium
Fang 114 C3 Chiang Mai, NW Thailand
Fanning Island see Tabuaeran
Fanø 74 C3 island W Denmark
Farafangana 57 G4 Fianarantsoa, SE Madagascar
Farāh 100 D4 var. Farah, Fararud. Farāh, W Afghanistan
Farah Rud 100 D4 river W Afghanistan
Faranah 52 C4 Haute-Guinée, S Guinea
Fararud see Farāh
Farasan, Jaza'ir 99 A6 island group SW Saudi Arabia
Farewell, Cape 128 C4 headland South Island, New Zealand
Farewell, Cape see Nunap Isua
Fargo 23 F2 North Dakota, N USA
Farg'ona 101 F2 Rus. Fergana; prev. Novyy Margilan. Farg'ona Viloyati, E Uzbekistan
Faribault 23 F2 Minnesota, N USA
Farīdābād 112 D3 Haryāna, N India
Farkhor 101 E3 Rus. Parkhar. SW Tajikistan
Farmington 23 G5 Missouri, C USA
Farmington 26 C1 New Mexico, SW USA
Faro 70 B5 Faro, S Portugal
Faroe-Iceland Ridge 58 C1 undersea ridge NW Norwegian Sea
Faroe Islands 61 E5 var. Faero Islands, Dan. Færøerne, Faer. Føroyar. Self-governing territory of Denmark N Atlantic Ocean
Faroe-Shetland Trough 58 C2 trough NE Atlantic Ocean
Farquhar Group 57 G2 island group S Seychelles
Fars, Khalīj-e see Persian Gulf
Farvel, Kap see Nunap Isua
Fastiv 87 E2 Rus. Fastov. Kyyivs'ka Oblast', NW Ukraine
Fastov see Fastiv
Fauske 62 C3 Nordland, C Norway
Faventia see Faenza
Faxa Bay see Faxaflói
Faxaflói 60 D5 Eng. Faxa Bay. bay W Iceland
Faya 54 C2 prev. Faya-Largeau, Largeau. Borkou-Ennedi-Tibesti, N Chad
Faya-Largeau see Faya
Fayetteville 20 A1 Arkansas, C USA
Fayetteville 21 F1 North Carolina, SE USA
Fdérick see Fdérik
Fdérik 52 C2 var. Fdérick, Fr. Fort Gouraud. Tiris Zemmour, NW Mauritania
Fear, Cape 21 F2 headland Bald Head Island, North Carolina, SE USA
Fécamp 68 B3 Seine-Maritime, N France
Fédala see Mohammedia
Federal Capital Territory see Australian Capital Territory
Fehérgyarmat 77 E6 Szabolcs-Szatmár-Bereg, E Hungary
Fehértemplom see Bela Crkva
Fehmarn 72 C2 island N Germany
Fehmarn Belt 72 C2 Dan. Femern Bælt, Ger. Fehmarnbelt. strait Denmark /Germany
Fehmarnbelt Fehmarn Belt/Femer Bælt
Feijó 40 C2 Acre, W Brazil
Feilding 128 D4 Manawatu-Wanganui, North Island, New Zealand
Feira see Feira de Santana
Feira de Santana 41 G3 var. Feira. Bahia, E Brazil
Feketehalom see Codlea
Felanitx 71 G3 Mallorca, Spain, W Mediterranean Sea
Felicitas Julia see Lisboa
Felidhu Atoll 110 B4 atoll C Maldives
Felipe Carrillo Puerto 29 H4 Quintana Roo, SE Mexico
Felixstowe 67 E6 E England, United Kingdom
Fellin see Viljandi
Felsőbánya see Baia Sprie
Felsőmuzslya see Mužlja
Femunden 63 B5 lake S Norway
Fénérive see Fenoarivo Atsinanana
Fengcheng 106 D3 var. Feng-cheng, Fenghwangcheng. Liaoning, NE China
Feng-cheng see Fengcheng
Fenghwangcheng see Fengcheng
Fengtien see Shenyang, China
Fengtien see Liaoning, China
Fenoarivo Atsinanana 57 G3 Fr. Fénérive. Toamasina, E Madagascar
Fens, The 67 E6 wetland E England, United Kingdom
Feodosiya 87 F5 var. Kefe, It. Kaffa; anc. Theodosia. Respublika Krym, S Ukraine
Ferdinand see Montana, Bulgaria
Ferdinandsberg see Oţelu Roşu
Féres 82 D3 Anatolikí Makedonía kai Thráki, NE Greece
Fergana see Farg'ona
Fergus Falls 23 F2 Minnesota, N USA
Ferizaj 79 D5 Serb. Uroševac. C Kosovo
Ferkessédougou 52 D4 N Ivory Coast
Fermo 74 C4 anc. Firmum Picenum. Marche, C Italy
Fernandina, Isla 38 A5 var. Narborough Island. island Galápagos Islands, Ecuador, E Pacific Ocean
Fernando de Noronha 41 H2 island E Brazil
Fernando Po/Fernando Póo see Bioco, Isla de
Ferrara 74 C2 anc. Forum Alieni. Emilia-Romagna, N Italy
Ferreñafe 38 B3 Lambayeque, W Peru
Ferro see Hierro
Ferrol 70 B1 var. El Ferrol; prev. El Ferrol del Caudillo. Galicia, NW Spain
Fertő see Neusiedler See
Ferwerd see Ferwert
Ferwert 64 D1 Dutch. Ferwerd. Friesland, N Netherlands
Fès 48 C2 Eng. Fez. N Morocco
Feteşti 86 D5 Ialomiţa, SE Romania
Fethiye 94 B4 Muğla, SW Turkey
Fetlar 66 D1 island NE Scotland, United Kingdom
Feuilles, Rivière aux 16 D2 river Québec, E Canada
Feyẕābād 101 F3 var. Faizabad, Faizābād, Feyzābād, Fyzabad. Badakhshān, NE Afghanistan
Feyẕābād see Feyẕābād
Fez see Fès
Fianarantsoa 57 F3 Fianarantsoa, C Madagascar
Fianga 54 B4 Mayo-Kébbi, SW Chad
Fier 79 C6 var. Fieri. Fier, SW Albania
Fieri see Fier
Figeac 69 C5 Lot, S France
Figig see Figuig
Figueira da Foz 70 B3 Coimbra, W Portugal
Figueres 71 G2 Cataluña, E Spain
Figuig 48 D2 var. Figig. E Morocco
Fiji 123 E5 off. Republic of Fiji, Fij. Viti. country SW Pacific Ocean
Fiji, Republic of see Fiji
Filadelfia 30 D4 Guanacaste, W Costa Rica
Filiaşi 86 B5 Dolj, SW Romania
Filipstad 63 B6 Värmland, C Sweden
Finale Ligure 74 A3 Liguria, NW Italy
Finchley 67 A7 United Kingdom
Findlay 18 C4 Ohio, N USA
Finike 94 B4 Antalya, SW Turkey
Finland 62 D4 off. Republic of Finland, Fin. Suomen Tasavalta, Suomi. country N Europe
Finland, Gulf of 63 D6 Est. Soome Laht, Fin. Suomenlahti, Ger. Finnischer Meerbusen, Rus. Finskiy Zaliv, Swe. Finska Viken. gulf E Baltic Sea
Finland, Republic of see Finland
Finnischer Meerbusen see Finland, Gulf of
Finnmarksvidda 62 D2 physical region N Norway
Finska Viken/Finskiy Zaliv see Finland, Gulf of
Finsterwalde 72 D4 Brandenburg, E Germany
Fiordland 129 A7 physical region South Island, New Zealand
Fiorina 74 E1 NE San Marino
Firat Nehri see Euphrates
Firenze 74 C3 Eng. Florence; anc. Florentia. Toscana, C Italy
Firmum Picenum see Fermo
First State see Delaware
Fischbacher Alpen 73 E7 mountain range E Austria
Fischhausen see Primorsk
Fish 56 B4 var. Vis. river S Namibia
Fishguard 67 C6 Wel. Abergwaun. SW Wales, United Kingdom
Fisterra, Cabo 70 B1 headland NW Spain
Fitzroy Crossing 124 C3 Western Australia
Fitzroy River 124 C3 river Western Australia
Fiume see Rijeka
Flagstaff 26 B2 Arizona, SW USA
Flanders 65 A6 Dut. Vlaanderen, Fr. Flandre. cultural region Belgium/France
Flandre see Flanders
Flathead Lake 22 B1 lake Montana, NW USA
Flat Island 106 C8 island NE Spratly Islands
Flatts Village 20 B5 var. The Flatts Village. C Bermuda
Flensburg 72 B2 Schleswig-Holstein, N Germany
Flessingue see Vlissingen
Flickertail State see North Dakota
Flinders Island 127 C8 island Furneaux Group, Tasmania, SE Australia
Flinders Ranges 127 B6 mountain range South Australia
Flinders River 126 C3 river Queensland, NE Australia
Flin Flon 15 F5 Manitoba, C Canada
Flint 18 C3 Michigan, N USA
Flint Island 123 G4 island Line Islands, E Kiribati
Floreana, Isla see Santa María, Isla
Florence 20 C1 Alabama, S USA
Florence 21 F2 South Carolina, SE USA
Florence see Firenze
Florencia 36 B4 Caquetá, S Colombia
Florentia see Firenze
Flores 30 B1 Petén, N Guatemala
Flores 117 E5 island Nusa Tenggara, C Indonesia
Flores 70 A5 island Azores, Portugal, NE Atlantic Ocean
Flores, Laut see Flores Sea
Flores Sea 116 D5 Ind. Laut Flores. sea C Indonesia
Floriano 41 F2 Piauí, E Brazil
Florianópolis 41 F5 prev. Destêrro. state capital Santa Catarina, S Brazil
Florida 42 D4 Florida, S Uruguay
Florida 21 E4 off. State of Florida, also known as Peninsular State, Sunshine State. state SE USA
Florida Bay 21 E5 bay Florida, SE USA
Florida Keys 21 E5 island group Florida, SE USA
Florida, Straits of 32 B1 strait Atlantic Ocean/ Gulf of Mexico
Flórina 82 B4 var. Phlórina. Dytikí Makedonía, N Greece
Florissant 23 G4 Missouri, C USA
Floúda, Akrotírio 83 D7 headland Astypálaia, Kykládes, Greece, Aegean Sea
Flushing see Vlissingen
Flylân see Vlieland
Foča 78 C4 var. Srbinje. SE Bosnia and Herzegovina
Focşani 86 C4 Vrancea, E Romania
Fogaras/Fogarasch see Făgăraş
Foggia 75 D5 Puglia, SE Italy
Fogo 52 A3 island Ilhas de Sotavento, SW Cape Verde
Foix 69 B6 Ariège, S France
Folégandros 83 C7 island Kykládes, Greece, Aegean Sea
Foleyet 16 C4 Ontario, S Canada
Foligno 74 C4 Umbria, C Italy
Folkestone 67 E7 SE England, United Kingdom
Fond du Lac 18 B2 Wisconsin, N USA
Fonseca, Golfo de see Fonseca, Gulf of
Fonseca, Gulf of 30 C3 Sp. Golfo de Fonseca. gulf C Central America
Fontainebleau 68 C3 Seine-et-Marne, N France
Fontenay-le-Comte 68 B4 Vendée, NW France
Fontvieille 69 B8 SW Monaco Europe
Fonyód 77 C7 Somogy, W Hungary
Foochow see Fuzhou
Foochow see Fuzhou
Forchheim 73 C5 Bayern, SE Germany
Forel, Mont 60 D4 mountain SE Greenland
Forfar 66 C3 E Scotland, United Kingdom
Forge du Sud see Dudelange
Forlì 74 C3 anc. Forum Livii. Emilia-Romagna, N Italy
Formentera 71 G4 anc. Ophiusa, Lat. Frumentum. island Islas Baleares, Spain, W Mediterranean Sea
Formosa 42 D2 Formosa, NE Argentina
Formosa/Formo'sa see Taiwan
Formosa, Serra 41 E3 mountain range C Brazil
Formosa Strait see Taiwan Strait
Føroyar see Faroe Islands
Forrest City 20 B1 Arkansas, C USA
Fort Albany 16 C3 Ontario, C Canada
Fortaleza 39 F2 Pando, N Bolivia
Fortaleza 41 G2 prev. Ceará. state capital Ceará, NE Brazil

Fort-Archambault – Giżycko

Fort-Archambault see Sarh
Fort-Bayard see Zhanjiang
Fort-Cappolani see Tidjikja
Fort Charlet see Djanet
Fort-Chimo see Kuujjuaq
Fort Collins 22 D4 Colorado, C USA
Fort-Crampel see Kaga Bandoro
Fort Davis 27 E3 Texas, SW USA
Fort-de-France 33 H4 prev. Fort-Royal. dependent territory capital (Martinique) W Martinique
Fort Dodge 23 F3 Iowa, C USA
Fortescue River 124 A4 river Western Australia
Fort-Foureau see Kousséri
Fort Frances 16 B4 Ontario, S Canada
Fort Good Hope 15 E3 var. Rádeyilikóé. Northwest Territories, NW Canada
Fort Gouraud see Fdérik
Forth 66 C4 river C Scotland, United Kingdom
Forth, Firth of 66 C4 estuary E Scotland, United Kingdom
Fortín General Eugenio Garay see General Eugenio A. Garay
Fort Jameson see Chipata
Fort-Lamy see Ndjamena
Fort Lauderdale 21 F5 Florida, SE USA
Fort Liard 15 E4 var. Liard. Northwest Territories, W Canada
Fort Madison 23 G4 Iowa, C USA
Fort McMurray 15 E4 Alberta, C Canada
Fort McPherson 14 D3 var. McPherson. Northwest Territories, NW Canada
Fort Morgan 22 D4 Colorado, C USA
Fort Myers 21 E5 Florida, SE USA
Fort Nelson 15 E4 British Columbia, W Canada
Fort Peck Lake 22 C1 reservoir Montana, NW USA
Fort Pierce 21 F4 Florida, SE USA
Fort Providence 15 E4 var. Providence. Northwest Territories, W Canada
Fort-Repoux see Akjoujt
Fort Rosebery see Mansa
Fort Rousset see Owando
Fort-Royal see Fort-de-France
Fort St. John 15 E4 British Columbia, W Canada
Fort Scott 23 F5 Kansas, C USA
Fort Severn 16 C2 Ontario, C Canada
Fort-Shevchenko 92 A4 Mangistau, W Kazakhstan
Fort-Sibut see Sibut
Fort Simpson 15 E4 var. Simpson. Northwest Territories, W Canada
Fort Smith 15 E4 Northwest Territories, W Canada
Fort Smith 20 B1 Arkansas, C USA
Fort Stockton 27 E3 Texas, SW USA
Fort-Trinquet see Bîr Mogreïn
Fort Vermilion 15 E4 Alberta, W Canada
Fort Victoria see Masvingo
Fort Walton Beach 20 C3 Florida, SE USA
Fort Wayne 18 C4 Indiana, N USA
Fort William 66 C3 N Scotland, United Kingdom
Fort Worth 27 G2 Texas, SW USA
Fort Yukon 14 D3 Alaska, USA
Forum Alieni see Ferrara
Forum Livii see Forlì
Fossa Claudia see Chioggia
Fougamou 55 A6 Ngounié, C Gabon
Fougères 68 B3 Ille-et-Vilaine, NW France
Fou-hsin see Fuxin
Foulwind, Cape 129 B5 headland South Island, New Zealand
Foumban 54 A4 Ouest, NW Cameroon
Fou-shan see Fushun
Foveaux Strait 129 A8 strait S New Zealand
Foxe Basin 15 G3 sea Nunavut, N Canada
Fox Glacier 129 B6 West Coast, South Island, New Zealand
Fraga 71 F2 Aragón, NE Spain
Fram Basin 133 C3 var. Amundsen Basin. undersea basin Arctic Ocean
France 68 B4 off. French Republic, It./Sp. Francia; prev. Gaul, Gaule, Lat. Gallia. country W Europe
Franceville 55 B6 var. Massoukou, Masuku. Haut-Ogooué, E Gabon
Francfort see Frankfurt am Main
Franche-Comté 68 D4 cultural region E France
Francia see France
Francis Case, Lake 23 E3 reservoir South Dakota, N USA
Francisco Escárcega 29 G4 Campeche, SE Mexico
Francistown 56 D3 North East, NE Botswana
Franconian Jura see Fränkische Alb
Frankenalb see Fränkische Alb
Frankenstein/Frankenstein in Schlesien see Ząbkowice Śląskie
Frankfort 18 C5 state capital Kentucky, S USA
Frankfort on the Main see Frankfurt am Main
Frankfurt see Frankfurt am Main, Germany
Frankfurt see Słubice, Poland
Frankfurt am Main 73 B5 var. Frankfurt, Fr. Francfort; prev. Eng. Frankfort on the Main. Hessen, SW Germany
Frankfurt an der Oder 72 D3 Brandenburg, E Germany
Fränkische Alb 73 C6 var. Frankenalb, Eng. Franconian Jura. mountain range S Germany
Franklin 20 C1 Tennessee, S USA
Franklin D. Roosevelt Lake 24 C1 reservoir Washington, NW USA
Franz Josef Land 92 D1 Eng. Franz Josef Land. island group N Russia
Franz Josef Land see Frantsa-Iosifa, Zemlya
Fraserburgh 66 D3 NE Scotland, United Kingdom
Fraser Island 126 E4 var. Great Sandy Island. island Queensland, E Australia
Frauenbach see Baia Mare
Frauenburg see Saldus, Latvia
Fredericksburg 19 E5 Virginia, NE USA
Fredericton 17 F4 province capital New Brunswick, SE Canada
Frederikshåb see Paamiut
Fredrikshald see Halden
Fredrikstad 63 B6 Østfold, S Norway
Freeport 32 C1 Grand Bahama Island, N The Bahamas
Freeport 27 H4 Texas, SW USA
Free State see Maryland
Freetown 52 C4 country capital (Sierra Leone) W Sierra Leone
Freiburg see Freiburg im Breisgau, Germany
Freiburg im Breisgau 73 A6 var. Freiburg, Fr. Fribourg-en-Brisgau. Baden-Württemberg, SW Germany
Freiburg in Schlesien see Świebodzice
Fremantle 125 A6 Western Australia
Fremont 23 F4 Nebraska, C USA
French Guiana 37 H3 var. Guiana, Guyane. French overseas department N South America
French Guinea see Guinea
French Polynesia 121 F4 French overseas territory S Pacific Ocean
French Republic see France
French Somaliland see Djibouti
French Southern and Antarctic Lands 119 B7 Fr. Terres Australes et Antarctiques Françaises. French overseas territory S Indian Ocean
French Sudan see Mali
French Territory of the Afars and Issas see Djibouti
French Togoland see Togo
Fresnillo 28 D3 var. Fresnillo de González Echeverría. Zacatecas, C Mexico
Fresnillo de González Echeverría see Fresnillo
Fresno 25 C6 California, W USA
Frías 42 C3 Catamarca, N Argentina
Fribourg-en-Brisgau see Freiburg im Breisgau
Friedek-Mistek see Frýdek-Místek
Friedrichshafen 73 B7 Baden-Württemberg, S Germany
Friendly Islands see Tonga
Frisches Haff see Vistula Lagoon
Frobisher Bay 60 B3 inlet Baffin Island, Nunavut, NE Canada
Frobisher Bay see Iqaluit
Frohavet 62 B4 sound C Norway
Frome, Lake 127 B6 salt lake South Australia
Frontera 29 G4 Tabasco, SE Mexico
Frontignan 69 C6 Hérault, S France
Frostviken see Kvarnbergsvattnet
Frøya 62 A4 island W Norway
Frumentum see Formentera
Frunze see Bishkek
Frýdek-Místek 77 C5 Ger. Friedek-Mistek. Moravskoslezský Kraj, E Czech Republic (Czechia)
Fu-chien see Fujian
Fu-chou see Fuzhou
Fuengirola 70 D5 Andalucía, S Spain
Fuerte Olimpo 42 D2 var. Olimpo. Alto Paraguay, NE Paraguay
Fuerte, Río 26 C5 river C Mexico
Fuerteventura 48 B3 island Islas Canarias, Spain, NE Atlantic Ocean
Fuhkien see Fujian
Fu-hsin see Fuxin
Fuji 109 D6 var. Huzi. Shizuoka, Honshū, S Japan
Fujian 106 D6 var. Fu-chien, Fuhkien, Fukien, Min, Fujian Sheng. province SE China
Fujian Sheng see Fujian
Fuji, Mount/Fujiyama see Fuji-san
Fuji-san 109 C6 var. Fujiyama, Eng. Mount Fuji. mountain Honshū, SE Japan
Fukang 104 C2 Xinjiang Uygur Zizhiqu, W China
Fukien see Fujian
Fukui 109 C6 var. Hukui. Fukui, Honshū, SW Japan
Fukuoka 109 A7 var. Hukuoka, hist. Najima. Fukuoka, Kyūshū, SW Japan
Fukushima 108 D4 var. Hukusima. Fukushima, Honshū, C Japan
Fulda 73 B5 Hessen, C Germany
Funafuti Atoll 123 E3 atoll and capital (Tuvalu) C Tuvalu
Funchal 48 A2 Madeira, Portugal, NE Atlantic Ocean
Fundy, Bay of 17 F5 bay Canada/USA
Fünen see Fyn
Fünfkirchen see Pécs
Furnes see Veurne
Fürth 73 C5 Bayern, S Germany
Furukawa 108 D4 var. Hurukawa, Ōsaki. Miyagi, Honshū, C Japan
Fusan see Busan
Fushë Kosovë 79 D5 Serb. Kosovo Polje. C Kosovo
Fushun 106 D3 var. Fou-shan, Fu-shun. Liaoning, NE China
Fu-shun see Fushun
Fusin see Fuxin
Füssen 73 C7 Bayern, S Germany
Futog 78 D3 Vojvodina, NW Serbia
Futuna, Île 123 E4 island S Wallis and Futuna
Fuxin 106 D3 var. Fou-hsin, Fu-hsin, Fusin. Liaoning, NE China
Fuzhou 106 D6 var. Foochow, Fu-chou. province capital Fujian, SE China
Fyn 63 B8 Ger. Fünen. island C Denmark
FYR Macedonia/FYROM see Macedonia
Fyzabad see Feyẕābād

G

Gaafu Alifu Atoll see North Huvadhu Atoll
Gaalkacyo 51 E5 var. Galka'yo, It. Galcaio. Mudug, C Somalia
Gabela 56 B2 Cuanza Sul, W Angola
Gaberones see Gaborone
Gabès 49 E2 var. Qābis. E Tunisia
Gabès, Golfe de 49 F2 Ar. Khalīj Qābis. gulf E Tunisia
Gabon 55 B6 off. Gabonese Republic. country C Africa
Gabonese Republic see Gabon
Gaborone 56 C4 prev. Gaberones. country capital (Botswana) South East, SE Botswana
Gabrovo 82 D2 Gabrovo, N Bulgaria
Gadag 110 C1 Karnātaka, W India
Gades/Gadier/Gadir/Gadire see Cádiz
Gadsden 20 D2 Alabama, S USA
Gaeta 75 C5 Lazio, C Italy
Gaeta, Golfo di 75 C5 var. Gulf of Gaeta. gulf C Italy
Gaeta, Gulf of see Gaeta, Golfo di
Gäfle see Gävle
Gafsa 49 E2 var. Qafşah. W Tunisia
Gagnoa 52 D5 C Ivory Coast
Gagra 95 E1 NW Georgia
Gaillac 69 C6 var. Gaillac-sur-Tarn. Tarn, S France
Gaillac-sur-Tarn see Gaillac
Gaillimh see Galway
Gaillimhe, Cuan na see Galway Bay
Gainesville 21 E3 Florida, SE USA
Gainesville 20 D2 Georgia, SE USA
Gainesville 27 G2 Texas, SW USA
Lake Gairdner 127 A6 salt lake South Australia
Gaizina Kalns see Gaiziņkalns
Gaiziņkalns 84 C3 var. Gaizina Kalns. mountain E Latvia
Galán, Cerro 42 B3 mountain NW Argentina
Galanta 77 C6 Hung. Galánta. Trnavský Kraj, W Slovakia
Galapagos Fracture Zone 131 E3 tectonic feature E Pacific Ocean
Galápagos Islands 131 F3 var. Islas de los Galápagos, Sp. Archipiélago de Colón, Eng. Galapagos Islands, Tortoise Islands. island group Ecuador, E Pacific Ocean
Galapagos Islands see Galápagos Islands
Galápagos, Islas de los see Galápagos Islands
Galapagos Rise 131 F3 undersea rise E Pacific Ocean
Galashiels 66 C4 SE Scotland, United Kingdom
Galaţi 86 D4 Ger. Galatz. Galaţi, E Romania
Galatz see Galaţi
Galcaio see Gaalkacyo
Galesburg 18 B3 Illinois, N USA
Galicia 70 B1 anc. Gallaecia. autonomous community NW Spain
Galicia Bank 58 B4 undersea bank E Atlantic Ocean
Galilee, Sea of see Tiberias, Lake
Galka'yo see Gaalkacyo
Galkynyş 100 D3 prev. Rus. Deynau, Dyanev, Turkm. Dänew. Lebap Welaýaty, NE Turkmenistan
Gallaecia see Galicia
Galle 110 D4 prev. Point de Galle. Southern Province, SW Sri Lanka
Gallego Rise 131 F3 undersea rise E Pacific Ocean
Gallegos see Río Gallegos
Gallia see France
Gallipoli 75 E6 Puglia, SE Italy
Gällivare 62 C3 Lapp. Váhtjer. Norrbotten, N Sweden
Gallup 26 C1 New Mexico, SW USA
Galtat-Zemmour 48 B3 C Western Sahara
Galveston 27 H4 Texas, SW USA
Galway 67 A5 Ir. Gaillimh. W Ireland
Galway Bay 67 A6 Ir. Cuan na Gaillimhe. bay W Ireland
Gámas see Kaamanen
Gambell 14 C2 Saint Lawrence Island, Alaska, USA
Gambia see Gambia, The
Gambia 52 C3 Fr. Gambie. river W Africa
Gambia, Republic of The see Gambia, The
Gambia, The 52 B3 off. Republic of The Gambia var. Gambia. country W Africa
Gambie see Gambia, The
Gambier, Îles 121 G4 island group E French Polynesia
Gamboma 55 B6 Plateaux, E Congo
Gamlakarleby see Kokkola
Gan 110 B5 Addu Atoll, C Maldives
Gan see Gansu, China
Gan see Jiangxi, China
Ganaane see Juba
Gäncä 95 G2 Rus. Gyandzha; prev. Kirovabad, Yelisavetpol. W Azerbaijan
Gand see Gent
Gandajika 55 D7 Kasai-Oriental, S Dem. Rep. Congo
Gander 17 G3 Newfoundland and Labrador, SE Canada
Gāndhīdhām 112 C4 Gujarāt, W India
Gandia 71 F3 prev. Gandía. País Valenciano, E Spain
Gandía see Gandia
Ganges 113 F3 Ben. Padma. river Bangladesh/India
Ganges Cone see Ganges Fan
Ganges Fan 118 D3 var. Ganges Cone. undersea fan N Bay of Bengal
Ganges, Mouths of the 113 G4 delta Bangladesh/India
Gangra see Çankırı
Gangtok 113 F3 state capital Sikkim, N India
Gansos, Lago dos see Goose Lake
Gansu 106 B4 var. Gan, Gansu Sheng, Kansu. province N China
Gansu Sheng see Gansu
Gantsevichi see Hantsavichy
Ganzhou 106 D6 Jiangxi, S China
Gao 53 E3 Gao, E Mali
Gaocheng see Litang
Gaoual 52 C4 N Guinea
Gaoxiong see Kaohsiung
Gap 69 D5 anc. Vapincum. Hautes-Alpes, SE France
Gaplaňgyr Platosy 100 C2 Rus. Plato Kaplangky. ridge Turkmenistan/Uzbekistan
Gar see Gar Xincun
Garabil Belentligi 100 D3 Rus. Vozvyshennost' Karabil'. mountain range S Turkmenistan
Garabogaz Aylagy 100 B2 Rus. Zaliv Kara-Bogaz-Gol. bay NW Turkmenistan
Garachiné 31 G5 Darién, SE Panama
Garagum 100 C3 var. Garagumy, Qara Qum, Eng. Black Sand Desert, Kara Kum; prev. Peski Karakumy. desert C Turkmenistan
Garagum Canal 100 D3 var. Kara Kum Canal, Rus. Karagumskiy Kanal, Karakumskiy Kanal. canal C Turkmenistan
Garagumy see Garagum
Gara Khitrino 82 D2 Shumen, NE Bulgaria
Gárassavon see Kaaresuvanto
Garda, Lago di 74 C2 var. Benaco, Eng. Lake Garda, Ger. Gardasee. lake NE Italy
Garda, Lake see Garda, Lago di
Gardasee see Garda, Lago di
Garden City 23 E5 Kansas, C USA
Garden State, The see New Jersey
Gardēz 101 E4 prev. Gardīz. E Afghanistan
Gardīz see Gardēz
Gardner Island see Nikumaroro
Garegegasnjárga see Karigasniemi
Gargždai 84 B3 Klaipėda, W Lithuania
Garissa 51 D6 Coast, E Kenya
Garland 27 G2 Texas, SW USA
Garoe see Garoowe
Garonne 69 B5 anc. Garumna. river S France
Garoowe 51 E5 var. Garoe. Nugaal, N Somalia
Garoua 54 B4 var. Garua. Nord, N Cameroon
Garrygala see Magtymguly
Garry Lake 15 F3 lake Nunavut, N Canada
Garsen 51 D6 Coast, S Kenya
Garua see Garoua
Garumna see Garonne
Garwolin 76 D4 Mazowieckie, E Poland
Gar Xincun 104 A4 prev. Gar. Xizang Zizhiqu, W China
Gary 18 B3 Indiana, N USA
Garzón 36 B4 Huila, S Colombia
Gasan-Kuli see Esenguly
Gascogne 69 B6 Eng. Gascony. cultural region S France
Gascony see Gascogne
Gascoyne River 125 A5 river Western Australia
Gaspé 17 F3 Québec, SE Canada
Gaspé, Péninsule de 17 E4 var. Péninsule de la Gaspésie. peninsula Québec, SE Canada
Gaspésie, Péninsule de la see Gaspé, Péninsule de
Gastonia 21 E1 North Carolina, SE USA
Gastoúni 83 B6 Dytikí Elláda, S Greece
Gatchina 88 B4 Leningradskaya Oblast', NW Russia
Gatineau 16 D4 Québec, SE Canada
Gatooma see Kadoma
Gatún, Lake 31 F4 reservoir C Panama
Gauhāti see Guwāhāti
Gauja 84 D3 Ger. Aa. river Estonia/Latvia
Gaul/Gaule see France
Gauteng see Johannesburg, South Africa
Gāvbandī 98 D4 Hormozgān, S Iran
Gávdos 83 C8 island SE Greece
Gavere 65 B6 Oost-Vlaanderen, NW Belgium
Gävle 63 C6 var. Gäfle; prev. Gefle. Gävleborg, C Sweden
Gawler 127 B6 South Australia
Gaya 113 F3 Bihār, N India
Gaya see Kyjov
Gayndah 127 E5 Queensland, E Australia
Gaysin see Haysyn
Gaza 97 A6 Ar. Ghazzah, Heb. 'Azza. NE Gaza Strip
Gaz-Achak see Gazojak
Gazandzhyk/Gazanjyk see Bereket
Gaza Strip 97 A7 Ar. Qita Ghazzah. disputed region SW Asia
Gaziantep 94 D4 var. Gazi Antep; prev. Aintab, Antep. Gaziantep, S Turkey
Gazi Antep see Gaziantep
Gazimağusa 80 D5 var. Famagusta, Gk. Ammóchostos. E Cyprus
Gazimağusa Körfezi 80 C5 var. Famagusta Bay, Gk. Kólpos Ammóchostos. bay E Cyprus
Gazli 100 D2 Buxoro Viloyati, C Uzbekistan
Gazojak 100 D2 Rus. Gaz-Achak. Lebap Welaýaty, NE Turkmenistan
Gbanga 52 D5 var. Gbarnga. N Liberia
Gbarnga see Gbanga
Gdańsk 76 C2 Fr. Dantzig, Ger. Danzig. Pomorskie, N Poland
Gdan'skaya Bukhta/Gdańsk, Gulf of see Danzig, Gulf of
Gdańska, Zakota see Danzig, Gulf of
Gdingen see Gdynia
Gdynia 76 C2 Ger. Gdingen. Pomorskie, N Poland
Gedaref 50 C4 var. Al Qaḍārif, El Gedaref. Gedaref, E Sudan
Gediz 94 B3 Kütahya, W Turkey
Gediz Nehri 94 A3 river W Turkey
Geel 65 C5 var. Gheel. Antwerpen, N Belgium
Geelong 127 C7 Victoria, SE Australia
Ge'e'mu see Golmud
Gefle see Gävle
Geilo 63 A5 Buskerud, S Norway
Gejiu 106 B6 var. Kochiu. Yunnan, S China
Gëkdepe see Gökdepe
Gela 75 C7 prev. Terranova di Sicilia. Sicilia, Italy, C Mediterranean Sea
Geldermalsen 64 C4 Gelderland, C Netherlands
Geleen 65 D6 Limburg, SE Netherlands
Gelib see Jilib
Gellinsor 51 E5 Mudug, C Somalia
Gembloux 65 C6 Namur, Belgium
Gemena 55 C5 Equateur, NW Dem. Rep. Congo
Gem of the Mountains see Idaho
Gemona del Friuli 74 D2 Friuli-Venezia Giulia, NE Italy
Gem State see Idaho
Genalē Wenz see Juba
Genck see Genk
General Alvear 42 B4 Mendoza, W Argentina
General Carrera, Lago see Buenos Aires, Lago
General Eugenio A. Garay 42 C1 var. Fortín General Eugenio Garay; prev. Yrendagüé. Nueva Asunción, NW Paraguay
General José F.Uriburu see Zárate
General Machado see Camacupa
General Santos 117 F3 off. General Santos City. Mindanao, S Philippines
General Santos City see General Santos
Gênes see Genova
Geneva see Genève
Geneva, Lake 73 A7 Fr. Lac de Genève, Lac Léman, le Léman, Ger. Genfer See. lake France/Switzerland
Genève 73 A7 Eng. Geneva, Ger. Genf, It. Ginevra. Genève, SW Switzerland
Genève, Lac de see Geneva, Lake
Genf see Genève
Genfer See see Geneva, Lake
Genichesk see Heniches'k
Genk 65 D6 var. Genck. Limburg, NE Belgium
Gennep 64 D4 Limburg, SE Netherlands
Genoa see Genova
Genoa, Gulf of 74 A3 Eng. Gulf of Genoa. gulf NW Italy
Genoa, Gulf of see Genova, Golfo di
Genova 80 D1 Eng. Genoa; anc. Genua, Fr. Gênes. Liguria, NW Italy
Genovesa, Isla 38 B5 var. Tower Island. island Galápagos Islands, Ecuador, E Pacific Ocean
Gent 65 B5 Eng. Ghent, Fr. Gand. Oost-Vlaanderen, NW Belgium
Genua see Genova
Geok-Tepe see Gökdepe
George 56 C5 Western Cape, S South Africa
George 60 A4 river Newfoundland and Labrador/Québec, E Canada
George, Lake 21 E3 lake Florida, SE USA
Georgenburg see Jurbarkas
Georges Bank 13 D5 undersea bank W Atlantic Ocean
George Sound 129 A7 sound South Island, New Zealand
Georges River 126 D2 river New South Wales, E Australia
Georgetown 37 F2 country capital (Guyana) N Guyana
George Town 32 C2 Great Exuma Island, C The Bahamas
George Town 32 B3 var. Georgetown. dependent territory capital (Cayman Islands) Grand Cayman, SW Cayman Islands
George Town 116 B3 var. Penang, Pinang. Pinang, Peninsular Malaysia
Georgetown 21 F2 South Carolina, SE USA
Georgetown see George Town
George V Land 132 C4 physical region Antarctica
Georgia 95 F2 off. Republic of Georgia, Geor. Sak'art'velo, Rus. Gruzinskaya SSR, Gruziya. country SW Asia
Georgia 20 D2 off. State of Georgia, also known as Empire State of the South, Peach State. state SE USA
Georgian Bay 18 D2 lake bay Ontario, S Canada
Georgia, Republic of see Georgia
Georgia, Strait of 24 A1 strait British Columbia, W Canada
Georgi Dimitrov see Kostenets
Georgiu-Dezh see Liski
Georg von Neumayer 132 A2 German research station Antarctica
Gera 72 C4 Thüringen, E Germany
Geráki 83 B6 Pelopónnisos, S Greece
Geraldine 129 B6 Canterbury, South Island, New Zealand
Geraldton 125 A6 Western Australia
Geral, Serra 35 D5 mountain range S Brazil
Gerede 94 C2 Bolu, N Turkey
Gereshk 100 D5 Helmand, SW Afghanistan
Gering 22 D3 Nebraska, C USA
German East Africa see Tanzania
Germanicopolis see Çankırı
Germanicum, Mare/German Ocean see North Sea
German Southwest Africa see Namibia
Germany 72 B4 off. Federal Republic of Germany, Bundesrepublik Deutschland, Ger. Deutschland. country N Europe
Germany, Federal Republic of see Germany
Geroliménas 83 B7 Pelopónnisos, S Greece
Gerona see Girona
Gerpinnes 65 C7 Hainaut, S Belgium
Gerunda see Girona
Gerze 94 D2 Sinop, N Turkey
Gesoriacum see Boulogne-sur-Mer
Gessoriacum see Boulogne-sur-Mer
Getafe 70 D3 Madrid, C Spain
Gevaş 95 F3 Van, SE Turkey
Gevgeli see Gevgelija
Gevgelija 79 E6 var. Đevđelija, Djevdjelija, Turk. Gevgeli. SE Macedonia
Ghaba see Al Ghābah
Ghana 53 E5 off. Republic of Ghana. country W Africa
Ghanzi 56 C3 var. Khanzi. Ghanzi, W Botswana
Gharandal 97 B7 Al 'Aqabah, SW Jordan
Gharbt, Jabal al see Liban, Jebel
Ghardaïa 48 D2 N Algeria
Gharvān see Gharyān
Gharyān 49 F2 var. Gharvān. NW Libya
Ghawdex see Gozo
Ghaznī 101 E4 var. Ghazni. Ghaznī, E Afghanistan
Ghazzah see Gaza
Gheel see Geel
Ghent see Gent
Gheorgheni 86 C4 prev. Gheorghieni, Sîn-Miclăuş, Ger. Niklasmarkt, Hung. Gyergyószentmiklós. Harghita, C Romania
Gheorghieni see Gheorgheni
Ghōriān 100 D4 prev. Ghūrīān. Herāt, W Afghanistan
Ghŭdara 101 F3 var. Gudara, Rus. Kudara. SE Tajikistan
Ghurdaqah see Al Ghurdaqah
Ghūrīān see Ghōriān
Giamame see Jamaame
Giannitsá 82 B4 var. Yiannitsá. Kentrikí Makedonía, N Greece
Gibraltar 71 G4 UK dependent territory SW Europe
Gibraltar, Bay of 71 G5 bay Gibraltar/Spain Europe Mediterranean Sea Atlantic Ocean
Gibraltar, Détroit de/Gibraltar, Estrecho de see Gibraltar, Strait of
Gibraltar, Strait of 70 C5 Fr. Détroit de Gibraltar, Sp. Estrecho de Gibraltar. strait Atlantic Ocean/Mediterranean Sea
Gibson Desert 125 B5 desert Western Australia
Giedraičiai 85 C5 Utena, E Lithuania
Giessen 73 B5 Hessen, W Germany
Gifu 109 C6 var. Gihu. Gifu, Honshū, SW Japan
Giganta, Sierra de la 28 B3 mountain range NW Mexico
Gihu see Gifu
G'ijduvon 100 D2 Rus. Gizhduvon. Buxoro Viloyati, C Uzbekistan
Gijón 70 D1 var. Xixón. Asturias, NW Spain
Gila River 26 A2 river Arizona, SW USA
Gilbert Islands see Tungaru
Gilbert River 126 C3 river Queensland, NE Australia
Gilf Kebir Plateau see Haḍabat al Jilf al Kabīr
Gillette 22 D3 Wyoming, C USA
Gilolo see Halmahera, Pulau
Gilroy 25 B6 California, W USA
Gimie, Mount 33 F1 mountain C Saint Lucia
Gimma see Jīma
Ginevra see Genève
Gingin 125 A6 Western Australia
Giohar see Jawhar
Gipeswic see Ipswich
Girardot 36 B3 Cundinamarca, C Colombia
Giresun 95 E2 var. Kerasunt; anc. Cerasus, Pharnacia. Giresun, NE Turkey
Girgenti see Agrigento
Girin see Jilin
Girne 80 C5 Gk. Kerýneia, Kyrenia. N Cyprus
Giron see Kiruna
Girona 71 G2 var. Gerona; anc. Gerunda. Cataluña, NE Spain
Gisborne 128 E3 Gisborne, North Island, New Zealand
Gissar Range 101 E3 Rus. Gissarskiy Khrebet. mountain range Tajikistan/Uzbekistan
Gissarskiy Khrebet see Gissar Range
Githio see Gýtheio
Giulianova 74 D4 Abruzzi, C Italy
Giumri see Gyumri
Giurgiu 86 C5 Giurgiu, S Romania
Giza 50 B1 var. Al Jīzah, El Giza, Gizeh. N Egypt
Gizhduvon see G'ijduvon
Giżycko 76 D2 Ger. Lötzen. Warmińsko-Mazurskie, NE Poland

Gjakovë – Haast

Gjakovë 79 D5 Serb. Đakovica. W Kosovo
Gjilan 79 D5 Serb. Gnjilane. E Kosovo
Gjinokastër see Gjirokastër
Gjirokastër 79 C7 var. Gjirokastra; prev. Gjinokastër, Gk. Argyrokastron, It. Argirocastro. Gjirokastër, S Albania
Gjirokastra see Gjirokastër
Gjoa Haven 15 F3 var. Uqsuqtuuq. King William Island, Nunavut, NW Canada
Gjøvik 63 B5 Oppland, S Norway
Glace Bay 17 G4 Cape Breton Island, Nova Scotia, SE Canada
Gladstone 126 E4 Queensland, E Australia
Gláma 63 B5 var. Glommen. river S Norway
Glasgow 66 C4 S Scotland, United Kingdom
Glavinitsa 82 D1 prev. Pravda, Dogrular. Silistra, NE Bulgaria
Glavn'a Morava see Velika Morava
Glazov 89 D5 Udmurtskaya Respublika, NW Russia
Gleiwitz see Gliwice
Glendale 26 B2 Arizona, SW USA
Glendive 22 D2 Montana, NW USA
Glens Falls 19 F3 New York, NE USA
Glevum see Gloucester
Glina 78 B3 var. Banijska Palanka. Sisak-Moslavina, NE Croatia
Glittertind 63 A5 mountain S Norway
Gliwice 77 C5 Ger. Gleiwitz. Śląskie, S Poland
Globe 26 B2 Arizona, SW USA
Globino see Hlobyne
Glogau see Głogów
Głogów 76 B4 Ger. Glogau, Glogow. Dolnośląskie, SW Poland
Glogow see Głogów
Glomma see Gláma
Glommen see Gláma
Gloucester 67 D6 hist. Caer Glou, Lat. Glevum. C England, United Kingdom
Głowno 76 D4 Łódź, C Poland
Glubokoye see Hlybokaye
Glukhov see Hlukhiv
Gnesen see Gniezno
Gniezno 76 C3 Ger. Gnesen. Weilkopolskie, C Poland
Gnjilane see Gjilan
Gobabis 56 B3 Omaheke, E Namibia
Gobi 104 D3 desert China/Mongolia
Gobō 109 C6 Wakayama, Honshū, SW Japan
Godāvari 102 B3 var. Godavari. river C India
Godavari see Godāvari
Godhavn see Qeqertarsuaq
Godhra 112 C4 Gujarāt, W India
Göding see Hodonín
Godoy Cruz 42 B4 Mendoza, W Argentina
Godthaab/Godthåb see Nuuk
Godwin Austen, Mount see K2
Goede Hoop, Kaap de see Good Hope, Cape of
Goeie Hoop, Kaap die see Good Hope, Cape of
Goeree 64 B4 island SW Netherlands
Goes 65 B5 Zeeland, SW Netherlands
Goettingen see Göttingen
Gogebic Range 18 B1 hill range Michigan/Wisconsin, N USA
Goiânia 41 E3 prev. Goyania. state capital Goiás, C Brazil
Goiás 41 E3 off. Estado de Goiás; prev. Goiaz, Goyaz. state/region C Brazil
Goiás, Estado de see Goiás
Goiaz see Goiás
Goidhoo Atoll see Horsburgh Atoll
Gojōme 108 D4 Akita, Honshū, NW Japan
Gökçeada 82 D4 var. Imroz Adasī, Gk. Imbros. island NW Turkey
Gökdepe 100 C3 Rus. Gëkdepe, Geok-Tepe. Ahal Welaýaty, C Turkmenistan
Göksun 94 D4 Kahramanmaraş, C Turkey
Gol 63 A5 Buskerud, S Norway
Golan Heights 97 B5 Ar. Al Jawlān, Heb. HaGolan. mountain range SW Syria
Golaya Pristan see Hola Prystan'
Gołdap 76 E2 Ger. Goldap. Warmińsko-Mazurskie, NE Poland
Gold Coast 127 E5 cultural region Queensland, E Australia
Golden Bay 128 C4 bay South Island, New Zealand
Golden State, The see California
Goldingen see Kuldīga
Goldsboro 21 F1 North Carolina, SE USA
Goleniów 76 B3 Ger. Gollnow. Zachodnio-pomorskie, NW Poland
Gollnow see Goleniów
Golmo see Golmud
Golmud 104 D4 var. Ge'e'mu, Golmo, Chin. Ko-erh-mu. Qinghai, C China
Golovanevsk see Holovanivs'k
Golub-Dobrzyń 76 C3 Kujawski-pomorskie, C Poland
Goma 55 E6 Nord-Kivu, NE Dem. Rep. Congo
Gombi 53 H4 Adamawa, E Nigeria
Gombroon see Bandar-e 'Abbās
Gomel' see Homyel'
Gomera 48 A3 island Islas Canarias, Spain, NE Atlantic Ocean
Gómez Palacio 28 D3 Durango, C Mexico
Gonaïves 32 D3 var. Les Gonaïves. N Haiti
Gonâve, Île de la 32 D3 island C Haiti
Gondar see Gonder
Gonder 50 C4 var. Gondar. Āmara, NW Ethiopia
Gondia 113 E4 Mahārāshtra, C India
Gonggar 104 C5 var. Gyixong. Xizang Zizhiqu, W China
Gongola 53 G4 river E Nigeria
Gongtang see Damxung
Gonni/Gónnos see Gónnoi
Gónnoi 82 B4 var. Gonni, Gónnos; prev. Dereli. Thessalía, C Greece
Good Hope, Cape of 56 B5 Afr. Kaap de Goede Hoop, Kaap die Goeie Hoop. headland SW South Africa
Goodland 22 D4 Kansas, C USA
Goondiwindi 127 D5 Queensland, E Australia
Goor 64 E3 Overijssel, E Netherlands
Goose Green 43 D7 var. Prado del Ganso. East Falkland, Falkland Islands
Goose Lake 24 B4 var. Lago dos Gansos. lake California/Oregon, W USA
Gopher State see Minnesota
Göppingen 73 B6 Baden-Württemberg, SW Germany
Góra Kalwaria 92 D4 Mazowieckie, C Poland
Gorakhpur 113 E3 Uttar Pradesh, N India
Gorany see Harany
Goražde 78 C4 Federacija Bosna I Hercegovina, SE Bosnia and Herzegovina
Gorbovichi see Harbavichy
Goré 54 C4 Logone-Oriental, S Chad
Gorē 51 C5 Oromīya, C Ethiopia
Gore 129 B7 Southland, South Island, New Zealand
Gorgān 98 D2 var. Astarabad, Astrabad, Gurgan, prev. Asterābād; anc. Hyrcania. Golestān, N Iran
Gori 95 F2 C Georgia
Gorinchem 64 C4 var. Gorkum. Zuid-Holland, C Netherlands
Goris 95 G3 SE Armenia
Gorki see Horki
Gor'kiy see Nizhniy Novgorod
Gorkum see Gorinchem
Görlitz 72 D4 Sachsen, E Germany
Görlitz see Zgorzelec
Gorlovka see Horlivka
Gorna Dzhumaya see Blagoevgrad
Gornja Mužlja see Mužlja
Gornji Milanovac 78 C4 Serbia, C Serbia
Gorodets see Haradzyets
Gorodishche see Horodyshche
Gorodnya see Horodnya
Gorodok see Haradok
Gorodok/Gorodok Yagellonski see Horodok
Gorontalo 117 E4 Sulawesi, C Indonesia
Gorontalo, Teluk see Tomini, Gulf of
Gorssel 64 D3 Gelderland, E Netherlands
Goryn see Horyn'
Gorzów Wielkopolski 76 B3 Ger. Landsberg, Landsberg an der Warthe. Lubuskie, W Poland
Gosford 127 D6 New South Wales, SE Australia
Goshogawara 108 D3 var. Gosyogawara. Aomori, Honshū, C Japan
Gospić 78 A3 Lika-Senj, C Croatia
Gostivar 79 D6 W Macedonia
Gosyogawara see Goshogawara
Göteborg 63 B7 Eng. Gothenburg. Västra Götaland, S Sweden
Gotel Mountains 53 G5 mountain range E Nigeria
Gotha 72 C4 Thüringen, C Germany
Gothenburg see Göteborg
Gotland 63 C7 island SE Sweden
Goto-retto 109 A7 island group SW Japan
Gotska Sandön 84 B1 island SE Sweden
Gōtsu 109 B6 var. Gôtu. Shimane, Honshū, SW Japan
Göttingen 72 B4 var. Goettingen. Niedersachsen, C Germany
Gottschee see Kočevje
Gottwaldov see Zlín
Gôtu see Gōtsu
Gouda 64 C4 Zuid-Holland, C Netherlands
Gough Fracture Zone 45 C6 tectonic feature S Atlantic Ocean
Gough Island 47 B8 island Tristan da Cunha, S Atlantic Ocean
Gouin, Réservoir 16 D4 reservoir Québec, SE Canada
Goulburn 127 D6 New South Wales, SE Australia
Goundam 53 E3 Tombouctou, NW Mali
Gouré 53 G3 Zinder, SE Niger
Goverla, Gora see Hoverla, Hora
Governador Valadares 41 F4 Minas Gerais, SE Brazil
Govi Altayn Nuruu 105 E3 mountain range S Mongolia
Goya 42 D3 Corrientes, NE Argentina
Goyania see Goiânia
Goyaz see Goiás
Goz Beïda 54 C3 Ouaddaï, SE Chad
Gozo 75 C8 var. Ghawdex. island N Malta
Graciosa 70 A5 var. Ilha Graciosa. island Azores, Portugal, NE Atlantic Ocean
Graciosa, Ilha see Graciosa
Gradačac 78 C3 Federacija Bosna I Hercegovina, N Bosnia and Herzegovina
Gradaús, Serra dos 41 E3 mountain range C Brazil
Gradiška see Bosanska Gradiška
Grafton 127 E5 New South Wales, SE Australia
Grafton 23 E1 North Dakota, N USA
Graham Land 132 A2 physical region Antarctica
Grajewo 76 E3 Podlaskie, NE Poland
Grampian Mountains 66 C3 mountain range C Scotland, United Kingdom
Gran see Esztergom, Hungary
Granada 30 D3 Granada, SW Nicaragua
Granada 70 D5 Andalucía, S Spain
Gran Canaria 48 A3 var. Grand Canary. island Islas Canarias, Spain, NE Atlantic Ocean
Gran Chaco 42 D2 var. Chaco. lowland plain South America
Grand Bahama Island 32 B1 island N The Bahamas
Grand Banks of Newfoundland 12 E4 undersea basin NW Atlantic Ocean
Grand Bassa see Buchanan
Grand Canary see Gran Canaria
Grand Canyon 26 A1 canyon Arizona, SW USA
Grand Canyon State see Arizona
Grand Cayman 32 B3 island SW Cayman Islands
Grand Duchy of Luxembourg see Luxembourg
Grande, Bahía 43 B7 bay S Argentina
Grande-Comore see Ngazidja
Grande de Chiloé, Isla see Chiloé, Isla de
Grande Prairie 15 E4 Alberta, W Canada
Grand Erg Occidental 48 D3 desert W Algeria
Grand Erg Oriental 49 E3 desert Algeria/Tunisia
Rio Grande 29 E2 var. Río Bravo, Sp. Río Bravo del Norte, Bravo del Norte. river Mexico/USA
Grande Terre 33 G3 island E West Indies
Grand Falls 17 G3 Newfoundland, Newfoundland and Labrador, SE Canada
Grand Forks 23 E1 North Dakota, N USA
Grandichi see Hrandzichy
Grand Island 23 E4 Nebraska, C USA
Grand Junction 22 C4 Colorado, C USA
Grand Paradis see Gran Paradiso
Grand Rapids 18 C3 Michigan, N USA
Grand Rapids 23 F1 Minnesota, N USA
Grand-Saint-Bernard, Col du see Great Saint Bernard Pass
Grand-Santi 37 G3 W French Guiana
Granite State see New Hampshire
Gran Lago see Nicaragua, Lago de
Gran Malvina see West Falkland
Gran Paradiso 74 A2 Fr. Grand Paradis. mountain NW Italy
Gran San Bernardo, Passo di see Great Saint Bernard Pass
Gran Santiago see Santiago
Grants 26 C2 New Mexico, SW USA
Grants Pass 24 B4 Oregon, NW USA
Granville 68 B3 Manche, N France
Gratianopolis see Grenoble
Gratz see Graz
Graudenz see Grudziądz
Graulhet 69 C6 Tarn, S France
Grave 64 D4 Noord-Brabant, SE Netherlands
Grayling 14 C2 Alaska, USA
Graz 73 E7 prev. Gratz. Steiermark, SE Austria
Great Abaco 32 C1 var. Abaco Island. island N The Bahamas
Great Alfold see Great Hungarian Plain
Great Ararat see Büyükağrı Dağı
Great Australian Bight 125 D7 bight S Australia
Great Barrier Island 128 D2 island N New Zealand
Great Barrier Reef 126 D2 reef Queensland, NE Australia
Great Basin 25 C5 basin W USA
Great Bear Lake 15 E3 Fr. Grand Lac de l'Ours. lake Northwest Territories, NW Canada
Great Belt 63 B8 var. Store Bælt, Eng. Great Belt, Storebelt. channel Baltic Sea/Kattegat
Great Belt see Storebælt
Great Bend 23 E5 Kansas, C USA
Great Britain see Britain
Great Dividing Range 126 D4 mountain range NE Australia
Greater Antilles 32 D3 island group West Indies
Greater Caucasus 95 G2 mountain range Azerbaijan/Georgia/Russia Asia/Europe
Greater Sunda Islands 102 D5 var. Sunda Islands. island group Indonesia
Great Exhibition Bay 128 C1 inlet North Island, New Zealand
Great Exuma Island 32 C2 island C The Bahamas
Great Falls 22 B1 Montana, NW USA
Great Grimsby see Grimsby
Great Hungarian Plain 77 C7 var. Great Alfold, Plain of Hungary, Hung. Alföld. plain SE Europe
Great Inagua 32 D2 var. Inagua Islands. island S The Bahamas
Great Indian Desert see Thar Desert
Great Khingan Range see Da Hinggan Ling
Great Lake see Tônlé Sap
Great Lakes 13 C5 lakes Ontario, Canada/USA
Great Lakes State see Michigan
Great Meteor Seamount see Great Meteor Tablemount
Great Meteor Tablemount 44 B3 var. Great Meteor Seamount. seamount E Atlantic Ocean
Great Nicobar 111 G3 island Nicobar Islands, India, NE Indian Ocean
Great Plain of China 103 E2 plain E China
Great Plains 23 E3 var. High Plains. plains Canada/USA
Great Rift Valley 51 C5 var. Rift Valley. depression Asia/Africa
Great Ruaha 51 C7 river S Tanzania
Great Saint Bernard Pass 74 A1 Fr. Col du Grand-Saint-Bernard, It. Passo del Gran San Bernardo. pass Italy/Switzerland
Great Salt Desert see Kavīr, Dasht-e
Great Salt Lake 22 A3 salt lake Utah, W USA
Great Salt Lake Desert 22 A4 plain Utah, W USA
Great Sand Sea 49 H3 desert Egypt/Libya
Great Sandy Desert 124 C4 desert Western Australia
Great Sandy Desert see Ar Rub 'al Khālī
Great Sandy Island see Fraser Island
Great Slave Lake 15 E4 Fr. Grand Lac des Esclaves. lake Northwest Territories, NW Canada
Great Socialist People's Libyan Arab Jamahiriya see Libya
Great Sound 20 A5 sound Bermuda, NW Atlantic Ocean
Great Victoria Desert 125 C5 desert South Australia/Western Australia
Great Wall of China 106 C4 ancient monument N China Asia
Great Yarmouth 67 E6 var. Yarmouth. E England, United Kingdom
Grebenka see Hrebinka
Gredos, Sierra de 70 D3 mountain range W Spain
Greece 83 A5 off. Hellenic Republic, Gk. Ellás; anc. Hellas. country SE Europe
Greeley 22 D4 Colorado, C USA
Green Bay 18 B2 Wisconsin, N USA
Green Bay 18 B2 lake bay Michigan/Wisconsin, N USA
Greeneville 21 E1 Tennessee, S USA
Greenland 60 D3 Dan. Grønland, Inuit Kalaallit Nunaat. Danish self-governing territory NE North America
Greenland Sea 61 F2 sea Arctic Ocean
Green Mountains 19 G2 mountain range Vermont, NE USA
Green Mountain State see Vermont
Greenock 66 C4 W Scotland, United Kingdom
Green River 22 B3 Wyoming, C USA
Green River 18 C5 river Kentucky, C USA
Green River 22 B4 river Utah, W USA
Greensboro 21 F1 North Carolina, SE USA
Greenville 20 B2 Mississippi, S USA
Greenville 21 F1 North Carolina, SE USA
Greenville 21 E1 South Carolina, SE USA
Greenville 27 G2 Texas, SW USA
Greenwich 67 B8 United Kingdom
Greenwood 20 B2 Mississippi, S USA
Greenwood 21 E2 South Carolina, SE USA
Gregory Range 126 C3 mountain range Queensland, E Australia
Greifenberg/Greifenberg in Pommern see Gryfice
Greifswald 72 D2 Mecklenburg-Vorpommern, NE Germany
Grenada 20 C2 Mississippi, S USA
Grenada 33 G5 country SE West Indies
Grenadines, The 33 H4 island group Grenada/St Vincent and the Grenadines
Grenoble 69 D5 anc. Cularo, Gratianopolis. Isère, E France
Gresham 24 B3 Oregon, NW USA
Grevená 82 B4 Dytikí Makedonía, N Greece
Grevenmacher 65 E8 Grevenmacher, E Luxembourg
Greymouth 129 B5 West Coast, South Island, New Zealand
Grey Range 127 C5 mountain range New South Wales/Queensland, E Australia
Greytown see San Juan del Norte
Griffin 20 D2 Georgia, SE USA
Grimari 54 C4 Ouaka, C Central African Republic
Grimsby 67 E5 prev. Great Grimsby. E England, United Kingdom
Grobin see Grobiņa
Grobiņa 84 B3 Ger. Grobin. Liepāja, W Latvia
Gródek Jagielloński see Horodok
Grodno see Hrodna
Grodzisk Wielkopolski 76 B3 Wielkopolskie, C Poland
Groesbeek 64 D4 Gelderland, SE Netherlands
Grójec 76 D4 Mazowieckie, C Poland
Groningen 64 E1 Groningen, NE Netherlands
Grønland see Greenland
Groote Eylandt 126 B2 island Northern Territory, N Australia
Grootfontein 56 B3 Otjozondjupa, N Namibia
Groot Karasberge 56 B4 mountain range S Namibia
Gros Islet 33 F1 N Saint Lucia
Grossa, Isola see Dugi Otok
Grossbetschkerek see Zrenjanin
Grosse Morava see Velika Morava
Grosser Sund see Suur Väin
Grosseto 74 B4 Toscana, C Italy
Grossglockner 73 C7 mountain W Austria
Grosskanizsa see Nagykanizsa
Gross-Karol see Carei
Grosskikinda see Kikinda
Grossmichel see Michalovce
Gross-Schlatten see Abrud
Grosswardein see Oradea
Groznyy 89 B8 Chechenskaya Respublika, SW Russia
Grudovo see Sredets
Grudziądz 76 C3 Ger. Graudenz. Kujawsko-pomorskie, C Poland
Grums 63 B6 Värmland, C Sweden
Grünberg/Grünberg in Schlesien see Zielona Góra
Grüneberg see Zielona Góra
Gruzinskaya SSR/Gruziya see Georgia
Gryazi 89 B6 Lipetskaya Oblast', W Russia
Gryfice 76 B2 Ger. Greifenberg, Greifenberg in Pommern. Zachodnio-pomorskie, NW Poland
Guabito 31 E4 Bocas del Toro, NW Panama
Guadalajara 28 D4 Jalisco, C Mexico
Guadalajara 71 E3 Ar. Wad Al-Hajarah; anc. Arriaca. Castilla-La Mancha, C Spain
Guadalcanal 122 C3 island C Solomon Islands
Guadalquivir 70 D4 river W Spain
Guadalupe 28 D3 Zacatecas, C Mexico
Guadalupe Peak 26 D3 mountain Texas, SW USA
Guadalupe River 27 G4 river SW USA
Guadarrama, Sierra de 71 E2 mountain range C Spain
Guadeloupe 33 H3 French overseas department E West Indies
Guadiana 70 C4 river Portugal/Spain
Guadix 71 E4 Andalucía, S Spain
Guaimaca 30 C2 Francisco Morazán, C Honduras
Guajira, Península de la 36 B1 peninsula N Colombia
Gualaco 30 D2 Olancho, C Honduras
Gualán 30 B2 Zacapa, C Guatemala
Gualdicciolo 74 D1 NW San Marino
Gualeguaychú 42 D4 Entre Ríos, E Argentina
Guam 122 B1 US unincorporated territory W Pacific Ocean
Guamúchil 28 C3 Sinaloa, C Mexico
Guanabacoa 32 B2 La Habana, W Cuba
Guanajuato 29 E4 Guanajuato, C Mexico
Guanare 36 C2 Portuguesa, N Venezuela
Guanare, Río 36 D2 river W Venezuela
Guangdong 106 C6 var. Guangdong Sheng, Kuang-tung, Kwangtung, Yue. province S China
Guangdong Sheng see Guangdong
Guangju see Gwangju
Guangxi see Guangxi Zhuangzu Zizhiqu
Guangxi Zhuangzu Zizhiqu 106 C6 var. Guangxi, Gui, Kuang-hsi, Kwangsi, Eng. Kwangsi Chuang Autonomous Region. autonomous region S China
Guangyuan 106 B5 var. Kuang-yuan, Kwangyuan. Sichuan, C China
Guangzhou 106 C6 var. Kuang-chou, Kwangchow, Eng. Canton. province capital Guangdong, S China
Guantánamo 32 D3 Guantánamo, SE Cuba
Guantánamo, Bahía de 32 D3 Eng. Guantanamo Bay. US military base SE Cuba
Guantanamo Bay see Guantánamo, Bahía de
Guaporé, Rio 40 D3 var. Río Iténez. river Bolivia/Brazil
Guarda 70 C3 Guarda, N Portugal
Guarumal 31 F5 Veraguas, S Panama
Guasave 28 C3 Sinaloa, C Mexico
Guatemala 30 A2 off. Republic of Guatemala. country Central America
Guatemala Basin 13 B7 undersea basin E Pacific Ocean
Guatemala City see Ciudad de Guatemala
Guatemala, Republic of see Guatemala
Guaviare 34 B2 off. Comisaría Guaviare. province S Colombia
Guaviare, Comisaría see Guaviare
Guaviare, Río 36 D3 river E Colombia
Guayanas, Macizo de las see Guiana Highlands
Guayaquil 38 A2 var. Santiago de Guayaquil. Guayas, SW Ecuador
Guayaquil, Golfo de 38 A2 var. Gulf of Guayaquil. gulf SW Ecuador
Guayaquil, Gulf of see Guayaquil, Golfo de
Guaymas 28 B2 Sonora, NW Mexico
Gubadag 100 C2 Turkm. Tel'man; prev. Tel'mansk. Daşoguz Welaýaty, N Turkmenistan
Guben 72 D4 var. Wilhelm-Pieck-Stadt. Brandenburg, E Germany
Gudara see Ghŭdara
Gudauta 95 E1 NW Georgia
Guéret 68 C4 Creuse, C France
Guernsey 67 D8 British Crown Dependency Channel Islands, NW Europe
Guerrero Negro 28 A2 Baja California Sur, NW Mexico
Gui see Guangxi Zhuangzu Zizhiqu
Guiana see French Guiana
Guiana Highlands 40 D1 var. Macizo de las Guayanas. mountain range N South America
Guiba see Juba
Guidder see Guider
Guider 54 B4 var. Guidder. Nord, N Cameroon
Guidimouni 53 G3 Zinder, S Niger
Guildford 67 D7 SE England, United Kingdom
Guilin 106 C6 var. Kuei-lin, Kweilin. Guangxi Zhuangzu Zizhiqu, S China
Guimarães 70 B2 var. Guimaráes. Braga, N Portugal
Guimaráes see Guimarães
Guinea 52 C4 off. Republic of Guinea, var. Guinée; prev. French Guinea, People's Revolutionary Republic of Guinea. country W Africa
Guinea Basin 47 A5 undersea basin E Atlantic Ocean
Guinea-Bissau 52 B4 off. Republic of Guinea-Bissau, Fr. Guinée-Bissau, Port. Guiné-Bissau; prev. Portuguese Guinea. country W Africa
Guinea-Bissau, Republic of see Guinea-Bissau
Guinea, Gulf of 46 B4 Fr. Golfe de Guinée. gulf E Atlantic Ocean
Guinea, People's Revolutionary Republic of see Guinea
Guinea, Republic of see Guinea
Guiné-Bissau see Guinea-Bissau
Guinée see Guinea
Guinée-Bissau see Guinea-Bissau
Guinée, Golfe de see Guinea, Gulf of
Güiria 37 E1 Sucre, NE Venezuela
Guiyang 106 B6 var. Kuei-Yang, Kuei-yang, Kueyang, Kweiyang; prev. Kweichu. province capital Guizhou, S China
Guizhou 106 B6 Guangdong, SE China
Gujarāt 112 C4 var. Gujerat. cultural region W India
Gujerat see Gujarāt
Gujrānwāla 112 D2 Punjab, NE Pakistan
Gujrāt 112 D2 Punjab, E Pakistan
Gulbarga see Kalaburagi
Gulbene 84 D3 Ger. Alt-Schwanenburg. Gulbene, NE Latvia
Gulf of Liaotung see Liaodong Wan
Gulfport 20 C3 Mississippi, S USA
Gulf, The see Persian Gulf
Gulistan see Guliston
Guliston 101 E2 Rus. Gulistan. Sirdaryo Viloyati, E Uzbekistan
Gulja see Yining
Gulkana 14 D3 Alaska, USA
Gulu 51 B6 N Uganda
Gulyantsi 82 C1 Pleven, N Bulgaria
Guma see Pishan
Gumbinnen see Gusev
Gumpolds see Humpolec
Gümülcine/Gümüljina see Komotiní
Gümüşane see Gümüşhane
Gümüşhane 95 E3 var. Gümüşane, Gumushkhane. Gümüşhane, NE Turkey
Gumushkhane see Gümüşhane
Güney Doğu Toroslar 95 E4 mountain range SE Turkey
Gunnbjørn Fjeld 60 D4 var. Gunnbjörns Bjerge. mountain C Greenland
Gunnbjörns Bjerge see Gunnbjørn Fjeld
Gunnedah 127 D6 New South Wales, SE Australia
Gunnison 22 C5 Colorado, C USA
Gurbansoltan Eje 100 C2 prev. Ýylanly, Rus. Il'yaly. Daşoguz Welaýaty, N Turkmenistan
Gurbantünggüt Shamo 104 B2 desert W China
Gurgan see Gorgān
Guri, Embalse de 37 E2 reservoir E Venezuela
Gurkfeld see Krško
Gurktaler Alpen 73 D7 mountain range S Austria
Gürün 94 D3 Sivas, C Turkey
Gur'yev/Gur'yevskaya Oblast' see Atyrau
Gusau 53 G4 Zamfara, NW Nigeria
Gusev 84 B4 Ger. Gumbinnen. Kaliningradskaya Oblast', W Russia
Gustavus 14 D4 Alaska, USA
Güstrow 72 C3 Mecklenburg-Vorpommern, NE Germany
Guta/Gúta see Kolárovo
Gütersloh 72 B4 Nordrhein-Westfalen, W Germany
Gutta see Kolárovo
Guttstadt see Dobre Miasto
Guwāhāti 113 G3 prev. Gauhāti. Assam, NE India
Guyana 37 F3 off. Co-operative Republic of Guyana; prev. British Guiana. country N South America
Guyana, Co-operative Republic of see Guyana
Guyane see French Guiana
Guymon 27 E1 Oklahoma, C USA
Güzelyurt Körfezi 80 C5 Gk. Kólpos Mórfu, Morphou. W Cyprus
Gvardeysk 84 A4 Ger. Tapaiu. Kaliningradskaya Oblast', W Russia
Gwādar 112 A3 var. Gwadur. Baluchistān, SW Pakistan
Gwadur see Gwādar
Gwalior 112 D3 Madhya Pradesh, C India
Gwanda 56 D3 Matabeleland South, SW Zimbabwe
Gwangju 107 E4 off. Gwangju Gwang-yeoksi, prev. Kwangju, var. Guangju, Kwangchu, Jap. Kōshū. SW South Korea
Gwangju Gwang-yeoksi see Gwangju
Gwy see Wye
Gyandzha see Gäncä
Gyangzê 104 C5 Xizang Zizhiqu, W China
Gyaring Co 104 C5 lake W China
Gyêgu see Yushu
Gyergyószentmiklós see Gheorgheni
Gyixong see Gonggar
Gympie 127 E5 Queensland, E Australia
Gyomaendrőd 77 D7 Békés, SE Hungary
Gyöngyös 77 D6 Heves, NE Hungary
Győr 77 C6 Ger. Raab, Lat. Arrabona. Győr-Moson-Sopron, NW Hungary
Gýtheio 83 B6 var. Githio; prev. Ýithion. Pelopónnisos, S Greece
Gyulafehérvár see Alba Iulia
Gyumri 95 F2 var. Giumri, Rus. Kumayri; prev. Aleksandropol', Leninakan. W Armenia
Gyzyrlabat see Serdar

H

Haabai see Ha'apai Group
Haacht 65 C6 Vlaams Brabant, C Belgium
Haaksbergen 64 E3 Overijssel, E Netherlands
Ha'apai Group 123 F4 var. Haabai. island group C Tonga
Haapsalu 84 D2 Ger. Hapsal. Läänemaa, W Estonia
Ha'Arava see 'Arabah, Wādī al
Haarlem 64 C3 prev. Harlem. Noord-Holland, W Netherlands
Haast 129 B6 West Coast, South Island, New Zealand

Hachijo-jima – Houlton

Hachijo-jima *109 D6 island* Izu-shotō, SE Japan
Hachinohe *108 D3* Aomori, Honshū, C Japan
Haḍabat al Jilf al Kabīr *50 A2 var.* Gilf Kebir Plateau. *plateau* SW Egypt
Hadama *see* Nazrēt
Hadejia *53 G4* Jigawa, N Nigeria
Hadejia *53 G3 river* N Nigeria
Hadera *97 A6 var.* Khadera; *prev.* Ḏadera. Haifa, C Israel
Ḏadera *see* Hadera
Hadhdhunmathi Atoll *110 A5 atoll* S Maldives
Ha Đông *114 D3 var.* Hadong. Ha Tây, N Vietnam
Hadong *see* Ha Đông
Hadramaut *see* Ḥaḍramawt
Ḥaḍramawt *99 C6 Eng.* Hadramaut. *mountain range* S Yemen
Hadrianopolis *see* Edirne
Haerbin/Haerhpin/Ha-erh-pin *see* Harbin
Hafnia *see* Denmark
Hafnia *see* København
Hafren *see* Severn
Hafun, Ras *see* Xaafuun, Raas
Hagåtña *122 B1* , *var.* Agaña. *dependent territory capital* (Guam) NW Guam
Hagerstown *19 E4* Maryland, NE USA
Ha Giang *114 D3* Ha Giang, N Vietnam
Hagios Evstrátios *see* Ágios Efstrátios
HaGolan *see* Golan Heights
Hagondange *68 D3* Moselle, NE France
Haguenau *68 E3* Bas-Rhin, NE France
Haibowan *see* Wuhai
Haicheng *106 D3* Liaoning, NE China
Haidarabad *see* Hyderābād
Haifa *see* Hefa
Haifa, Bay of *see* Mifrats Hefa
Haifong *see* Hai Phong
Haikou *106 C7 var.* Hai-k'ou, Hoihow, *Fr.* Hoï-Hao. *province capital* Hainan, S China
Hai-k'ou *see* Haikou
Ḥā'il *98 B4* Ḥā'il, NW Saudi Arabia
Hailuoto *62 D4 Swe.* Karlö. *island* W Finland
Hainan *106 B7 var.* Hainan Sheng, Qiong. *province* S China
Hainan Dao *106 C7 island* S China
Hainan Sheng *see* Hainan
Hainasch *see* Ainaži
Haines *14 D4* Alaska, USA
Hainichen *72 D4* Sachsen, E Germany
Hai Phong *114 D3 var.* Haifong, Haiphong. N Vietnam
Haiphong *see* Hai Phong
Haiti *32 D3 off.* Republic of Haiti. *country* C West Indies
Haiti, Republic of *see* Haiti
Haiya *50 C3* Red Sea, NE Sudan
Hajdúhadház *77 D6* Hajdú-Bihar, E Hungary
Hajîne *see* Abū Ḥardān
Hajnówka *76 E3 Ger.* Hermhausen. Podlaskie, NE Poland
Hakodate *108 D3* Hokkaidō, NE Japan
Hal *see* Halle
Ḥalab *96 B2 Eng.* Aleppo, *Fr.* Alep; *anc.* Beroea. Ḥalab, NW Syria
Hala'ib Triangle *50 C3 region* Egypt/Sudan
Ḥalāniyāt, Juzur al *99 D6 var.* Jazā'ir Bin Ghalfān, *Eng.* Kuria Muria Islands. *island group* S Oman
Halberstadt *72 C4* Sachsen-Anhalt, C Germany
Halden *63 B6 prev.* Fredrikshald. Østfold, S Norway
Halfmoon Bay *129 A8 var.* Oban. Stewart Island, Southland, New Zealand
Haliacmon *see* Aliákmonas
Halifax *17 F4 province capital* Nova Scotia, SE Canada
Halkida *see* Chalkída
Halle *65 B6 Fr.* Hal. Vlaams Brabant, C Belgium
Halle *72 C4 var.* Halle an der Saale. Sachsen-Anhalt, C Germany
Halle an der Saale *see* Halle
Halle-Neustadt *72 C4* Sachsen-Anhalt, C Germany
Halley *132 B2 UK research station* Antarctica
Hall Islands *120 B2 island group* C Micronesia
Halls Creek *124 C3* Western Australia
Halmahera, Laut *117 F3 Eng.* Halmahera Sea; *sea* E Indonesia
Halmahera, Pulau *117 F3 prev.* Djailolo, Gilolo, Jailolo. *island* E Indonesia
Halmahera Sea *see* Halmahara, Laut
Halmstad *63 B7* Halland, S Sweden
Ha Long *114 E3 prev.* Hông Gai; *var.* Hon Gai, Hongay. Quang Ninh, N Vietnam
Hälsingborg *see* Helsingborg
Hamada *109 B6* Shimane, Honshū, SW Japan
Hamadān *98 C3 anc.* Ecbatana. Hamadān, W Iran
Ḥamāh *96 B3 var.* Hama; *anc.* Epiphania, *Bibl.* Hamath. Ḥamāh, W Syria
Hamamatsu *109 D6 var.* Hamamatu. Shizuoka, Honshū, S Japan
Hamamatu *see* Hamamatsu
Hamar *63 B5 prev.* Storhammer. Hedmark, S Norway
Hamath *see* Ḥamāh
Hamburg *72 B3* Hamburg, N Germany
Hamd, Wadi al *98 A4 dry watercourse* W Saudi Arabia
Hämeenlinna *63 D5 Swe.* Tavastehus. Kanta-Häme, S Finland
HaMela h, Yam *see* Dead Sea
Hamersley Range *124 A4 mountain range* Western Australia
Hamhŭng *107 E3* C North Korea
Hami *104 C3 var.* Ha-mi, *Uigh.* Kumul, Qomul. Xinjiang Uygur Zizhiqu, NW China
Ha-mi *see* Hami
Hamilton *20 A5 dependent territory capital* (Bermuda) C Bermuda
Hamilton *16 D5* Ontario, S Canada
Hamilton *128 D3* Waikato, North Island, New Zealand
Hamilton *66 C4* S Scotland, United Kingdom
Hamilton *20 C2* Alabama, S USA
Hamim, Wadi al *49 G2 river* NE Libya
Hamïs Musait *see* Khamīs Mushayt
Hamm *72 B4 var.* Hamm in Westfalen. Nordrhein-Westfalen, W Germany
Ḥammāmāt, Khalīj al *see* Hammamet, Golfe de
Hammamet, Golfe de *80 D3 Ar.* Khalīj al Ḥammāmāt. *gulf* NE Tunisia
Ḥammār, Hawr al *98 C3 lake* SE Iraq
Hamm in Westfalen *see* Hamm
Hampden *129 B7* Otago, South Island, New Zealand
Hampstead *67 A7* Maryland, USA
Ħamrun *80 B5* C Malta
Hāmūn, Daryācheh-ye *see* Şāberī, Hāmūn-e/Sīstān, Daryācheh-ye
Hamwih *see* Southampton
Hâncești *see* Hîncești
Hancewicze *see* Hantsavichy
Handan *106 C4 var.* Han-tan. Hebei, E China
Haneda *108 A2* (Tōkyō) Tōkyō, Honshū, S Japan
HaNegev *97 A7 Eng.* Negev. *desert* S Israel
Hanford *25 C6* California, W USA
Hangayn Nuruu *104 D2 mountain range* C Mongolia
Hang-chou/Hangchow *see* Hangzhou
Hangö *see* Hanko
Hangzhou *106 D5 var.* Hang-chou, Hangchow. *province capital* Zhejiang, SE China
Hania *see* Chaniá
Hanka, Lake *see* Khanka, Lake
Hanko *63 D6 Swe.* Hangö. Uusimaa, SW Finland
Han-kou/Han-k'ou/Hankow *see* Wuhan
Hanmer Springs *129 C5* Canterbury, South Island, New Zealand
Hannibal *23 G4* Missouri, C USA
Hannover *72 B3 Eng.* Hanover. Niedersachsen, NW Germany
Hanöbukten *63 B7 bay* S Sweden
Ha Nôi *114 D3 Eng.* Hanoi, *Fr.* Hanoï. *country capital* (Vietnam) N Vietnam
Hanover *see* Hannover
Han Shui *105 E4 river* C China
Han-tan *see* Handan
Hantsavichy *85 B6 Pol.* Hancewicze, *Rus.* Gantsevichi. Brestskaya Voblasts', SW Belarus
Hanyang *see* Wuhan
Hanzhong *106 B5* Shaanxi, C China
Hāora *113 F4 prev.* Howrah. West Bengal, NE India
Haparanda *62 D4* Norrbotten, N Sweden
Hapsal *see* Haapsalu
Haradok *85 E5 Rus.* Gorodok. Vitsyebskaya Voblasts', N Belarus
Haradzyets *85 B6 Rus.* Gorodets. Brestskaya Voblasts', SW Belarus
Haramachi *108 D4* Fukushima, Honshū, E Japan
Harany *85 D5 Rus.* Gorany. Vitsyebskaya Voblasts', N Belarus
Harare *56 D3 prev.* Salisbury. *country capital* (Zimbabwe) Mashonaland East, NE Zimbabwe
Harbavichy *85 E6 Rus.* Gorbovichi. Mahilyowskaya Voblasts', E Belarus
Harbel *52 C5* W Liberia
Harbin *107 E2 var.* Haerbin, Ha-erh-pin, Kharbin; *prev.* Haerhpin, Pingkiang, Pinkiang. *province capital* Heilongjiang, NE China
Hardangerfjorden *63 A6 fjord* S Norway
Hardangervidda *63 A6 plateau* S Norway
Hardenberg *64 E3* Overijssel, E Netherlands
Harelbeke *65 A6 var.* Harlebeke. West-Vlaanderen, W Belgium
Harem *see* Ḥārim
Haren *64 E2* Groningen, NE Netherlands
Hārer *51 D5* E Ethiopia
Hargeisa *see* Hargeysa
Hargeysa *51 D5 var.* Hargeisa. Woqooyi Galbeed, NW Somalia
Hariana *see* Haryāna
Hari, Batang *116 B4 prev.* Djambi. *river* Sumatera, W Indonesia
Ḥārim *96 B2 var.* Harem. Idlib, W Syria
Harima-nada *109 B6 sea* S Japan
Harirud *101 E4 var.* Tedzhen, *Turkm.* Tejen. *river* Afghanistan/Iran
Harlan *23 F3* Iowa, C USA
Harlebeke *see* Harelbeke
Harlem *see* Haarlem
Harlingen *64 D2 Fris.* Harns. Friesland, N Netherlands
Harlingen *27 G5* Texas, SW USA
Harlow *67 E6* E England, United Kingdom
Harney Basin *24 B4 basin* Oregon, NW USA
Härnösand *63 C5 var.* Hernösand. Västernorrland, C Sweden
Harns *see* Harlingen
Harper *52 D5 var.* Cape Palmas. NE Liberia
Harricana *16 D3 river* Québec, SE Canada
Harris *66 B3 physical region* NW Scotland, United Kingdom
Harrisburg *19 E4 state capital* Pennsylvania, NE USA
Harrisonburg *19 E4* Virginia, NE USA
Harrison, Cape *17 F2 headland* Newfoundland and Labrador, E Canada
Harris Ridge *see* Lomonosov Ridge
Harrogate *67 D5* N England, United Kingdom
Hârșova *86 D5 prev.* Hîrșova. Constanța, SE Romania
Harstad *62 C2* Troms, N Norway
Hartford *19 G3 state capital* Connecticut, NE USA
Hartlepool *67 D5* N England, United Kingdom
Harunabad *see* Eslāmābād
Har Us Gol *104 C2 lake* Hovd, W Mongolia
Har Us Nuur *104 C2 lake* NW Mongolia
Harwich *67 E6* E England, United Kingdom
Haryāna *112 D2 var.* Hariana. *cultural region* N India
Hashemite Kingdom of Jordan *see* Jordan
Hasselt *65 C6* Limburg, NE Belgium
Hassetché *see* Al Ḥasakah
Hasta Colonia/Hasta Pompeia *see* Asti
Hastings *128 E4* Hawke's Bay, North Island, New Zealand
Hastings *67 E7* SE England, United Kingdom
Hastings *23 E4* Nebraska, C USA
Hațeg *86 B4 Ger.* Wallenthal, *Hung.* Hátszeg; *prev.* Hatzeg, Hötzing. Hunedoara, SW Romania
Hátszeg *see* Hațeg
Hattem *64 D3* Gelderland, E Netherlands
Hatteras, Cape *21 G1 headland* North Carolina, SE USA
Hatteras Plain *13 D6 abyssal plain* Atlantic Ocean
Hattiesburg *20 C3* Mississippi, S USA
Hatton Bank *see* Hatton Ridge
Hatton Ridge *58 B2 var.* Hatton Bank. *undersea ridge* N Atlantic Ocean
Hat Yai *115 C7 var.* Ban Hat Yai. Songkhla, SW Thailand
Hatzeg *see* Hațeg
Hatzfeld *see* Jimbolia
Haugesund *63 A6* Rogaland, S Norway
Haukeligrend *63 A6* Telemark, S Norway
Haukivesi *63 E5 lake* SE Finland
Hauraki Gulf *128 D2 gulf* North Island, N New Zealand
Hauroko, Lake *129 A7 lake* South Island, New Zealand
Haut Atlas *48 C2 Eng.* High Atlas. *mountain range* C Morocco
Hautes Fagnes *65 D6 Ger.* Hohes Venn. *mountain range* E Belgium
Hauts Plateaux *48 D2 plateau* Algeria/Morocco
Hauzenberg *73 D6* Bayern, SE Germany
Havana *13 D6* Illinois, N USA
Havana *see* La Habana
Havant *67 D7* S England, United Kingdom
Havelock *21 F1* North Carolina, SE USA
Havelock North *128 E4* Hawke's Bay, North Island, New Zealand
Haverfordwest *67 C6* SW Wales, United Kingdom
Havířov *77 C5* Moravskoslezský Kraj, E Czech Republic (Czechia)
Havre *22 C1* Montana, NW USA
Havre *see* le Havre
Havre-St-Pierre *17 F3* Québec, E Canada
Hawaii *25 A8 off.* State of Hawaii, *also known as* Aloha State, Paradise of the Pacific, *var.* Hawai'i. *state* USA, C Pacific Ocean
Hawai'i *25 B8 var.* Hawaii. *island* Hawai'ian Islands, USA, C Pacific Ocean
Hawai'ian Islands *130 D2 prev.* Sandwich Islands. *island group* Hawaii, USA
Hawaiian Ridge *130 H4 undersea ridge* N Pacific Ocean
Hawea, Lake *129 B6 lake* South Island, New Zealand
Hawera *128 D4* Taranaki, North Island, New Zealand
Hawick *66 C4* SE Scotland, United Kingdom
Hawke Bay *128 E4 bay* North Island, New Zealand
Hawkeye State *see* Iowa
Hawlēr *see* Arbīl
Hawthorne *25 C6* Nevada, W USA
Hay *127 C6* New South Wales, SE Australia
HaYarden *see* Jordan
Hayastani Hanrapetut'yun *see* Armenia
Hayes *16 B2 river* Manitoba, C Canada
Hay River *15 E4* Northwest Territories, W Canada
Hays *23 E5* Kansas, C USA
Haysyn *86 D3 Rus.* Gaysin. Vinnyts'ka Oblast', C Ukraine
Hazar *100 B2 prev. Rus.* Cheleken. Balkan Welaýaty, W Turkmenistan
Heard and McDonald Islands *119 B7 Australian external territory* S Indian Ocean
Hearst *16 C4* Ontario, S Canada
Heart of Dixie *see* Alabama
Heathrow *67 A8* (London) SE England, United Kingdom
Hebei *106 C4 var.* Hebei Sheng, Hopeh, Hopei, Ji; *prev.* Chihli. *province* E China
Hebei Sheng *see* Hebei
Hebron *97 A6 var.* Al Khalīl, El Khalil, *Heb.* Hevron; *anc.* Kiriath-Arba. S West Bank
Heemskerk *64 C3* Noord-Holland, W Netherlands
Heerde *64 D3* Gelderland, E Netherlands
Heerenveen *64 D2 Fris.* It Hearrenfean. Friesland, N Netherlands
Heerhugowaard *64 C2* Noord-Holland, NW Netherlands
Heerlen *65 D6* Limburg, SE Netherlands
Heerwegen *see* Polkowice
Hefa *97 A5 var.* Haifa, *hist.* Caiffa, Caiphas; *anc.* Sycaminum. Haifa, N Israel
Ḥefa, Mifraẕ *see* Mifrats Hefa
Hefei *106 D5 var.* Hofei, *hist.* Luchow. *province capital* Anhui, E China
Hegang *107 E2* Heilongjiang, NE China
Hei *see* Heilongjiang
Heide *72 B2* Schleswig-Holstein, N Germany
Heidelberg *73 B5* Baden-Württemberg, SW Germany
Heidenheim *see* Heidenheim an der Brenz
Heidenheim an der Brenz *73 B6 var.* Heidenheim. Baden-Württemberg, S Germany
Hei-ho *see* Nagqu
Heilbronn *73 B6* Baden-Württemberg, SW Germany
Heiligenbeil *see* Mamonovo
Heilongjiang *106 D2 var.* Hei, Heilongjiang Sheng, Hei-lung-chiang, Heilungkiang. *province* NE China
Heilong Jiang *see* Amur
Heilongjiang Sheng *see* Heilongjiang
Heiloo *64 C3* Noord-Holland, NW Netherlands
Heilsberg *see* Lidzbark Warmiński
Hei-lung-chiang/Heilungkiang *see* Heilongjiang
Heimdal *63 B5* Sør-Trøndelag, S Norway
Heinaste *see* Ainaži
Hekimhan *94 D3* Malatya, C Turkey
Helena *22 B2 state capital* Montana, NW USA
Helensville *128 D2* Auckland, North Island, New Zealand
Helgoland Bay *see* Helgoländer Bucht
Helgoländer Bucht *72 A2 var.* Helgoland Bay, Heligoland Bight. *bay* NW Germany
Heligoland Bight *see* Helgoländer Bucht
Heliopolis *see* Baalbek
Hellas *see* Greece
Hellenic Republic *see* Greece
Hellevoetsluis *64 B4* Zuid-Holland, SW Netherlands
Hellín *71 E4* Castilla-La Mancha, C Spain
Darya-ye Helmand *100 D5 var.* Rūd-e Hīrmand. *river* Afghanistan/Iran
Helmantica *see* Salamanca
Helmond *65 D5* Noord-Brabant, S Netherlands
Helsingborg *63 B7 prev.* Hälsingborg. Skåne, S Sweden
Helsingfors *see* Helsinki
Helsinki *63 D6 Swe.* Helsingfors. *country capital* (Finland) Etelä-Suomi, S Finland
Heltau *see* Cisnădie
Helvetia *see* Switzerland
Henan *106 C5 var.* Henan Sheng, Honan, Yu. *province* C China
Henderson *18 B5* Kentucky, S USA
Henderson *25 D7* Nevada, W USA
Henderson *27 H3* Texas, SW USA
Hendū Kosh *see* Hindu Kush
Hengchow *see* Hengyang
Hengduan Shan *106 A5 mountain range* SW China
Hengelo *64 E3* Overijssel, E Netherlands
Hengnan *see* Hengyang
Hengyang *106 C6 var.* Hengnan, Heng-yang; *prev.* Hengchow. Hunan, S China
Heng-yang *see* Hengyang
Heniches'k *87 F4 Rus.* Genichesk. Khersons'ka Oblast', S Ukraine
Hennebont *68 A3* Morbihan, NW France
Henrique de Carvalho *see* Saurimo
Henzada *see* Hinthada
Herakleion *see* Irákleio
Herāt *100 D4 var.* Herat; *anc.* Aria. Herāt, W Afghanistan
Heredia *31 E4* Heredia, C Costa Rica
Hereford *27 E2* Texas, SW USA
Herford *72 B4* Nordrhein-Westfalen, NW Germany
Héristal *see* Herstal
Herk-de-Stad *65 C6* Limburg, NE Belgium
Herlen Gol/Herlen He *see* Kerulen
Hermannstadt *see* Sibiu
Hermansverk *63 A5* Sogn Og Fjordane, S Norway
Hermhausen *see* Hajnówka
Hermiston *24 C2* Oregon, NW USA
Hermon, Mount *97 B5 Ar.* Jabal ash Shaykh. *mountain* S Syria
Hermosillo *28 B2* Sonora, NW Mexico
Hermoupolis *see* Ermoúpoli
Hernösand *see* Härnösand
Herrera del Duque *70 D3* Extremadura, W Spain
Herselt *65 C5* Antwerpen, C Belgium
Herstal *65 D6 Fr.* Héristal. Liège, E Belgium
Herzogenbusch *see* 's-Hertogenbosch
Hesse *see* Hessen
Hessen *73 B5 Eng./Fr.* Hesse. *state* C Germany
Hevron *see* Hebron
Heydebrech *see* Kędzierzyn-Kozle
Heydekrug *see* Šilutė
Heywood Islands *124 C3 island group* Western Australia
Hibbing *23 F1* Minnesota, N USA
Hibernia *see* Ireland
Hidalgo del Parral *28 C2 var.* Parral. Chihuahua, N Mexico
Hida-sanmyaku *109 C5 mountain range* Honshū, S Japan
Hierosolyma *see* Jerusalem
Hierro *48 A3 var.* Ferro. *island* Islas Canarias, Spain, NE Atlantic Ocean
High Atlas *see* Haut Atlas
High Plains *see* Great Plains
High Point *21 E1* North Carolina, SE USA
Hiiumaa *84 C2 Ger.* Dagden, *Swe.* Dagö. *island* W Estonia
Hikurangi *128 D2* Northland, North Island, New Zealand
Hildesheim *72 B4* Niedersachsen, N Germany
Hilla *see* Al Ḥillah
Hillaby, Mount *33 G1 mountain* N Barbados
Hill Bank *30 C1* Orange Walk, N Belize
Hillegom *64 C3* Zuid-Holland, W Netherlands
Hilo *25 B8* Hawaii, USA, C Pacific Ocean
Hilton Head Island *21 E2* South Carolina, SE USA
Hilversum *64 C3* Noord-Holland, C Netherlands
Himalaya/Himalaya Shan *see* Himalayas
Himalayas *113 E2 var.* Himalaya, *Chin.* Himalaya Shan. *mountain range* S Asia
Himeji *109 C6 var.* Himezi. Hyōgo, Honshū, SW Japan
Himezi *see* Himeji
Ḥimș *96 B4 var.* Homs; *anc.* Emesa. Ḥimș, C Syria
Hîncești *86 D4 var.* Hâncești; *prev.* Kotovsk. C Moldova
Hinchinbrook Island *126 D3 island* Queensland, NE Australia
Hinds *129 C6* Canterbury, South Island, New Zealand
Hindu Kush *101 F4 Per.* Hendū Kosh. *mountain range* Afghanistan/Pakistan
Hinesville *21 E3* Georgia, SE USA
Hinnøya *62 C3 Lapp.* Iinnasuolu. *island* C Norway
Hinson Bay *20 A5 bay* W Bermuda, W Atlantic Ocean
Hinthada *114 B4 var.* Henzada. Ayeyarwady, SW Myanmar (Burma)
Hios *see* Chíos
Hirfanlı Barajı *94 C3 reservoir* C Turkey
Hīrmand, Rūd-e *see* Helmand, Daryā-ye
Hirosaki *108 D3* Aomori, Honshū, C Japan
Hiroshima *109 B6 var.* Hirosima. Hiroshima, Honshū, SW Japan
Hirschberg/Hirschberg im Riesengebirge/Hirschberg in Schlesien *see* Jelenia Góra
Hirson *68 D3* Aisne, N France
Hîrșova *see* Hârșova
Hispalis *see* Sevilla
Hispana/Hispania *see* Spain
Hispaniola *34 B1 island* Dominion Republic/Haiti
Hitachi *109 D5 var.* Hitati. Ibaraki, Honshū, S Japan
Hitati *see* Hitachi
Hitra *62 A4 prev.* Hitteren. *island* S Norway
Hitteren *see* Hitra
Hjälmaren *63 C6 Eng.* Lake Hjalmar. *lake* C Sweden
Hjalmar, Lake *see* Hjälmaren
Hjørring *63 B7* Nordjylland, N Denmark
Hkakabo Razi *114 B1 mountain* Myanmar (Burma)/China
Hlobyne *87 F2 Rus.* Globino. Poltavs'ka Oblast', NE Ukraine
Hlukhiv *87 F1 Rus.* Glukhov. Sums'ka Oblast', NE Ukraine
Hlybokaye *85 D5 Rus.* Glubokoye. Vitsyebskaya Voblasts', N Belarus
Hoa Binh *114 D3* Hoa Binh, N Vietnam
Hoang Lien Son *114 D3 mountain range* N Vietnam
Hobart *127 C8 prev.* Hobarton, Hobart Town. *state capital* Tasmania, SE Australia
Hobarton/Hobart Town *see* Hobart
Hobbs *27 E3* New Mexico, SW USA
Hobro *63 A7* Nordjylland, N Denmark
Hô Chi Minh *115 E6 var.* Ho Chi Minh City; *prev.* Saigon. S Vietnam
Ho Chi Minh City *see* Hô Chi Minh
Hodeida *see* Al Ḥudaydah
Hódmezővásárhely *77 D7* Csongrád, SE Hungary
Hodna, Chott El *80 C4 var.* Chott el-Hodna, *Ar.* Shatt al-Hodna. *salt lake* N Algeria
Hodna, Chott el-/Hodna, Shatt al- *see* Hodna, Chott El
Hodonín *77 C5 Ger.* Göding. Jihomoravský Kraj, SE Czech Republic (Czechia)
Hoei *see* Huy
Hoey *see* Huy
Hof *73 C5* Bayern, SE Germany
Hofei *see* Hefei
Hōfu *109 B7* Yamaguchi, Honshū, SW Japan
Hofuf *see* Al Hufūf
Hogoley Islands *see* Chuuk Islands
Hohensalza *see* Inowrocław
Hohenstadt *see* Zábřeh
Hohes Venn *see* Hautes Fagnes
Hohe Tauern *73 C7 mountain range* W Austria
Hohhot *105 F3 var.* Huhehot, Huhohaote, *Mong.* Kukukhoto; *prev.* Kweisui, Kwesui. Nei Mongol Zizhiqu, N China
Hôi An *115 E5 prev.* Faifo. Quang Nam-Đa Năng, C Vietnam
Hoï-Hao/Hoihow *see* Haikou
Hokianga Harbour *128 C2 inlet* SE Tasman Sea
Hokitika *129 B5* West Coast, South Island, New Zealand
Hokkaido *108 C2 prev.* Ezo, Yeso, Yezo. *island* NE Japan
Hola Prystan' *87 E4 Rus.* Golaya Pristan. Khersons'ka Oblast', S Ukraine
Holbrook *26 B2* Arizona, SW USA
Holetown *33 G1 prev.* Jamestown. W Barbados
Holguín *32 C2* Holguín, SE Cuba
Hollabrunn *73 E6* Niederösterreich, NE Austria
Holland *see* Netherlands
Hollandia *see* Jayapura
Holly Springs *20 C1* Mississippi, S USA
Holman *15 E3* Victoria Island, Northwest Territories, N Canada
Holmsund *62 D4* Västerbotten, N Sweden
Holon *97 A6 var.* Kholon; *prev.* Ḏolon. Tel Aviv, C Israel
Ḏolon *see* Holon
Holovanivs'k *87 E3 Rus.* Golovanevsk. Kirovohrads'ka Oblast', C Ukraine
Holstebro *63 A7* Ringkøbing, W Denmark
Holsteinborg/Holsteinsborg/Holstenborg/Holstensborg *see* Sisimiut
Holyhead *67 C5 Wel.* Caer Gybi. NW Wales, United Kingdom
Hombori *53 E3* Mopti, S Mali
Homs *see* Al Khums, Libya
Homs *see* Ḥimș
Homyel' *85 D7 Rus.* Gomel'. Homyel'skaya Voblasts', SE Belarus
Honan *see* Luoyang, China
Honan *see* Henan, China
Hondo *27 F4* Texas, SW USA
Hondo *see* Honshū
Honduras *30 C2 off.* Republic of Honduras. *country* Central America
Honduras, Golfo de *see* Honduras, Gulf of
Honduras, Gulf of *30 C2 Sp.* Golfo de Honduras. *gulf* W Caribbean Sea
Honduras, Republic of *see* Honduras
Hønefoss *63 B6* Buskerud, S Norway
Honey Lake *25 B5 lake* California, W USA
Hon Gai *see* Ha Long
Hongay *see* Ha Long
Hông Gai *see* Ha Long
Hông Hà, Sông *see* Red River
Hong Kong *106 A1* Hong Kong, S China
Hong Kong Island *106 B2 island* S China Asia
Honiara *122 C3 country capital* (Solomon Islands) Guadalcanal, C Solomon Islands
Honjō *108 D4 var.* Honzyô, Yurihonjō. Akita, Honshū, C Japan
Honolulu *25 A8 state capital* O'ahu, Hawaii, USA, C Pacific Ocean
Honshu *109 E5 var.* Hondo, Honsyû. *island* SW Japan
Honsyû *see* Honshū
Honte *see* Westerschelde
Honzyô *see* Honjō
Hoogeveen *64 E2* Drenthe, NE Netherlands
Hoogezand-Sappemeer *64 E2* Groningen, NE Netherlands
Hoorn *64 C2* Noord-Holland, NW Netherlands
Hoosier State *see* Indiana
Hopa *95 E2* Artvin, NE Turkey
Hope *14 C3* British Columbia, SW Canada
Hopedale *17 F2* Newfoundland and Labrador, NE Canada
Hopeh/Hopei *see* Hebei
Hopkinsville *18 B5* Kentucky, S USA
Horasan *95 F3* Erzurum, NE Turkey
Horizon Deep *130 D4 trench* W Pacific Ocean
Horki *85 E6 Rus.* Gorki. Mahilyowskaya Voblasts', E Belarus
Horlivka *87 G3 Rom.* Adâncata, *Rus.* Gorlovka. Donets'ka Oblast', E Ukraine
Hormoz, Tangeh-ye *see* Hormuz, Strait of
Hormuz, Strait of *98 D4 var.* Strait of Ormuz, *Per.* Tangeh-ye Hormoz. *strait* Iran/Oman
Cape Horn *43 C8 Eng.* Cape Horn. *headland* S Chile
Horn, Cape *see* Hornos, Cabo de
Hornsby *126 E1* New South Wales, SE Australia
Horodnya *87 E1 Rus.* Gorodnya. Chernihivs'ka Oblast', NE Ukraine
Horodok *86 B2 Pol.* Gródek Jagielloński, *Rus.* Gorodok, Gorodok Yagellonski. L'vivs'ka Oblast', NW Ukraine
Horodyshche *87 E2 Rus.* Gorodishche. Cherkas'ka Oblast', C Ukraine
Horoshiri-dake *108 D2 var.* Horosiri Dake. *mountain* Hokkaidō, N Japan
Horosiri Dake *see* Horoshiri-dake
Horsburgh Atoll *110 A4 var.* Goidhoo Atoll. *atoll* N Maldives
Horseshoe Bay *20 A5 bay* W Bermuda W Atlantic Ocean
Horseshoe Seamounts *58 A4 seamount range* E Atlantic Ocean
Horsham *127 B7* Victoria, SE Australia
Horst *65 D5* Limburg, SE Netherlands
Horten *63 B6* Vestfold, S Norway
Horyn' *85 B7 Rus.* Goryn. *river* NW Ukraine
Hosingen *65 D7* Diekirch, NE Luxembourg
Hospitalet *see* L'Hospitalet de Llobregat
Hotan *104 B4 var.* Khotan, *Chin.* Ho-t'ien. Xinjiang Uygur Zizhiqu, NW China
Ho-t'ien *see* Hotan
Hoting *62 C4* Jämtland, C Sweden
Hot Springs *20 B1* Arkansas, C USA
Hötzing *see* Hațeg
Houayxay *114 C3 var.* Ban Houayxay. Bokèo, N Laos
Houghton *18 B1* Michigan, N USA
Houilles *69 B5* Yvelines, Île-de-France, N France Europe
Houlton *19 H1* Maine, NE USA

Jaisalmer – Karakax

Jaisalmer 112 C3 Rājasthān, NW India
Jajce 78 B3 Federacija Bosna I Hercegovina, W Bosnia and Herzegovina
Jakarta 116 C5 prev. Djakarta, Dut. Batavia. country capital (Indonesia) Jawa, C Indonesia
Jakobstad 62 D4 Fin. Pietarsaari. Österbotten, W Finland
Jakobstadt see Jēkabpils
Jalālābād 101 F4 var. Jalalabad, Jelalabad. Nangarhār, E Afghanistan
Jalal-Abad see Dzhalal-Abad, Dzhalal-Abadskaya Oblast', Kyrgyzstan
Jalandhar 112 D2 prev. Jullundur. Punjab, N India
Jalapa 30 D3 Nueva Segovia, NW Nicaragua
Jalpa 28 D4 Zacatecas, C Mexico
Jālū 49 G3 var. Jūlā. NE Libya
Jaluit Atoll 122 D2 var. Jālwōj. atoll Ralik Chain, S Marshall Islands
Jālwōj see Jaluit Atoll
Jamaame 51 D6 It. Giamame; prev. Margherita. Jubbada Hoose, S Somalia
Jamaica 32 A4 country W West Indies
Jamaica 34 A1 island W West Indies
Jamaica Channel 32 D3 channel Haiti/Jamaica
Jamālpur 113 F3 Bihār, NE India
Jambi 116 B4 var. Telanaipura; prev. Djambi. Sumatera, W Indonesia
Jamdena see Yamdena, Pulau
James Bay 16 C3 bay Ontario/Québec, E Canada
James River 23 E2 river North Dakota/South Dakota, N USA
James River 19 E5 river Virginia, NE USA
Jamestown 19 E3 New York, NE USA
Jamestown 23 E2 North Dakota, N USA
Jamestown see Holetown
Jammu 112 D2 prev. Jummoo. state capital Jammu and Kashmir, NW India
Jammu and Kashmīr 112 D1 disputed region India/Pakistan
Jāmnagar 112 C4 prev. Navanagar. Gujarāt, W India
Jamshedpur 113 F4 Jhārkhand, NE India
Jamuna see Brahmaputra
Janaúba 41 F3 Minas Gerais, SE Brazil
Janesville 18 B3 Wisconsin, N USA
Janina see Ioánnina
Janischken see Joniškis
Jankovac see Jánoshalma
Jan Mayen 61 F4 constituent part of Norway. island N Atlantic Ocean
Jánoshalma 77 C7 SCr. Jankovac. Bács-Kiskun, S Hungary
Janów see Ivanava, Belarus
Janow/Janów see Jonava, Lithuania
Janów Poleski see Ivanava
Japan 108 C4 var. Nippon, Jap. Nihon. country E Asia
Japan, Sea of 108 A4 var. East Sea, Rus. Yapanskoye More. sea NW Pacific Ocean
Japan Trench 103 F1 trench NW Pacific Ocean
Japen see Yapen, Pulau
Japiim 40 C2 var. Máncio Lima. Acre, W Brazil
Japurá, Rio 40 C2 var. Río Caquetá, Yapurá. river Brazil/Colombia
Japurá, Rio see Caquetá, Río
Jaqué 31 G5 Darién, SE Panama
Jaquemel see Jacmel
Jarablos see Jarābulus
Jarābulus 96 C2 var. Jarablos, Jerablus, Fr. Djérablous. Ḩalab, N Syria
Jarbah, Jazīrat see Jerba, Île de
Jardines de la Reina, Archipiélago de los 32 B2 island group C Cuba
Jarīd, Shaṭṭ al see Jerid, Chott el
Jarocin 76 C4 Wielkopolskie, C Poland
Jaroslau see Jarosław
Jarosław 77 E5 Ger. Jaroslau, Rus. Yaroslav. Podkarpackie, SE Poland
Jarqo'rg'on 101 E3 Rus. Dzharkurgan. Surkhondaryo Viloyati, S Uzbekistan
Jarvis Island 123 G2 US unincorporated territory C Pacific Ocean
Jasło 77 D5 Podkarpackie, SE Poland
Jastrzębie-Zdrój 77 C5 Śląskie, S Poland
Jataí 41 E3 Goiás, C Brazil
Jativa see Xàtiva
Jauf see Al Jawf
Jaunpiebalga 84 D3 Gulbene, NE Latvia
Jaunpur 113 E3 Uttar Pradesh, N India
Java 130 A3 South Dakota, N USA
Javalambre 71 E3 mountain E Spain
Java Sea 116 D4 Ind. Laut Jawa. sea W Indonesia
Java Trench 102 D5 var. Sunda Trench. trench E Indian Ocean
Jawa, Laut see Java Sea
Jawhar 51 D6 var. Jowhar, It. Giohar. Shabeellaha Dhexe, S Somalia
Jaworów see Yavoriv
Jaya, Puncak 117 G4 prev. Puntjak Carstensz, Puntjak Sukarno. mountain Papua, E Indonesia
Jayapura 117 H4 var. Djajapura, Dut. Hollandia; prev. Kotabaru, Sukarnapura. Papua, E Indonesia
Jay Dairen see Dalian
Jayhawker State see Kansas
Jaz Murian, Hamun-e 98 E4 lake SE Iran
Jebba 53 F4 Kwara, W Nigeria
Jebel, Bahr el see White Nile
Jeble see Jablah
Jedda see Jiddah
Jędrzejów 76 D4 Ger. Endersdorf. Świętokrzyskie, C Poland
Jefferson City 23 G5 state capital Missouri, C USA
Jega 53 F4 Kebbi, NW Nigeria
Jehol see Chengde
Jeju-do 107 E4 Jap. Saishū; prev. Cheju-do, Quelpart. island S South Korea
Jeju Strait 107 E4 var. Jeju-haehyŏp; prev. Cheju-Strait. strait S South Korea
Jeju-haehyŏp see Jeju Strait
Jēkabpils 84 D4 Ger. Jakobstadt. Jēkabpils, S Latvia
Jelalabad see Jalālābād
Jelenia Góra 76 B4 Ger. Hirschberg, Hirschberg im Riesengebirge, Hirschberg in Riesengebirge, Hirschberg in Schlesien. Dolnośląskie, SW Poland
Jelgava 84 C3 Ger. Mitau. Jelgava, C Latvia
Jemappes 65 B6 Hainaut, S Belgium
Jember 116 D5 prev. Djember. Jawa, C Indonesia
Jena 72 C4 Thüringen, C Germany
Jengish Chokusu see Tömür Feng
Jenin 97 A6 N West Bank
Jerablus see Jarābulus
Jerada 48 D2 NE Morocco
Jérémie 32 D3 SW Haiti
Jerez see Jeréz de la Frontera, Spain
Jerez de la Frontera 70 C5 var. Jerez; prev. Xeres. Andalucía, SW Spain
Jerez de los Caballeros 70 C4 Extremadura, W Spain
Jericho see Arīḩā
Jerid, Chott el 49 E2 var. Shaṭṭ al Jarīd. salt lake SW Tunisia
Jersey 67 D8 British Crown Dependency Channel Islands, NW Europe
Jerusalem 81 H4 Ar. Al Quds, Al Quds ash Sharīf, Heb. Yerushalayim; anc. Hierosolyma. country capital (Israel) Jerusalem, NE Israel
Jesenice 73 D7 Ger. Assling. NW Slovenia
Jesselton see Kota Kinabalu
Jessore 113 G4 Khulna, W Bangladesh
Jesús María 42 C3 Córdoba, C Argentina
Jeypore see Jaipur, Rājasthān, India
Jhānsi 112 D3 Uttar Pradesh, N India
Jhārkhand 113 F4 cultural region NE India
Jhelum 112 C2 Punjab, NE Pakistan
Ji see Hebei, China
Ji see Jilin, China
Jiangmen 106 C6 Guangdong, S China
Jiangsu 106 D4 var. Chiang-su, Jiangsu Sheng, Kiangsu, Su. province E China
Jiangsu see Nanjing
Jiangsu Sheng see Jiangsu
Jiangxi 106 C6 var. Chiang-hsi, Gan, Jiangxi Sheng, Kiangsi. province S China
Jiangxi Sheng see Jiangxi
Jiaxing 106 D5 Zhejiang, SE China
Jiayi see Chiayi
Jibuti see Djibouti
Jiddah 99 A5 Eng. Jedda. Makkah, W Saudi Arabia
Jih-k'a-tse see Xigazê
Jihlava 77 B5 Ger. Iglau, Pol. Igława. Vysocina, S Czech Republic (Czechia)
Jilib 51 D6 It. Gelib. Jubbada Dhexe, S Somalia
Jilin 106 E3 var. Chi-lin, Girin, Kirin; prev. Yungki, Yunki. Jilin, NE China
Jilin 106 E3 var. Chi-lin, Girin, Ji, Jilin Sheng, Kirin. province NE China
Jilin Sheng see Jilin
Jilong see Keelung
Jīma 51 C5 var. Jimma, It. Gimma. Oromīya, C Ethiopia
Jimbolia 86 A4 Ger. Hatzfeld, Hung. Zsombolya. Timiş, W Romania
Jiménez 28 D2 Chihuahua, N Mexico
Jimma see Jīma
Jimsar 104 C3 Xinjiang Uygur Zizhiqu, NW China
Jin see Shanxi
Jin see Tianjin Shi
Jinan 106 C4 var. Chinan, Chi-nan, Tsinan. province capital Shandong, E China
Jingdezhen 106 C5 Jiangxi, S China
Jinghong 106 A6 var. Yunjinghong. Yunnan, SW China
Jinhua 106 D5 Zhejiang, SE China
Jining 105 F3 Shandong, E China
Jinja 51 C6 S Uganda
Jinotega 30 D3 Jinotega, NW Nicaragua
Jinotepe 30 D3 Carazo, SW Nicaragua
Jinsen see Incheon
Jinzhong 106 C4 var. Yuci. Shanxi, C China
Jinzhou 106 D3 var. Chin-chou, Chinchow; prev. Chinhsien. Liaoning, NE China
Jirgalanta see Hovd
Jisr ash Shadadi see Ash Shadādah
Jiu 86 B5 Ger. Schil, Schyl, Hung. Zsil, Zsily. river S Romania
Jiujiang 106 C5 Jiangxi, S China
Jixi 107 E2 Heilongjiang, NE China
Jīzān 99 B6 var. Qīzān. Jīzān, SW Saudi Arabia
Jizzax 101 E2 Rus. Dzhizak. Jizzax Viloyati, C Uzbekistan
João Belo see Xai-Xai
João Pessoa 41 G2 prev. Paraíba. state capital Paraíba, E Brazil
Joazeiro see Juazeiro
Job'urg see Johannesburg
Jo-ch'iang see Ruoqiang
Jodhpur 112 C3 Rājasthān, NW India
Joensuu 63 E5 Pohjois-Karjala, SE Finland
Jōetsu 109 C5 var. Zyôetu. Niigata, Honshū, C Japan
Jogjakarta see Yogyakarta
Johannesburg 56 D4 var. Egoli, Erautini, Gauteng, abbrev. Job'urg. Gauteng, NE South Africa
Johannisburg see Pisz
John Day River 24 C3 river Oregon, NW USA
John o'Groats 66 C2 N Scotland, United Kingdom
Johnston Atoll 121 E1 US unincorporated territory C Pacific Ocean
Johor Baharu see Johor Bahru
Johor Bahru 116 B3 var. Johor Baharu, Johore Bahru. Johor, Peninsular Malaysia
Johore Bahru see Johor Bahru
Johore Strait 116 A1 strait Johor, Peninsular Malaysia, Malaysia/Singapore Asia Andaman Sea/South China Sea
Joinvile see Joinville
Joinville 41 E4 var. Joinvile. Santa Catarina, S Brazil
Jokkmokk 62 C3 Lapp. Dálvvadis. Norrbotten, N Sweden
Jokyakarta see Yogyakarta
Joliet 18 B3 Illinois, N USA
Jonava 84 B4 Ger. Janow, Pol. Janów. Kaunas, C Lithuania
Jonesboro 20 B1 Arkansas, C USA
Joniškis 84 C3 Ger. Janischken. Šiauliai, N Lithuania
Jönköping 63 B7 Jönköping, S Sweden
Jonquière 17 E4 Québec, SE Canada
Joplin 23 F5 Missouri, C USA
Jordan 97 B6 off. Hashemite Kingdom of Jordan, Ar. Al Mamlaka al Urdunīya al Hashemīyah, Al Urdunn; prev. Transjordan. country SW Asia
Jordan 97 B5 Ar. Urdunn, Heb. HaYarden. river SW Asia
Jorhāt 113 H3 Assam, NE India
Jos 53 G4 Plateau, C Nigeria
Joseph Bonaparte Gulf 124 D2 gulf N Australia
Jos Plateau 53 G4 plateau C Nigeria
Jotunheimen 63 A5 mountain range S Norway
Joûnié 96 A4 var. Juniyah. W Lebanon
Joure 64 D2 Fris. De Jouwer. Friesland, N Netherlands
Joutseno 63 E5 Etelä-Kariala, SE Finland
Jowhar see Jawhar
J.Storm Thurmond Reservoir see Clark Hill Lake
Juan Aldama 28 D3 Zacatecas, C Mexico
Juan de Fuca, Strait of 24 A1 strait Canada/USA
Juan Fernández, Islas 35 A6 Eng. Juan Fernandez Islands. island group W Chile
Juan Fernandez Islands see Juan Fernández, Islas
Juazeiro 41 G2 prev. Joazeiro. Bahia, E Brazil
Juazeiro do Norte 41 G2 Ceará, E Brazil
Juba 51 B5 var. Jūbā. country capital (South Sudan) Bahr el Gabel, S South Sudan
Juba 51 D6 Amh. Genalē Wenz, It. Guiba, Som. Ganaane, Webi Jubba. river Ethiopia/Somalia
Jubba, Webi see Juba
Jubbulpore see Jabalpur
Júcar 71 E3 var. Jucar. river C Spain
Juchitán 29 F5 var. Juchitán de Zaragosa. Oaxaca, SE Mexico
Juchitán de Zaragosa see Juchitán
Judayyidat Hāmir 98 B3 Al Anbār, S Iraq
Judenburg 73 D7 Steiermark, C Austria
Jugoslavija see Serbia
Juigalpa 30 D3 Chontales, S Nicaragua
Juiz de Fora 41 F4 Minas Gerais, SE Brazil
Jujuy see San Salvador de Jujuy
Jūlā see Jālū, Libya
Julia Beterrae see Béziers
Juliaca 39 E4 Puno, SE Peru
Juliana Top 37 G3 mountain C Suriname
Julianehåb see Qaqortoq
Julio Briga see Bragança
Juliobriga see Logroño
Juliomagus see Angers
Jullundur see Jalandhar
Jumilla 71 E4 Murcia, SE Spain
Jummoo see Jammu
Jumna see Yamuna
Jumporn see Chumphon
Junction City 23 F4 Kansas, C USA
Juneau 14 D4 state capital Alaska, USA
Junín 42 C4 Buenos Aires, E Argentina
Juniyah see Joûnié
Junkseylon see Phuket
Jur 51 B5 river C Sudan
Jura 68 D4 cultural region E France
Jura 73 A7 var. Jura Mountains. mountain range France/Switzerland
Jura 66 B4 island SW Scotland, United Kingdom
Jura Mountains see Jura
Jurbarkas 84 B4 Ger. Georgenburg, Jurburg. Tauragė, W Lithuania
Jurburg see Jurbarkas
Jūrmala 84 C3 Rīga, C Latvia
Juruá, Rio 40 C2 var. Río Yuruá. river Brazil/Peru
Juruena, Rio 40 D3 river W Brazil
Jutiapa 30 B2 Jutiapa, S Guatemala
Juticalpa 30 D2 Olancho, C Honduras
Jutland 63 A7 Den. Jylland. peninsula W Denmark
Juvavum see Salzburg
Juventud, Isla de la 32 A2 var. Isla de Pinos, Eng. Isle of Youth; prev. The Isle of the Pines. island W Cuba
Južna Morava 79 E5 Ger. Südliche Morava. river SE Serbia
Jwaneng 56 C4 Southern, S Botswana
Jylland see Jutland
Jyrgalan see Dzhergalan
Jyväskylä 63 D5 Keski-Suomi, C Finland

K

K2 104 A4 Chin. Qogir Feng, Eng. Mount Godwin Austen. mountain China/Pakistan
Kaafu Atoll see Male' Atoll
Kaaimanston 37 G3 Sipaliwini, N Suriname
Kaakhka see Kaka
Kaala see Caála
Kaamanen 62 D2 Lapp. Gámas. Lappi, N Finland
Kaapstad see Cape Town
Kaaresuvanto 62 C3 Lapp. Gárassavon. Lappi, N Finland
Kabale 51 B6 SW Uganda
Kabinda 55 D7 Kasai-Oriental, SE Dem. Rep. Congo
Kabinda see Cabinda
Kābol see Kābul
Kabompo 56 C2 river W Zambia
Kābul 101 E4 prev. Kābol. country capital (Afghanistan) Kābul, E Afghanistan
Kabul 101 E4 var. Daryā-ye Kābul. river Afghanistan/Pakistan
Kābul, Daryā-ye see Kabul
Kabwe 56 D2 Central, C Zambia
Kachchh, Gulf of 112 B4 var. Gulf of Cutch, Gulf of Kutch. gulf W India
Kachchh, Rann of 112 B4 var. Rann of Kachh, Rann of Kutch. salt marsh India/Pakistan
Kachh, Rann of see Kachchh, Rann of
Kadan Kyun 115 B5 prev. King Island. island Myeik Archipelago, S Myanmar (Burma)
Kadavu 123 E4 prev. Kandavu. island S Fiji
Kadiyivka 87 H3 Rus. Stakhanov. Luhans'ka Oblast', E Ukraine
Kadoma 56 D3 prev. Gatooma. Mashonaland West, C Zimbabwe
Kadugli 50 B4 Southern Kordofan, S Sudan
Kaduna 53 G4 Kaduna, C Nigeria
Kadzhi-Say 101 G2 Kir. Kajisay. Issyk-Kul'skaya Oblast', NE Kyrgyzstan
Kaédi 52 C3 Gorgol, S Mauritania
Kaffa see Feodosiya
Kafue 56 D2 Lusaka, SE Zambia
Kafue 56 C2 river C Zambia
Kaga Bandoro 54 C4 prev. Fort-Crampel. Nana-Grébizi, C Central African Republic
Kagan see Kogon
Kāgheṭ 52 D1 var. Karet. physical region N Mauritania
Kagi see Chiayi
Kagoshima 109 B8 var. Kagosima. Kagoshima, Kyūshū, SW Japan
Kagoshima-wan 109 A8 bay SW Japan
Kagosima see Kagoshima
Kagul see Cahul
Darya-ye Kahmard 101 E4 prev. Darya-i-surkhab. river NE Afghanistan
Kahramanmaraş 94 D4 var. Kahraman Maraş, Maraş, Marash. Kahramanmaraş, S Turkey
Kaiapoi 129 C6 Canterbury, South Island, New Zealand
Kaifeng 106 C4 Henan, C China
Kai, Kepulauan 117 F4 prev. Kei Islands. island group Maluku, SE Indonesia
Kaikohe 128 C2 Northland, North Island, New Zealand
Kaikoura 129 C5 Canterbury, South Island, New Zealand
Kaikoura Peninsula 129 C5 peninsula South Island, New Zealand
Kainji Lake see Kainji Reservoir
Kainji Reservoir 53 F4 var. Kainji Lake. reservoir W Nigeria
Kaipara Harbour 128 C2 harbour North Island, New Zealand
Kairouan 49 E2 var. Al Qayrawān. E Tunisia
Kaisaria see Kayseri
Kaiserslautern 73 A5 Rheinland-Pfalz, SW Germany
Kaišiadorys 85 B5 Kaunas, S Lithuania
Kaitaia 128 C2 Northland, North Island, New Zealand
Kajaani 62 E4 Swe. Kajana. Kainuu, C Finland
Kajan see Kayan, Sungai
Kajana see Kajaani
Kajisay see Kadzhi-Say
Kaka 100 C2 Rus. Kaakhka. Ahal Welaýaty, S Turkmenistan
Kake 14 D4 Kupreanof Island, Alaska, USA
Kakhovka 87 F4 Khersons'ka Oblast', S Ukraine
Kakhovs'ke Vodoskhovyshche 87 F4 Rus. Kakhovskoye Vodokhranilishche. reservoir SE Ukraine
Kakhovskoye Vodokhranilishche see Kakhovs'ke Vodoskhovyshche
Kākināda 110 D1 prev. Cocanada. Andhra Pradesh, E India
Kakshaal-Too, Khrebet see Kokshaal-Tau
Kaktovik 14 D2 Alaska, USA
Kalaallit Nunaat see Greenland
Kalaburagi 110 C1 prev. Gulbarga. Karnātaka, C India
Kalahari Desert 56 B4 desert Southern Africa
Kalaikhum see Qal'aikhum
Kalámai see Kalámata
Kalamariá 82 B4 Kentrikí Makedonía, N Greece
Kalamás 82 A4 var. Thiamis; prev. Thýamis. river W Greece
Kalámata 83 B6 prev. Kalámai. Pelopónnisos, S Greece
Kalamazoo 18 C3 Michigan, N USA
Kalambaka see Kalampáka
Kálamos 83 C5 Attikí, C Greece
Kalampáka 82 B4 var. Kalambaka. Thessalía, C Greece
Kalanchak 87 F4 Khersons'ka Oblast', S Ukraine
Kalarash see Călăraşi
Kalasin 114 D4 var. Muang Kalasin. Kalasin, E Thailand
Kālat 112 B2 var. Kelat, Khelat. Baluchistān, SW Pakistan
Kalāt see Qalāt
Kalbarri 125 A5 Western Australia
Kalecik 94 C3 Ankara, N Turkey
Kalemie 55 E6 prev. Albertville. Katanga, SE Dem. Rep. Congo
Kale Sultanie see Çanakkale
Kalgan see Zhangjiakou
Kalgoorlie 125 B6 Western Australia
Kalima 55 D6 Maniema, E Dem. Rep. Congo
Kalimantan 116 D4 Eng. Indonesian Borneo. geopolitical region Borneo, C Indonesia
Kalinin see Tver'
Kaliningrad 84 A4 Kaliningradskaya Oblast', W Russia
Kaliningrad see Kaliningradskaya Oblast'
Kaliningradskaya Oblast' 84 B4 var. Kaliningrad. province and enclave W Russia
Kalinkavichy 85 C7 Rus. Kalinkovichi. Homyel'skaya Voblasts', SE Belarus
Kalinkovichi see Kalinkavichy
Kalisch/Kalish see Kalisz
Kalispell 22 B1 Montana, NW USA
Kalisz 76 C4 Ger. Kalisch, Rus. Kalish; anc. Calisia. Wielkopolskie, C Poland
Kalix 62 D4 Norrbotten, N Sweden
Kalixälven 62 D3 river N Sweden
Kallaste 84 E3 Ger. Krasnogor. Tartumaa, SE Estonia
Kallavesi 63 E5 lake SE Finland
Kalloní 83 D5 Lésvos, E Greece
Kalmar 63 C7 var. Calmar. Kalmar, S Sweden
Kalmthout 65 C5 Antwerpen, N Belgium
Kalpáki 82 A4 Ípeiros, W Greece
Kalpeni Island 110 B3 island Lakshadweep, India, N Indian Ocean
Kaltdorf see Pruszków
Kaluga 89 B5 Kaluzhskaya Oblast', W Russia
Kalush 86 C2 Pol. Kałusz. Ivano-Frankivs'ka Oblast', W Ukraine
Kałusz see Kalush
Kalutara 110 D4 Western Province, SW Sri Lanka
Kalvarija 85 B5 Pol. Kalwaria. Marijampolė, S Lithuania
Kalwaria see Kalvarija
Kalyān 112 C5 Mahārāshtra, W India
Kálymnos 83 D6 var. Kálimnos. island Dodekánisa, Greece, Aegean Sea
Kama 88 D4 river NW Russia
Kamarang 37 F3 W Guyana
Kambryk see Cambrai
Kamchatka see Kamchatka, Poluostrov
Kamchatka, Poluostrov 93 G3 Eng. Kamchatka. peninsula E Russia
Kamenets-Podol'skiy see Kam"yanets'-Podil's'kyy
Kamenka Dneprovskaya see Kam"yanka-Dniprovs'ka
Kamenskoye see **Kam"yans'ke**
Kamensk-Shakhtinskiy 89 B6 Rostovskaya Oblast', SW Russia
Kamina 55 D7 Katanga, S Dem. Rep. Congo
Kamishli see Al Qāmishlī
Kamloops 15 E5 British Columbia, SW Canada
Kammu Seamount 130 C2 guyot N Pacific Ocean
Kampala 51 B6 country capital (Uganda) S Uganda
Kampong Cham 115 D6 Khmer. Kâmpóng Cham. Kampong Cham, C Cambodia
Kâmpóng Cham see Kampong Cham
Kampong Chhang 115 D6 Khmer. Kâmpóng Chhnăng. Kampong Chhang, C Cambodia
Kâmpóng Chhnăng see Kampong Chhang
Kampong Speu 115 D6 Khmer. Kâmpóng Spœ. Kampong Speu, S Cambodia
Kâmpóng Spœ see Kampong Speu
Kampong Thom 115 D5 Khmer. Kâmpóng Thum, prev. Trâpeăng Vêng. Kampong Thom, C Cambodia
Kâmpóng Thum see Kampong Thom
Kâmpóng Trâbêk 115 D5 prev. Phumĭ Kâmpóng Trâbêk, Phum Kompong Trabek. Kampong Thom, C Cambodia
Kampot 115 D6 Khmer. Kâmpôt. Kampot, SW Cambodia
Kâmpôt see **Kampot**
Kampuchea see Cambodia
Kampuchea, Democratic see Cambodia
Kampuchea, People's Democratic Republic of see Cambodia
Kam"yanets'-Podil's'kyy 86 C3 Rus. Kamenets-Podol'skiy. Khmel'nyts'ka Oblast', W Ukraine
Kam"yanka-Dniprovs'ka 87 F3 Rus. Kamenka Dneprovskaya. Zaporiz'ka Oblast', SE Ukraine
Kam"yans'ke 87 F3 Rus. Dniprodzerzhyns'k, prev. Kamenskoye. Dnipropetrovs'ka Oblast', E Ukraine
Kamyshin 89 B6 Volgogradskaya Oblast', SW Russia
Kanaky see New Caledonia
Kananga 55 D6 prev. Luluabourg. Kasai-Occidental, S Dem. Rep. Congo
Kanara see Karnātaka
Kanash 89 C5 Chuvashskaya Respublika, W Russia
Kanazawa 109 C5 Ishikawa, Honshū, SW Japan
Kanbe 114 B4 Yangon, SW Myanmar (Burma)
Kānchīpuram 110 C2 prev. Conjeeveram. Tamil Nādu, SE India
Kandahār 101 E5 Per. Qandahār. Kandahār, S Afghanistan
Kandalakša see Kandalaksha
Kandalaksha 88 B2 var. Kandalakša, Fin. Kantalahti. Murmanskaya Oblast', NW Russia
Kandangan 116 D4 Borneo, C Indonesia
Kandau see Kandava
Kandava 84 C3 Ger. Kandau. Tukums, W Latvia
Kandavu see Kadavu
Kandi 53 F4 N Benin
Kandy 110 D3 Central Province, C Sri Lanka
Kane Fracture Zone 44 B4 fracture zone NW Atlantic Ocean
Kāne'ohe 25 A8 var. Kaneohe. O'ahu, Hawaii, USA, C Pacific Ocean
Kanestron, Akrotírio see Palioúri, Akrotírio
Kanëv see Kaniv
Kanevskoye Vodokhranilishche see Kanivs'ke Vodoskhovyshche
Kangān see Bandar-e Kangān
Kangaroo Island 127 A7 island South Australia
Kangertittivaq 61 E4 Dan. Scoresby Sund. fjord E Greenland
Kangikajik 61 E4 var. Kap Brewster. headland E Greenland
Kaniv 87 E2 Rus. Kanëv. Cherkas'ka Oblast', C Ukraine
Kanivs'ke Vodoskhovyshche 87 E2 Rus. Kanevskoye Vodokhranilishche. reservoir C Ukraine
Kanjiža 78 D2 Ger. Altkanischa, Hung. Magyarkanizsa, Ókanizsa; prev. Stara Kanjiža. Vojvodina, N Serbia
Kankaanpää 63 D5 Satakunta, SW Finland
Kankakee 18 B3 Illinois, N USA
Kankan 52 D4 E Guinea
Kannur 110 B2 var. Cannanore. Kerala, SW India
Kano 53 G4 Kano, N Nigeria
Kānpur 113 E3 Eng. Cawnpore. Uttar Pradesh, N India
Kansas 23 F5 off. State of Kansas, also known as Jayhawker State, Sunflower State. state C USA
Kansas City 23 F4 Kansas, C USA
Kansas City 23 F4 Missouri, C USA
Kansas River 23 F5 river Kansas, C USA
Kansk 93 E4 Krasnoyarskiy Kray, S Russia
Kansu see Gansu
Kantalahti see Kandalaksha
Kántanos 83 C7 Kríti, Greece, E Mediterranean Sea
Kantemirovka 89 B6 Voronezhskaya Oblast', W Russia
Kantipur see Kathmandu
Kanton 123 F3 var. Abariringa, Canton Island; prev. Mary Island. atoll Phoenix Islands, C Kiribati
Kanye 56 C4 Southern, SE Botswana
Kaohsiung 106 D6 var. Gaoxiong, Jap. Takao, Takow. S Taiwan
Kaolack 52 B3 var. Kaolak. W Senegal
Kaolak see Kaolack
Kaolan see Lanzhou
Kaoma 56 C2 Western, W Zambia
Kapelle 65 B5 Zeeland, SW Netherlands
Kapellen 65 C5 Antwerpen, N Belgium
Kapka, Massif du 54 C2 mountain range E Chad
Kaplangky, Plato see Gaplaňyr Platosy
Kapoeas see Kapuas, Sungai
Kapoeta 51 C5 E Equatoria, SE South Sudan
Kaposvár 77 C7 Somogy, SW Hungary
Kappeln 72 B2 Schleswig-Holstein, N Germany
Kaproncza see Koprivnica
Kapstad see Cape Town
Kapsukas see Marijampolė
Kaptsevichy 85 C7 Rus. Koptsevichi. Homyel'skaya Voblasts', SE Belarus
Kapuas, Sungai 116 C4 prev. Kapoeas. river Borneo, C Indonesia
Kapuskasing 16 C4 Ontario, S Canada
Kapyl' 85 C6 Rus. Kopyl'. Minskaya Voblasts', C Belarus
Kara-Balta 101 F2 Chuyskaya Oblast', N Kyrgyzstan
Karabil', Vozvyshennost' see Garabil Belentligi
Kara-Bogaz-Gol, Zaliv see Garabogaz Aylagy
Karabük 94 C2 Karabük, NW Turkey
Karāchi 112 B3 Sind, SE Pakistan
Karácsonkő see Piatra-Neamţ
Karadeniz see Black Sea
Karadeniz Boğazi see İstanbul Boğazı
Karaferiye see Véroia
Karagandy 92 C4 prev. Karaganda, Kaz. Qaraghandy. Karagandy, C Kazakhstan
Karaginskiy, Ostrov 93 H2 island E Russia
Karagumskiy Kanal see Garagum Kanaly
Karak see Al Karak
Kara-Kala see Magtymguly
Karakax see Moyu

Karakılısse – Kitab

Karakılısse see Ağrı
Karakol *101 G2* prev. Przheval'sk. Issyk-Kul'skaya Oblast', NE Kyrgyzstan
Karakol *101 G2* var. Karakolka. Issyk-Kul'skaya Oblast', NE Kyrgyzstan
Karakolka see Karakol
Karakoram Range *112 D1* mountain range C Asia
Karaköse see Ağrı
Karakul' see Qarokŭl, Tajikistan
Kara Kum see Garagum
Kara Kum Canal/Karakumskiy Kanal see Garagum Kanaly
Karakumy, Peski see Garagum
Karamai see Karamay
Karaman *94 C4* Karaman, S Turkey
Karamay *104 B2* var. Karamai, Kelamayi; prev. Chin. K'o-la-ma-i. Xinjiang Uygur Zizhiqu, NW China
Karamea Bight *129 B5* gulf South Island, New Zealand
Karapelit *82 E1* Rom. Stejarul. Dobrich, NE Bulgaria
Kara-Say *101 G2* Issyk-Kul'skaya Oblast', NE Kyrgyzstan
Karasburg *56 B4* Karas, S Namibia
Kara Sea see Karskoye More
Kara Strait see Karskiye Vorota, Proliv
Karatau *92 C5* Kaz. Qarataū. Zhambyl, S Kazakhstan
Karavás *83 B7* Kýthira, S Greece
Karbalā' *98 B3* var. Kerbala, Kerbela. Karbalā', S Iraq
Kardeljevo see Ploče
Kardhítsa see Kardítsa
Kardítsa *83 B5* var. Kardhítsa. Thessalía, C Greece
Kärdla *84 C2* Ger. Kertel. Hiiumaa, W Estonia
Karet see Kâgheţ
Kargı *94 C2* Çorum, N Turkey
Kargilik see Yecheng
Kariba *56 D2* Mashonaland West, N Zimbabwe
Kariba, Lake *56 D3* reservoir Zambia/Zimbabwe
Karibib *56 B3* Erongo, C Namibia
Karies see Karyés
Karigasniemi *62 D2* Lapp. Garegegasnjárga. Lappi, N Finland
Karimata, Selat *116 C4* strait W Indonesia
Karīmnagar *112 D5* Telangana, C India
Karin *50 D4* Sahil, N Somalia
Kariot see Ikaría
Káristos see Kárystos
Karkinits'ka Zatoka *87 E4* Rus. Karkinitskiy Zaliv. gulf S Ukraine
Karkinitskiy Zaliv see Karkinits'ka Zatoka
Karkūk see Kirkūk
Karleby see Kokkola
Karl-Marx-Stadt see Chemnitz
Karlö see Hailuoto
Karlovac *78 B3* Ger. Karlstadt, Hung. Károlyváros. Karlovac, C Croatia
Karlovy Vary *77 A5* Ger. Karlsbad; prev. Eng. Carlsbad. Karlovarský Kraj, W Czech Republic (Czechia)
Karlsbad see Karlovy Vary
Karlsburg see Alba Iulia
Karlskrona *63 C7* Blekinge, S Sweden
Karlsruhe *73 B6* var. Carlsruhe. Baden-Württemberg, SW Germany
Karlstad *63 B6* Värmland, C Sweden
Karlstadt see Karlovac
Karnāl *112 D2* Haryāna, N India
Karnātaka *110 C1* var. Kanara; prev. Maisur, Mysore. cultural region W India
Karnobat *82 D2* Burgas, E Bulgaria
Karnul see Kurnool
Karol see Carei
Károly-Fehérvár see Alba Iulia
Károlyváros see Karlovac
Karpaten see Carpathian Mountains
Kárpathos *83 E7* Kárpathos, SE Greece
Kárpathos *83 E7* It. Scarpanto; anc. Carpathos, Carpathus. island SE Greece
Karpaty see Carpathian Mountains
Karpenísi *83 B5* prev. Karpenísion. Stereá Ellás, C Greece
Karpenísion see Karpenísi
Karpilovka see Aktsyabrski
Kars *95 F2* var. Qars. Kars, NE Turkey
Karsau see Kārsava
Kārsava *84 D4* Ger. Karsau; prev. Rus. Korsovka. Ludza, E Latvia
Karshi see Qarshi, Uzbekistan
Karskiye Vorota, Proliv *88 E2* Eng. Kara Strait. strait N Russia
Karskoye More *92 D2* Eng. Kara Sea. sea Arctic Ocean
Karyés *82 C4* var. Karies. Ágion Óros, N Greece
Kárystos *83 C6* var. Káristos. Évvoia, C Greece
Kasai *55 C6* var. Cassai, Kassai. river Angola/Dem. Rep. Congo
Kasaji *55 D7* Katanga, S Dem. Rep. Congo
Kasama *56 D1* Northern, N Zambia
Kasan see Koson
Kāsaragod *110 B2* Kerala, SW India
Kaschau see Košice
Kāshān *98 C3* Eşfahān, C Iran
Kashgar see Kashi
Kashi *104 A3* Chin. Kaxgar, K'o-shih, Uigh. Kashgar. Xinjiang Uygur Zizhiqu, NW China
Kasi see Vārānasi
Kasongo *55 D6* Maniema, E Dem. Rep. Congo
Kasongo-Lunda *55 C7* Bandundu, SW Dem. Rep. Congo
Kásos *83 D7* island S Greece
Kaspīy Mangy Oypaty see Caspian Depression
Kaspiysk *89 B8* Respublika Dagestan, SW Russia
Kaspiyskoye More/Kaspiy Tengizi see Caspian Sea
Kassa see Košice
Kassai see Kasai
Kassala *50 C4* Kassala, E Sudan
Kassel *72 B4* prev. Cassel. Hessen, C Germany
Kasserine *49 E2* var. Al Qaşrayn. W Tunisia
Kastamonu *94 C2* var. Castamoni, Kastamuni. Kastamonu, N Turkey
Kastamuni see Kastamonu
Kastaneá *82 B4* Kentrikí Makedonía, N Greece
Kastélli see Kíssamos
Kastoriá *82 B4* Dytikí Makedonía, N Greece
Kástro *83 C6* Sífnos, Kykládes, Greece, Aegean Sea
Kastsyukovichy *85 E7* Rus. Kostyukovichi. Mahilyowskaya Voblasts', E Belarus
Kastsyukowka *85 D7* Rus. Kostyukovka. Homyel'skaya Voblasts', SE Belarus
Kasulu *51 B7* Kigoma, W Tanzania
Kasumiga-ura *109 D5* lake Honshū, S Japan
Katahdin, Mount *19 G1* mountain Maine, NE USA
Katalla *14 C3* Alaska, USA
Katana see Qaţanā
Katanning *125 B7* Western Australia
Katawaz see Zarghūn Shahr
Katchall Island *111 F3* island Nicobar Islands, India, NE Indian Ocean
Kateríni *82 B4* Kentrikí Makedonía, N Greece
Katha *114 B2* Sagaing, N Myanmar (Burma)
Katherine *126 A2* Northern Territory, N Australia
Kathmandu *102 C3* prev. Kantipur. country capital (Nepal) Central, C Nepal
Katikati *128 D3* Bay of Plenty, North Island, New Zealand
Katima Mulilo *56 C3* Caprivi, NE Namibia
Katiola *52 D4* C Ivory Coast
Káto Achaïa *83 B5* var. Kato Ahaia, Káto Akhaía. Dytikí Ellás, S Greece
Kato Ahaia/Káto Akhaía see Káto Achaïa
Katoúna *83 A5* Dytikí Ellás, C Greece
Katowice *77 C5* Ger. Kattowitz. Śląskie, S Poland
Katsina *53 G3* Katsina, N Nigeria
Kattakurgan see Kattaqo'rg'on
Kattaqo'rg'on *101 E2* Rus. Kattakurgan. Samarqand Viloyati, C Uzbekistan
Kattavía *83 E7* Ródos, Dodekánisa, Greece, Aegean Sea
Kattegat *63 B7* Dan. Kattegatt. strait N Europe
Kattegatt see Kattegat
Kattowitz see Katowice
Kaua'i *25 A7* var. Kauai. island Hawai'ian Islands, Hawaii, USA, C Pacific Ocean
Kauai see Kaua'i
Kauen see Kaunas
Kaufbeuren *73 C6* Bayern, S Germany
Kaunas *84 B4* Ger. Kauen, Pol. Kowno; prev. Rus. Kovno. Kaunas, C Lithuania
Kavadar see Kavadarci
Kavadarci *79 E6* Turk. Kavadar. C Macedonia
Kavaja see Kavajë
Kavajë *79 C6* It. Cavaia, Kavaja. Tiranë, W Albania
Kavakli see Topolovgrad
Kavála *82 C3* prev. Kaválla. Anatolikí Makedonía kai Thráki, NE Greece
Kāvali *110 D2* Andhra Pradesh, E India
Kavála see Kavála
Kavango see Cubango/Okavango
Kavaratti Island *110 A3* island Lakshadweep, Lakshadweep, SW India Asia N Indian Ocean
Kavarna *82 E2* Dobrich, NE Bulgaria
Kavengo see Cubango/Okavango
Kavir, Dasht-e *98 D3* var. Great Salt Desert. salt pan N Iran
Kavkaz see Caucasus
Kawagoe *109 D5* Saitama, Honshū, S Japan
Kawasaki *108 A2* Kanagawa, Honshū, S Japan
Kawerau *128 E3* Bay of Plenty, North Island, New Zealand
Kaxgar see Kashi
Kaya *53 E3* C Burkina
Kayan *114 B4* Yangon, SW Burma (Myanmar)
Kayan, Sungai *116 D3* prev. Kajan. river Borneo, C Indonesia
Kayes *52 C3* Kayes, W Mali
Kayseri *94 D3* var. Kaisaria; anc. Caesarea Mazaca, Mazaca. Kayseri, C Turkey
Kazach'ye *93 F2* Respublika Sakha (Yakutiya), NE Russia
Kazakhskaya SSR/Kazakh Soviet Socialist Republic see Kazakhstan
Kazakhstan *92 B4* off. Republic of Kazakhstan, var. Kazakstan, Kaz. Qazaqstan, Qazaqstan Respublikasy; prev. Kazakh Soviet Socialist Republic, Rus. Kazakhskaya SSR. country C Asia
Kazakhstan, Republic of see Kazakhstan
Kazakh Uplands *92 C4* Eng. Kazakh Uplands, Kirghiz Steppe, Kaz. Saryarqa. uplands C Kazakhstan
Kazakh Uplands see Kazakhskiy Melkosopochnik
Kazakstan see Kazakhstan
Kazan' *89 C5* Respublika Tatarstan, W Russia
Kazandzhik see Bereket
Kazanlik see Kazanlŭk
Kazanlŭk *82 D2* prev. Kazanlik. Stara Zagora, C Bulgaria
Kazatin see Kozyatyn
Kazbegi see Kazbek
Kazbek *95 F1* var. Kazbegi, Geor. Mqinvartsveri. mountain N Georgia
Kāzerūn *98 D4* Fārs, S Iran
Kazi Magomed see Qazimämmäd
Kazvin see Qazvīn
Kéa *83 C6* var. Tziá, Kéos; anc. Ceos. island Kykládes, Greece, Aegean Sea
Kea, Mauna *25 B8* mountain Hawaii, USA
Kéamu see Aneityum
Kearney *23 E4* Nebraska, C USA
Keban Barajı *95 E3* reservoir C Turkey
Kebkabiya *50 A4* Northern Darfur, W Sudan
Kebnekaise *62 C3* mountain N Sweden
Kecskemét *77 D7* Bács-Kiskun, C Hungary
Kediri *116 D5* Jawa, C Indonesia
Kędzierzyn-Kozle *77 C5* Ger. Heydebrech. Opolskie, S Poland
Keelung *106 D6* var. Chilung, Jilong; Jap. Kirun, Kirun'; prev. Sp. Santissima Trinidad. N Taiwan
Keetmanshoop *56 B4* Karas, S Namibia
Kefallinía see Kefallonía
Kefallonía *83 A5* var. Kefallinía. island Iónia Nisiá, Greece, C Mediterranean Sea
Kefe see Feodosiya
Kegel see Keila
Kehl *73 A6* Baden-Württemberg, SW Germany
Kei Islands see Kai, Kepulauan
Keijō see Seoul
Keila *84 D2* Ger. Kegel. Harjumaa, NW Estonia
Keïta *53 F3* Tahoua, C Niger
Keith *127 B7* South Australia
Këk-Art *101 G2* prev. Alaykel', Alay-Kuu. Oshskaya Oblast', SW Kyrgyzstan
Kékes *77 C6* mountain N Hungary
Kelamayi see Karamay
Kelang see Klang
Kelat see Kālat
Kelifskiy Uzboy see Kelif Uzboÿy
Kelif Uzboÿy *100 D3* Rus. Kelifskiy Uzboy. salt marsh E Turkmenistan
Kelkit Çayı *95 E3* river N Turkey
Kelmė *84 B4* Šiaulai, C Lithuania
Kélo *54 B4* Tandjilé, SW Chad
Kelowna *15 E5* British Columbia, SW Canada
Kelso *24 B2* Washington, NW USA
Keltsy see Kielce
Keluang *116 B3* var. Kluang. Johor, Peninsular Malaysia, Malaysia
Kem' *88 B3* Respublika Kareliya, NW Russia
Kemah *95 E3* Erzincan, E Turkey
Kemaman see Cukai
Kemerovo *92 D4* prev. Shcheglovsk. Kemerovskaya Oblast', C Russia
Kemi *62 D4* Lappi, NW Finland
Kemijärvi *62 D3* Swe. Kemiträsk. Lappi, N Finland
Kemijoki *62 D3* river NW Finland
Kemin *101 G2* prev. Bystrovka. Chuyskaya Oblast', N Kyrgyzstan
Kemins Island see Nikumaroro
Kemiträsk see Kemijärvi
Kemmuna *80 A5* var. Comino. island C Malta
Kempele *62 D4* Pohjois-Pohjanmaa, C Finland
Kempten *73 B7* Bayern, S Germany
Kendal *67 D5* NW England, United Kingdom
Kendari *117 E4* Sulawesi, C Indonesia
Kenedy *27 G4* Texas, SW USA
Kenema *52 C4* SE Sierra Leone
Këneurgench see Köneürgenç
Kenge *55 C6* Bandundu, SW Dem. Rep. Congo
Kengtung *114 C3* pev. Keng Tung. Shan State, E Myanmar (Burma)
Keng Tung see Kengtung
Kénitra *48 C2* prev. Port-Lyautey. NW Morocco
Kennett *23 H5* Missouri, C USA
Kennewick *24 C2* Washington, NW USA
Kenora *16 A3* Ontario, S Canada
Kenosha *18 B3* Wisconsin, N USA
Kentau *92 B5* Yuzhnyy Kazakhstan, S Kazakhstan
Kentucky *18 C5* off. Commonwealth of Kentucky, also known as Bluegrass State. state C USA
Kentucky Lake *18 B5* reservoir Kentucky/Tennessee, S USA
Kentung see Keng Tung
Kenya *51 C6* off. Republic of Kenya. country E Africa
Kenya, Mount see Kirinyaga
Kenya, Republic of see Kenya
Keokuk *23 G4* Iowa, C USA
Kéos see Keá
Kępno *76 C4* Wielkopolskie, C Poland
Keppel Island see Niuatoputapu
Kerak see Al Karak
Kerala *110 C2* cultural region S India
Kerasunt see Giresun
Keratéa *83 C6* var. Keratea. Attikí, C Greece
Keratea see Keratéa
Kerbala/Kerbela see Karbalā'
Kerch *87 G5* Rus. Kerch'. Respublika Krym, SE Ukraine
Kerch' see Kerch
Kerchens'ka Protska/Kerchenskiy Proliv see Kerch Strait
Kerch Strait *87 G4* var. Bosporus Cimmerius, Enikale Strait, Rus. Kerchenskiy Proliv, Ukr. Kerchens'ka Protska. strait Black Sea/Sea of Azov
Keremitlik see Lyulyakovo
Kerguelen *119 C7* island C French Southern and Antarctic Lands
Kerguelen Plateau *119 C7* undersea plateau S Indian Ocean
Kerí *83 A6* Zákynthos, Iónia Nisiá, Greece, C Mediterranean Sea
Kerikeri *128 D2* Northland, North Island, New Zealand
Kerkenah, Îles de *80 D4* var. Kerkenna Islands, Ar. Juzur Qarqannah. island group E Tunisia
Kerkenna Islands see Kerkenah, Îles de
Kerki see Atamyrat
Kérkira see Kérkyra
Kerkrade *65 D6* Limburg, SE Netherlands
Kerkuk see Kirkūk
Kérkyra *82 A4* var. Kérkira, Eng. Corfu. Kérkyra, Iónia Nisiá, Greece, C Mediterranean Sea
Kermadec Islands *130 C4* island group New Zealand, SW Pacific Ocean
Kermadec Trench *121 E4* trench SW Pacific Ocean
Kermān *98 D3* var. Kirman; anc. Carmana. Kermān, C Iran
Kermānshāh *98 C3* var. Qahremānshahr; prev. Bākhtarān. Kermānshāhān, W Iran
Kerrville *27 F4* Texas, SW USA
Kertel see Kärdla
Kerulen *105 E2* Chin. Herlen He, Mong. Herlen Gol. river China/Mongolia
Keryneia see Girne
Kesennuma *108 D4* Miyagi, Honshū, C Japan
Keszthely *77 C7* Zala, SW Hungary
Ketchikan *14 D4* Revillagigedo Island, Alaska, USA
Kętrzyn *76 D2* Ger. Rastenburg. Warmińsko-Mazurskie, NE Poland
Kettering *67 D6* C England, United Kingdom
Kettering *18 C4* Ohio, N USA
Keupriya see Primorsko
Keuruu *63 D5* Keski-Suomi, C Finland
Keweenaw Peninsula *18 B1* peninsula Michigan, N USA
Key Largo *21 F5* Key Largo, Florida, SE USA
Keystone State see Pennsylvania
Key West *21 E5* Florida Keys, Florida, SE USA
Kezdivásárhely see Târgu Secuiesc
Khabarovsk *93 G4* Khabarovskiy Kray, SE Russia
Khachmas see Xaçmaz
Khadera see Hadera
Khairpur *112 B3* Sind, SE Pakistan
Khalīj as Suways *50 B2* var. Suez, Gulf of. gulf NE Egypt
Khalkidhikí see Chalkidikí
Khalkís see Chalkída
Khambhat, Gulf of *112 C4* Eng. Gulf of Cambay. gulf W India
Khamīs Mushayt *99 B6* var. Hamīs Musait. 'Asīr, SW Saudi Arabia
Khānābād *101 E3* Kunduz, NE Afghanistan
Khān al Baghdādī see Al Baghdādī
Khandwa *112 D4* Madhya Pradesh, C India
Khanh Hung see Soc Trăng
Khaniá see Chaniá
Khanka, Lake *107 E2* var. Hsing-K'ai Hu, Lake Hanka, Chin. Xingkai Hu, Rus. Ozero Khanka. lake China/Russia
Khanka, Ozero see Khanka, Lake
Khankendi see Xankändi
Khanthabouli *114 D4* prev. Savannakhét. Savannakhét, S Laos
Khanty-Mansiysk *92 C3* prev. Ostyako-Voguls'k. Khanty-Mansiyskiy Avtonomnyy Okrug-Yugra, C Russia
Khān Yūnis *97 A7* var. Khān Yūnus. S Gaza Strip
Khān Yūnus see Khān Yūnis
Khanzi see Ghanzi
Kharagpur *113 F4* West Bengal, NE India
Kharbin see Harbin
Kharkiv *87 G2* Rus. Khar'kov. Kharkivs'ka Oblast', NE Ukraine
Khar'kov see Kharkiv
Kharmanli *82 D3* Khaskovo, S Bulgaria
Khartoum *50 B4* var. El Khartûm, Khartum. country capital (Sudan) Khartoum, C Sudan
Khartum see Khartoum
Khasavyurt *89 B8* Respublika Dagestan, SW Russia
Khash, Dasht-e *100 D5* Eng. Khash Desert. desert SW Afghanistan
Khash Desert see Khāsh, Dasht-e
Khashim Al Qirba/Khashm al Qirbah see Khashm el Girba
Khashm el Girba *50 C4* var. Khashim Al Qirba, Khashm al Qirbah. Kassala, E Sudan
Khaskovo *82 D3* Khaskovo, S Bulgaria
Khaybar, Kowtal-e see Khyber Pass
Khaydarkan *101 F2* var. Khaydarken. Batkenskaya Oblast', SW Kyrgyzstan
Khaydarken see Khaydarkan
Khazar, Baḩr-e/Khazar, Daryā-ye see Caspian Sea
Khelat see Kālat
Kherson *87 E4* Khersons'ka Oblast', S Ukraine
Kheta *93 E2* river N Russia
Khíos see Chíos
Khirbet el 'Aujā et Taḥtā *97 E7* E West Bank, NE Asia
Khiva/Khiwa see Xiva
Khmel'nitskiy see Khmel 'nyts'kyy
Khmel 'nyts'kyy *86 C2* Rus. Khmel'nitskiy; prev. Proskurov. Khmel'nyts'ka Oblast', W Ukraine
Khodasy *85 E6* Rus. Khodosy. Mahilyowskaya Voblasts', E Belarus
Khodoriv *86 C2* Pol. Chodorów, Rus. Khodorov. L'vivs'ka Oblast', NW Ukraine
Khodorov see Khodoriv
Khodosy see Khodasy
Khodzhent see Khujand
Khoi see Khvoy
Khojend see Khujand
Khokand see Qo'qon
Kholm see Khulm
Kholm see Chełm
Kholon see Holon
Khoms see Al Khums
Khong Sedone see Muang Khôngxédôn
Khon Kaen *114 D4* var. Muang Khon Kaen. Khon Kaen, E Thailand
Khor *93 G4* Khabarovskiy Kray, SE Russia
Khorat see Nakhon Ratchasima
Khorog see Khorugh
Khorugh *101 F3* Rus. Khorog. S Tajikistan
Khōst *101 F4* prev. Khowst. Khōst, E Afghanistan
Khotan see Hotan
Khouribga *48 B2* C Morocco
Khovd see Hovd
Khowst see Khōst
Khoy see Khvoy
Khoyniki *85 D8* Homyel'skaya Voblasts', SE Belarus
Khrustal'nyy *87 H3* Rus. Krasnyy Luch. prev. Krindachevka. Luhans'ka Oblast', E Ukraine
Khudzhand see Khujand
Khujand *101 E2* var. Khodzhent, Khojend, Rus. Khudzhand; prev. Leninabad, Taj. Leninobod. N Tajikistan
Khulm *101 E3* var. Kholm, Tashqurghan. Balkh, N Afghanistan
Khulna *113 G4* Khulna, SW Bangladesh
Khums see Al Khums
Khust *86 B3* var. Husté, Cz. Chust, Hung. Huszt. Zakarpats'ka Oblast', W Ukraine
Khvoy *98 C2* var. Khoi, Khoy. Āžarbāyjān-e Bākhtarī, NW Iran
Khyber Pass *112 C1* var. Kowtal-e Khaybar. pass Afghanistan/Pakistan
Kiangmai see Chiang Mai
Kiang-ning see Nanjing
Kiangsi see Jiangxi
Kiangsu see Jiangsu
Kiáto *83 B6* prev. Kiáton. Pelopónnisos, S Greece
Kiáton see Kiáto
Kiayi see Chiayi
Kibangou *55 B6* Niari, SW Congo
Kibombo *55 D6* Maniema, E Dem. Rep. Congo
Kıbrıs/Kıbrıs Cumhuriyeti see Cyprus
Kičevo *79 D6* SW Macedonia
Kidderminster *67 D6* C England, United Kingdom
Kiel *72 B2* Schleswig-Holstein, N Germany
Kielce *76 D4* Rus. Keltsy. Świętokrzyskie, C Poland
Kieler Bucht *72 B2* bay N Germany
Kiev see Kyiv
Kiev Reservoir see Kyyivs'ke Vodoskhovyshche
Kiffa *52 C3* Assaba, S Mauritania
Kigali *51 B6* country capital (Rwanda) C Rwanda
Kigoma *51 B7* Kigoma, W Tanzania
Kihnu *84 C2* var. Kihnu Saar, Ger. Kühnö. island SW Estonia
Kihnu Saar see Kihnu
Kii-suido *109 C7* strait S Japan
Kikinda *78 D3* Ger. Grosskikinda, Hung. Nagykikinda; prev. Velika Kikinda. Vojvodina, N Serbia
Kikládhes see Kykládes
Kikwit *55 C6* Bandundu, W Dem. Rep. Congo
Kilien Mountains see Qilian Shan
Kilimane see Quelimane
Kilimanjaro *47 E5* region E Tanzania
Kilimanjaro *51 C7* var. Uhuru Peak. volcano NE Tanzania
Kilingi-Nõmme *84 D3* Ger. Kurkund. Pärnumaa, SW Estonia
Kilis *94 D4* Kilis, S Turkey
Kiliya *86 D4* Rom. Chilia-Nouă. Odes'ka Oblast', SW Ukraine
Kilkenny *67 B6* Ir. Cill Chainnigh. Kilkenny, S Ireland
Kilkís *82 B3* Kentrikí Makedonía, N Greece
Killarney *67 A6* Ir. Cill Airne. Kerry, SW Ireland
Killeen *27 G3* Texas, SW USA
Kilmain see Quelimane
Kilmarnock *66 C4* W Scotland, United Kingdom
Kilwa see Kilwa Kivinje
Kilwa Kivinje *51 C7* var. Kilwa. Lindi, SE Tanzania
Kimberley *56 C4* Northern Cape, C South Africa
Kimberley Plateau *124 C3* plateau Western Australia
Kimch'aek *107 E3* prev. Sŏngjin. E North Korea
Kími see Kými
Kinabalu, Gunung *116 D3* mountain East Malaysia
Kindersley *15 F5* Saskatchewan, S Canada
Kindia *52 C4* Guinée-Maritime, SW Guinea
Kindley Field *20 A4* air base E Bermuda
Kindu *55 D6* prev. Kindu-Port-Empain. Maniema, C Dem. Rep. Congo
Kindu-Port-Empain see Kindu
Kineshma *89 C5* Ivanovskaya Oblast', W Russia
King Abdullah Economic City *99 A5* W Saudi Arabia
King Charles Islands see Kong Karls Land
King Christian IX Land see Kong Christian IX Land
King Frederik VI Coast see Kong Frederik VI Kyst
King Frederik VIII Land see Kong Frederik VIII Land
King Island *127 B8* island Tasmania, SE Australia
King Island see Kadan Kyun
Kingissepp see Kuressaare
Kingman *26 A1* Arizona, SW USA
Kingman Reef *123 E2* US territory C Pacific Ocean
Kingsford Smith *126 E2* (Sydney) New South Wales, SE Australia
King's Lynn *67 E6* var. Bishop's Lynn, Kings Lynn, Lynn, Lynn Regis. E England, United Kingdom
Kings Lynn see King's Lynn
King Sound *124 B3* sound Western Australia
Kingsport *21 E1* Tennessee, S USA
Kingston *32 B5* country capital (Jamaica) E Jamaica
Kingston *16 D5* Ontario, SE Canada
Kingston *19 F3* New York, NE USA
Kingston upon Hull *67 D5* var. Hull. E England, United Kingdom
Kingston upon Thames *67 A8* SE England, United Kingdom
Kingstown *33 H4* country capital (Saint Vincent and the Grenadines) Saint Vincent, Saint Vincent and the Grenadines
Kingstown see Dún Laoghaire
Kingsville *27 G5* Texas, SW USA
King William Island *15 F3* island Nunavut, N Canada
Kinneret, Yam see Tiberias, Lake
Kinrooi *65 D5* Limburg, NE Belgium
Kinshasa *55 B6* prev. Léopoldville. country capital (Dem. Rep. Congo) Kinshasa, W Dem. Rep. Congo
Kintyre *66 B4* peninsula W Scotland, United Kingdom
Kinyeti *51 B5* mountain S South Sudan
Kiparissía see Kyparissía
Kipili *51 B7* Rukwa, W Tanzania
Kipushi *55 D8* Katanga, SE Dem. Rep. Congo
Kirdzhali see Kardzhali
Kirghizia see Kyrgyzstan
Kirghiz Range *101 F2* Rus. Kirgizskiy Khrebet; prev. Alexander Range. mountain range Kazakhstan/Kyrgyzstan
Kirghiz SSR see Kyrgyzstan
Kirghiz Steppe see Kazakhskiy Melkosopochnik
Kirgizskaya SSR see Kyrgyzstan
Kirgizskiy Khrebet see Kirghiz Range
Kiriath-Arba see Hebron
Kiribati *123 F2* off. Republic of Kiribati. country C Pacific Ocean
Kiribati, Republic of see Kiribati
Kırıkhan *94 D4* Hatay, S Turkey
Kırıkkale *94 C3* province C Turkey
Kirin see Jilin
Kirinyaga *51 C6* prev. Mount Kenya. volcano C Kenya
Kirishi *88 B4* var. Kirisi. Leningradskaya Oblast', NW Russia
Kirisi see Kirishi
Kiritimati *123 G2* prev. Christmas Island. atoll Line Islands, E Kiribati
Kirkenes *62 E2* Fin. Kirkkoniemi. Finnmark, N Norway
Kirk-Kilissa see Kırklareli
Kirkkoniemi see Kirkenes
Kirkland Lake *16 D4* Ontario, S Canada
Kırklareli *94 A2* prev. Kirk-Kilissa. Kırklareli, NW Turkey
Kirkpatrick, Mount *132 B3* mountain Antarctica
Kirksville *23 G4* Missouri, C USA
Kirkūk *98 B3* var. Karkūk, Kerkuk. At Ta'mīn, N Iraq
Kirkwall *66 C2* NE Scotland, United Kingdom
Kirkwood *23 G4* Missouri, C USA
Kir Moab/Kir of Moab see Al Karak
Kirov *89 C5* prev. Vyatka. Kirovskaya Oblast', NW Russia
Kirovabad see Gäncä
Kirovakan see Vanadzor
Kirovo-Chepetsk *89 D5* Kirovskaya Oblast', NW Russia
Kirovohrad see Kropyvnyts'kyy
Kirovo see Kropyvnyts'kyy
Kīrthar Range *112 B3* mountain range S Pakistan
Kiruna *62 C3* Lapp. Giron. Norrbotten, N Sweden
Kirun/Kirun' see Keelung
Kisalföld see Little Alföld
Kisangani *55 D5* prev. Stanleyville. Orientale, NE Dem. Rep. Congo
Kishinev see Chişinău
Kislovodsk *89 B7* Stavropol'skiy Kray, SW Russia
Kismaayo *51 D6* var. Chisimayu, Kismayu, It. Chisimaio. Jubbada Hoose, S Somalia
Kismayu see Kismaayo
Kíssamos *83 C7* prev. Kastélli. Kríti, Greece, E Mediterranean Sea
Kissidougou *52 C4* Guinée-Forestière, S Guinea
Kissimmee, Lake *21 E4* lake Florida, SE USA
Kistna see Krishna
Kisumu *51 C6* prev. Port Florence. Nyanza, W Kenya
Kisvárda *77 E6* Ger. Kleinwardein. Szabolcs-Szatmár-Bereg, E Hungary
Kita *52 D3* Kayes, W Mali
Kitab see Kitob

Kitakyūshū – Kuril Islands

Kitakyūshū *109 A7 var.* Kitakyûsyû. Fukuoka, Kyūshū, SW Japan
Kitakyûsyû *see* Kitakyūshū
Kitami *108 D2* Hokkaidō, NE Japan
Kitchener *16 C5* Ontario, S Canada
Kíthnos *see* Kýthnos
Kitimat *14 D4* British Columbia, SW Canada
Kitinen *62 D3 river* N Finland
Kitob *101 E3 Rus.* Kitab. Qashqadaryo Viloyati, S Uzbekistan
Kitwe *56 D2 var.* Kitwe-Nkana. Copperbelt, C Zambia
Kitwe-Nkana *see* Kitwe
Kitzbüheler Alpen *73 C7 mountain range* W Austria
Kivalina *14 C2* Alaska, USA
Kivalo *62 D3 ridge* C Finland
Kivertsi *86 C1 Pol.* Kiwerce, *Rus.* Kivertsy. Volyns'ka Oblast', NW Ukraine
Kivertsy *see* Kivertsi
Kivu, Lac *see* Kivu, Lake
Kivu, Lake *55 E6 Fr.* Lac Kivu. *lake* Rwanda/Dem. Rep. Congo
Kiwerce *see* Kivertsi
Kiyev *see* Kyiv
Kiyevskoye Vodokhranilishche *see* Kyyivs'ke Vodoskhovyshche
Kızıl Irmak *94 C3 river* C Turkey
Kizil Kum *see* Kyzyl Kum
Kizyl-Arvat *see* Serdar
Kjølen *see* Kölen
Kladno *77 A5* Středočeský, NW Czech Republic (Czechia)
Klagenfurt *73 D7 Slvn.* Celovec. Kärnten, S Austria
Klaipėda *84 B3 Ger.* Memel. Klaipeda, NW Lithuania
Klamath Falls *24 B4* Oregon, NW USA
Klamath Mountains *24 A4 mountain range* California/Oregon, W USA
Klang *116 B3 var.* Kelang; *prev.* Port Swettenham. Selangor, Peninsular Malaysia
Klarälven *63 B6 river* Norway/Sweden
Klatovy *77 A5 Ger.* Klattau. Plzeňský Kraj, W Czech Republic (Czechia)
Klattau *see* Klatovy
Klausenburg *see* Cluj-Napoca
Klazienaveen *64 E2* Drenthe, NE Netherlands
Kleines Ungarisches Tiefland *see* Little Alföld
Klein Karas *56 B4* Karas, S Namibia
Kleinwardein *see* Kisvárda
Kleisoúra *83 A5* Ípeiros, W Greece
Klerksdorp *56 D4* North-West, N South Africa
Klimavichy *85 E7 Rus.* Klimovichi. Mahilyowskaya Voblasts', E Belarus
Klimovichi *see* Klimavichy
Klintsy *89 A5* Bryanskaya Oblast', W Russia
Klisura *82 C2* Plovdiv, C Bulgaria
Ključ *78 B3* Federacija Bosna I Hercegovina, NW Bosnia and Herzegovina
Kłobuck *76 C4* Śląskie, S Poland
Klosters *73 B7* Graubünden, SE Switzerland
Kluang *see* Keluang
Kluczbork *76 C4 Ger.* Kreuzburg, Kreuzburg in Oberschlesien. Opolskie, S Poland
Klyuchevskaya Sopka, Vulkan *93 H3 volcano* E Russia
Knin *78 B4* Šibenik-Knin, S Croatia
Knjaževac *78 E4* Serbia, E Serbia
Knokke-Heist *65 A5* West-Vlaanderen, NW Belgium
Knoxville *20 D1* Tennessee, S USA
Knud Rasmussen Land *60 D1 physical region* N Greenland
Kobdo *see* Hovd
Kōbe *109 C6* Hyōgo, Honshū, SW Japan
København *63 B7 Eng.* Copenhagen; *anc.* Hafnia. *country capital* (Denmark) Sjælland, København, E Denmark
Kobenni *52 D3* Hodh el Gharbi, S Mauritania
Koblenz *73 A5 prev.* Coblenz, *Fr.* Coblence; *anc.* Confluentes. Rheinland-Pfalz, W Germany
Kobrin *see* Kobryn
Kobryn *85 A6 Rus.* Kobrin. Brestskaya Voblasts', SW Belarus
Kobuleti *95 F2 prev.* K'obulet'i. W Georgia
K'obulet'i *see* Kobuleti
Kočani *79 E6* NE Macedonia
Kočevje *73 D8 Ger.* Gottschee. S Slovenia
Koch Bihār *113 G3* West Bengal, NE India
Kochchi *see* Kochi
Kochi *110 C3 var.* Cochin, Kochchi. Kerela SW India
Kōchi *109 B7 var.* Kôti. Kōchi, Shikoku, SW Japan
Kochiu *see* Gejiu
Kodiak *14 C3* Kodiak Island, Alaska, USA
Kodiak Island *14 C3 island* Alaska, USA
Koedoes *see* Kudus
Koeln *see* Köln
Koepang *see* Kupang
Ko-erh-mu *see* Golmud
Koetai *see* Mahakam, Sungai
Koetaradja *see* Bandaaceh
Kōfu *109 D5 var.* Kôhu. Yamanashi, Honshū, S Japan
Kogarah *126 E2* New South Wales, E Australia
Kogon *100 D2 Rus.* Kagan. Buxoro Viloyati, C Uzbekistan
Kőhalom *see* Rupea
Kohīma *113 H3 state capital* Nāgāland, E India
Koh I Noh *see* Büyükağrı Dağı
Kohtla-Järve *84 E2* Ida-Virumaa, NE Estonia
Kôhu *see* Kōfu
Kokand *see* Qo'qon
Kokchetav *see* Kokshetau
Kokkola *62 D4 Swe.* Karleby; *prev. Swe.* Gamlakarleby. Österbotten, W Finland
Koko *53 F4* Kebbi, W Nigeria
Kokomo *18 C4* Indiana, N USA
Koko Nor *see* Qinghai, China
Koko Nor *see* Qinghai Hu, China
Kokrines *14 C2* Alaska, USA
Kokshaal-Tau *101 G2 Rus.* Khrebet Kakshaal-Too. *mountain range* China/Kyrgyzstan
Kokshetau *92 C4 Kaz.* Kökshetaū; *prev.* Kokchetav. Kokshetau, N Kazakhstan
Kökshetaū *see* Kokshetau
Koksijde *65 A5* West-Vlaanderen, W Belgium
Koksoak *16 D2 river* Québec, E Canada
Kokstad *56 D5* KwaZulu/Natal, E South Africa
Kolaka *117 E4* Sulawesi, C Indonesia
K'o-la-ma-i *see* Karamay
Kola Peninsula *see* Kol'skiy Poluostrov
Kolari *62 D3* Lappi, NW Finland
Kolárovo *77 C6 Ger.* Gutta; *prev.* Guta, *Hung.* Gúta. Nitriansky Kraj, SW Slovakia
Kolberg *see* Kołobrzeg
Kolda *52 C3* S Senegal
Kolding *63 A7* Vejle, C Denmark
Kölen *59 E1 Nor.* Kjølen. *mountain range* Norway/Sweden
Kolguyev, Ostrov *88 C2 island* NW Russia
Kolhāpur *110 B1* Mahārāshtra, SW India
Kolhumadulu *110 A5 var.* Thaa Atoll. *atoll* S Maldives
Kolín *77 B5 Ger.* Kolin. Střední Čechy, C Czech Republic (Czechia)
Kolka *84 C2* Talsi, NW Latvia
Kolkasrags *84 C2 prev. Eng.* Cape Domesnes. *headland* NW Latvia
Kolkata *113 G4 prev.* Calcutta. West Bengal, N India
Kollam *110 C3 var.* Quilon. Kerala, SW India
Kolmar *see* Colmar
Köln *72 A4 var.* Koeln, *Eng./Fr.* Cologne, *prev.* Cöln; *anc.* Colonia Agrippina, Oppidum Ubiorum. Nordrhein-Westfalen, W Germany
Koło *76 C3* Wielkopolskie, C Poland
Kołobrzeg *76 B2 Ger.* Kolberg. Zachodnio-pomorskie, NW Poland
Kolokani *52 D3* Koulikoro, W Mali
Kolomea *see* Kolomyya
Kolomna *89 B5* Moskovskaya Oblast', W Russia
Kolomyya *86 C3 Ger.* Kolomea. Ivano-Frankivs'ka Oblast', W Ukraine
Kolosjoki *see* Nikel'
Kolozsvár *see* Cluj-Napoca
Kolpa *78 A2 Ger.* Kulpa, *SCr.* Kupa. *river* Croatia/Slovenia
Kolpino *88 B4* Leningradskaya Oblast', NW Russia
Kólpos Mórfu *see* Güzelyurt Körfezi
Kol'skiy Poluostrov *88 C2 Eng.* Kola Peninsula. *peninsula* NW Russia
Kolwezi *55 D7* Katanga, S Dem. Rep. Congo
Kolyma *93 G2 river* NE Russia
Komatsu *109 C5 var.* Komatu. Ishikawa, Honshū, SW Japan
Komatu *see* Komatsu
Kommunizm, Qullai *see* Ismoili Somoní, Qullai
Komoé *53 E4 var.* Komoé Fleuve. *river* E Ivory Coast
Komoé Fleuve *see* Komoé
Komotau *see* Chomutov
Komotiní *82 D3 var.* Gümüljina, *Turk.* Gümülcine. Anatolikí Makedonía kai Thráki, NE Greece
Kompong Som *see* Sihanoukville
Komrat *see* Comrat
Komsomolets, Ostrov *93 E1 island* Severnaya Zemlya, N Russia
Komsomol'sk-na-Amure *93 G4* Khabarovskiy Kray, SE Russia
Kondolovo *82 E3* Burgas, E Bulgaria
Kondopoga *88 B3* Respublika Kareliya, NW Russia
Kondoz *see* Kunduz
Köneürgenç *100 C2 var.* Köneürgench, *Rus.* Këneurgench; *prev.* Kunya-Urgench. Daşoguz Welaýaty, N Turkmenistan
Kong Christian IX Land *60 D4 Eng.* King Christian IX Land. *physical region* SE Greenland
Kong Frederik IX Land *60 C3 physical region* SW Greenland
Kong Frederik VIII Land *61 E2 Eng.* King Frederik VIII Land. *physical region* NE Greenland
Kong Frederik VI Kyst *60 C4 Eng.* King Frederik VI Coast. *physical region* SE Greenland
Kong Karls Land *61 G2 Eng.* King Charles Islands. *island group* SE Svalbard
Kongo *see* Congo (river)
Kongolo *55 D6* Katanga, E Dem. Rep. Congo
Kongor *51 B5* Jonglei, E South Sudan
Kong Oscar Fjord *61 E3 fjord* E Greenland
Kongsberg *63 B6* Buskerud, S Norway
Kông, Tônle *115 E5 var.* Xê Kong. *river* Cambodia/Laos
Kong, Xê *see* Kông, Tônle
Königgrätz *see* Hradec Králové
Königshütte *see* Chorzów
Konin *76 C3 Ger.* Kuhnau. Weilkopolskie, C Poland
Koninkrijk der Nederlanden *see* Netherlands
Konispol *79 C7 var.* Konispoli. Vlorë, S Albania
Konispoli *see* Konispol
Kónitsa *82 A4* Ípeiros, W Greece
Konitz *see* Chojnice
Konjic *78 C4* Federacija Bosna I Hercegovina, S Bosnia and Herzegovina
Konosha *88 C4* Arkhangel'skaya Oblast', NW Russia
Konotop *87 F1* Sums'ka Oblast', NE Ukraine
Konstantinovka *see* Kostyantynivka
Konstanz *73 B7 var.* Constanz, *Eng.* Constance, *hist.* Kostnitz; *anc.* Constantia. Baden-Württemberg, S Germany
Konstanza *see* Constanța
Konya *94 C4 var.* Konieh, *prev.* Konia; *anc.* Iconium. Konya, C Turkey
Kopaonik *79 D5 mountain range* S Serbia
Kopar *see* Koper
Koper *73 D8 It.* Capodistria; *prev.* Kopar. SW Slovenia
Köpetdag Gershi *100 C3 mountain range* Iran/Turkmenistan
Köpetdag Gershi/Kopetdag, Khrebet *see* Koppeh Dāgh
Koppeh Dagh *98 D2 Rus.* Khrebet Kopetdag, *Turkm.* Köpetdag Gershi. *mountain range* Iran/Turkmenistan
Kopreinitz *see* Koprivnica
Koprivnica *78 B2 Ger.* Kopreinitz, *Hung.* Kaproncza. Koprivnica-Križevci, N Croatia
Köprülü *see* Veles
Koptsevichi *see* Kaptsevichy
Kopyl' *see* Kapyl'
Korat *see* Nakhon Ratchasima
Korat Plateau *114 D4 plateau* E Thailand
Korba *113 E4* Chhattīsgarh, C India
Korça *see* Korçë
Korçë *79 D6 var.* Korça, *Gk.* Korytsa, *It.* Corriza; *prev.* Koritsa. Korçë, SE Albania
Korčula *78 B4 It.* Curzola; *anc.* Corcyra Nigra. *island* S Croatia
Korea Bay *105 G3 bay* China/North Korea
Korea, Democratic People's Republic of *see* North Korea
Korea, Republic of *see* South Korea
Korea Strait *109 A7 Jap.* Chōsen-kaikyō, *Kor.* Taehan-haehyŏp. *channel* Japan/South Korea
Korhogo *52 D4* N Ivory Coast
Kórinthos *83 B6 anc.* Corinthus *Eng.* Corinth. Pelopónnisos, S Greece
Korinthiakós Kólpos *83 B5 Eng.* Gulf of Corinth; *anc.* Corinthiacus Sinus. *gulf* C Greece
Koritsa *see* Korçë
Kōriyama *109 D5* Fukushima, Honshū, C Japan
Korla *104 C3 Chin.* K'u-erh-lo. Xinjiang Uygur Zizhiqu, NW China
Körmend *77 B7* Vas, W Hungary
Koróni *83 B6* Pelopónnisos, S Greece
Koror *122 A2* (Palau) Oreor, N Palau
Kőrös *see* Križevci
Korosten' *86 D1* Zhytomyrs'ka Oblast', NW Ukraine
Koro Toro *54 C2* Borkou-Ennedi-Tibesti, N Chad
Korsovka *see* Kārsava
Kortrijk *65 A6 Fr.* Courtrai. West-Vlaanderen, W Belgium
Koryak Range *93 H2 var.* Koryakskiy Khrebet, *Eng.* Koryak Range. *mountain range* NE Russia
Koryak Range *see* Koryakskoye Nagor'ye
Koryakskiy Khrebet *see* Koryakskoye Nagor'ye
Koryazhma *88 C4* Arkhangel'skaya Oblast', NW Russia
Korytsa *see* Korçë
Kos *83 E6* Kos, Dodekánisa, Greece, Aegean Sea
Kos *83 E6 It.* Coo; *anc.* Cos. *island* Dodekánisa, Greece, Aegean Sea
Ko-saki *109 A7 headland* Nagasaki, Tsushima, SW Japan
Kościan *76 B4 Ger.* Kosten. Wielkopolskie, C Poland
Kościerzyna *76 C2* Pomorskie, NW Poland
Kosciusko, Mount *see* Kosciuszko, Mount
Kosciuszko, Mount *127 C7 prev.* Mount Kosciusko. *mountain* New South Wales, SE Australia
K'o-shih *see* Kashi
Koshikijima-retto *109 A8 var.* Kosikizima Rettô. *island group* SW Japan
Kōshū *see* Gwangju
Košice *77 D6 Ger.* Kaschau, *Hung.* Kassa. Košický Kraj, E Slovakia
Kosikizima Rettô *see* Koshikijima-rettō
Köslin *see* Koszalin
Koson *101 E3 Rus.* Kasan. Qashqadaryo Viloyati, S Uzbekistan
Kosovo *79 D5 prev.* Autonomous Province of Kosovo and Metohija. *country (not fully recognised)* SE Europe
Kosovo and Metohija, Autonomous Province of *see* Kosovo
Kosovo Polje *see* Fushë Kosovë
Kosovska Mitrovica *see* Mitrovicë
Kosrae *122 C2 prev.* Kusaie. *island* Caroline Islands, E Micronesia
Kossou, Lac de *52 D5 lake* C Ivory Coast
Kostanay *92 C4 var.* Kustanay, *Kaz.* Qostanay. Kostanay, N Kazakhstan
Kosten *see* Kościan
Kostenets *82 C2 prev.* Georgi Dimitrov. Sofiya, W Bulgaria
Kostnitz *see* Konstanz
Kostroma *88 B4* Kostromskaya Oblast', NW Russia
Kostyantynivka *87 G3 Rus.* Konstantinovka. Donets'ka Oblast', SE Ukraine
Kostyukovichi *see* Kastsyukovichy
Kostyukovka *see* Kastsyukowka
Koszalin *76 B2 Ger.* Köslin. Zachodnio-pomorskie, NW Poland
Kota *112 D3 prev.* Kotah. Rājasthān, N India
Kota Baharu *see* Kota Bharu
Kota Bahru *see* Kota Bharu
Kotabaru *see* Jayapura
Kota Bharu *116 B3 var.* Kota Baharu, Kota Bahru. Kelantan, Peninsular Malaysia
Kotaboemi *see* Kotabumi
Kotabumi *116 B4 prev.* Kotaboemi. Sumatera, W Indonesia
Kotah *see* Kota
Kota Kinabalu *116 D3 prev.* Jesselton. Sabah, East Malaysia
Kotel'nyy, Ostrov *93 E2 island* Novosibirskiye Ostrova, N Russia
Kotka *63 E5* Kymenlaakso, S Finland
Kotlas *88 C4* Arkhangel'skaya Oblast', NW Russia
Kotonu *see* Cotonou
Kotor *79 C5 It.* Cattaro. SW Montenegro
Kotovs'k *see* Podil's'k
Kotovsk *see* Hîncești
Kottbus *see* Cottbus
Kotto *54 D4 river* Central African Republic/Dem. Rep. Congo
Kotuy *93 E2 river* N Russia
Koudougou *53 E4* C Burkina
Koulamoutou *55 B6* Ogooué-Lolo, C Gabon
Koulikoro *52 D3* Koulikoro, SW Mali
Koumra *54 C4* Moyen-Chari, S Chad
Kourou *37 H3* N French Guiana
Kousseir *see* Al Quṣayr
Kousséri *54 B3 prev.* Fort-Foureau. Extrême-Nord, NE Cameroon
Koutiala *52 D4* Sikasso, S Mali
Kouvola *63 E5* Kymenlaakso, S Finland
Kovel' *86 C1 Pol.* Kowel. Volyns'ka Oblast', NW Ukraine
Kovno *see* Kaunas
Koweit *see* Kuwait
Kowel *see* Kovel'
Kowloon *106 A2* Hong Kong, S China
Kowno *see* Kaunas
Kozáni *82 B4* Dytikí Makedonía, N Greece
Kozara *78 B3 mountain range* NW Bosnia and Herzegovina
Kozarska Dubica *see* Bosanska Dubica
Kozhikode *110 C2 var.* Calicut. Kerala, SW India
Kozu-shima *109 D6 island* E Japan
Kozyatyn *86 D2 Rus.* Kazatin. Vinnyts'ka Oblast', C Ukraine
Kpalimé *53 E5 var.* Palimé. SW Togo
Krâchéh *see* Kratie
Kragujevac *78 D4* Serbia, C Serbia
Krainburg *see* Kranj
Kra, Isthmus of *115 B6 isthmus* Malaysia/Thailand
Krakau *see* Kraków
Kraków *77 D5 Eng.* Cracow, *Ger.* Krakau; *anc.* Cracovia. Małopolskie, S Poland
Krâlănh *115 D5* Siĕmréab, NW Cambodia
Kralendijk *33 E5 dependent territory capital* (Bonaire) Lesser Antilles, S Caribbean Sea
Kraljevo *78 D4 prev.* Rankovićevo. Serbia, C Serbia
Kramators'k *87 G3 Rus.* Kramatorsk. Donets'ka Oblast', SE Ukraine
Kramatorsk *see* Kramators'k
Kramfors *63 C5* Västernorrland, C Sweden
Kranéa *see* Kraniá
Kraniá *82 B4 var.* Kranéa. Dytikí Makedonía, N Greece
Kranj *73 D7 Ger.* Krainburg. NW Slovenia
Kranz *see* Zelenogradsk
Krāslava *84 D4* Krāslava, SE Latvia
Krasnaye *85 C5 Rus.* Krasnoye. Minskaya Voblasts', C Belarus
Krasnoarmeysk *89 C6* Saratovskaya Oblast', W Russia
Krasnodar *89 A7 prev.* Ekaterinodar, Yekaterinodar. Krasnodarskiy Kray, SW Russia
Krasnodon *see* Sorokyne
Krasnogor *see* Kallaste
Krasnogvardeyskoye *see* Krasnohvardiys'ke
Krasnohvardiys'ke *87 F4 Rus.* Krasnogvardeyskoye. Respublika Krym, S Ukraine
Krasnokamensk *93 F4* Chitinskaya Oblast', S Russia
Krasnokamsk *89 D5* Permskaya Oblast', W Russia
Krasnoperekops'k *see* Yany Kapu
Krasnostav *see* Krasnystaw
Krasnovodsk *see* Türkmenbaşy
Krasnovodskiy Zaliv *see* Türkmenbasy Aylagy
Krasnowodsk Aylagy *see* Türkmenbasy Aylagy
Krasnoyarsk *92 D4* Krasnoyarskiy Kray, S Russia
Krasnoye *see* Krasnaye
Krasnystaw *76 E4 Rus.* Krasnostav. Lubelskie, SE Poland
Krasnyy Kut *89 C6* Saratovskaya Oblast', W Russia
Krasnyy Luch *see* Khrustal'nyy
Kratie *115 D6 Khmer.* Krâchéh. Kratie, E Cambodia
Krâvanh, Chuŏr Phnum *115 C6 Eng.* Cardamom Mountains, *Fr.* Chaîne des Cardamomes. *mountain range* W Cambodia
Krefeld *72 A4* Nordrhein-Westfalen, W Germany
Kreisstadt *see* Krosno Odrzańskie
Kremenchug *see* Kremenchuk
Kremenchugskoye Vodokhranilishche/Kremenchuk Reservoir *see* Kremenchuts'ke Vodoskhovyshche
Kremenchuk *87 F2 Rus.* Kremenchug. Poltavs'ka Oblast', NE Ukraine
Kremenchuk Reservoir *87 F2 Eng.* Kremenchuk Reservoir, *Rus.* Kremenchugskoye Vodokhranilishche. *reservoir* C Ukraine
Kremenets' *86 C2 Pol.* Krzemieniec, *Rus.* Kremenets. Ternopil's'ka Oblast', W Ukraine
Kremennaya *see* Kreminna
Kreminna *87 G2 Rus.* Kremennaya. Luhans'ka Oblast', E Ukraine
Kresena *see* Kresna
Kresna *82 C3 var.* Kresena. Blagoevgrad, SW Bulgaria
Kretikon Delagos *see* Kritikó Pélagos
Kretinga *84 B3 Ger.* Krottingen. Klaipėda, NW Lithuania
Kreutz *see* Cristuru Secuiesc
Kreuz *see* Križevci, Croatia
Kreuz *see* Risti, Estonia
Kreuzburg/Kreuzburg in Oberschlesien *see* Kluczbork
Krichëv *see* Krychaw
Krievija *see* Russia
Krindachevka *see* Khrustal'nyy
Krishna *110 C1 prev.* Kistna. *river* C India
Krishnagiri *110 C2* Tamil Nādu, SE India
Kristiania *see* Oslo
Kristiansand *63 A6 var.* Christiansand. Vest-Agder, S Norway
Kristianstad *63 B7* Skåne, S Sweden
Kristiansund *62 A4 var.* Christiansund. Møre og Romsdal, S Norway
Kríti *83 C7 Eng.* Crete. *island* Greece, Aegean Sea
Kritikó Pélagos *83 D7 var.* Kretikon Delagos, *Eng.* Sea of Crete; *anc.* Mare Creticum. *sea* Greece, Aegean Sea
Krivoy Rog *see* Kryvyy Rih
Križevci *78 B2 Ger.* Kreuz, *Hung.* Kőrös. Varaždin, NE Croatia
Krk *78 A3 It.* Veglia; *anc.* Curieta. *island* NW Croatia
Kroatien *see* Croatia
Krolevets' *87 F1 Rus.* Krolevets. Sums'ka Oblast', NE Ukraine
Krolevets *see* Krolevets'
Królewska Huta *see* Chorzów
Kronach *73 C5* Bayern, E Germany
Kronstadt *see* Brașov
Kroonstad *56 D4* Free State, C South Africa
Kropotkin *89 A7* Krasnodarskiy Kray, SW Russia
Kropyvnyts'kyy *87 E3 Rus.* Kirovohrad; *prev.* Kirovo, Yelizavetgrad, Zinov'yevsk. Kirovohrads'ka Oblast', C Ukraine
Krosno *77 D5 Ger.* Krossen. Podkarpackie, SE Poland
Krosno Odrzańskie *76 B3 Ger.* Crossen, Kreisstadt. Lubuskie, W Poland
Krossen *see* Krosno
Krottingen *see* Kretinga
Krško *73 E8 Ger.* Gurkfeld; *prev.* Videm-Krško. E Slovenia
Krugloye *see* Kruhlaye
Kruhlaye *85 D6 Rus.* Krugloye. Mahilyowskaya Voblasts', E Belarus
Kruja *see* Krujë
Krujë *79 C6 var.* Kruja, *It.* Croia. Durrës, C Albania
Krummau *see* Český Krumlov
Krung Thep, Ao *115 C5 var.* Bight of Bangkok. *bay* S Thailand
Krung Thep Mahanakhon *see* Ao Krung Thep
Krupki *85 D6* Minskaya Voblasts', C Belarus
Krušné Hory *see* Erzgebirge
Krychaw *85 E7 Rus.* Krichëv. Mahilyowskaya Voblasts', E Belarus
Kryms'ki Hory *87 F5 mountain range* S Ukraine
Kryms'kyy Pivostriv *87 F5 Eng.* Crimea. (Ukrainian territory annexed by Russia since 2014). *peninsula* S Ukraine
Krynica *77 D5 Ger.* Tannenhof. Małopolskie, S Poland
Kryve Ozero *87 E3* Odes'ka Oblast', SW Ukraine
Kryvyy Rih *87 F3 Rus.* Krivoy Rog. Dnipropetrovs'ka Oblast', SE Ukraine
Krzemieniec *see* Kremenets'
Ksar al Kabir *see* Ksar-el-Kebir
Ksar al Soule *see* Er-Rachidia
Ksar-el-Kebir *48 C2 var.* Alcázar, Ksar al Kabir, Ksar-el-Kébir, *Ar.* Al-Kasr al-Kebir, Al-Qsar al-Kbir, *Sp.* Alcazarquivir. NW Morocco
Ksar-el-Kébir *see* Ksar-el-Kebir
Kuala Dungun *see* Dungun
Kuala Lumpur *116 B3 country capital* (Malaysia) Kuala Lumpur, Peninsular Malaysia
Kuala Terengganu *116 B3 var.* Kuala Trengganu. Terengganu, Peninsular Malaysia
Kualatungkal *116 B4* Sumatera, W Indonesia
Kuang-chou *see* Guangzhou
Kuang-hsi *see* Guangxi Zhuangzu Zizhiqu
Kuang-tung *see* Guangdong
Kuang-yuan *see* Guangyuan
Kuantan *116 B3* Pahang, Peninsular Malaysia
Kuba *see* Quba
Kuban' *87 G5 var.* Hypanis. *river* SW Russia
Kubango *see* Cubango/Okavango
Kuching *116 C3 prev.* Sarawak. Sarawak, East Malaysia
Kūchnay Darwāshān *100 D5 prev.* Kūchnay Darweyshān. Helmand, S Afghanistan
Kūchnay Darweyshān *see* Kūchnay Darwāshān
Kuçova *see* Kuçovë
Kuçovë *79 C6 var.* Kuçova; *prev.* Qyteti Stalin. Berat, C Albania
Kudara *see* Ghŭdara
Kudus *116 C5 prev.* Koedoes. Jawa, C Indonesia
Kuei-lin *see* Guilin
Kuei-Yang/Kuei-yang *see* Guiyang
K'u-erh-lo *see* Korla
Kueyang *see* Guiyang
Kugaaruk *15 G3 prev.* Pelly Bay. Nunavut, N Canada
Kugluktuk *31 E3 var.* Qurlurtuuq; *prev.* Coppermine. Nunavut, NW Canada
Kuhmo *62 E4* Kainuu, E Finland
Kuhnau *see* Konin
Kühnö *see* Kihnu
Kuibyshev *see* Kuybyshevskoye Vodokhranilishche
Kuito *56 B2 Port.* Silva Porto. Bié, C Angola
Kuji *108 D3 var.* Kuzi. Iwate, Honshū, C Japan
Kukës *79 D5 var.* Kukësi. Kukës, NE Albania
Kukësi *see* Kukës
Kukong *see* Shaoguan
Kukukhoto *see* Hohhot
Kula Kangri *113 G3 var.* Kulhakangri. *mountain* Bhutan/China
Kuldīga *84 B3 Ger.* Goldingen. Kuldīga, W Latvia
Kuldja *see* Yining
Kulhakangri *see* Kula Kangri
Kullorsuaq *60 D2 var.* Kuvdlorssuak. Kitaa, C Greenland
Kulm *see* Chełmno
Kulmsee *see* Chełmża
Kŭlob *101 F3 Rus.* Kulyab. SW Tajikistan
Kulpa *see* Kolpa
Kulu *94 C3* Konya, W Turkey
Kulunda Steppe *92 C4 Kaz.* Qulyndy Zhazyghy, *Rus.* Kulundinskaya Ravnina. *grassland* Kazakhstan/Russia
Kulundinskaya Ravnina *see* Kulunda Steppe
Kulyab *see* Kŭlob
Kum *see* Qom
Kuma *89 B7 river* SW Russia
Kumamoto *109 A7* Kumamoto, Kyūshū, SW Japan
Kumanova *see* Kumanovo
Kumanovo *79 E5 Turk.* Kumanova. N Macedonia
Kumasi *53 E5 prev.* Coomassie. C Ghana
Kumayri *see* Gyumri
Kumba *55 A5* Sud-Ouest, W Cameroon
Kumertau *89 D6* Respublika Bashkortostan, W Russia
Kumillā *see* Comilla
Kumo *53 G4* Gombe, E Nigeria
Kumon Range *114 B2 mountain range* N Myanmar (Burma)
Kumul *see* Hami
Kunashiri *see* Kunashir, Ostrov
Kunashir, Ostrov *108 E1 var.* Kunashiri. *island* Kuril'skiye Ostrova, SE Russia
Kunda *84 E2* Lääne-Virumaa, NE Estonia
Kunduz *101 E3 prev.* Kondoz. NE Afghanistan
Kunene *47 C6 var.* Kunene. *river* Angola/Namibia
Kunene *see* Cunene
Kungsbacka *63 B7* Halland, S Sweden
Kungur *89 D5* Permskaya Oblast', NW Russia
Kunlun Mountains *see* Kunlun Shan
Kunlun Shan *104 B4 Eng.* Kunlun Mountains. *mountain range* NW China
Kunming *106 B6 var.* K'un-ming; *prev.* Yunnan. *province capital* Yunnan, SW China
K'un-ming *see* Kunming
Kununurra *124 D3* Western Australia
Kunya-Urgench *see* Köneürgenç
Kuopio *63 E5* Pohjois-Savo, C Finland
Kupa *see* Kolpa
Kupang *117 E5 prev.* Koepang. Timor, C Indonesia
Kup"yans'k *87 G2 Rus.* Kupyansk. Kharkivs'ka Oblast', E Ukraine
Kupyansk *see* Kup"yans'k
Kür *see* Kura
Kura *95 H3 Az.* Kür, *Geor.* Mtkvari, *Turk.* Kura Nehri. *river* SW Asia
Kura Nehri *see* Kura
Kurashiki *109 B6 var.* Kurasiki. Okayama, Honshū, SW Japan
Kurasiki *see* Kurashiki
Kurdistan *95 F4 cultural region* SW Asia
Kŭrdzhali *82 D3 var.* Kirdzhali. Kŭrdzhali, S Bulgaria
Kure *109 B7* Hiroshima, Honshū, SW Japan
Küre Dağları *94 C2 mountain range* N Turkey
Kuressaare *84 C2 Ger.* Arensburg; *prev.* Kingissepp. Saaremaa, W Estonia
Kureyka *90 D2 river* N Russia
Kurgan-Tyube *see* Qŭrghonteppa
Kuria Muria Islands *see* Ḥalāniyāt, Juzur al
Kuril'skiye Ostrova *93 H4 Eng.* Kuril Islands. *island group* SE Russia
Kuril Islands *see* Kuril'skiye Ostrova

Kuril-Kamchatka Depression – Lewis, Isle of

Kuril-Kamchatka Depression see Kuril-Kamchatka Trench
Kuril-Kamchatka Trench *91 F3* var. Kuril-Kamchatka Depression. trench NW Pacific Ocean
Kuril'sk *108 E1* Jap. Shana. Kuril'skiye Ostrova, Sakhalinskaya Oblats', SE Russia
Ku-ring-gai *126 E1* New South Wales, E Australia
Kurisches Haff see Courland Lagoon
Kurkund see Kilingi-Nõmme
Kurnool *110 C1* var. Karnul. Andhra Pradesh, S India
Kursk *89 A6* Kurskaya Oblast', W Russia
Kurskiy Zaliv see Courland Lagoon
Kuršumlija *79 D5* Serbia, S Serbia
Kurtbunar see Tervel
Kurtitsch/Kürtös see Curtici
Kuruktag *104 C3* mountain range NW China
Kurume *109 A7* Fukuoka, Kyūshū, SW Japan
Kurupukari *37 F3* C Guyana
Kusaie see Kosrae
Kushiro *108 D2* var. Kusiro. Hokkaidō, NE Japan
Kushka see Serhetabat
Kusiro see Kushiro
Kuskokwim Mountains *14 C3* mountain range Alaska, USA
Kustanay see Kostanay
Küstence/Küstendje see Constanța
Kütahya *94 B3* prev. Kutaia. Kütahya, W Turkey
Kutai see Mahakam, Sungai
Kutaisi *95 F2* prev. K'ut'aisi *95 F2* W Georgia W Georgia
K'ut'aisi see Kutaisi
Kūt al 'Amārah see Al Kūt
Kut al Imara see Al Kūt
Kutaradja/Kutaraja see Bandaaceh
Kutch, Gulf of see Kachchh, Gulf of
Kutch, Rann of see Kachchh, Rann of
Kutina *78 B3* Sisak-Moslavina, NE Croatia
Kutno *76 C3* Łódzkie, C Poland
Kuujjuaq *17 E2* prev. Fort-Chimo. Québec, E Canada
Kuusamo *62 E3* Pohjois-Pohjanmaa, E Finland
Kuvango see Cubango
Kuvdlorssuak see Kullorsuaq
Kuwait *98 C4* off. State of Kuwait, var. Dawlat al Kuwait, Koweit, Kuweit. country SW Asia
Kuwait see Al Kuwayt
Kuwait City see Al Kuwayt
Kuwait, Dawlat al see Kuwait
Kuwait, State of see Kuwait
Kuwajleen see Kwajalein Atoll
Kuwayt *98 C3* Maysān, E Iraq
Kuweit see Kuwait
Kuybyshev see Samara
Kuybyshev Reservoir see Kuybyshevskoye Vodokhranilishche
Kuybyshevskoye Vodokhranilishche *89 C5* var. Kuibyshev, Eng. Kuybyshev Reservoir. reservoir W Russia
Kuytun *104 B2* Xinjiang Uygur Zizhiqu, NW China
Kuzi see Kuji
Kuznetsk *89 B6* Penzenskaya Oblast', W Russia
Kuźnica *76 E2* Białystok, NE Poland Europe
Kvaløya *62 C2* island N Norway
Kvarnbergsvattnet *62 B4* var. Frostviken. lake N Sweden
Kvarner *78 A3* var. Carnaro, It. Quarnero. gulf W Croatia
Kvitøya *61 G1* island NE Svalbard
Kwajalein Atoll *122 C1* var. Kuwajleen. atoll Ralik Chain, C Marshall Islands
Kwando see Cuando
Kwangchow see Guangzhou
Kwangchu see Gwangju
Kwangju see Gwangju
Kwango *55 C7* Port. Cuango. river Angola/Dem. Rep. Congo
Kwangsi/Kwangsi Chuang Autonomous Region see Guangxi Zhuangzu Zizhiqu
Kwangtung see Guangdong
Kwangyuan see Guangyuan
Kwanza see Cuanza
Kweichu see Guiyang
Kweilin see Guilin
Kweisui see Hohhot
Kweiyang see Guiyang
Kwekwe *56 D3* prev. Que Que. Midlands, C Zimbabwe
Kwesui see Hohhot
Kwidzyń *76 C2* Ger. Marienwerder. Pomorskie, N Poland
Kwigillingok *14 C3* Alaska, USA
Kwilu *55 C6* river W Dem. Rep. Congo
Kwito see Cuito
Kyabé *54 C4* Moyen-Chari, S Chad
Kyaikkami *115 B5* prev. Amherst. Mon State, S Myanmar (Burma)
Kyaiklat *114 B4* Ayeyarwady, SW Myanmar (Burma)
Kyaikto *114 B4* Mon State, S Myanmar (Burma)
Kyakhta *93 E5* Respublika Buryatiya, S Russia
Kyaukse *114 B3* Mandalay, C Myanmar (Burma)
Kyiv *87 E2* var. Kyyiv, Eng. Kiev, Rus. Kiyev. country capital (Ukraine) Kyyivs'ka Oblast', N Ukraine
Kyjov *77 C5* Ger. Gaya. Jihomoravský Kraj, SE Czech Republic (Czechia)
Kykládes *83 D6* var. Kikládhes, Eng. Cyclades. island group SE Greece
Kými *83 C5* prev. Kími. Évvoia, C Greece
Kyŏngsŏng see Seoul
Kyōto *109 C6* Kyōto, Honshū, SW Japan
Kyparissía *83 B6* var. Kiparissía. Pelopónnisos, S Greece
Kypros see Cyprus
Kyrá Panagía *83 C5* island Vóreies Sporádes, Greece, Aegean Sea
Kyrenia see Girne
Kyrgyz Republic see Kyrgyzstan
Kyrgyzstan *101 F2* off. Kyrgyz Republic, var. Kirghizia; prev. Kirgizskaya SSR, Kirghiz SSR, Republic of Kyrgyzstan. country C Asia
Kyrgyzstan, Republic of see Kyrgyzstan
Kythira *83 C7* var. Kíthira, It. Cerigo, Lat. Cythera. island S Greece
Kýthnos *83 C6* Kńythnos, Kykládes, Greece, Aegean Sea
Kythnos *83 C6* var. Kíthnos, Thermiá, It. Termia; anc. Cythnos. island Kykládes, Greece, Aegean Sea
Kythréa see Değirmenlik
Kyushu *109 B7* var. Kyūsyū. island SW Japan
Kyushu-Palau Ridge *103 F3* var. Kyusyu-Palau Ridge. undersea ridge W Pacific Ocean
Kyustendil *82 B2* anc. Pautalia. Kyustendil, W Bulgaria
Kyûsyû see Kyūshū
Kyusyu-Palau Ridge see Kyushu-Palau Ridge
Kyyiv see Kyiv
Kyyivs'ke Vodoskhovyshche *87 E1* Eng. Kiev Reservoir, Rus. Kiyevskoye Vodokhranilishche. reservoir N Ukraine
Kyzyl *92 D4* Respublika Tyva, C Russia
Kyzyl Kum *100 D2* var. Kizil Kum, Qizil Qum, Uzb. Qizilqum. desert Kazakhstan/Uzbekistan
Kyzylrabot see Qizilrabot
Kyzyl-Suu *101 G2* prev. Pokrovka. Issyk-Kul'skaya Oblast', NE Kyrgyzstan
Kzylorda *92 B5* var. Kzyl-Orda, Qizil Orda, Qyzylorda; prev. Perovsk. Kyzylorda, S Kazakhstan
Kzyl-Orda see Kzylorda

L

Laaland see Lolland
La Algaba *70 C4* Andalucía, S Spain
Laarne *65 B5* Oost-Vlaanderen, NW Belgium
La Asunción *37 E1* Nueva Esparta, NE Venezuela
Laatokka see Ladozhskoye, Ozero
Laâyoune *48 B3* var. Aaiún. country capital (Western Sahara) NW Western Sahara
La Banda Oriental see Uruguay
la Baule-Escoublac *68 A4* Loire-Atlantique, NW France
Labé *52 C4* NW Guinea
Labe see Elbe
Laborca see Laborec
Laborec *77 E5* Hung. Laborca. river E Slovakia
Labrador *17 F2* cultural region Newfoundland and Labrador, SW Canada
Labrador Basin *12 E3* var. Labrador Sea Basin. undersea basin Labrador Sea
Labrador Sea *60 A4* sea NW Atlantic Ocean
Labrador Sea Basin see Labrador Basin
Labudalin see Ergun
Labutta *115 A5* Ayeyarwady, SW Myanmar (Burma)
Laç *79 C6* var. Laci. Lezhë, C Albania
La Calera *42 B4* Valparaíso, C Chile
La Carolina *70 D4* Andalucía, S Spain
Laccadive Islands *110 A3* Eng. Laccadive Islands. island group India, N Indian Ocean
Laccadive Islands/Laccadive Minicoy and Amindivi Islands, the see Lakshadweep
La Ceiba *30 D2* Atlántida, N Honduras
Lachanás *82 B3* Kentrikí Makedonía, N Greece
La Chaux-de-Fonds *73 A7* Neuchâtel, W Switzerland
Lachlan River *127 C6* river New South Wales, SE Australia
Laci see Laç
la Ciotat *69 D6* anc. Citharista. Bouches-du-Rhône, SE France
Lacobriga see Lagos
La Concepción *31 E5* var. Concepción. Chiriquí, W Panama
La Concepción *36 C1* Zulia, NW Venezuela
La Condamine *69 C8* W Monaco
Laconia *19 G2* New Hampshire, NE USA
La Crosse *18 A2* Wisconsin, N USA
La Cruz *30 D4* Guanacaste, NW Costa Rica
Ladoga, Lake see Ladozhskoye, Ozero
Ladozhskoye, Ozero *88 B3* Eng. Lake Ladoga, Fin. Laatokka. lake NW Russia
Ladysmith *18 B2* Wisconsin, N USA
Lae *122 B3* Morobe, W Papua New Guinea
La Esperanza *30 C2* Intibucá, SW Honduras
Lafayette *18 C4* Indiana, N USA
Lafayette *20 B3* Louisiana, S USA
La Fé *32 A2* Pinar del Río, W Cuba
Lafia *53 G4* Nassarawa, C Nigeria
la Flèche *68 B4* Sarthe, NW France
Lagdo, Lac de *54 B4* lake N Cameroon
Laghouat *48 D2* N Algeria
Lagos *53 F5* Lagos, SW Nigeria
Lagos *70 B5* anc. Lacobriga. Faro, S Portugal
Lagos de Moreno *29 E4* Jalisco, SW Mexico
Lagouira *48 A4* SW Western Sahara
La Grande *24 C3* Oregon, NW USA
La Guaira *44 B4* Distrito Federal, N Venezuela
Lagunas *42 B1* Tarapacá, N Chile
Lagunillas *39 G4* Santa Cruz, SE Bolivia
La Habana *32 B2* var. Havana. country capital (Cuba) Ciudad de La Habana, W Cuba
Lahat *116 B4* Sumatera, W Indonesia
La Haye see 's-Gravenhage
Laholm *63 B7* Halland, S Sweden
Lahore *112 D2* Punjab, NE Pakistan
Lahr *73 A6* Baden-Württemberg, S Germany
Lahti *63 D5* Swe. Lahtis. Päijät-Häme, S Finland
Lahtis see Lahti
Laï *54 B4* prev. Behagle, De Behagle. Tandjilé, S Chad
Laibach see Ljubljana
Lai Châu *114 D3* Lai Châu, N Vietnam
Laila see Laylā
La Junta *22 D5* Colorado, C USA
Lake Charles *20 A3* Louisiana, S USA
Lake City *21 E3* Florida, SE USA
Lake District *67 C5* physical region NW England, United Kingdom
Lake Havasu City *26 A2* Arizona, SW USA
Lake Jackson *27 H4* Texas, SW USA
Lakeland *21 E4* Florida, SE USA
Lakeside *25 C8* California, W USA
Lake State see Michigan
Lakewood *22 D4* Colorado, C USA
Lakhnau see Lucknow
Lakonikós Kólpos *83 B7* gulf S Greece
Lakselv *62 D2* Lapp. Leavdnja. Finnmark, N Norway
la Laon see Laon
Lalibela *50 C4* Āmara, Ethiopia
La Libertad *30 B1* Petén, N Guatemala
La Ligua *42 B4* Valparaíso, C Chile
Lalín *70 C1* Galicia, NW Spain
Lalitpur *113 F3* Central, C Nepal
La Louvière *65 B6* Hainaut, S Belgium
la Maddalena *74 A4* Sardegna, Italy, C Mediterranean Sea
la Manche see English Channel
Lamar *22 D5* Colorado, C USA
La Marmora, Punta *75 A5* mountain Sardegna, Italy, C Mediterranean Sea
La Massana *69 A8* La Massana, W Andorra Europe
Lambaréné *55 A6* Moyen-Ogooué, W Gabon
Lamego *70 C2* Viseu, N Portugal
Lamesa *27 E3* Texas, SW USA
Lamezia Terme *75 D6* Calabria, SE Italy
Lamía *83 B5* Stereá Ellás, C Greece
Lamoni *23 F4* Iowa, C USA
Lampang *114 C4* var. Muang Lampang. Lampang, NW Thailand
Lámpeia *83 B6* Dytikí Ellás, S Greece
Lanbi Kyun *115 B6* prev. Sullivan Island. island Myeik Archipelago, S Myanmar (Burma)
Lancang Jiang see Mekong
Lancaster *67 D5* NW England, United Kingdom
Lancaster *25 C7* California, W USA
Lancaster *19 F4* Pennsylvania, NE USA
Lancaster Sound *15 F2* sound Nunavut, N Canada
Lan-chou/Lan-chow/Lanchow see Lanzhou
Landao see Lantau Island
Landen *65 C6* Vlaams Brabant, C Belgium
Lander *22 C3* Wyoming, C USA
Landerneau *68 A3* Finistère, NW France
Landes *69 B5* cultural region SW France
Land of Enchantment see New Mexico
The Land of Opportunity see Arkansas
Land of Steady Habits see Connecticut
Land of the Midnight Sun see Alaska
Landsberg see Gorzów Wielkopolski, Lubuskie, Poland
Landsberg an der Warthe see Gorzów Wielkopolski
Land's End *67 B8* headland SW England, United Kingdom
Landshut *73 C6* Bayern, SE Germany
Langar *101 E2* Rus. Lyangar. Navoiy Viloyati, C Uzbekistan
Langfang *106 D4* Hebei, E China
Langkawi, Pulau *115 B7* island Peninsular Malaysia
Langres *68 D4* Haute-Marne, N France
Langsa *116 A3* Sumatera, W Indonesia
Lang Shan *105 E3* mountain range N China
Lang Sơn *114 D3* var. Langson. Lang Sơn, N Vietnam
Langson see Lang Sơn
Lang Suan *115 B6* Chumphon, SW Thailand
Languedoc *69 C6* cultural region S France
Länkäran *95 H3* Rus. Lenkoran'. S Azerbaijan
Lansing *18 C3* state capital Michigan, N USA
Lanta, Ko *115 B7* island S Thailand
Lantau Island *106 A2* Cant. Tai Yue Shan, Chin. Landao. island Hong Kong, S China
Lan-ts'ang Chiang see Mekong
Lantung, Gulf of see Liaodong Wan
Lanzarote *48 B3* island Islas Canarias, Spain, NE Atlantic Ocean
Lanzhou *106 B4* var. Lan-chou, Lanchow, Lan-chow; prev. Kaolan. province capital Gansu, C China
Lao Cai *114 D3* Lao Cai, N Vietnam
Laodicea/Laodicea ad Mare see Al Lādhiqīyah
Laoet see Laut, Pulau
Laojunmiao *106 A3* prev. Yumen. Gansu, N China
Laon *68 D3* var. la Laon; anc. Laudunum. Aisne, N France
Lao People's Democratic Republic see Laos
La Orchila, Isla *36 D1* island N Venezuela
La Oroya *38 C3* Junín, C Peru
Laos *114 D4* off. Lao People's Democratic Republic. country SE Asia
La Palma *31 G5* Darién, SE Panama
La Palma *48 A3* island Islas Canarias, Spain, NE Atlantic Ocean
La Paz *39 F4* var. La Paz de Ayacucho. country capital (Bolivia-seat of government) La Paz, W Bolivia
La Paz *28 B3* Baja California Sur, NW Mexico
La Paz, Bahía de *28 B3* bay NW Mexico
La Paz de Ayacucho see La Paz
La Pérouse Strait *108 D1* Jap. Sōya-kaikyō, Rus. Proliv Laperuza. strait Japan/Russia
Laperuza, Proliv see La Pérouse Strait
Lápithos see Lapta
Lapland *62 D3* Fin. Lappi, Swe. Lappland. cultural region N Europe
La Plata *42 D4* Buenos Aires, E Argentina
La Plata see Sucre
La Pola *70 D1* var. Pola de Lena. Asturias, N Spain
Lappeenranta *63 E5* Swe. Villmanstrand. Etelä-Karjala, SE Finland
Lappi/Lappland see Lapland
Lappo see Lapua
Lapta *80 C5* Gk. Lápithos. NW Cyprus
Laptev Sea see Laptevykh, More
Laptevykh, More *93 E2* Eng. Laptev Sea. sea Arctic Ocean
Lapua *63 D5* Swe. Lappo. Etelä-Pohjanmaa, W Finland
Lapurdum see Bayonne
Łapy *76 E3* Podlaskie, NE Poland
La Quiaca *42 C2* Jujuy, N Argentina
L'Aquila *74 C4* var. Aquila, Aquila degli Abruzzi. Abruzzo, C Italy
Laracha *70 B1* Galicia, NW Spain
Laramie *22 C4* Wyoming, C USA
Laramie Mountains *22 C3* mountain range Wyoming, C USA
Laredo *71 E1* Cantabria, N Spain
Laredo *27 F5* Texas, SW USA
La Réunion see Réunion
Largeau see Faya
Largo *21 E4* Florida, SE USA
Largo, Cayo *32 B2* island W Cuba
Lario see Como, Lago di
La Rioja *42 C3* La Rioja, NW Argentina
La Rioja *71 E2* autonomous community N Spain
Lárisa *82 B4* var. Larissa. Thessalía, C Greece
Larissa see Lárisa
Lārkāna *112 B3* var. Larkhana. Sind, SE Pakistan
Larkhana see Lārkāna
Larnaca see Lárnaka
Lárnaka *80 C5* var. Larnaca, Larnax. SE Cyprus
Larnax see Lárnaka
la Rochelle *68 B4* anc. Rupella. Charente-Maritime, W France
la Roche-sur-Yon *68 B4* prev. Bourbon Vendée, Napoléon-Vendée. Vendée, NW France
La Roda *71 E3* Castilla-La Mancha, C Spain
La Romana *33 E3* E Dominican Republic
Larvotto *69 C8* N Monaco Europe
La-sa see Lhasa
Las Cabezas de San Juan *70 C5* Andalucía, S Spain
Las Cruces *26 D3* New Mexico, SW USA
La See d'Urgel see La Seu d'Urgell
La Serena *42 B3* Coquimbo, C Chile
La Seu d'Urgell *71 G1* prev. La See d'Urgel, Seo de Urgel. Cataluña, NE Spain
la Seyne-sur-Mer *69 D6* Var, SE France
Lashio *114 B3* Shan State, E Myanmar (Burma)
Lashkar Gāh *100 D5* var. Lash-Kar-Gar'. Helmand, S Afghanistan
Lash-Kar-Gar' see Lashkar Gāh
La Sila *75 D6* mountain range SW Italy
La Sirena *30 D3* Región Autónoma Atlántico Sur, E Nicaragua
Łask *76 C4* Łódzkie, C Poland
Las Lomitas *42 D2* Formosa, N Argentina
La Solana *71 E4* Castilla-La Mancha, C Spain
Las Palmas *48 A3* var. Las Palmas de Gran Canaria. Gran Canaria, Islas Canarias, Spain, NE Atlantic Ocean
Las Palmas de Gran Canaria see Las Palmas
La Spezia *74 B3* Liguria, NW Italy
Lassa see Lhasa
Las Tablas *31 F5* Los Santos, S Panama
Last Frontier, The see Alaska
Las Tunas *32 C2* var. Victoria de las Tunas. Las Tunas, E Cuba
La Suisse see Switzerland
Las Vegas *25 D7* Nevada, W USA
Latacunga *38 B1* Cotopaxi, C Ecuador
Latakia see Al Lādhiqīyah
la Teste *69 B5* Gironde, SW France
Latina *75 C5* prev. Littoria. Lazio, C Italy
La Tortuga, Isla *37 E1* var. Isla Tortuga. island N Venezuela
La Tuque *17 E4* Québec, SE Canada
Latvia *84 C3* off. Republic of Latvia, Ger. Lettland, Latv. Latvija, Latvijas Republika; prev. Latvian SSR, Rus. Latviyskaya SSR. country NE Europe
Latvian SSR/Latvija/Latvijas Republika/Latviyskaya SSR see Latvia
Latvia, Republic of see Latvia
Laudunum see Laon
Laudus see St-Lô
Lauenburg/Lauenburg in Pommern see Lębork
Lau Group *123 E4* island group E Fiji
Lauis see Lugano
Launceston *127 C8* Tasmania, SE Australia
La Unión *30 C2* Olancho, C Honduras
La Unión *71 F4* Murcia, SE Spain
Laurel *20 C3* Mississippi, S USA
Laurel *22 C2* Montana, NW USA
Laurentian Highlands see Laurentian Mountains
Laurentian Mountains *17 E3* var. Laurentian Highlands, Fr. Les Laurentides. plateau Newfoundland and Labrador/Québec, Canada
Laurentides, Les see Laurentian Mountains
Lauria *75 D6* Basilicata, S Italy
Laurinburg *21 F1* North Carolina, SE USA
Lauru see Choiseul
Lausanne *73 A7* It. Losanna. Vaud, SW Switzerland
Laut, Pulau *116 D4* prev. Laoet. island Borneo, C Indonesia
Laval *16 D4* Québec, SE Canada
Laval *68 B3* Mayenne, NW France
La Vall d'Uixó *71 F3* var. Vall D'Uxó. País Valenciano, E Spain
La Vega *33 E3* var. Concepción de la Vega. C Dominican Republic
La Vila Joíosa see Villajoyosa
Lávrio *83 C6* prev. Lávrion. Attikí, C Greece
Lávrion see Lávrio
Lawrence *19 G3* Massachusetts, NE USA
Lawrenceburg *20 C1* Tennessee, S USA
Lawton *27 F2* Oklahoma, C USA
La Yarada *39 E4* Tacna, SW Peru
Laylā *99 C5* var. Laila. Ar Riyāḍ, C Saudi Arabia
Lazarev Sea *132 B1* sea Antarctica
Lázaro Cárdenas *29 E5* Michoacán, SW Mexico
Leal see Lihula
Leamhcán see Lucan
Leamington *16 C5* Ontario, S Canada
Leavdnja see Lakselv
Lebak *117 E3* Mindanao, S Philippines
Lebanese Republic see Lebanon
Lebanon *23 G5* Missouri, C USA
Lebanon *19 G2* New Hampshire, NE USA
Lebanon *24 B3* Oregon, NW USA
Lebanon *96 A4* off. Lebanese Republic, Ar. Al Lubnān, Fr. Liban. country SW Asia
Lebanon, Mount see Liban, Jebel
Lebap *100 D2* Lebapskiy Velayat, NE Turkmenistan
Lebedin see Lebedyn
Lebedyn *87 F2* Rus. Lebedin. Sums'ka Oblast', NE Ukraine
Lębork *76 C2* var. Lębórk, Ger. Lauenburg, Lauenburg in Pommern. Pomorskie, N Poland
Lebrija *70 C5* Andalucía, S Spain
Lebu *43 A5* Bío Bío, C Chile
le Cannet *69 D6* Alpes-Maritimes, SE France
Le Cap see Cap-Haïtien
Lecce *75 E6* Puglia, SE Italy
Lechainá *83 B6* var. Lehena, Lekhainá. Dytikí Ellás, S Greece
Ledo Salinarius see Lons-le-Saunier
Leduc *15 E5* Alberta, SW Canada
Leech Lake *23 F2* lake Minnesota, N USA
Leeds *67 D5* N England, United Kingdom
Leek *64 E2* Groningen, NE Netherlands
Leer *72 A3* Niedersachsen, NW Germany
Leeuwarden *64 D1* Fris. Ljouwert. Friesland, N Netherlands
Leeuwin, Cape *120 A5* headland Western Australia
Leeward Islands *33 G3* island group E West Indies
Leeward Islands see Sotavento, Ilhas de
Lefkáda *83 A5* prev. Levkás. Lefkáda, Iónia Nisiá, Greece, C Mediterranean Sea
Lefkáda *83 A5* It. Santa Maura, prev. Levkás; anc. Leucas. island Iónia Nisiá, Greece, C Mediterranean Sea
Lefká Óri *83 C7* mountain range Kriti, Greece, E Mediterranean Sea
Lefkímmi *83 A5* var. Levkímmi. Kérkyra, Iónia Nisiá, Greece, C Mediterranean Sea
Lefkosía/Lefkoşa see Nicosia
Legaceaster see Chester
Legaspi see Legazpi City
Legazpi City *117 E2* var. Legaspi. C Philippines
Leghorn see Livorno
Legnica *76 B4* Ger. Liegnitz. Dolnośląskie, SW Poland
le Havre *68 B3* Eng. Havre; prev. le Havre-de-Grâce. Seine-Maritime, N France
le Havre-de-Grâce see le Havre
Lehena see Lechainá
Leicester *67 D6* Lat. Batae Coritanorum. C England, United Kingdom
Leiden *64 B3* prev. Leyden; anc. Lugdunum Batavorum. Zuid-Holland, W Netherlands
Leie *68 D2* Fr. Lys. river Belgium/France
Leinster *67 B6* Ir. Cúige Laighean. cultural region E Ireland
Leipsic see Leipzig
Leipsoí *83 E6* island Dodekánisa, Greece, Aegean Sea
Leipzig *72 C4* Pol. Lipsk, hist. Leipsic; anc. Lipsia. Sachsen, E Germany
Leiria *70 B3* anc. Collipo. Leiria, C Portugal
Leirvik *63 A6* Hordaland, S Norway
Lek *64 C4* river SW Netherlands
Lekhainá see Lechainá
Lekhchevo see Boychinovtsi
Leksand *63 C5* Dalarna, C Sweden
Lel'chitsy see Lyel'chytsy
le Léman see Geneva, Lake
Lelystad *64 D3* Flevoland, C Netherlands
Léman, Lac see Geneva, Lake
le Mans *68 B3* Sarthe, NW France
Lemberg see L'viv
Lemesós *80 C5* var. Limassol. SW Cyprus
Lemhi Range *24 D3* mountain range Idaho, C USA North America
Lemnos see Límnos
Lemovices see Limoges
Lena *93 F3* river NE Russia
Lena Tablemount *119 B7* seamount S Indian Ocean
Len Dao *106 C8* island S Spratly Islands
Lengshuitan see Yongzhou
Leninabad see Khujand
Leninakan see Gyumri
Lenine *87 G5* Rus. Lenino. Respublika Krym, S Ukraine
Leningor see Ridder
Leningrad see Sankt-Peterburg
Leningradskaya *132 B4* Russian research station Antarctica
Lenino see Lenine, Ukraine
Leninobod see Khujand
Leninogorsk see Ridder
Leninpol' *101 F2* Talasskaya Oblast', NW Kyrgyzstan
Lenin-Turkmenski see Türkmenabat
Lenkoran' see Länkäran
Lenti *77 B7* Zala, SW Hungary
Lentia see Linz
Leoben *73 E7* Steiermark, C Austria
León *29 E4* var. León de los Aldamas. Guanajuato, C Mexico
León *30 C3* León, NW Nicaragua
León *70 D1* Castilla-León, NW Spain
León de los Aldamas see León
Leonídi see Leonídio
Leonídio *83 B6* var. Leonídi. Pelopónnisos, S Greece
Léopold II, Lac see Mai-Ndombe, Lac
Léopoldville see Kinshasa
Lepe *70 C4* Andalucía, S Spain
Lepel' see Lyepyel'
le Portel *68 C2* Pas-de-Calais, N France
Le Puglie see Puglia
le Puy *69 C5* prev. le Puy-en-Velay, hist. Anicium, Podium Anicensis. Haute-Loire, C France
le Puy-en-Velay see le Puy
Léré *54 B4* Mayo-Kébbi, SW Chad
Lérida see Lleida
Lerma *70 D2* Castilla-León, N Spain
Leros *83 D6* island Dodekánisa, Greece, Aegean Sea
Lerrnayin Gharabakh see Nagornyy-Karabakh
Lerwick *66 D1* NE Scotland, United Kingdom
Lesbos see Lésvos
Les Cayes see Cayes
Les Gonaïves see Gonaïves
Leshan *106 B5* Sichuan, C China
les Herbiers *68 B4* Vendée, NW France
Lesh/Leshi see Lezhë
Lesina see Hvar
Leskovac *79 E5* Serbia, SE Serbia
Lesnoy *92 C3* Sverdlovskaya Oblast', C Russia
Lesotho *56 D4* off. Kingdom of Lesotho; prev. Basutoland. country S Africa
Lesotho, Kingdom of see Lesotho
les Sables-d'Olonne *68 B4* Vendée, NW France
Lesser Antarctica see West Antarctica
Lesser Antilles *33 G4* island group E West Indies
Lesser Caucasus *95 F2* Rus. Malyy Kavkaz. mountain range SW Asia
Lesser Khingan Range see Xiao Hinggan Ling
Lesser Sunda Islands *117 E5* Eng. Lesser Sunda Islands. island group C Indonesia
Lesser Sunda Islands see Nusa Tenggara
Lésvos *94 A3* anc. Lesbos. island E Greece
Leszno *76 B4* Ger. Lissa. Wielkopolskie, C Poland
Lethbridge *15 E5* Alberta, SW Canada
Lethem *37 F3* S Guyana
Leti, Kepulauan *117 F5* island group E Indonesia
Letpadan *114 B4* Bago, SW Myanmar (Burma)
Letsôk-aw Kyun *115 B6* var. Letsutan Island; prev. Domel Island. island Myeik Archipelago, S Myanmar (Burma)
Letsutan Island see Letsôk-aw Kyun
Lettland see Latvia
Lëtzebuerg see Luxembourg
Leucas see Lefkáda
Leuven *65 C6* Fr. Louvain, Ger. Löwen. Vlaams Brabant, C Belgium
Leuze see Leuze-en-Hainaut
Leuze-en-Hainaut *65 B6* var. Leuze. Hainaut, SW Belgium
Léva see Levice
Levanger *62 B4* Nord-Trøndelag, C Norway
Levelland *27 E2* Texas, SW USA
Leverkusen *72 A4* Nordrhein-Westfalen, W Germany
Levice *77 C6* Ger. Lewentz, Hung. Léva, Lewenz. Nitriansky Kraj, SW Slovakia
Levin *128 D4* Manawatu-Wanganui, North Island, New Zealand
Levkás see Lefkáda
Levkímmi see Lefkímmi
Lewentz/Lewenz see Levice
Lewis, Isle of *66 B2* island NW Scotland, United Kingdom

Lewis Range – Lyusino

Lewis Range 22 B1 mountain range Montana, NW USA
Lewiston 24 C2 Idaho, NW USA
Lewiston 19 G2 Maine, NE USA
Lewistown 22 C1 Montana, NW USA
Lexington 18 C5 Kentucky, S USA
Lexington 23 E4 Nebraska, C USA
Leyden see Leiden
Leyte 117 F2 island C Philippines
Leżajsk 77 E5 Podkarpackie, SE Poland
Lezha see Lezhë
Lezhë 79 C6 var. Lezha; prev. Lesh, Leshi. Lezhë, NW Albania
Lhasa 104 C5 var. La-sa, Lassa. Xizang Zizhiqu, W China
Lhaviyani Atoll see Faadhippolhu Atoll
Lhazê 104 C5 var. Quxar. Xizang Zizhiqu, China E Asia
L'Hospitalet de Llobregat 71 G2 var. Hospitalet. Cataluña, NE Spain
Liancourt Rocks 109 A5 Jap. Takeshima, Kor. Dokdo. island group Japan/South Korea
Lianyungang 106 D4 var. Xinpu. Jiangsu, E China
Liao see Liaoning
Liaodong Wan 105 G3 Eng. Gulf of Lantung, Gulf of Liaotung. gulf NE China
Liao He 103 E1 river NE China
Liaoning 106 D3 var. Liao, Liaoning Sheng, Shengking, hist. Fengtien, Shenking. province NE China
Liaoning Sheng see Liaoning
Liaoyuan 107 E3 var. Dongliao, Shuang-liao, Jap. Chengchiatun. Jilin, NE China
Liard see Fort Liard
Liban see Lebanon
Liban, Jebel 96 B4 Ar. Jabal al Gharbt, Jabal Lubnān, Eng. Mount Lebanon. mountain range C Lebanon
Libau see Liepāja
Libby 22 A1 Montana, NW USA
Liberal 23 E5 Kansas, C USA
Liberalitas Julia see Évora
Liberec 76 B4 Ger. Reichenberg. Liberecký Kraj, N Czech Republic (Czechia)
Liberia 30 D4 Guanacaste, NW Costa Rica
Liberia 52 C5 off. Republic of Liberia. country W Africa
Liberia, Republic of see Liberia
Libian Desert see Libyan Desert
Lībīyah, Aş Şaḩrā' al see Libyan Desert
Libourne 69 B5 Gironde, SW France
Libreville 55 A5 country capital (Gabon) Estuaire, NW Gabon
Libya 49 F3 off. Libya, Ar. Al Jamāhīrīyah al 'Arabīyah al Lībīyah ash Sha'bīyah al Ishtirākīy; prev. Libyan Arab Republic, Great Socialist People's Libyan Arab Jamahiriya. country N Africa
Libyan Arab Republic see Libya
Libyan Desert 49 H4 var. Libian Desert, Ar. Aş Şaḩrā' al Lībīyah. desert N Africa
Libyan Plateau 81 F4 var. Aḑ Ḑiffah. plateau Egypt/Libya
Lichtenfels 73 C5 Bayern, SE Germany
Lichtenvoorde 64 E4 Gelderland, E Netherlands
Lichuan 106 C5 Hubei, C China
Lida 85 B5 Hrodzyenskaya Voblasts', W Belarus
Lidhorikíon see Lidoríki
Lidköping 63 B6 Västra Götaland, S Sweden
Lidokhorikion see Lidoríki
Lidoríki 83 B5 prev. Lidhorikíon, Lidokhorikion. Stereá Elláda, C Greece
Lidzbark Warmiński 76 D2 Ger. Heilsberg. Olsztyn, N Poland
Liechtenstein 72 D1 off. Principality of Liechtenstein. country C Europe
Liechtenstein, Principality of see Liechtenstein
Liège 65 D6 Dut. Luik, Ger. Lüttich. Liège, E Belgium
Liegnitz see Legnica
Lienz 73 D7 Tirol, W Austria
Liepāja 84 B3 Ger. Libau. Liepāja, W Latvia
Lietuva see Lithuania
Lievenhof see Līvāni
Liezen 73 D7 Steiermark, C Austria
Liffey 67 B6 river Ireland
Lifou 122 D5 island Îles Loyauté, E New Caledonia
Liger see Loire
Ligure, Appennino 74 A2 Eng. Ligurian Mountains. mountain range NW Italy
Ligure, Mar see Ligurian Sea
Ligurian Mountains see Ligure, Appennino
Ligurian Sea 74 A3 Fr. Mer Ligurienne, It. Mar Ligure. sea N Mediterranean Sea
Ligurienne, Mer see Ligurian Sea
Lihu'e 25 A7 var. Lihue. Kaua'i, Hawaii, USA
Lihue see Lihu'e
Lihula 84 D2 Ger. Leal. Läänemaa, W Estonia
Liivi Laht see Riga, Gulf of
Likasi 55 D7 prev. Jadotville. Shaba, SE Dem. Rep. Congo
Likenes 63 A6 Vest-Agder, S Norway
Lille 68 C2 var. l'Isle, Dut. Rijssel, Flem. Ryssel, prev. Lisle; anc. Insula. Nord, N France
Lillehammer 63 B5 Oppland, S Norway
Lillestrøm 63 B6 Akershus, S Norway
Lilongwe 57 E2 country capital (Malawi) Central, W Malawi
Lilybaeum see Marsala
Lima 38 C4 country capital (Peru) Lima, W Peru
Limanowa 77 D5 Małopolskie, S Poland
Limassol see Lemesós
Limerick 67 A6 Ir. Luimneach. Limerick, SW Ireland
Límín Vathéos see Sámos
Límnos 81 F3 anc. Lemnos. island E Greece
Limoges 69 C5 anc. Augustoritum Lemovicensium, Lemovices. Haute-Vienne, C France
Limón 31 E4 var. Puerto Limón. Limón, E Costa Rica
Limón 30 D2 Colón, NE Honduras
Limonum see Poitiers
Limousin 69 C5 cultural region C France
Limoux 69 C6 Aude, S France
Limpopo 56 D3 var. Crocodile. river S Africa
Linares 42 B4 Maule, C Chile
Linares 29 E3 Nuevo León, NE Mexico
Linares 70 D4 Andalucía, S Spain
Lincoln 67 D5 anc. Lindum, Lindum Colonia. E England, United Kingdom
Lincoln 19 H2 Maine, NE USA
Lincoln 23 F4 state capital Nebraska, C USA
Lincoln Sea 12 D2 sea Arctic Ocean
Linden 37 F3 E Guyana
Líndhos see Líndos
Lindi 51 D8 Lindi, SE Tanzania
Líndos 83 E7 var. Líndhos. Ródos, Dodekánisa, Greece, Aegean Sea
Lindum/Lindum Colonia see Lincoln
Line Islands 123 G3 island group E Kiribati
Lingeh see Bandar-e Lengeh
Lingen 72 A3 var. Lingen an der Ems. Niedersachsen, NW Germany
Lingen an der Ems see Lingen
Lingga, Kepulauan 116 B4 island group W Indonesia
Linköping 63 C6 Östergötland, S Sweden
Linz 73 D6 anc. Lentia. Oberösterreich, N Austria
Lion, Golfe du 69 C7 Eng. Gulf of Lion, Gulf of Lions; anc. Sinus Gallicus. gulf S France
Lion, Gulf of/Lions, Gulf of see Lion, Golfe du
Liozno see Lyozna
Lipari 75 D6 island Isole Eolie, S Italy
Lipari Islands/Lipari, Isole see Eolie, Isole
Lipetsk 89 B5 Lipetskaya Oblast', W Russia
Lipno 76 C3 Kujawsko-pomorskie, C Poland
Lipova 86 A4 Hung. Lippa. Arad, W Romania
Lipovets see Lypovets'
Lippa see Lipova
Lipsia/Lipsk see Leipzig
Lira 51 B6 N Uganda
Lisala 55 C5 Equateur, N Dem. Rep. Congo
Lisboa 70 B4 Eng. Lisbon; anc. Felicitas Julia, Olisipo. country capital (Portugal) Lisboa, W Portugal
Lisbon see Lisboa
Lisichansk see Lysychans'k
Lisieux 68 B3 anc. Noviomagus. Calvados, N France
Liski 89 B6 prev. Georgiu-Dezh. Voronezhskaya Oblast', W Russia
Lisle/l'Isle see Lille
Lismore 127 E5 New South Wales, SE Australia
Lissa see Vis, Croatia
Lissa see Leszno, Poland
Lisse 64 C3 Zuid-Holland, W Netherlands
Litang 106 A5 var. Gaocheng. Sichuan, C China
Lītani, Nahr el 97 B5 var. Nahr al Litant. river C Lebanon
Litant, Nahr al see Lītani, Nahr el
Litauen see Lithuania
Lithgow 127 D6 New South Wales, SE Australia
Lithuania 84 B4 off. Republic of Lithuania, Ger. Litauen, Lith. Lietuva, Pol. Litwa, Rus. Litva; prev. Lithuanian SSR, Rus. Litovskaya SSR. country NE Europe
Lithuanian SSR see Lithuania
Lithuania, Republic of see Lithuania
Litóchoro 82 B4 var. Litohoro, Litókhoron. Kentrikí Makedonía, N Greece
Litohoro/Litókhoron see Litóchoro
Litovskaya SSR see Lithuania
Little Alföld 77 C6 Ger. Kleines Ungarisches Tiefland, Hung. Kisalföld, Slvk. Podunajská Rovina. plain Hungary/Slovakia
Little Andaman 111 F2 island Andaman Islands, India, NE Indian Ocean
Little Barrier Island 128 D2 island N New Zealand
Little Bay 71 H5 bay Alboran Sea, Mediterranean Sea
Little Cayman 32 B3 island E Cayman Islands
Little Falls 23 F2 Minnesota, N USA
Littlefield 27 E2 Texas, SW USA
Little Inagua 32 D2 var. Inagua Islands. island S The Bahamas
Little Minch, The 66 B3 strait NW Scotland, United Kingdom
Little Missouri River 22 D2 river NW USA
Little Nicobar 111 G3 island Nicobar Islands, India, NE Indian Ocean
Little Rhody see Rhode Island
Little Rock 20 B1 state capital Arkansas, C USA
Little Saint Bernard Pass 69 D5 Fr. Col du Petit St-Bernard, It. Colle del Piccolo San Bernardo. pass France/Italy
Little Sound 20 A5 bay Bermuda, NW Atlantic Ocean
Littleton 22 D4 Colorado, C USA
Littoria see Latina
Litva/Litwa see Lithuania
Liu-chou/Liuchow see Liuzhou
Liuzhou 106 C6 var. Liu-chou, Liuchow. Guangxi Zhuangzu Zizhiqu, S China
Livanátai see Livanátes
Livanátes 83 B5 prev. Livanátai. Stereá Elláda, C Greece
Līvāni 84 D4 Ger. Lievenhof. Preiļi, SE Latvia
Liverpool 17 F5 Nova Scotia, SE Canada
Liverpool 67 C5 NW England, United Kingdom
Livingston 22 B2 Montana, NW USA
Livingston 27 H3 Texas, SW USA
Livingstone 56 C3 var. Maramba. Southern, S Zambia
Livingstone Mountains 129 A7 mountain range South Island, New Zealand
Livno 78 B4 Federicija Bosna I Hercegovina, SW Bosnia and Herzegovina
Livojoki 62 D4 river C Finland
Livonia 18 D3 Michigan, N USA
Livorno 74 B3 Eng. Leghorn. Toscana, C Italy
Lixian Jiang see Black River
Lixoúri 83 A5 prev. Lixoúrion. Kefallinía, Iónia Nisiá, Greece, C Mediterranean Sea
Lixoúrion see Lixoúri
Lizarra see Estella
Ljouwert see Leeuwarden
Ljubelj see Loibl Pass
Ljubljana 73 D7 Ger. Laibach, It. Lubiana; anc. Aemona, Emona. country capital (Slovenia) C Slovenia
Ljungby 63 B7 Kronoberg, S Sweden
Ljusdal 63 C5 Gävleborg, C Sweden
Ljusnan 63 C5 river C Sweden
Llanelli 67 C6 prev. Llanelly. SW Wales, United Kingdom
Llanelly see Llanelli
Llanes 70 D1 Asturias, N Spain
Llanos 36 D2 physical region Colombia/Venezuela
Lleida 71 F2 Cast. Lérida; anc. Ilerda. Cataluña, NE Spain
Llucmajor 71 G3 Mallorca, Spain, W Mediterranean Sea
Loaita Island 106 C8 island W Spratly Islands
Loanda see Luanda
Lobamba 56 D4 country capital (Swaziland- royal and legislative) NW Swaziland
Lobatse 56 C4 var. Lobatsi. Kgatleng, SE Botswana
Lobatsi see Lobatse
Löbau 72 D4 Sachsen, E Germany
Lobito 56 B2 Benguela, W Angola
Lob Nor see Lop Nur
Lobositz see Lovosice
Loburi see Lop Buri
Locarno 73 B8 Ger. Luggarus. Ticino, S Switzerland
Lochem 64 E3 Gelderland, E Netherlands
Lockport 19 E3 New York, NE USA
Lodja 55 D6 Kasai-Oriental, C Dem. Rep. Congo
Lodwar 51 C6 Rift Valley, NW Kenya
Łódź 76 D4 Rus. Lodz. Łódź, C Poland
Loei 114 C4 var. Loey, Muang Loei. Loei, C Thailand
Loey see Loei
Lofoten 62 B3 var. Lofoten Islands. island group C Norway
Lofoten Islands see Lofoten
Logan 22 B3 Utah, W USA
Logan, Mount 14 D3 mountain Yukon, W Canada
Logroño 71 E1 anc. Vareia, Lat. Juliobriga. La Rioja, N Spain
Loibl Pass 73 D7 Ger. Loiblpass, Slvn. Ljubelj. pass Austria/Slovenia
Loiblpass see Loibl Pass
Loikaw 114 B4 Kayah State, C Myanmar (Burma)
Loire 68 B4 var. Liger. river C France
Loja 38 B2 Loja, S Ecuador
Lokitaung 51 C5 Rift Valley, NW Kenya
Lokoja 53 G4 Kogi, C Nigeria
Loksa 84 E2 Ger. Loxa. Harjumaa, NW Estonia
Lolland 63 B8 prev. Laaland. island S Denmark
Lom 82 C1 prev. Lom-Palanka. Montana, NW Bulgaria
Lomami 55 D6 river C Dem. Rep. Congo
Lomas 38 D4 Arequipa, SW Peru
Lomas de Zamora 42 D4 Buenos Aires, E Argentina
Lombamba 56 D4 country capital (Swaziland - royal and legislative) NW Swaziland
Lombardia 74 B2 Eng. Lombardy. region N Italy
Lombardy see Lombardia
Lombok, Pulau 116 D5 island Nusa Tenggara, C Indonesia
Lomé 53 F5 country capital (Togo) S Togo
Lomela 55 D6 Kasai-Oriental, C Dem. Rep. Congo
Lommel 65 C5 Limburg, N Belgium
Lomond, Loch 66 B4 lake C Scotland, United Kingdom
Lomonosov Ridge 133 B3 var. Harris Ridge, Rus. Khrebet Homonsova. undersea ridge Arctic Ocean
Lomonsova, Khrebet see Lomonosov Ridge
Lom-Palanka see Lom
Lompoc 25 B7 California, W USA
Lom Sak 114 C4 var. Muang Lom Sak. Phetchabun, C Thailand
Łomża 76 D3 Rus. Lomzha. Podlaskie, NE Poland
Lomzha see Łomża
Loncoche 43 B5 Araucanía, C Chile
Londinium see London
London 67 A7 anc. Augusta, Lat. Londinium. country capital (United Kingdom) SE England, United Kingdom
London 16 C5 Ontario, S Canada
London 18 C5 Kentucky, S USA
Londonderry 66 B4 var. Derry, Ir. Doire. NW Northern Ireland, United Kingdom
Londonderry, Cape 124 C2 cape Western Australia
Londrina 41 E4 Paraná, S Brazil
Lone Star State see Texas
Long Bay 21 F2 bay W Jamaica
Long Beach 25 C7 California, W USA
Longford 67 B5 Ir. An Longfort. Longford, C Ireland
Long Island 32 D2 island C The Bahamas
Long Island 19 G4 island New York, NE USA
Longlac 16 C3 Ontario, S Canada
Longmont 22 D4 Colorado, C USA
Longreach 126 C4 Queensland, E Australia
Long Strait 93 G1 Eng. Long Strait. strait NE Russia
Long Strait see Longa, Proliv
Longview 27 H3 Texas, SW USA
Longview 24 B2 Washington, NW USA
Long Xuyên 115 D6 var. Longxuyen. An Giang, S Vietnam
Longxuyen see Long Xuyên
Longyan 106 D6 Fujian, SE China
Longyearbyen 61 G2 dependent territory capital (Svalbard) Spitsbergen, W Svalbard
Lons-le-Saunier 68 D4 anc. Ledo Salinarius. Jura, E France
Lop Buri 115 C5 var. Loburi. Lop Buri, C Thailand
Lop Nor see Lop Nur
Lop Nur 104 C3 var. Lob Nor, Lop Nor, Lo-pu Po. seasonal lake NW China
Loppersum 64 E1 Groningen, NE Netherlands
Lo-pu Po see Lop Nur
Lorca 71 E4 Ar. Lurka; anc. Eliocroca, Lat. Illurco. Murcia, S Spain
Lord Howe Island 120 C4 island E Australia
Lord Howe Rise 120 C4 undersea rise SW Pacific Ocean
Loreto 28 B3 Baja California Sur, NW Mexico
Lorient 68 A3 prev. l'Orient. Morbihan, NW France
l'Orient see Lorient
Lorn, Firth of 66 B4 inlet W Scotland, United Kingdom
Loro Sae see East Timor
Lörrach 73 A7 Baden-Württemberg, S Germany
Lorraine 68 D3 cultural region NE France
Los Alamos 26 C1 New Mexico, SW USA
Los Amates 30 B2 Izabal, E Guatemala
Los Ángeles 43 B5 Bío Bío, C Chile
Los Angeles 25 C7 California, W USA
Losanna see Lausanne
Lošinj 78 A3 Ger. Lussin, It. Lussino. island W Croatia
Loslau see Wodzisław Śląski
Los Mochis 28 C3 Sinaloa, C Mexico
Losonc/Losontz see Lučenec
Los Roques, Islas 36 D1 island group N Venezuela
Lot 69 B5 cultural region S France
Lot 69 B5 river S France
Lotagipi Swamp 51 C5 wetland Kenya/ South Sudan
Lötzen see Giżycko
Loualaba see Lualaba
Louangnamtha 114 C3 var. Luong Nam Tha. Louang Namtha, N Laos
Louangphabang 102 D3 var. Louangphrabang, Luang Prabang. Louangphabang, N Laos
Louangphrabang see Louangphabang
Loubomo see Dolisie
Loudéac 68 A3 Côtes d'Armor, NW France
Loudi 106 C5 Hunan, S China
Louga 52 B3 NW Senegal
Louisiade Archipelago 122 B4 island group SE Papua New Guinea
Louisiana 20 A2 off. State of Louisiana, also known as Creole State, Pelican State. state S USA
Louisville 18 C5 Kentucky, S USA
Louisville Ridge 121 E4 undersea ridge S Pacific Ocean
Loup River 23 E4 river Nebraska, C USA
Lourdes 69 B6 Hautes-Pyrénées, S France
Lourenço Marques see Maputo
Louth 67 E5 E England, United Kingdom
Loutrá 82 C4 Kentrikí Makedonía, N Greece
Louvain see Leuven
Louvain-la Neuve 65 C6 Walloon Brabant, C Belgium
Louviers 68 C3 Eure, N France
Lovech 82 C2 Lovech, N Bulgaria
Loveland 22 D4 Colorado, C USA
Lovosice 76 A4 Ger. Lobositz. Ústecký Kraj, NW Czech Republic (Czechia)
Lóvua 56 C1 Moxico, E Angola
Lowell 19 G3 Massachusetts, NE USA
Löwen see Leuven
Lower California see Baja California
Lower Hutt 129 D5 Wellington, North Island, New Zealand
Lower Lough Erne 67 A5 lake SW Northern Ireland, United Kingdom
Lower Red Lake 23 F1 lake Minnesota, N USA
Lower Rhine see Neder Rijn
Lower Tunguska see Nizhnyaya Tunguska
Lowestoft 67 E6 E England, United Kingdom
Loxa see Loksa
Lo-yang see Luoyang
Loyauté, Îles 122 D5 island group S New Caledonia
Loyev see Loyew
Loyew 85 D8 Rus. Loyev. Homyel'skaya Voblasts', SE Belarus
Loznica 78 C3 Serbia, W Serbia
Lu see Shandong, China
Lualaba 55 D6 Fr. Loualaba. river SE Dem. Rep. Congo
Luanda 56 A1 var. Loanda, Port. São Paulo de Loanda. country capital (Angola) Luanda, NW Angola
Luang Prabang see Louangphabang
Luang, Thale 115 C7 lagoon S Thailand
Luangua, Rio see Luangwa
Luangwa 51 B8 var. Aruângua, Rio Luangua. river Mozambique/Zambia
Luanshya 56 D2 Copperbelt, C Zambia
Luarca 70 C1 Asturias, N Spain
Lubaczów 77 E5 var. Lúbaczów. Podkarpackie, SE Poland
L'uban' 76 B4 Leningradskaya Oblast', Russia
Lubānas Ezers see Lubāns
Lubango 56 B2 Port. Sá da Bandeira. Huíla, SW Angola
Lubāns 84 D4 var. Lubānas Ezers. lake E Latvia
Lubao 55 D6 Kasai-Oriental, C Dem. Rep. Congo
Lübben 72 D4 Brandenburg, E Germany
Lübbenau 72 D4 Brandenburg, E Germany
Lubbock 27 E2 Texas, SW USA
Lübeck 72 C2 Schleswig-Holstein, N Germany
Lubelska, Wyżyna 76 E4 plateau SE Poland
Lüben see Lubin
Lubiana see Ljubljana
Lubin 76 B4 Ger. Lüben. Dolnośląskie, SW Poland
Lublin 76 E4 Rus. Lyublin. Lubelskie, E Poland
Lubliniec 76 C4 Śląskie, S Poland
Lubnān, Jabal see Liban, Jebel
Lubny 87 F2 Poltavs'ka Oblast', NE Ukraine
Lubsko 76 B4 Ger. Sommerfeld. Lubuskie, W Poland
Lubumbashi 55 E8 prev. Élisabethville. Shaba, SE Dem. Rep. Congo
Lubutu 55 D6 Maniema, E Dem. Rep. Congo
Luca see Lucca
Lucan 67 B5 Ir. Leamhcán. Dublin, E Ireland
Lucanian Mountains see Lucano, Appennino
Lucano, Appennino 75 D5 Eng. Lucanian Mountains. mountain range S Italy
Lucapa 56 C1 var. Lukapa. Lunda Norte, NE Angola
Lucca 74 B3 anc. Luca. Toscana, C Italy
Lucea 32 A4 W Jamaica
Lucena 117 E1 off. Lucena City. Luzon, N Philippines
Lucena 70 D4 Andalucía, S Spain
Lucena City see Lucena
Lučenec 77 D6 Ger. Losontz, Hung. Losonc. Banskobystrický Kraj, C Slovakia
Lucentum see Alicante
Lucerna/Lucerne see Luzern
Luchow see Hefei
Luck see Luts'k
Lucknow 113 E3 var. Lakhnau. state capital Uttar Pradesh, N India
Lüda see Dalian
Luda Kamchiya 82 D2 river E Bulgaria
Ludasch see Luduş
Lüderitz 56 B4 prev. Angra Pequena. Karas, SW Namibia
Ludhiāna 112 D2 Punjab, N India
Ludington 18 C2 Michigan, N USA
Ludsan see Ludza
Luduş 86 B4 Ger. Ludasch, Hung. Marosludas. Mureş, C Romania
Ludvika 63 C6 Dalarna, C Sweden
Ludwigsburg 73 B6 Baden-Württemberg, SW Germany
Ludwigsfelde 72 D3 Brandenburg, NE Germany
Ludwigshafen 73 B5 var. Ludwigshafen am Rhein. Rheinland-Pfalz, W Germany
Ludwigshafen am Rhein see Ludwigshafen
Ludwigslust 72 C3 Mecklenburg-Vorpommern, N Germany
Ludza 84 D4 Ger. Ludsan. Ludza, E Latvia
Luebo 55 C6 Kasai-Occidental, SW Dem. Rep. Congo
Luena 56 C2 var. Lwena, Port. Luso. Moxico, E Angola
Lufira 55 E7 river SE Dem. Rep. Congo
Lufkin 27 H3 Texas, SW USA
Luga 88 A4 Leningradskaya Oblast', NW Russia
Lugano 73 B8 Ger. Lauis. Ticino, S Switzerland
Lugansk see Luhans'k
Lugdunum see Lyon
Lugdunum Batavorum see Leiden
Lugenda, Rio 57 E2 river N Mozambique
Luggarus see Locarno
Lugh Ganana see Luuq
Lugo 70 C1 anc. Lugus Augusti. Galicia, NW Spain
Lugoj 86 A4 Ger. Lugosch, Hung. Lugos. Timiş, W Romania
Lugos/Lugosch see Lugoj
Lugus Augusti see Lugo
Luguvallium/Luguvallum see Carlisle
Luhans'k 87 H3 Rus. Lugansk; prev. Voroshilovgrad. Luhans'ka Oblast', E Ukraine
Luimneach see Limerick
Lukapa see Lucapa
Lukenie 55 C6 river C Dem. Rep. Congo
Lukovit 82 C2 Lovech, N Bulgaria
Łuków 76 E4 Ger. Bogendorf. Lubelskie, E Poland
Lukuga 55 D7 river SE Dem. Rep. Congo
Luleå 62 D4 Norrbotten, N Sweden
Luleälven 62 C3 river N Sweden
Lulonga 55 C5 river NW Dem. Rep. Congo
Lulua 55 D7 river S Dem. Rep. Congo
Luluabourg see Kananga
Lumber State see Maine
Lumbo 57 F2 Nampula, NE Mozambique
Lumsden 129 A7 Southland, South Island, New Zealand
Lund 63 B7 Skåne, S Sweden
Lüneburg 72 C3 Niedersachsen, N Germany
Lunga, Isola see Dugi Otok
Lungkiang see Qiqihar
Lungué-Bungo 56 C2 var. Lungwebungu. river Angola/Zambia
Lungwebungu see Lungué-Bungo
Luninets see Luninyets
Łuniniec see Luninyets
Luninyets 85 B7 Pol. Łuniniec, Rus. Luninets. Brestskaya Voblasts', SW Belarus
Lunteren 64 D4 Gelderland, C Netherlands
Luong Nam Tha see Louangnamtha
Luoyang 106 C4 var. Honan, Lo-yang. Henan, C China
Lupatia see Altamura
Lúrio 57 F2 Nampula, NE Mozambique
Lúrio, Rio 57 E2 river NE Mozambique
Lurka see Lorca
Lusaka 56 D2 country capital (Zambia) Lusaka, SE Zambia
Lushnja see Lushnjë
Lushnjë 79 C6 var. Lushnja. Fier, C Albania
Luso see Luena
Lussin/Lussino see Lošinj
Lūt, Baḩrat/Lut, Bahret see Dead Sea
Lut, Dasht-e 98 D3 var. Kavīr-e Lūt. desert E Iran
Lutetia/Lutetia Parisiorum see Paris
Lūt, Kavīr-e see Lūt, Dasht-e
Luton 67 D6 E England, United Kingdom
Łutselk'e 15 F4 prev. Snowdrift. Northwest Territories, W Canada
Luts'k 86 C1 Pol. Łuck, Rus. Lutsk. Volyns'ka Oblast', NW Ukraine
Lutsk see Luts'k
Lüttich see Liège
Lutzow-Holm Bay 132 C2 var. Lutzow-Holm Bay. bay Antarctica
Lutzow-Holm Bay see Lützow Holmbukta
Luuq 51 D6 It. Lugh Ganana. Gedo, SW Somalia
Luvua 55 D7 river SE Dem. Rep. Congo
Luwego 51 C8 river S Tanzania
Luxembourg 65 D8 country capital (Luxembourg) Luxembourg, S Luxembourg
Luxembourg 65 D8 off. Grand Duchy of Luxembourg, var. Lëtzebuerg, Luxemburg. country NW Europe
Luxemburg see Luxembourg
Luxor 50 B2 Ar. Al Uqşur. E Egypt
Luza 88 C4 Kirovskaya Oblast', NW Russia
Luz, Costa de la 70 C5 coastal region SW Spain
Luzern 73 B7 Fr. Lucerne, It. Lucerna. Luzern, C Switzerland
Luzon 117 E1 island N Philippines
Luzon Strait 103 E3 strait Philippines/Taiwan
L'viv 86 B2 Ger. Lemberg, Pol. Lwów, Rus. L'vov. L'vivs'ka Oblast', W Ukraine
L'vov see L'viv
Lwena see Luena
Lwów see L'viv
Lyakhavichy 85 B6 Rus. Lyakhovichi. Brestskaya Voblasts', SW Belarus
Lyakhovichi see Lyakhavichy
Lyallpur see Faisalābād
Lyangar see Langar
Lyck see Ełk
Lycksele 62 C4 Västerbotten, N Sweden
Lycopolis see Asyūţ
Lyel'chytsy 85 C7 Rus. Lel'chitsy. Homyel'skaya Voblasts', SE Belarus
Lyepyel' 85 D5 Rus. Lepel'. Vitsyebskaya Voblasts', N Belarus
Lyme Bay 67 C7 bay S England, United Kingdom
Lynchburg 19 E5 Virginia, NE USA
Lynn see King's Lynn
Lynn Lake 15 F4 Manitoba, C Canada
Lynn Regis see King's Lynn
Lyon 69 D5 Eng. Lyons; anc. Lugdunum. Rhône, E France
Lyons see Lyon
Lyozna 85 E6 Rus. Liozno. Vitsyebskaya Voblasts', NE Belarus
Lypovets' 86 D2 Rus. Lipovets. Vinnyts'ka Oblast', C Ukraine
Lys see Leie
Lysychans'k 87 H3 Rus. Lisichansk. Luhans'ka Oblast', E Ukraine
Lyttelton 129 C6 South Island, New Zealand
Lyublin see Lublin
Lyubotin see Lyubotyn
Lyubotyn 87 G2 Rus. Lyubotin. Kharkivs'ka Oblast', E Ukraine
Lyulyakovo 82 E2 prev. Keremitlik. Burgas, E Bulgaria
Lyusina 85 B6 Rus. Lyusino. Brestskaya Voblasts', SW Belarus
Lyusino see Lyusina

Maale – Marshall Islands

M

Maale *110 B4 var.* Male'. *country capital* (Maldives) Male' Atoll, C Maldive
Ma'ān *97 B7* Ma'ān, SW Jordan
Maardu *84 D2 Ger.* Maart. Harjumaa, NW Estonia
Ma'aret-en-Nu'man *see* Ma'arrat an Nu'mān
Ma'arrat an Nu'mān *96 B3 var.* Ma'aret-en-Nu'man, *Fr.* Maarret enn Naamâne. Idlib, NW Syria
Maarret enn Naamâne *see* Ma'arrat an Nu'mān
Maart *see* Maardu
Maas *see* Meuse
Maaseik *65 D5 prev.* Maeseyck. Limburg, NE Belgium
Maastricht *65 D6 var.* Maestricht; *anc.* Traiectum ad Mosam, Traiectum Tungorum. Limburg, SE Netherlands
Macao *107 C6 Port.* Macau. Guangdong, SE China
Macapá *41 E1 state capital* Amapá, N Brazil
Macarsca *see* Makarska
Macassar *see* Makassar
Macău *see* Makó, Hungary
Macau *see* Macao
MacCluer Gulf *see* Berau, Teluk
Macdonnell Ranges *124 D4 mountain range* Northern Territory, C Australia
Macedonia *79 D6 off.* Republic of Macedonia, *var.* the former Yugoslav Republic of Macedonia (used by UN), *Mac.* Makedonija, *abbrev.* FYR Macedonia, FYROM. *country* SE Europe
Macedonia, FYR *see* Macedonia
Macedonia, Republic of *see* Macedonia
Macedonia, the former Yugoslav Republic of *see* Macedonia
Maceió *41 G3 state capital* Alagoas, E Brazil
Machachi *38 B1* Pichincha, C Ecuador
Machala *38 B2* El Oro, SW Ecuador
Machanga *57 E3* Sofala, E Mozambique
Machilīpatnam *110 D1 var.* Bandar Masulipatnam. Andhra Pradesh, E India
Machiques *36 C2* Zulia, NW Venezuela
Macías Nguema Biyogo *see* Bioco, Isla de
Măcin *86 D5* Tulcea, SE Romania
Mackay *126 D4* Queensland, NE Australia
Mackay, Lake *124 C4 salt lake* Northern Territory/Western Australia
Mackenzie *15 E3 river* Northwest Territories, NW Canada
Mackenzie Bay *132 D3 bay* Antarctica
Mackenzie Mountains *14 D3 mountain range* Northwest Territories, NW Canada
Macleod, Lake *124 A4 lake* Western Australia
Macomb *18 A4* Illinois, N USA
Macomer *75 A5* Sardegna, Italy, C Mediterranean Sea
Mâcon *69 D5 anc.* Matisco, Matisco Ædourum. Saône-et-Loire, C France
Macon *20 D2* Georgia, SE USA
Macon *23 G4* Missouri, C USA
Macquarie Ridge *132 C5 undersea ridge* SW Pacific Ocean
Macuspana *29 G4* Tabasco, SE Mexico
Ma'dabā *97 B6 var.* Mādabā, Madeba; *anc.* Medeba. Ma'dabā, NW Jordan
Mādabā *see* Ma'dabā
Madagascar *57 F3 off.* Democratic Republic of Madagascar, *Malg.* Madagasikara; *prev.* Malagasy Republic. *country* W Indian Ocean
Madagascar *57 F3 island* W Indian Ocean
Madagascar Basin *47 E7 undersea basin* W Indian Ocean
Madagascar, Democratic Republic of *see* Madagascar
Madagascar Plateau *47 E7 var.* Madagascar Ridge, Madagascar Rise, *Rus.* Madagaskarskiy Khrebet. *undersea plateau* W Indian Ocean
Madagascar Rise/Madagascar Ridge *see* Madagascar Plateau
Madagasikara *see* Madagascar
Madagaskarskiy Khrebet *see* Madagascar Plateau
Madang *122 B3* Madang, N Papua New Guinea
Madanīyīn *see* Médenine
Mađarska *see* Hungary
Made *64 C4* Noord-Brabant, S Netherlands
Madeba *see* Ma'dabā
Madeira *48 A2 var.* Ilha de Madeira. *island* Madeira, Portugal, NE Atlantic Ocean
Madeira, Ilha de *see* Madeira
Madeira Plain *44 C3 abyssal plain* E Atlantic Ocean
Madeira, Rio *40 D2 var.* Río Madera. *river* Bolivia/Brazil
Madeleine, Îles de la *17 F4 Eng.* Magdalen Islands. *island group* Québec, E Canada
Madera *25 B6* California, W USA
Madera, Río *see* Madeira, Rio
Madhya Pradesh *113 E4 prev.* Central Provinces and Berar. *cultural region* C India
Madīnat ath Thawrah *96 C2 var.* Ath Thawrah. Ar Raqqah, N Syria
Madioen *see* Madiun
Madison *23 F3* South Dakota, N USA
Madison *18 B3 state capital* Wisconsin, N USA
Madiun *116 D5 prev.* Madioen. Jawa, C Indonesia
Madoera *see* Madura, Pulau
Madona *84 D4 Ger.* Modohn. Madona, E Latvia
Madras *see* Chennai
Madras *see* Tamil Nādu
Madre de Dios, Río *39 E3 river* Bolivia/Peru
Madre del Sur, Sierra *29 E5 mountain range* S Mexico
Madre, Laguna *29 F3 lagoon* NE Mexico
Madre, Laguna *27 G5 lagoon* Texas, SW USA
Madre Occidental, Sierra *28 C3 var.* Western Sierra Madre. *mountain range* C Mexico
Madre Oriental, Sierra *29 E3 var.* Eastern Sierra Madre. *mountain range* C Mexico
Madre, Sierra *30 B2 var.* Sierra de Soconusco. *mountain range* Guatemala/Mexico
Madrid *70 D3 country capital* (Spain) Madrid, C Spain
Madura *see* Madurai
Madurai *110 C3 prev.* Madura, Mathurai. Tamil Nādu, S India
Madura, Pulau *116 D5 prev.* Madoera. *island* C Indonesia
Maebashi *109 D5 var.* Maebasi, Mayebashi. Gunma, Honshū, S Japan
Maebasi *see* Maebashi
Mae Nam Khong *see* Mekong
Mae Nam Nan *114 C4 river* NW Thailand
Mae Nam Yom *114 C4 river* W Thailand
Maeseyck *see* Maaseik
Maestricht *see* Maastricht
Maéwo *122 D4 prev.* Aurora. *island* C Vanuatu
Mafia *51 D7 island* E Tanzania
Mafraq/Muḥāfaẓat al Mafraq *see* Al Mafraq
Magadan *93 G3* Magadanskaya Oblast', E Russia
Magallanes *see* Punta Arenas
Magallanes, Estrecho de *see* Magellan, Strait of
Magangué *36 B2* Bolívar, N Colombia
Magdalena *39 F3* Beni, N Bolivia
Magdalena *28 B1* Sonora, NW Mexico
Magdalena. Isla *28 B3 island* NW Mexico
Magdalena, Río *36 B2 river* C Colombia
Magdalen Islands *see* Madeleine, Îles de la
Magdeburg *72 C4* Sachsen-Anhalt, C Germany
Magelang *116 C5* Jawa, C Indonesia
Magellan, Strait of *43 B8 Sp.* Estrecho de Magallanes. *strait* Argentina/Chile
Magerøy *see* Magerøya
Magerøya *62 D1 var.* Magerøy, *Lapp.* Máhkarávju. *island* N Norway
Maggiore, Lago *see* Maggiore, Lake
Maggiore, Lake *74 B1 It.* Lago Maggiore. *lake* Italy/Switzerland
Maglaj *78 C3* Federacija Bosna I Hercegovina, N Bosnia and Herzegovina
Maglie *75 E6* Puglia, SE Italy
Magna *22 B4* Utah, W USA
Magnesia *see* Manisa
Magnitogorsk *92 B4* Chelyabinskaya Oblast', C Russia
Magnolia State *see* Mississippi
Magṭa' Laḥjar *52 C3 var.* Magta Lahjar, Magta' Lahjar, Magtá Lahjar. Brakna, SW Mauritania
Magtymguly *100 C3 prev.* Garrygala; *Rus.* Kara-gala. W Turkmenistan
Magway *114 A3 var.* Magwe. Magway, W Myanmar (Burma)
Magyar-Becse *see* Bečej
Magyarkanizsa *see* Kanjiža
Magyarország *see* Hungary
Mahajanga *57 F2 var.* Majunga. Mahajanga, NW Madagascar
Mahakam, Sungai *116 D4 var.* Koetai, Kutai. *river* Borneo, C Indonesia
Mahalapye *56 D3 var.* Mahalatswe. Central, SE Botswana
Mahalatswe *see* Mahalapye
Māhān *98 D3* Kermān, E Iran
Mahanādi *113 F4 river* E India
Mahārāshtra *112 D5 cultural region* W India
Mahbés *see* El Mahbas
Mahbūbnagar *112 D5* Telangana, C India
Mahdia *49 F2 var.* Al Mahdīyah, Mehdia. NE Tunisia
Mahé *57 H1 island* Inner Islands, NE Seychelles
Mahia Peninsula *128 E4 peninsula* North Island, New Zealand
Mahilyow *85 D6 Rus.* Mogilëv. Mahilyowskaya Voblasts', E Belarus
Máhkarávju *see* Magerøya
Mahmūd-e 'Erāqī *see* Maḥmūd-e Rāqī
Maḥmūd-e Rāqī *101 E4 var.* Mahmūd-e 'Erāqī. Kāpīsā, NE Afghanistan
Mahón *see* Maó
Mähren *see* Moravia
Mährisch-Weisskirchen *see* Hranice
Maicao *36 C1* La Guajira, N Colombia
Mai Ceu/Mai Chio *see* Maych'ew
Maīdān Shahr *101 E4 prev.* Meydān Shahr. Vardak, E Afghanistan
Maidstone *67 E7* SE England, United Kingdom
Maiduguri *53 H4* Borno, NE Nigeria
Mailand *see* Milano
Maīmanah *100 D3 var.* Meymaneh, Maymana. Fāryāb, NW Afghanistan
Main *73 B5 river* C Germany
Mai-Ndombe, Lac *55 C6 prev.* Lac Léopold II. *lake* W Dem. Rep. Congo
Maine *19 G2 off.* State of Maine, *also known as* Lumber State, Pine Tree State. *state* NE USA
Maine *68 B3 cultural region* NW France
Maine, Gulf of *19 H2 gulf* NE USA
Mainland *66 C2 island* N Scotland, United Kingdom
Mainland *66 D1 island* NE Scotland, United Kingdom
Mainz *73 B5 Fr.* Mayence. Rheinland-Pfalz, SW Germany
Maio *52 A3 var.* Mayo. *island* Ilhas de Sotavento, SE Cape Verde
Maisur *see* Mysūru, India
Maisur *see* Karnātaka, India
Maitri *132 C2 Indian research station* Antarctica
Maizhokunggar *104 C5* Xizang Zizhiqu, W China
Majorca *see* Mallorca
Mājro *see* Majuro Atoll
Majunga *see* Mahajanga
Majuro Atoll *122 D2 var.* Mājro. *atoll and capital* (Marshall Islands) Ratak Chain, SE Marshall Islands
Makale *see* Mek'elē
Makarov Basin *133 B3 undersea basin* Arctic Ocean
Makarska *78 B4 It.* Macarsca. Split-Dalmacija, SE Croatia
Makasar *see* Makassar
Makasar, Selat *see* Makassar Straits
Makassar *117 E4 var.* Macassar, Makasar; *prev.* Ujungpandang. Sulawesi, C Indonesia
Makassar Straits *116 D4 Ind.* Makasar Selat. *strait* C Indonesia
Makay *57 F3 var.* Massif du Makay. *mountain range* SW Madagascar
Makay, Massif du *see* Makay
Makedonija *see* Macedonia
Makeni *52 C4* C Sierra Leone
Makeyevka *see* Makiyivka
Makhachkala *92 A4 prev.* Petrovsk-Port. Respublika Dagestan, SW Russia
Makin *122 D2 prev.* Pitt Island. *atoll* Tungaru, W Kiribati
Makira *see* San Cristobal
Makiyivka *87 G3 Rus.* Makeyevka; *prev.* Dmitriyevsk. Donets'ka Oblast', E Ukraine
Makkah *99 A5 Eng.* Mecca. Makkah, W Saudi Arabia
Makkovik *17 F2* Newfoundland and Labrador, NE Canada
Makó *77 D7 Rom.* Macău. Csongrád, SE Hungary
Makoua *55 B5* Cuvette, C Congo
Makran Coast *98 E4 coastal region* SE Iran
Makrany *85 A6 Rus.* Mokrany. Brestskaya Voblasts', SW Belarus
Mākū *98 B2* Āžarbāyjān-e Gharbī, NW Iran
Makurdi *53 G4* Benue, C Nigeria
Mala *see* Malaita, Solomon Islands
Malabār Coast *110 B3 coast* SW India
Malabo *55 A5 prev.* Santa Isabel. *country capital* (Equatorial Guinea) Isla de Bioco, NW Equatorial Guinea
Malaca *see* Málaga
Malacca, Strait of *116 B3 Ind.* Selat Malaka. *strait* Indonesia/Malaysia
Malacka *see* Malacky
Malacky *77 C6 Hung.* Malacka. Bratislavský Kraj, W Slovakia
Maladzyechna *85 C5 Pol.* Molodeczno, *Rus.* Molodechno. Minskaya Voblasts', C Belarus
Málaga *70 D5 anc.* Malaca. Andalucía, S Spain
Malagarasi River *51 B7 river* W Tanzania Africa
Malagasy Republic *see* Madagascar
Malaita *122 C3 var.* Mala. *island* N Solomon Islands
Malakal *51 B5* Upper Nile, NE South Sudan
Malakula *see* Malekula
Malang *116 D5* Jawa, C Indonesia
Malange *see* Malanje
Malanje *56 B1 var.* Malange. Malanje, NW Angola
Mälaren *63 C6 lake* C Sweden
Malatya *95 E4 anc.* Melitene. Malatya, SE Turkey
Mala Vyska *87 E3 Rus.* Malaya Viska. Kirovohrads'ka Oblast', S Ukraine
Malawi *57 E1 off.* Republic of Malaŵi; *prev.* Nyasaland, Nyasaland Protectorate. *country* S Africa
Malawi, Lake *see* Nyasa, Lake
Malaŵi, Republic of *see* Malawi
Malaya Viska *see* Mala Vyska
Malay Peninsula *102 D4 peninsula* Malaysia/Thailand
Malaysia *116 B3 off.* Malaysia, *var.* Federation of Malaysia; *prev.* the separate territories of Federation of Malaya, Sarawak and Sabah (North Borneo) and Singapore. *country* SE Asia
Malaysia, Federation of *see* Malaysia
Malbork *76 C2 Ger.* Marienburg, Marienburg in Westpreussen. Pomorskie, N Poland
Malchin *72 C3* Mecklenburg-Vorpommern, N Germany
Malden *23 H5* Missouri, C USA
Malden Island *123 G3 prev.* Independence Island. *atoll* E Kiribati
Maldives *110 A4 off.* Maldivian Divehi, Republic of Maldives. *country* N Indian Ocean
Maldives, Republic of *see* Maldives
Maldivian Divehi *see* Maldives
Male' *see* Maale
Male' Atoll *110 B4 var.* Kaafu Atoll. *atoll* C Maldives
Malekula *122 D4 var.* Malakula; *prev.* Mallicolo. *island* W Vanuatu
Malesína *83 C5* Stereá Ellás, E Greece
Malheur Lake *24 C3 lake* Oregon, NW USA
Mali *53 E3 off.* Republic of Mali, *Fr.* République du Mali; *prev.* French Sudan, Sudanese Republic. *country* W Africa
Malik, Wadi al *see* Milk, Wadi el
Mali Kyun *115 B5 var.* Tavoy Island. *island* Myeik Archipelago, S Myanmar (Burma)
Malin *see* Malyn
Malindi *51 D7* Coast, SE Kenya
Malines *see* Mechelen
Mali, Republic of *see* Mali
Mali, République du *see* Mali
Malkiye *see* Al Mālikīyah
Malko Tŭrnovo *82 E3* Burgas, E Bulgaria
Mallaig *66 B3* N Scotland, United Kingdom
Mallawī *50 B2 var.* Mallawi. C Egypt
Mallawi *see* Mallawī
Mallicolo *see* Malekula
Mallorca *71 G3 Eng.* Majorca; *anc.* Baleares Major. *island* Islas Baleares, Spain, W Mediterranean Sea
Malmberget *62 C3 Lapp.* Malmivaara. Norrbotten, N Sweden
Malmédy *65 D6* Liège, E Belgium
Malmivaara *see* Malmberget
Malmö *63 B7* Skåne, S Sweden
Maļoeļap *see* Maloelap Atoll
Maloelap Atoll *122 D1 var.* Maļoeļap. *atoll* E Marshall Islands
Małopolska, Wyżyna *76 D4 plateau* S Poland
Malozemel'skaya Tundra *88 D3 physical region* NW Russia
Malta *84 D4* Rēzekne, SE Latvia
Malta *22 C1* Montana, NW USA
Malta *75 C8 off.* Republic of Malta. *country* C Mediterranean Sea
Malta *75 C8 island* Malta, C Mediterranean Sea
Malta, Canale di *see* Malta Channel
Malta Channel *75 C8 It.* Canale di Malta. *strait* Italy/Malta
Malta, Republic of *see* Malta
Maluku *117 F4 Dut.* Molukken, *Eng.* Moluccas; *prev.* Spice Islands. *island group* E Indonesia
Maluku, Laut *see* Molucca Sea
Malung *63 B6* Dalarna, C Sweden
Malventum *see* Benevento
Malvina, Isla Gran *see* West Falkland
Malvinas, Islas *see* Falkland Islands
Malyn *86 D2 Rus.* Malin. Zhytomyrs'ka Oblast', N Ukraine
Malyy Kavkaz *see* Lesser Caucasus
Mamberamo, Sungai *117 H4 river* Papua, E Indonesia
Mambij *see* Manbij
Mamonovo *84 A4 Ger.* Heiligenbeil. Kaliningradskaya Oblast', W Russia
Mamoré, Rio *39 F3 river* Bolivia/Brazil
Mamou *52 C4* W Guinea
Mamoudzou *57 F2 dependent territory capital* (Mayotte) C Mayotte
Mamuno *56 C3* Ghanzi, W Botswana
Manacor *71 G3* Mallorca, Spain, W Mediterranean Sea
Manado *117 F3 prev.* Menado. Sulawesi, C Indonesia
Managua *30 D3 country capital* (Nicaragua) Managua, W Nicaragua
Managua, Lake *30 C3 var.* Xolotlán. *lake* W Nicaragua
Manakara *57 G4* Fianarantsoa, SE Madagascar
Manama *see* Al Manāmah
Mananjary *57 G3* Fianarantsoa, SE Madagascar
Manáos *see* Manaus
Manapouri, Lake *129 A7 lake* South Island, New Zealand
Manar *see* Mannar
Manas, Gora *101 E2 mountain* Kyrgyzstan/Uzbekistan
Manaus *40 D2 prev.* Manáos. *state capital* Amazonas, NW Brazil
Manavgat *94 B4* Antalya, SW Turkey
Manbij *96 C2 var.* Mambij, *Fr.* Membidj. Ḥalab, N Syria
Manchester *67 D5 Lat.* Mancunium. NW England, United Kingdom
Manchester *19 G3* New Hampshire, NE USA
Man-chou-li *see* Manzhouli
Manchurian Plain *103 E1 plain* NE China
Máncio Lima *see* Japiim
Mancunium *see* Manchester
Mand *see* Mand, Rūd-e
Mandalay *114 B3* Mandalay, C Myanmar (Burma)
Mandan *23 E2* North Dakota, N USA
Mandeville *32 B5* C Jamaica
Mándra *83 C6* Attikí, C Greece
Rud-e Mand *98 D4 var.* Mand. *river* S Iran
Mandurah *125 A6* Western Australia
Manduria *75 E5* Puglia, SE Italy
Mandya *110 C2* Karnātaka, C India
Manfredonia *75 D5* Puglia, SE Italy
Mangai *55 C6* Bandundu, W Dem. Rep. Congo
Mangaia *123 G5 island group* S Cook Islands
Mangalia *86 D5 anc.* Callatis. Constanța, SE Romania
Mangalmé *54 C3* Guéra, SE Chad
Mangalore *see* Mangalūru
Mangalūru *110 B2 prev.* Mangalore. Karnātaka, W India
Mangaung *see* Bloemfontein
Mango *see* Sansanné-Mango, Togo
Mangoky *57 F3 river* W Madagascar
Manhattan *23 F4* Kansas, C USA
Manicouagan, Réservoir *16 D3 lake* Québec, E Canada
Manihiki *123 G4 atoll* N Cook Islands
Manihiki Plateau *121 E3 undersea plateau* C Pacific Ocean
Maniitsoq *60 C3 var.* Manîtsoq, *Dan.* Sukkertoppen. Kitaa, S Greenland
Manila *117 E1 off.* City of Manila. *country capital* (Philippines) Luzon, N Philippines
Manila, City of *see* Manila
Manisa *94 A3 var.* Manissa, *prev.* Saruhan; *anc.* Magnesia. Manisa, W Turkey
Manissa *see* Manisa
Manitoba *15 F5 province* S Canada
Manitoba, Lake *15 F5 lake* Manitoba, S Canada
Manitoulin Island *16 C4 island* Ontario, S Canada
Manîtsoq *see* Maniitsoq
Manizales *36 B3* Caldas, W Colombia
Manjimup *125 A7* Western Australia
Mankato *23 F3* Minnesota, N USA
Manlleu *71 G2* Cataluña, NE Spain
Manly *126 E1* Iowa, C USA
Manmād *112 C5* Mahārāshtra, W India
Mannar *110 C3 var.* Manar. Northern Province, NW Sri Lanka
Mannar, Gulf of *110 C3 gulf* India/Sri Lanka
Mannheim *73 B5* Baden-Württemberg, SW Germany
Manokwari *117 G4* New Guinea, E Indonesia
Manono *55 E7* Shaba, SE Dem. Rep. Congo
Manosque *69 D6* Alpes-de-Haute-Provence, SE France
Manra *123 F3 prev.* Sydney Island. *atoll* Phoenix Islands, C Kiribati
Mansa *56 D2 prev.* Fort Rosebery. Luapula, N Zambia
Mansel Island *15 G3 island* Nunavut, NE Canada
Mansfield *18 D4* Ohio, N USA
Manta *38 A2* Manabí, W Ecuador
Manteca *25 B6* California, W USA
Mantoue *see* Mantova
Mantova *74 B2 Eng.* Mantua, *Fr.* Mantoue. Lombardia, NW Italy
Mantua *see* Mantova
Manuae *123 G4 island* S Cook Islands
Manukau *see* Manurewa
Manurewa *128 D3 var.* Manukau. Auckland, North Island, New Zealand
Manzanares *71 E3* Castilla-La Mancha, C Spain
Manzanillo *32 C3* Granma, E Cuba
Manzanillo *28 D4* Colima, SW Mexico
Manzhouli *105 F1 var.* Man-chou-li. Nei Mongol Zizhiqu, N China
Mao *54 B3* Kanem, W Chad
Maó *71 H3 Cast.* Mahón, *Eng.* Port Mahon; *anc.* Portus Magonis. Menorca, Spain, W Mediterranean Sea
Maoke, Pegunungan *117 H4 Dut.* Sneeuw-gebergte, *Eng.* Snow Mountains. *mountain range* Papua, E Indonesia
Maoming *106 C6* Guangdong, S China
Mapmaker Seamounts *103 H2 seamount range* N Pacific Ocean
Maputo *56 D4 prev.* Lourenço Marques. *country capital* (Mozambique) Maputo, S Mozambique
Marabá *41 F2* Pará, NE Brazil
Maracaibo *36 C1* Zulia, NW Venezuela
Maracaibo, Gulf of *see* Venezuela, Golfo de
Maracaibo, Lago de *36 C2 var.* Lake Maracaibo. *inlet* NW Venezuela
Maracaibo, Lake *see* Maracaibo, Lago de
Maracay *36 D2* Aragua, N Venezuela
Marada *see* Marādah
Marādah *49 G3 var.* Marada. N Libya
Maradi *53 G3* Maradi, S Niger
Maragha *see* Marāgheh
Marāgheh *98 C2 var.* Maragha. Āžarbāyjān-e Khāvarī, NW Iran
Marajó, Baía de *41 F1 bay* N Brazil
Marajó, Ilha de *41 E1 island* N Brazil
Marakesh *see* Marrakech
Maramba *see* Livingstone
Maranhão *41 F2 off.* Estado do Maranhão. *state/region* E Brazil
Maranhão, Estado do *see* Maranhão
Marañón, Río *38 B2 river* N Peru
Marathon *16 C4* Ontario, S Canada
Marathón *see* Marathónas
Marathónas *83 C5 prev.* Marathón. Attikí, C Greece
Marbella *70 D5* Andalucía, S Spain
Marble Bar *124 B4* Western Australia
Marburg *see* Marburg an der Lahn, Germany
Marburg *see* Maribor, Slovenia
Marburg an der Lahn *72 B4 hist.* Marburg. Hessen, W Germany
March *see* Morava
Marche *74 C3 Eng.* Marches. *region* C Italy
Marche *69 C5 cultural region* C France
Marche-en-Famenne *65 C7* Luxembourg, SE Belgium
Marchena, Isla *38 B5 var.* Bindloe Island. *island* Galápagos Islands, Ecuador, E Pacific Ocean
Marches *see* Marche
Mar Chiquita, Laguna *42 C3 lake* C Argentina
Marcounda *see* Markounda
Mardān *112 C1* North-West Frontier Province, N Pakistan
Mar del Plata *43 D5* Buenos Aires, E Argentina
Mardin *95 E4* Mardin, SE Turkey
Maré *122 D5 island* Îles Loyauté, E New Caledonia
Marea Neagră *see* Black Sea
Mareeba *126 D3* Queensland, NE Australia
Marek *see* Dupnitsa
Marganets *see* Marhanets'
Margarita, Isla de *37 E1 island* N Venezuela
Margate *67 E7 prev.* Mergate. SE England, United Kingdom
Margherita *see* Jamaame
Margherita, Lake *51 C5 Eng.* Lake Margherita, *It.* Abbaia. *lake* SW Ethiopia
Margherita, Lake *see* Ābaya Hāyk'
Marghita *86 B3 Hung.* Margitta. Bihor, NW Romania
Margitta *see* Marghita
Marhanets' *87 F3 Rus.* Marganets. Dnipropetrovs'ka Oblast', E Ukraine
María Cleofas, Isla *28 C4 island* C Mexico
Maria Island *127 C8 island* Tasmania, SE Australia
María Madre, Isla *28 C4 island* C Mexico
María Magdalena, Isla *28 C4 island* C Mexico
Mariana Trench *103 G4 trench* W Pacific Ocean
Mariánské Lázně *77 A5 Ger.* Marienbad. Karlovarský Kraj, W Czech Republic (Czechia)
Marías, Islas *28 C4 island group* C Mexico
Maria-Theresiopel *see* Subotica
Maribor *73 E7 Ger.* Marburg. NE Slovenia
Marica *see* Maritsa
Maridi *51 B5* W Equatoria, S South Sudan
Marie Byrd Land *132 A3 physical region* Antarctica
Marie-Galante *33 G4 var.* Ceyre to the Caribs. *island* SE Guadeloupe
Marienbad *see* Mariánské Lázně
Marienburg *see* Alūksne, Latvia
Marienburg *see* Malbork, Poland
Marienburg in Westpreussen *see* Malbork
Marienhausen *see* Viļaka
Mariental *56 B4* Hardap, SW Namibia
Marienwerder *see* Kwidzyń
Mariestad *63 B6* Västra Götaland, S Sweden
Marietta *20 D2* Georgia, SE USA
Marijampolė *84 B4 prev.* Kapsukas. Marijampolė, S Lithuania
Marília *41 E4* São Paulo, S Brazil
Marín *70 B1* Galicia, NW Spain
Mar'ina Gorka *see* Mar"ina Horka
Mar"ina Horka *85 C6 Rus.* Mar'ina Gorka. Minskaya Voblasts', C Belarus
Maringá *41 E4* Paraná, S Brazil
Marion *23 G3* Iowa, C USA
Marion *18 D4* Ohio, N USA
Marion, Lake *21 E2 reservoir* South Carolina, SE USA
Mario Zucchelli *132 C4 Italian research station* Antarctica
Mariscal Estigarribia *42 D2* Boquerón, NW Paraguay
Maritsa *82 D3 var.* Marica, *Gk.* Évros, *Turk.* Meriç; *anc.* Hebrus. *river* SW Europe
Maritzburg *see* Pietermaritzburg
Mariupol' *87 G4 prev.* Zhdanov. Donets'ka Oblast', SE Ukraine
Marka *51 D6 var.* Merca. Shabeellaha Hoose, S Somalia
Markham, Mount *132 B4 mountain* Antarctica
Markounda *54 C4 var.* Marcounda. Ouham, NW Central African Republic
Marktredwitz *73 C5* Bayern, E Germany
Marlborough *126 D4* Queensland, E Australia
Marmanda *see* Marmande
Marmande *69 B5 anc.* Marmanda. Lot-et-Garonne, SW France
Sea of Marmara *94 A2 Eng.* Sea of Marmara. *sea* NW Turkey
Marmara, Sea of *see* Marmara Denizi
Marmaris *94 A4* Muğla, SW Turkey
Marne *68 C3 cultural region* N France
Marne *68 D3 river* N France
Maro *54 C4* Moyen-Chari, S Chad
Maroantsetra *57 G2* Toamasina, NE Madagascar
Maromokotro *57 G2 mountain* N Madagascar
Maroni *37 G3 Dut.* Marowijne. *river* French Guiana/Suriname
Maroshevíz *see* Toplița
Marosludas *see* Luduș
Marosvásárhely *see* Târgu Mureș
Marotiri *121 F4 var.* Îlots de Bass, Morotiri. *island group* Îles Australes, SW French Polynesia
Maroua *54 B3* Extrême-Nord, N Cameroon
Marowijne *see* Maroni
Marquesas Fracture Zone *131 E3 fracture zone* E Pacific Ocean
Marquette *18 B1* Michigan, N USA
Marrakech *48 C2 var.* Marakesh, *Eng.* Marrakesh; *prev.* Morocco. W Morocco
Marrakesh *see* Marrakech
Marrawah *127 C8* Tasmania, SE Australia
Marree *127 B5* South Australia
Marsá al Burayqah *49 G3 var.* Al Burayqah. N Libya
Marsabit *51 C6* Eastern, N Kenya
Marsala *75 B7 anc.* Lilybaeum. Sicilia, Italy, C Mediterranean Sea
Marsberg *72 B4* Nordrhein-Westfalen, W Germany
Marseille *69 D6 Eng.* Marseilles; *anc.* Massilia. Bouches-du-Rhône, SE France
Marseilles *see* Marseille
Marshall *23 F2* Minnesota, N USA
Marshall *27 H2* Texas, SW USA
Marshall Islands *122 C1 off.* Republic of the Marshall Islands. *country* W Pacific Ocean

Marshall Islands, Republic of the – Misti, Volcán

Marshall Islands, Republic of the see Marshall Islands
Marshall Seamounts 103 H3 seamount range SW Pacific Ocean
Marsh Harbour 32 C1 Great Abaco, W The Bahamas
Martaban see Mottama
Martha's Vineyard 19 G3 island Massachusetts, NE USA
Martigues 69 D6 Bouches-du-Rhône, SE France
Martin 77 C5 Ger. Sankt Martin, Hung. Turócszentmárton; prev. Turčianský Svätý Martin. Žilinský Kraj, N Slovakia
Martinique 33 G4 French overseas department E West Indies
Martinique Channel see Martinique Passage
Martinique Passage 33 G4 var. Dominica Channel, Martinique Channel. channel Dominica/Martinique
Marton 128 D4 Manawatu-Wanganui, North Island, New Zealand
Martos 70 D4 Andalucía, S Spain
Marungu 55 E7 mountain range SE Dem. Rep. Congo
Mary 100 D3 prev. Merv. Mary Welaýaty, S Turkmenistan
Maryborough 127 D4 Queensland, E Australia
Maryborough see Port Laoise
Mary Island see Kanton
Maryland 19 E5 off. State of Maryland, also known as America in Miniature, Cockade State, Free State, Old Line State. state NE USA
Maryland, State of see Maryland
Maryville 23 F4 Missouri, C USA
Maryville 20 D1 Tennessee, S USA
Masai Steppe 51 C7 grassland NW Tanzania
Masaka 51 B6 SW Uganda
Masallı 95 H3 Rus. Masally. S Azerbaijan
Masally see Masallı
Masasi 51 C8 Mtwara, SE Tanzania
Masawa/Massawa see Mits'iwa
Masaya 30 D3 Masaya, W Nicaragua
Mascarene Basin 119 B5 undersea basin W Indian Ocean
Mascarene Islands 57 H4 island group W Indian Ocean
Mascarene Plain 119 B5 abyssal plain W Indian Ocean
Mascarene Plateau 119 B5 undersea plateau W Indian Ocean
Maseru 56 D4 country capital (Lesotho) W Lesotho
Mas-ha 97 D7 W West Bank Asia
Mashhad 98 E2 var. Meshed. Khorāsān-Razavī, NE Iran
Masindi 51 B6 W Uganda
Masīra see Maṣīrah, Jazīrat
Masira, Gulf of see Maṣīrah, Khalīj
Maṣīrah, Jazīrat 99 E5 var. Masīra. island E Oman
Maṣīrah, Khalīj 99 E5 var. Gulf of Masira. bay E Oman
Masis see Büyükağrı Dağı
Maskat see Masqaṭ
Mason City 23 F3 Iowa, C USA
Masqaṭ 99 E5 var. Maskat, Eng. Muscat. country capital (Oman) NE Oman
Massa 74 B3 Toscana, C Italy
Massachusetts 19 G3 off. Commonwealth of Massachusetts, also known as Bay State, Old Bay State, Old Colony State. state NE USA
Massenya 54 B3 Chari-Baguirmi, SW Chad
Massif Central 69 C5 plateau C France
Massilia see Marseille
Massoukou see Franceville
Mastanli see Momchilgrad
Masterton 129 D5 Wellington, North Island, New Zealand
Masty 85 B5 Rus. Mosty. Hrodzyenskaya Voblasts', W Belarus
Masuda 109 B6 Shimane, Honshū, SW Japan
Masuku see Franceville
Masvingo 56 D3 prev. Fort Victoria, Nyanda, Victoria. Masvingo, SE Zimbabwe
Maṣyāf 96 B3 Fr. Misiaf. Ḥamāh, C Syria
Matadi 55 B6 Bas-Congo, W Dem. Rep. Congo
Matagalpa 30 D3 Matagalpa, C Nicaragua
Matale 110 D3 Central Province, C Sri Lanka
Matam 52 C3 NE Senegal
Matamata 128 D3 Waikato, North Island, New Zealand
Matamoros 28 D3 Coahuila, NE Mexico
Matamoros 29 E2 Tamaulipas, C Mexico
Matane 17 E4 Québec, SE Canada
Matanzas 32 B2 Matanzas, NW Cuba
Matara 110 D4 Southern Province, S Sri Lanka
Mataram 116 D5 Pulau Lombok, C Indonesia
Mataró 71 G2 anc. Illuro. Cataluña, E Spain
Mataura 129 B7 Southland, South Island, New Zealand
Mataura 129 B7 river South Island, New Zealand
Mata Uta see Matā'utu
Matā'utu 123 E4 var. Mata Uta. dependent territory capital (Wallis and Futuna) Île Uvea, Wallis and Futuna
Matera 75 E5 Basilicata, S Italy
Mathurai see Madurai
Matianus see Orūmīyeh, Daryācheh-ye
Matías Romero 29 F5 Oaxaca, SE Mexico
Matisco/Matisco Ædourum see Mâcon
Mato Grosso 41 E3 off. Estado de Mato Grosso; prev. Matto Grosso. state/region W Brazil
Mato Grosso do Sul 41 E4 off. Estado de Mato Grosso do Sul. state/region S Brazil
Mato Grosso do Sul, Estado de see Mato Grosso do Sul
Mato Grosso, Estado de see Mato Grosso
Mato Grosso, Planalto de 34 C4 plateau C Brazil
Matosinhos 70 B2 prev. Matozinhos. Porto, NW Portugal
Matozinhos see Matosinhos
Matsue 109 B6 var. Matsuye, Matue. Shimane, Honshū, SW Japan
Matsumoto 109 C5 var. Matumoto. Nagano, Honshū, S Japan
Matsuyama 109 B7 var. Matuyama. Ehime, Shikoku, SW Japan
Matsuye see Matsue
Matterhorn 73 A8 It. Monte Cervino. mountain Italy/Switzerland
Matthews Ridge 37 F2 N Guyana
Matthew Town 32 D2 Great Inagua, S The Bahamas
Matto Grosso see Mato Grosso
Matucana 38 C4 Lima, W Peru
Matue see Matsue
Matumoto see Matsumoto
Maturín 37 E2 Monagas, NE Venezuela
Matuyama see Matsuyama
Mau 113 E3 var. Maunāth Bhanjan. Uttar Pradesh, N India
Maui 25 B8 island Hawaii, USA, C Pacific Ocean
Maun 56 C3 North-West, C Botswana
Maunāth Bhanjan see Mau
Mauren 72 E1 NE Liechtenstein Europe
Maurice see Mauritius
Mauritania 52 C2 off. Islamic Republic of Mauritania, Ar. Mūrītānīyah. country W Africa
Mauritania, Islamic Republic of see Mauritania
Mauritius 57 H3 off. Republic of Mauritius, Fr. Maurice. country W Indian Ocean
Mauritius 119 B5 island W Indian Ocean
Mauritius, Republic of see Mauritius
Mawlamyaing see Mawlamyine
Mawlamyine 114 B4 var. Mawlamyaing, Moulmein. Mon State, S Myanmar (Burma)
Mawson 132 D2 Australian research station Antarctica
Mayadin see Al Mayādīn
Mayaguana 32 D2 island SE The Bahamas
Mayaguana Passage 32 D2 passage SE The Bahamas
Mayagüez 33 F3 W Puerto Rico
Mayamey 98 D2 Semnān, N Iran
Maya Mountains 30 B2 Sp. Montañas Mayas. mountain range Belize/Guatemala
Mayas, Montañas see Maya Mountains
Maych'ew 50 C4 var. Mai Chio, It. Mai Ceu. Tigray, N Ethiopia
Mayebashi see Maebashi
Mayence see Mainz
Mayfield 129 B6 Canterbury, South Island, New Zealand
Maykop 89 A7 Respublika Adygeya, SW Russia
Maymana see Maīmanah
Maymyo see Pyin-Oo-Lwin
Mayo see Maio
Mayor Island 128 D3 island NE New Zealand
Mayor Pablo Lagerenza see Capitán Pablo Lagerenza
Mayotte 57 F2 French territorial collectivity E Africa
May Pen 32 B5 C Jamaica
Mayyit, Al Baḥr al see Dead Sea
Mazabuka 56 D2 Southern, S Zambia
Mazaca see Kayseri
Mazagan see El-Jadida
Mazār-e Sharīf 101 E3 var. Mazār-i Sharif. Balkh, N Afghanistan
Mazār-i Sharif see Mazār-e Sharīf
Mazatlán 28 C3 Sinaloa, C Mexico
Mažeikiai 84 B3 Telšiai, NW Lithuania
Mazirbe 84 C2 Talsi, NW Latvia
Mazra'a see Al Mazra'ah
Mazury 76 D3 physical region NE Poland
Mazyr 85 C7 Rus. Mozyr'. Homyel'skaya Voblasts', SE Belarus
Mbabane 56 D4 country capital (Swaziland - administrative) NW Swaziland
Mbacké see Mbaké
Mbaïki 55 C5 var. M'Baiki. Lobaye, SW Central African Republic
M'Baiki see Mbaïki
Mbaké 52 B3 var. Mbacké. W Senegal
Mbala 56 D1 prev. Abercorn. Northern, NE Zambia
Mbale 51 C6 E Uganda
Mbandaka 55 C5 prev. Coquilhatville. Equateur, NW Dem. Rep. Congo
M'Banza Congo 56 B1 var. Mbanza Congo; prev. São Salvador, São Salvador do Congo. Dem. Rep. Congo, NW Angola
Mbanza-Ngungu 55 B6 Bas-Congo, W Dem. Rep. Congo
Mbarara 51 B6 SW Uganda
Mbé 54 B4 Nord, N Cameroon
Mbeya 51 C7 Mbeya, SW Tanzania
Mbomou/M'Bomu/Mbomu see Bomu
Mbour 52 B3 W Senegal
Mbuji-Mayi 55 D7 prev. Bakwanga. Kasai-Oriental, S Dem. Rep. Congo
McAlester 27 G2 Oklahoma, C USA
McAllen 27 G5 Texas, SW USA
McCamey 27 E3 Texas, SW USA
M'Clintock Channel 15 F2 channel Nunavut, N Canada
McComb 20 B3 Mississippi, S USA
McCook 23 E4 Nebraska, C USA
McKean Island 123 E3 island Phoenix Islands, C Kiribati
McKinley, Mount see Denali
McKinley Park 14 C3 Alaska, USA
McMinnville 24 B3 Oregon, NW USA
McMurdo 132 B4 US research station Antarctica
McPherson 23 E5 Kansas, C USA
McPherson see Fort McPherson
Mdantsane 56 D5 Eastern Cape, SE South Africa
Mead, Lake 25 D6 reservoir Arizona/Nevada, W USA
Mecca see Makkah
Mechelen 65 C5 Eng. Mechlin, Fr. Malines. Antwerpen, C Belgium
Mechlin see Mechelen
Mecklenburger Bucht 72 C2 bay N Germany
Mecsek 77 C7 mountain range SW Hungary
Medan 116 B3 Sumatera, E Indonesia
Medeba see Ma'dabā
Medellín 36 B3 Antioquia, NW Colombia
Médenine 49 F2 var. Madanīyīn. SE Tunisia
Medeshamstede see Peterborough
Medford 24 B4 Oregon, NW USA
Medgidia 86 D5 Constanța, SE Romania
Medgyes see Mediaș
Mediaș 86 B4 Ger. Mediasch, Hung. Medgyes. Sibiu, C Romania
Mediasch see Mediaș
Medicine Hat 15 F5 Alberta, SW Canada
Medina see Al Madīnah
Medinaceli 71 E2 Castilla-León, N Spain
Medina del Campo 70 D2 Castilla-León, N Spain
Medinat Israel see Israel
Mediolanum see Saintes, France
Mediolanum see Milano, Italy
Mediomatrica see Metz
Mediterranean Sea 80 D3 Fr. Mer Méditerranée. sea Africa/Asia/Europe
Méditerranée, Mer see Mediterranean Sea
Médoc 69 B5 cultural region SW France
Medvezh'yegorsk 88 B3 Respublika Kareliya, NW Russia
Meekatharra 125 B5 Western Australia
Meemu Atoll see Mulakatholhu
Meerssen 65 D6 var. Mersen. Limburg, SE Netherlands
Meerut 112 D2 Uttar Pradesh, N India
Megáli Préspa, Límni see Prespa, Lake
Meghālaya 91 G3 cultural region NE India
Mehdia see Mahdia
Meheso see Mī'ēso
Me Hka see Nmai Hka
Mehrīz 98 D3 Yazd, C Iran
Mehtar Lām 101 F4 var. Mehtarlām, Meterlam, Metharlam, Metharlam. Laghmān, E Afghanistan
Mehtarlām see Mehtar Lām
Meiktila 114 B3 Mandalay, C Myanmar (Burma)
Méjico see Mexico
Mejillones 42 B2 Antofagasta, N Chile
Mek'elē 50 C4 var. Makale. Tigray, N Ethiopia
Mékhé 52 B3 NW Senegal
Mekong 102 D3 var. Lan-ts'ang Chiang, Cam. Mékôngk, Chin. Lancang Jiang, Lao. Mènam Khong, Th. Mae Nam Khong, Tib. Dza Chu, Vtn. Sông Tiền Giang. river SE Asia
Mékôngk see Mekong
Mekong, Mouths of the 115 E6 delta S Vietnam
Melaka 116 B3 var. Malacca. Melaka, Peninsular Malaysia
Melaka, Selat see Malacca, Strait of
Melanesia 122 D3 island group W Pacific Ocean
Melanesian Basin 120 C2 undersea basin W Pacific Ocean
Melbourne 127 C7 state capital Victoria, SE Australia
Melbourne 21 E4 Florida, SE USA
Meleda see Mljet
Melghir, Chott 49 E2 var. Chott Melrhir. salt lake E Algeria
Melilla 58 B5 anc. Rusaddir, Russadir. Melilla, Spain, N Africa
Melilla 48 D2 enclave Spain, N Africa
Melita 15 F5 Manitoba, S Canada
Melita see Mljet
Melitene see Malatya
Melitopol' 87 F4 Zaporiz'ka Oblast', SE Ukraine
Melle 65 B5 Oost-Vlaanderen, NW Belgium
Mellerud 63 B6 Västra Götaland, S Sweden
Mellieħa 80 B5 E Malta
Mellizo Sur, Cerro 43 A7 mountain S Chile
Melo 42 E4 Cerro Largo, NE Uruguay
Melodunum see Melun
Melrhir, Chott see Melghir, Chott
Melsungen 72 B4 Hessen, C Germany
Melun 68 C3 anc. Melodunum. Seine-et-Marne, N France
Melville Bay/Melville Bugt see Qimusseriarsuaq
Melville Island 124 D2 island Northern Territory, N Australia
Melville Island 15 E2 island Parry Islands, Northwest Territories, NW Canada
Melville, Lake 17 F2 lake Newfoundland and Labrador, E Canada
Melville Peninsula 15 G3 peninsula Nunavut, NE Canada
Melville Sound see Viscount Melville Sound
Membidj see Manbij
Memel see Neman, NE Europe
Memel see Klaipeda, Lithuania
Memmingen 73 B6 Bayern, S Germany
Memphis 20 C1 Tennessee, S USA
Menaam see Menaldum
Menado see Manado
Ménaka 53 F3 Goa, E Mali
Menaldum 64 D1 Fris. Menaam. Friesland, N Netherlands
Mènam Khong see Mekong
Mendaña Fracture Zone 131 F4 fracture zone E Pacific Ocean
Mende 69 C5 anc. Mimatum. Lozère, S France
Mendeleyev Ridge 133 B2 undersea ridge Arctic Ocean
Mendocino Fracture Zone 130 D2 fracture zone NE Pacific Ocean
Mendoza 42 B4 Mendoza, W Argentina
Menemen 94 A3 İzmir, W Turkey
Menengiyn Tal 105 F2 plain E Mongolia
Menongue 56 B2 var. Vila Serpa Pinto, Port. Serpa Pinto. Cuando Cubango, C Angola
Menorca 71 H3 Eng. Minorca; anc. Balearis Minor. island Islas Baleares, Spain, W Mediterranean Sea
Mentawai, Kepulauan 116 A4 island group W Indonesia
Meppel 64 D2 Drenthe, NE Netherlands
Meran see Merano
Merano 74 C1 Ger. Meran. Trentino-Alto Adige, N Italy
Merca see Marka
Mercedes 42 D3 Corrientes, NE Argentina
Mercedes 42 D4 Soriano, SW Uruguay
Meredith, Lake 27 E1 reservoir Texas, SW USA
Merefa 87 G2 Kharkivs'ka Oblast', E Ukraine
Mergate see Margate
Mergui see Myeik
Mergui Archipelago see Myeik Archipelago
Mérida 29 H3 Yucatán, SW Mexico
Mérida 70 C4 anc. Augusta Emerita. Extremadura, W Spain
Mérida 36 C2 Mérida, W Venezuela
Meridian 20 C2 Mississippi, S USA
Mérignac 69 B5 Gironde, SW France
Merín, Laguna see Mirim Lagoon
Merkulovichi see Myerkulavichy
Merowe 50 B3 Northern, N Sudan
Merredin 125 B6 Western Australia
Mersen see Meerssen
Mersey 67 D5 river NW England, United Kingdom
Mersin 94 C4 var. İçel. İçel, S Turkey
Mērsrags 84 C3 Talsi, NW Latvia
Meru 51 C6 Eastern, C Kenya
Merv see Mary
Merzifon 94 D2 Amasya, N Turkey
Merzig 73 A5 Saarland, SW Germany
Mesa 26 B2 Arizona, SW USA
Meseritz see Międzyrzecz
Meshed see Mashhad
Mesopotamia 35 C5 var. Mesopotamia Argentina. physical region NE Argentina
Mesopotamia Argentina see Mesopotamia
Messalo, Rio 57 E2 var. Mualo. river NE Mozambique
Messana/Messene see Messina
Messina see Musina
Messina, Strait of see Messina, Stretto di
Messina, Stretto di 75 D7 Eng. Strait of Messina. strait SW Italy
Messíni 83 B6 Pelopónnisos, S Greece
Mesta see Néstos
Mestghanem see Mostaganem
Mestia 95 F1 var. Mestiya. N Georgia
Mestiya see Mestia
Mestre 74 C2 Veneto, NE Italy
Metairie 20 B3 Louisiana, S USA
Metán 42 C2 Salta, N Argentina
Metapán 30 B2 Santa Ana, NW El Salvador
Meta, Río 36 D3 river Colombia/Venezuela
Meterlam see Mehtar Lām
Methariam/Metharlam see Mehtar Lām
Metis see Metz
Metković 78 B4 Dubrovnik-Neretva, SE Croatia
Métsovo 82 B4 prev. Métsovon. Ípeiros, C Greece
Métsovon see Métsovo
Metz 68 D3 anc. Divodurum Mediomatricum, Mediomatrica, Metis. Moselle, NE France
Meulaboh 116 A3 Sumatera, W Indonesia
Meuse 65 C6 Dut. Maas. river W Europe
Mexcala, Río see Balsas, Río
Mexicali 28 A1 Baja California Norte, NW Mexico
Mexicanos, Estados Unidos see Mexico
México 29 E4 var. Ciudad de México, Eng. Mexico City. country capital (Mexico) México, C Mexico
Mexico 23 G4 Missouri, C USA
Mexico 28 C3 off. United Mexican States, var. Méjico, México, Sp. Estados Unidos Mexicanos. country N Central America
México see Mexico
México City see México
México, Golfo de see Mexico, Gulf of
Mexico, Gulf of 29 F2 Sp. Golfo de México. gulf W Atlantic Ocean
Meyadine see Al Mayādīn
Meydān Shahr see Maīdān Shahr
Meymaneh see Maīmanah
Mezen' 88 D3 river NW Russia
Mezőtúr 77 D7 Jász-Nagykun-Szolnok, E Hungary
Mgarr 80 A5 Gozo, N Malta
Miadzioł Nowy see Myadzyel
Miahuatlán 29 F5 var. Miahuatlán de Porfirio Díaz. Oaxaca, SE Mexico
Miahuatlán de Porfirio Díaz see Miahuatlán
Miami 21 F5 Florida, SE USA
Miami 27 G1 Oklahoma, C USA
Miami Beach 21 F5 Florida, SE USA
Mīāneh 98 C2 var. Miyāneh. Āžarbāyjān-e Sharqī, NW Iran
Mianyang 106 B5 Sichuan, C China
Miastko 76 C2 Ger. Rummelsburg in Pommern. Pomorskie, N Poland
Mi Chai see Nong Khai
Michalovce 77 E5 Ger. Grossmichel, Hung. Nagymihály. Košický Kraj, E Slovakia
Michigan 18 C1 off. State of Michigan, also known as Great Lakes State, Lake State, Wolverine State. state N USA
Michigan, Lake 18 C2 lake N USA
Michurin see Tsarevo
Michurinsk 89 B5 Tambovskaya Oblast', W Russia
Micoud 33 F2 SE Saint Lucia
Micronesia 122 B1 off. Federated States of Micronesia. country W Pacific Ocean
Micronesia 122 C1 island group W Pacific Ocean
Micronesia, Federated States of see Micronesia
Mid-Atlantic Cordillera see Mid-Atlantic Ridge
Mid-Atlantic Ridge 44 C3 var. Mid-Atlantic Cordillera, Mid-Atlantic Rise, Mid-Atlantic Swell. undersea ridge Atlantic Ocean
Mid-Atlantic Rise/Mid-Atlantic Swell see Mid-Atlantic Ridge
Middelburg 65 B5 Zeeland, SW Netherlands
Middelharnis 64 B4 Zuid-Holland, SW Netherlands
Middelkerke 65 A5 West-Vlaanderen, W Belgium
Middle America Trench 13 B7 trench E Pacific Ocean
Middle Andaman 111 F2 island Andaman Islands, India, NE Indian Ocean
Middle Atlas 48 C2 Eng. Middle Atlas. mountain range N Morocco
Middle Atlas see Moyen Atlas
Middleburg Island see 'Eua
Middle Congo see Congo (Republic of)
Middlesboro 18 C5 Kentucky, S USA
Middlesbrough 67 D5 N England, United Kingdom
Middletown 19 F4 New Jersey, NE USA
Middletown 19 F3 New York, NE USA
Mid-Indian Basin 119 C5 undersea basin N Indian Ocean
Mid-Indian Ridge 119 C5 var. Central Indian Ridge. undersea ridge C Indian Ocean
Midland 16 D5 Ontario, S Canada
Midland 18 C3 Michigan, N USA
Midland 27 E3 Texas, SW USA
Mid-Pacific Mountains 130 C2 var. Mid-Pacific Seamounts. seamount range NW Pacific Ocean
Mid-Pacific Seamounts see Mid-Pacific Mountains
Midway Islands 130 D2 US unincorporated territory C Pacific Ocean
Miechów 77 D5 Małopolskie, S Poland
Międzyrzec Podlaski 76 E3 Lubelskie, E Poland
Międzyrzecz 76 B3 Ger. Meseritz. Lubuskie, W Poland
Mielec 77 D5 Podkarpackie, SE Poland
Miercurea-Ciuc 86 C4 Ger. Szeklerburg, Hung. Csíkszereda. Harghita, C Romania
Mieres del Camín 70 D1 var. Mieres del Camino. Asturias, NW Spain
Mieres del Camino see Mieres del Camín
Mī'ēso 51 D5 var. Meheso, Miesso. Oromīya, C Ethiopia
Miesso see Mī'ēso
Mifrats Hefa 97 A5 Eng. Bay of Haifa; prev. MifrazḤefa. bay N Israel
Miguel Asua 28 D3 var. Miguel Auza. Zacatecas, C Mexico
Miguel Auza see Miguel Asua
Mijdrecht 64 C3 Utrecht, C Netherlands
Mikashevichi see Mikashevichy
Mikashevichy 85 C7 Pol. Mikaszewicze, Rus. Mikashevichi. Brestskaya Voblasts', SW Belarus
Mikaszewicze see Mikashevichy
Mikhaylovgrad see Montana
Mikhaylovka 89 B6 Volgogradskaya Oblast', SW Russia
Míkonos see Mýkonos
Mikre 82 C2 Lovech, N Bulgaria
Mikun' 88 D4 Respublika Komi, NW Russia
Mikuni-sanmyaku 109 D5 mountain range Honshū, N Japan Asia
Mikura-jima 109 D6 island E Japan
Milagro 38 B2 Guayas, SW Ecuador
Milan see Milano
Milange 57 E2 Zambézia, NE Mozambique
Milano 74 B2 Eng. Milan, Ger. Mailand; anc. Mediolanum. Lombardia, N Italy
Milas 94 A4 Muğla, SW Turkey
Milashavichy 85 C7 Rus. Milashevichi. Homyel'skaya Voblasts', SE Belarus
Milashevichi see Milashavichy
Mildura 127 C6 Victoria, SE Australia
Mile see Mili Atoll
Miles 127 D5 Queensland, E Australia
Miles City 22 C2 Montana, NW USA
Milford see Milford Haven
Milford Haven 67 C6 prev. Milford. SW Wales, United Kingdom
Milford Sound 129 A6 Southland, South Island, New Zealand
Milford Sound 129 A6 inlet South Island, New Zealand
Milḥ, Baḥr al see Razāzah, Buḥayrat ar
Mili Atoll 122 D2 var. Mile. atoll Ratak Chain, SE Marshall Islands
Mil'kovo 93 H3 Kamchatskaya Oblast', E Russia
Milk River 15 E5 Alberta, SW Canada
Milk River 22 C1 river Montana, NW USA
Milk, Wadi el 66 B4 var. Wadi al Malik. river C Sudan
Milledgeville 21 E2 Georgia, SE USA
Mille Lacs Lake 23 F2 lake Minnesota, N USA
Millennium Island 121 F3 prev. Caroline Island, Thornton Island. atoll Line Islands, E Kiribati
Millerovo 89 B6 Rostovskaya Oblast', SW Russia
Mílos 83 C7 island Kykládes, Greece, Aegean Sea
Milton 129 B7 Otago, South Island, New Zealand
Milton Keynes 67 D6 SE England, United Kingdom
Milwaukee 18 B3 Wisconsin, N USA
Mimatum see Mende
Min see Fujian
Mīnā' Qābūs 118 B3 NE Oman
Minas Gerais 41 F3 off. Estado de Minas Gerais. state/region E Brazil
Minas Gerais, Estado de see Minas Gerais
Minatitlán 29 F4 Veracruz-Llave, E Mexico
Minbu 114 A3 Magway, W Myanmar (Burma)
Minch, The 66 B3 var. North Minch. strait NW Scotland, United Kingdom
Mindanao 117 F2 island S Philippines
Mindanao Sea see Bohol Sea
Mindelheim 73 C6 Bayern, S Germany
Mindello see Mindelo
Mindelo 52 A2 var. Mindello; prev. Porto Grande. São Vicente, N Cape Verde
Minden 72 B4 anc. Minthun. Nordrhein-Westfalen, NW Germany
Mindoro 117 E2 island N Philippines
Mindoro Strait 117 E2 strait W Philippines
Mineral Wells 27 F2 Texas, SW USA
Mingäçevir 95 G2 Rus. Mingechaur, Mingechevir. C Azerbaijan
Mingechaur/Mingechevir see Mingäçevir
Mingora see Saidu Sharif
Minho 70 B2 former province N Portugal
Minho 70 B2 Sp. Miño. river Portugal/Spain
Minho, Rio see Miño
Minicoy Island 110 B3 island SW India
Minius see Miño
Minna 53 G4 Niger, C Nigeria
Minneapolis 23 F2 Minnesota, N USA
Minnesota 23 F2 off. State of Minnesota, also known as Gopher State, New England of the West, North Star State. state N USA
Miño 70 B2 var. Mino, Minius, Port. Rio Minho. river Portugal/Spain
Miño see Minho, Rio
Minorca see Menorca
Minot 23 E1 North Dakota, N USA
Minsk 85 C6 country capital (Belarus) Minskaya Voblasts', C Belarus
Minskaya Wzvyshsha 85 C6 mountain range C Belarus
Mińsk Mazowiecki 76 D3 var. Nowo-Minsk. Mazowieckie, C Poland
Minthun see Minden
Minto, Lac 16 D2 lake Québec, C Canada
Minya see Al Minyā
Miraflores 28 C3 Baja California Sur, NW Mexico
Miranda de Ebro 71 E1 La Rioja, N Spain
Mirgorod see Myrhorod
Miri 116 D3 Sarawak, East Malaysia
Mirim Lagoon 41 E5 var. Lake Mirim, Sp. Laguna Merín. lagoon Brazil/Uruguay
Mirim, Lake see Mirim Lagoon
Mírina see Mýrina
Mīrjāveh 98 E4 Sīstān va Balūchestān, SE Iran
Mirny 132 C3 Russian research station Antarctica
Mirnyy 93 F3 Respublika Sakha (Yakutiya), NE Russia
Mīrpur Khās 112 B3 Sind, SE Pakistan
Mirtoan Sea see Mirtóo Pélagos
Mirtóo Pélagos 83 C6 Eng. Mirtoan Sea; anc. Myrtoum Mare. sea S Greece
Misiaf see Maṣyāf
Miskito Coast see Mosquito Coast
Miskitos, Cayos 31 E2 island group NE Nicaragua
Miskolc 77 D6 Borsod-Abaúj-Zemplén, NE Hungary
Misool, Pulau 117 F4 island Maluku, E Indonesia
Miṣrātah 49 F2 var. Misurata. NW Libya
Mission 27 G5 Texas, SW USA
Mississippi 20 B2 off. State of Mississippi, also known as Bayou State, Magnolia State. state SE USA
Mississippi Delta 20 B4 delta Louisiana, S USA
Mississippi River 13 C6 river C USA
Missoula 22 B1 Montana, NW USA
Missouri 23 F5 off. State of Missouri, also known as Bullion State, Show Me State. state C USA
Missouri River 23 E3 river C USA
Mistassini, Lac 16 D3 lake Québec, SE Canada
Mistelbach an der Zaya 73 E6 Niederösterreich, NE Austria
Misti, Volcán 39 E4 volcano S Peru

Nãbulus – Ningbo

Parramatta – Pontoise

Parramatta 126 D1 New South Wales, SE Australia
Parras 28 D3 var. Parras de la Fuente. Coahuila, NE Mexico
Parras de la Fuente see Parras
Parsons 23 F5 Kansas, C USA
Pasadena 25 C7 California, W USA
Pasadena 27 H4 Texas, SW USA
Paşcani 86 C3 Hung. Páskán. Iaşi, NE Romania
Pasco 24 C2 Washington, NW USA
Pascua, Isla de 131 F4 var. Rapa Nui, Easter Island. island E Pacific Ocean
Pasewalk 72 D3 Mecklenburg-Vorpommern, NE Germany
Pashkeni see Bolyarovo
Pasinler 95 F3 Erzurum, NE Turkey
Páskán see Paşcani
Pasłęk 76 D2 Ger. Preußisch Holland. Warmińsko-Mazurskie, NE Poland
Pasni 112 A3 Baluchistān, SW Pakistan
Paso de Indios 43 B6 Chubut, S Argentina
Passarowitz see Požarevac
Passau 73 D6 Bayern, SE Germany
Passo Fundo 41 E5 Rio Grande do Sul, S Brazil
Pastavy 85 C5 Pol. Postawy, Rus. Postovy. Vitsyebskaya Voblasts', NW Belarus
Pastaza, Río 38 B2 river Ecuador/Peru
Pasto 36 A4 Nariño, SW Colombia
Pasvalys 84 C4 Panevėžys, N Lithuania
Patagonia 35 B7 physical region Argentina/Chile
Patalung see Phatthalung
Patani see Pattani
Patavium see Padova
Patea 128 D4 Taranaki, North Island, New Zealand
Paterson 19 F3 New Jersey, NE USA
Pathein 114 A4 var. Bassein. Ayeyarwady, SW Myanmar (Burma)
Pátmos 83 D6 island Dodekánisa, Greece, Aegean Sea
Patna 113 F3 var. Azimabad. state capital Bihār, N India
Patnos 95 F3 Ağrı, E Turkey
Patos, Lagoa dos 41 E5 lagoon S Brazil
Pátra 83 B5 Eng. Patras; prev. Pátrai. Dytikí Ellás, S Greece
Pátrai/Patras see Pátra
Pattani 115 C7 var. Patani. Pattani, SW Thailand
Pattaya 115 C5 Chon Buri, S Thailand
Patuca, Río 30 D2 river E Honduras
Pau 69 B6 Pyrénées-Atlantiques, SW France
Paulatuk 15 E3 Northwest Territories, NW Canada
Paungde 114 B4 Bago, C Myanmar (Burma)
Pautalia see Kyustendil
Pavia 74 B2 anc. Ticinum. Lombardia, N Italy
Pāvilosta 84 B3 Liepāja, W Latvia
Pavlikeni 82 D2 Veliko Tŭrnovo, N Bulgaria
Pavlodar 92 C4 Pavlodar, NE Kazakhstan
Pavlograd see Pavlohrad
Pavlohrad 87 G3 Rus. Pavlograd. Dnipropetrovs'ka Oblast', E Ukraine
Pawai, Pulau 116 A2 island SW Singapore Asia
Pawn 114 B3 river C Myanmar (Burma)
Pax Augusta see Badajoz
Pax Julia see Beja
Paxoí 83 A5 island Iónia Nisiá, Greece, C Mediterranean Sea
Payo Obispo see Chetumal
Paysandú 42 D4 Paysandú, W Uruguay
Pazar 95 E2 Rize, NE Turkey
Pazardzhik 82 C3 prev. Tatar Pazardzhik. Pazardzhik, SW Bulgaria
Peace Garden State see North Dakota
Peach State see Georgia
Pearl Islands 31 G5 Eng. Pearl Islands. island group SE Panama
Pearl Islands see Perlas, Archipiélago de las
Pearl Lagoon see Perlas, Laguna de
Pearl River 20 B3 river Louisiana/Mississippi, S USA
Pearsall 27 F4 Texas, SW USA
Peawanuk 16 C2 Ontario, S Canada
Peć see Pejë
Pechora 88 D3 Respublika Komi, NW Russia
Pechora 88 D3 river NW Russia
Pechora Sea see Pechorskoye More
Pechorskoye More 88 D2 Eng. Pechora Sea. sea NW Russia
Pecos 27 E3 Texas, SW USA
Pecos River 27 E3 river New Mexico/Texas, SW USA
Pécs 77 C7 Ger. Fünfkirchen, Lat. Sopianae. Baranya, SW Hungary
Pedra Lume 52 A3 Sal, NE Cape Verde
Pedro Cays 32 C3 island group Greater Antilles, S Jamaica North America N Caribbean Sea W Atlantic Ocean
Pedro Juan Caballero 42 D2 Amambay, E Paraguay
Peer 65 D5 Limburg, NE Belgium
Pegasus Bay 129 C6 bay South Island, New Zealand
Pegu see Bago
Pehuajó 42 C4 Buenos Aires, E Argentina
Pei-ching see Beijing/Beijing Shi
Peine 72 B3 Niedersachsen, C Germany
Pei-p'ing see Beijing/Beijing Shi
Peipsi Järv/Peipus-See see Peipus, Lake
Peipus, Lake 84 E3 Est. Peipsi Järv, Ger. Peipus-See, Rus. Chudskoye Ozero. lake Estonia/Russia
Peiraías 83 C6 prev. Piraiévs, Eng. Piraeus. Attikí, C Greece
Pejë 79 D5 Serb. Peć. W Kosovo
Pèk see Phônsaven
Pekalongan 116 C4 Jawa, C Indonesia
Pekanbaru 116 B3 var. Pakanbaru. Sumatera, W Indonesia
Pekin 18 B4 Illinois, N USA
Peking see Beijing/Beijing Shi
Pelagie 75 B8 island group SW Italy
Pelagosa see Palagruža
Pelican State see Louisiana
Pelly Bay see Kugaaruk
Pélmonostor see Beli Manastir
Peloponnese 83 B6 var. Morea, Eng. Peloponnese; anc. Peloponnesus. peninsula S Greece
Peloponnese/Peloponnesus see Pelopónnisos
Pematangsiantar 116 B3 Sumatera, W Indonesia
Pemba 57 F2 prev. Port Amelia, Porto Amélia. Cabo Delgado, NE Mozambique
Pemba 51 D7 island E Tanzania
Pembroke 16 D4 Ontario, SE Canada
Penang see Pinang, Pulau, Peninsular Malaysia
Penang see George Town
Penas, Golfo de 43 A7 gulf S Chile
Penderma see Bandırma
Pendleton 24 C3 Oregon, NW USA
Pend Oreille, Lake 24 D2 lake Idaho, NW USA
Peneius see Pineiós
Peng-pu see Bengbu
Penibético, Sistema see Béticos, Sistemas
Peniche 70 B3 Leiria, W Portugal
Peninsular State see Florida
Penninae, Alpes/Pennine, Alpi see Pennine Alps
Pennine Alps 73 A8 Fr. Alpes Pennines, It. Alpi Pennine, Lat. Alpes Penninae. mountain range Italy/Switzerland
Pennine Chain see Pennines
Pennines 67 D5 var. Pennine Chain. mountain range N England, United Kingdom
Pennines, Alpes see Pennine Alps
Pennsylvania 19 E4 off. Commonwealth of Pennsylvania, also known as Keystone State. state NE USA
Penobscot River 19 G2 river Maine, NE USA
Penong 127 A6 South Australia
Penonomé 31 F5 Coclé, C Panama
Penrhyn 123 G3 atoll N Cook Islands
Penrhyn Basin 121 F3 undersea feature C Pacific Ocean
Penrith 126 D1 New South Wales, SE Australia
Penrith 67 D5 NW England, United Kingdom
Pensacola 20 C3 Florida, SE USA
Pentecost 122 D4 Fr. Pentecôte. island C Vanuatu
Pentecôte see Pentecost
Penza 89 C6 Penzenskaya Oblast', W Russia
Penzance 67 C7 SW England, United Kingdom
Peoria 18 B4 Illinois, N USA
Perchtoldsdorf 73 E6 Niederösterreich, NE Austria
Percival Lakes 124 C4 lakes Western Australia
Perdido, Monte 71 F1 mountain NE Spain
Perece Vela Basin see West Mariana Basin
Pereira 36 B3 Risaralda, W Colombia
Peremyshl see Przemyśl
Pergamino 42 C4 Buenos Aires, E Argentina
Périgueux 69 C5 anc. Vesuna. Dordogne, SW France
Perito Moreno 43 B6 Santa Cruz, S Argentina
Perlas, Laguna de 31 E3 Eng. Pearl Lagoon. lagoon E Nicaragua
Perleberg 72 C3 Brandenburg, N Germany
Perlepe see Prilep
Perm' 92 C3 prev. Molotov. Permskaya Oblast', NW Russia
Pernambuco 41 G2 off. Estado de Pernambuco. state/region E Brazil
Pernambuco see Recife
Pernambuco Abyssal Plain see Pernambuco Plain
Pernambuco, Estado de see Pernambuco
Pernambuco Plain 45 C5 var. Pernambuco Abyssal Plain. undersea feature E Atlantic Ocean
Pernau see Pärnu
Pernauer Bucht see Pärnu Laht
Pērnava see Pärnu
Pernik 82 B2 prev. Dimitrovo. Pernik, W Bulgaria
Pernov see Pärnu
Perote 29 F4 Veracruz-Llave, E Mexico
Pérouse see Perugia
Perovsk see Kzylorda
Perpignan 69 C6 Pyrénées-Orientales, S France
Perryton 27 F1 Texas, SW USA
Perryville 23 H5 Missouri, C USA
Persia see Iran
Persian Gulf 98 C4 var. Gulf, The, Ar. Khalīj al 'Arabī, Per. Khalīj-e Fars. gulf SW Asia
Perth 125 A6 state capital Western Australia
Perth 66 C4 C Scotland, United Kingdom
Perth Basin 119 E6 undersea feature SE Indian Ocean
Peru 38 C3 off. Republic of Peru. country W South America
Peru see Beru
Peru Basin 45 A5 undersea feature E Pacific Ocean
Peru-Chile Trench 34 A4 undersea feature E Pacific Ocean
Perugia 74 C4 Fr. Pérouse; anc. Perusia. Umbria, C Italy
Perugia, Lake of see Trasimeno, Lago
Peru, Republic of see Peru
Perusia see Perugia
Péruwelz 65 B6 Hainaut, SW Belgium
Pervomays'k 87 E3 prev. Ol'viopol'. Mykolayivs'ka Oblast', S Ukraine
Pervyy Kuril'skiy Proliv 93 H3 strait E Russia
Pesaro 74 C3 anc. Pisaurum. Marche, C Italy
Pescara 74 D4 anc. Aternum, Ostia Aterni. Abruzzo, C Italy
Peshāwar 112 C1 North-West Frontier Province, N Pakistan
Peshkopi 79 C6 var. Peshkopia, Peshkopija. Dibër, NE Albania
Peshkopia/Peshkopija see Peshkopi
Pessac 69 B5 Gironde, SW France
Petach-Tikva see Petah Tikva
Petah Tikva 97 A6 var. Petach-Tikva, Petah Tiqva, Petakh Tikva; prev. Petaḥ Tiqwa. Tel Aviv, C Israel
Petaḥ Tiqwa see Petah Tikva
Petakh Tikva/Petah Tiqva see Petah Tikva
Pétange 65 D8 Luxembourg, SW Luxembourg
Petchaburi see Phetchaburi
Peterborough 127 B6 South Australia
Peterborough 16 D5 Ontario, SE Canada
Peterborough 67 E6 prev. Medeshamstede. E England, United Kingdom
Peterhead 66 D3 NE Scotland, United Kingdom
Peter I Øy 132 A3 Norwegian dependency Antarctica
Petermann Bjerg 61 E3 mountain C Greenland
Petersburg 19 E5 Virginia, NE USA
Peters Mine 37 F3 var. Peter's Mine. N Guyana
Petit St-Bernard, Col du see Little Saint Bernard Pass
Peto 29 H4 Yucatán, SE Mexico
Petoskey 18 C2 Michigan, N USA
Petra see Wādī Mūsā
Petrich 82 C3 Blagoevgrad, SW Bulgaria
Petrikau see Piotrków Trybunalski
Petrikov see Pyetrykaw
Petrinja 78 B3 Sisak-Moslavina, C Croatia
Petroaleksandrovsk see To'rtkok'l
Petrodvorets 88 A4 Fin. Pietarhovi. Leningradskaya Oblast', NW Russia
Petrograd see Sankt-Peterburg
Petrokov see Piotrków Trybunalski
Petropavl see Petropavlovsk
Petropavlovsk 92 C4 Kaz. Petropavl. Severnyy Kazakhstan, N Kazakhstan
Petropavlovsk-Kamchatskiy 93 H3 Kamchatskaya Oblast', E Russian Federation
Petroşani 86 B4 var. Petroşeni, Ger. Petroschen, Hung. Petrozsény. Hunedoara, W Romania
Petroschen/Petroşeni see Petroşani
Petroskoi see Petrozavodsk
Petrovgrad see Zrenjanin
Petrovsk-Port see Makhachkala
Petrozavodsk 92 B2 Fin. Petroskoi. Respublika Kareliya, NW Russia
Petrozsény see Petroşani
Pettau see Ptuj
Pevek 93 G1 Chukotskiy Avtonomnyy Okrug, NE Russia
Pezinok 77 C6 Ger. Bösing, Hung. Bazin. Bratislavský Kraj, W Slovakia
Pforzheim 73 B6 Baden-Württemberg, SW Germany
Pfungstadt 73 B5 Hessen, W Germany
Phangan, Ko 115 C6 island SW Thailand
Phang-Nga 115 B7 var. Pang-Nga, Phangnga. Phangnga, SW Thailand
Phangnga see Phang-Nga
Phan Rang/Phanrang see Phan Rang-Thap Cham
Phan Rang-Thap Cham 115 E6 var. Phanrang, Phan Rang, Phan Rang Thap Cham. Ninh Thuận, S Vietnam
Phan Thiêt 115 E6 Binh Thuận, S Vietnam
Pharnacia see Giresun
Pharus see Hvar
Phatthalung 115 C7 var. Padalung, Patalung. Phatthalung, SW Thailand
Phayao 114 C4 var. Muang Phayao. Phayao, NW Thailand
Phenix City 20 D2 Alabama, S USA
Phet Buri see Phetchaburi
Phetchaburi 115 C5 var. Bejraburi, Petchaburi, Phet Buri. Phetchaburi, SW Thailand
Philadelphia 19 F4 Pennsylvania, NE USA
Philadelphia see 'Ammān
Philippine Basin 103 F3 undersea feature W Pacific Ocean
Philippine Islands 117 E1 island group W Pacific Ocean
Philippines 117 E1 off. Republic of the Philippines. country SE Asia
Philippine Sea 103 F3 sea W Pacific Ocean
Philippines, Republic of the see Philippines
Philippine Trench 120 A1 undersea feature W Philippine Sea
Philippopolis see Plovdiv
Phitsanulok 114 C4 var. Bisnulok, Muang Phitsanulok, Pitsanulok. Phitsanulok, C Thailand
Phlórina see Flórina
Phnom Penh 115 D6 Khmer. Phnum Pénh. country capital (Cambodia) Phnom Penh, S Cambodia
Phnum Pénh see Phnom Penh
Phoenix 26 B2 state capital Arizona, SW USA
Phoenix Islands 123 E3 island group C Kiribati
Phôngsali 114 C3 var. Phong Saly. Phôngsali, N Laos
Phong Saly see Phôngsali
Phônsaven 114 D4 var. Xieng Khouang; prev. Pèk, Xiangkhoang. Xiangkhoang, N Laos
Phrae 114 C4 var. Muang Phrae, Prae. Phrae, NW Thailand
Phra Nakhon Si Ayutthaya see Ayutthaya
Phra Thong, Ko 115 B6 island SW Thailand
Phuket 115 B7 var. Bhuket, Puket, Mal. Ujung Salang; prev. Junkseylon, Salang. Phuket, SW Thailand
Phuket, Ko 115 B7 island SW Thailand
Phumĭ Kâmpóng Trâbêk see Kâmpóng Trâbêk
Phumĭ Sâmraông see Sâmraông
Phum Kompong Trabek see Kâmpóng Trâbêk
Phum Samrong see Sâmraông
Phu Vinh see Tra Vinh
Phyu 114 B4 var. Hpyu, Pyu. Bago, C Myanmar (Burma)
Piacenza 74 B2 Fr. Paisance; anc. Placentia. Emilia-Romagna, N Italy
Piatra-Neamţ 86 C4 Hung. Karácsonkő. Neamţ, NE Romania
Piauhy see Piauí
Piauí 41 F2 off. Estado do Piauí; prev. Piauhy. state/region E Brazil
Piauí, Estado do see Piauí
Picardie 68 C3 Eng. Picardy. cultural region N France
Picardy see Picardie
Piccolo San Bernardo, Colle di see Little Saint Bernard Pass
Pichilemu 42 B4 Libertador, C Chile
Pico 70 A5 var. Ilha do Pico. island Azores, Portugal, NE Atlantic Ocean
Pico, Ilha do see Pico
Picos 41 F2 Piauí, E Brazil
Picton 129 C5 Marlborough, South Island, New Zealand
Piedmont see Piemonte
Piedras Negras 29 E2 var. Ciudad Porfirio Díaz. Coahuila, NE Mexico
Pielavesi 62 D4 lake C Finland
Pielinen 62 E4 var. Pielisjärvi. lake E Finland
Pielisjärvi see Pielinen
Piemonte 74 A2 Eng. Piedmont. region NW Italy
Pierre 23 E3 state capital South Dakota, N USA
Piešt'any 77 C6 Ger. Pistyan, Hung. Pöstyén. Tranavský Kraj, W Slovakia
Pietarhovi see Petrodvorets
Pietari see Sankt-Peterburg
Pietarsaari see Jakobstad
Pietermaritzburg 56 C5 var. Maritzburg. KwaZulu/Natal, E South Africa
Pietersburg see Polokwane
Pigs, Bay of see Cochinos, Bahía de
Pihkva Järv see Pskov, Lake
Pijijiapán 29 G5 Chiapas, SE Mexico
Pikes Peak 22 C5 mountain Colorado, C USA
Pikeville 18 D5 Kentucky, S USA
Pikinni see Bikini Atoll
Piła 76 B3 Ger. Schneidemühl. Wielkopolskie, C Poland
Pilar 42 D3 var. Villa del Pilar. Ñeembucú, S Paraguay
Pilcomayo, Rio 35 C5 river C South America
Pilos see Pýlos
Pilsen see Plzeň
Pilzno see Plzeň
Pinang see Pinang, Pulau, Peninsular Malaysia
Pinang see George Town
Pinang, Pulau 116 B3 var. Penang, Pinang; prev. Prince of Wales Island. island Peninsular Malaysia
Pinar del Río 32 A2 Pinar del Río, W Cuba
Píndhos/Píndhos Óros see Píndos
Píndos 82 A4 var. Píndhos Óros, Eng. Pindus Mountains; prev. Píndhos. mountain range C Greece
Pindus Mountains see Píndos
Pine Bluff 20 B2 Arkansas, C USA
Pine Creek 124 D2 Northern Territory, N Australia
Pinega 88 C3 river NW Russia
Pineiós 82 B4 var. Piniós; anc. Peneius. river C Greece
Pineland 27 H3 Texas, SW USA
Píhes, Akrotírio 82 C4 var. Akrotírio Pínnes. headland N Greece
Pines, The Isle of the see Juventud, Isla de la
Pine Tree State see Maine
Pingdingshan 106 C4 Henan, C China
Pingkiang see Harbin
Ping, Mae Nam 114 B4 river W Thailand
Piniós see Pineiós
Pinkiang see Harbin
Pínnes, Akrotírio see Píhes, Akrotírio
Pinos, Isla de see Juventud, Isla de la
Pinotepa Nacional 29 F5 var. Santiago Pinotepa Nacional. Oaxaca, SE Mexico
Pinsk 85 B7 Pol. Pińsk. Brestskaya Voblasts', SW Belarus
Pinta, Isla 38 A5 var. Abingdon. island Galápagos Islands, Ecuador, E Pacific Ocean
Piombino 74 B3 Toscana, C Italy
Pioneer Mountains 24 D3 mountain range Montana, N USA North America
Pionerskiy 84 A4 Ger. Neukuhren. Kaliningradskaya Oblast', W Russia
Piotrków Trybunalski 76 D4 Ger. Petrikau, Rus. Petrokov. Łodzkie, C Poland
Piraeus/Piraiévs see Peiraías
Pírgos see Pýrgos
Pirineos see Pyrenees
Piripiri 41 F2 Piauí, E Brazil
Pirna 72 D4 Sachsen, E Germany
Pirot 79 E5 Serbia, SE Serbia
Piryatin see Pyryatyn
Pisa 74 B3 var. Pisae. Toscana, C Italy
Pisae see Pisa
Pisaurum see Pesaro
Pisco 38 D4 Ica, SW Peru
Písek 77 A5 Budějovický Kraj, S Czech Republic (Czechia)
Pishan 104 A3 var. Guma. Xinjiang Uygur Zizhiqu, NW China
Pishpek see Bishkek
Pistoia 74 B3 anc. Pistoria, Pistoriæ. Toscana, C Italy
Pistoria/Pistoriæ see Pistoia
Pistyan see Piešt'any
Pisz 76 D3 Ger. Johannisburg. Warmińsko-Mazurskie, NE Poland
Pita 52 C4 NW Guinea
Pitalito 36 B4 Huila, S Colombia
Pitcairn Group of Islands see Pitcairn, Henderson, Ducie & Oeno Islands
Pitcairn Island 121 G4 island S Pitcairn Group of Islands
Pitcairn, Henderson, Ducie & Oeno Islands 121 G4 var. Pitcairn Group of Islands. UK dependent territory C Pacific Ocean
Piteå 62 D4 Norrbotten, N Sweden
Piteşti 86 B5 Argeş, S Romania
Pitsanulok see Phitsanulok
Pitt Island see Makin
Pittsburg 23 F5 Kansas, C USA
Pittsburgh 19 E4 Pennsylvania, NE USA
Pittsfield 19 F3 Massachusetts, NE USA
Piura 38 B2 Piura, NW Peru
Pivdennyy Buh 87 E3 Rus. Yuzhnyy Bug. river S Ukraine
Placentia see Piacenza
Placetas 32 B2 Villa Clara, C Cuba
Plainview 27 E2 Texas, SW USA
Pláka 83 C7 Kykládes, Greece, Aegean Sea
Planeta Rica 36 B2 Córdoba, NW Colombia
Planken 72 E1 C Liechtenstein Europe
Plano 27 G2 Texas, SW USA
Plasencia 70 C3 Extremadura, W Spain
Plate, River 42 D4 var. River Plate. estuary Argentina/Uruguay
Plate, River see Plata, Río de la
Platinum 14 C3 Alaska, USA
Plattensee see Balaton
Platte River 23 E4 river Nebraska, C USA
Plattsburgh 19 F2 New York, NE USA
Plauen 73 C5 var. Plauen im Vogtland. Sachsen, E Germany
Plauen im Vogtland see Plauen
Pļaviņas 84 D4 Ger. Stockmannshof. Aizkraukle, S Latvia
Plây Cu 115 E5 var. Pleiku. Gia Lai, C Vietnam
Pleasant Island see Nauru
Pleiku see Plây Cu
Plenty, Bay of 128 E3 bay North Island, New Zealand
Plérin 68 A3 Côtes d'Armor, NW France
Plesetsk 88 C3 Arkhangel'skaya Oblast', NW Russia
Pleshchenitsy see Plyeshchanitsy
Pleskau see Pskov
Pleskauer See see Pskov, Lake
Pleskava see Pskov
Pleszew 76 C4 Wielkopolskie, C Poland
Pleven 82 C2 prev. Plevna. Pleven, N Bulgaria
Plevlja/Plevlje see Pljevlja
Plevna see Pleven
Pljevlja 78 C4 prev. Plevlja, Plevlje. N Montenegro
Plocce see Ploče
Ploče 78 B4 It. Plocce; prev. Kardeljevo. Dubrovnik-Neretva, SE Croatia
Płock 76 D3 Ger. Plozk. Mazowieckie, C Poland
Plöcken Pass 73 C7 Ger. Plöckenpass, It. Passo di Monte Croce Carnico. pass SW Austria
Plöckenpass see Plöcken Pass
Ploeşti see Ploieşti
Ploieşti 86 C5 prev. Ploeşti. Prahova, SE Romania
Plomári 83 D5 prev. Plomárion. Lésvos, E Greece
Plomárion see Plomári
Płońsk 76 D3 Mazowieckie, C Poland
Plovdiv 82 C3 prev. Eumolpias; anc. Evmolpia, Philippopolis, Lat. Trimontium. Plovdiv, C Bulgaria
Plozk see Płock
Plungė 84 B3 Telšiai, W Lithuania
Plyeshchanitsy 85 D5 Rus. Pleshchenitsy. Minskaya Voblasts', N Belarus
Plymouth 67 C7 SW England, United Kingdom
Plzeň 77 A5 Ger. Pilsen, Pol. Pilzno. Plzeňský Kraj, W Czech Republic (Czechia)
Po 58 D4 river N Italy
Pobedy, Pik see Tömür Feng
Po, Bocche del see Po, Foci del
Pocahontas 20 B1 Arkansas, C USA
Pocatello 24 E4 Idaho, NW USA
Pochinok 89 A5 Smolenskaya Oblast', W Russia
Pocking 73 D6 Bayern, SE Germany
Poděbrady 77 B5 Ger. Podiebrad. Středočeský Kraj, C Czech Republic (Czechia)
Podgorica 79 C5 prev. Titograd. country capital (Montenegro) S Montenegro
Podiebrad see Poděbrady
Podil's'k 86 D3 Rus. Kotovs'k. Odes'ka Oblast', SW Ukraine
Podil's'ka Vysochina 86 D3 plateau W Ukraine
Podium Anicensis see le Puy
Podol'sk 89 B5 Moskovskaya Oblast', W Russia
Podravska Slatina see Slatina
Podujevë 79 D5 Serb. Podujevo. N Kosovo
Podujevo see Podujevë
Podunajská Rovina see Little Alföld
Poetovio see Ptuj
Pogradec 79 D6 var. Pogradeci. Korçë, SE Albania
Pogradeci see Pogradec
Pohjanlahti see Bothnia, Gulf of
Pohnpei 122 C2 prev. Ponape Ascension Island. island E Micronesia
Poictiers see Poitiers
Poinsett, Cape 132 D4 headland Antarctica
Point de Galle see Galle
Pointe-à-Pitre 33 G3 Grande Terre, C Guadeloupe
Pointe-Noire 55 B6 Kouilou, S Congo
Point Lay 14 C2 Alaska, USA
Poitiers 68 B4 prev. Poictiers; anc. Limonum. Vienne, W France
Poitou 68 B4 cultural region W France
Pokharā 113 E3 Western, C Nepal
Pokrov 87 F3 Rus. Ordzhonikidze. Dnipropetrovs'ka Oblast', E Ukraine
Pokrovka see Kyzyl-Suu
Pokrovs'ke 87 G3 Rus. Pokrovskoye. Dnipropetrovs'ka Oblast', E Ukraine
Pokrovskoye see Pokrovs'ke
Pola see Pula
Pola de Lena see La Pola
Poland 76 B4 off. Republic of Poland, var. Polish Republic, Pol. Polska, Rzeczpospolita Polska; prev. Pol. Polska Rzeczpospolita Ludowa, The Polish People's Republic. country C Europe
Poland, Republic of see Poland
Polatlı 94 C3 Ankara, C Turkey
Polatsk 85 D5 Rus. Polotsk. Vitsyebskaya Voblasts', N Belarus
Pol-e Khomrī see Pul-e Khumrī
Poli see Pólis
Polikastro/Políkastron see Polýkastro
Polikrayshte see Dolna Oryakhovitsa
Pólis 80 C5 var. Poli. W Cyprus
Polish People's Republic, The see Poland
Polish Republic see Poland
Polkowice 76 B4 Ger. Heerwegen. Dolnośląskie, W Poland
Pollença 71 G3 Mallorca, Spain, W Mediterranean Sea
Pologi see Polohy
Polohy 87 G3 Rus. Pologi. Zaporiz'ka Oblast', SE Ukraine
Polokwane 56 D4 prev. Pietersburg. Limpopo, NE South Africa
Polonne 86 D2 Rus. Polonnoye. Khmel'nyts'ka Oblast', NW Ukraine
Polonnoye see Polonne
Polotsk see Polatsk
Polska/Polska, Rzeczpospolita/Polska Rzeczpospolita Ludowa see Poland
Polski Trŭmbesh 82 D2 prev. Polsko Kosovo. Ruse, N Bulgaria
Polsko Kosovo see Polski Trŭmbesh
Poltava 87 F2 Poltavs'ka Oblast', NE Ukraine
Poltoratsk see Aşgabat
Põlva 84 E3 Ger. Põlwe. Põlvamaa, SE Estonia
Põlwe see Põlva
Polyarnyy 88 C2 Murmanskaya Oblast', NW Russia
Polýkastro 82 B3 var. Polikastro; prev. Políkastron. Kentrikí Makedonía, N Greece
Polynesia 121 F4 island group C Pacific Ocean
Pomeranian Bay 72 D2 Ger. Pommersche Bucht, Pol. Zatoka Pomorska. bay Germany/Poland
Pomir, Dar"yoi see Pamir/Pāmīr, Daryā-ye
Pommersche Bucht see Pomeranian Bay
Pomorska, Zatoka see Pomeranian Bay
Pomorskiy Proliv 88 D2 strait NW Russia
Po, Mouth of the 74 C2 var. Bocche del Po. river NE Italy
Pompaelo see Pamplona
Pompano Beach 21 F5 Florida, SE USA
Ponape Ascension Island see Pohnpei
Ponca City 27 G1 Oklahoma, C USA
Ponce 33 F3 C Puerto Rico
Pondicherry 110 C2 var. Puducheri, Fr. Pondichéry. Pondicherry, SE India
Ponferrada 70 C1 Castilla-León, NW Spain
Poniatowa 76 E4 Lubelskie, E Poland
Pons Aelii see Newcastle upon Tyne
Pons Vetus see Pontevedra
Ponta Delgada 70 B5 São Miguel, Azores, Portugal, NE Atlantic Ocean
Ponta Grossa 41 E4 Paraná, S Brazil
Pontarlier 68 D4 Doubs, E France
Ponteareas 70 B2 Galicia, NW Spain
Ponte da Barca 70 B2 Viana do Castelo, N Portugal
Pontevedra 70 B1 anc. Pons Vetus. Galicia, NW Spain
Pontiac 18 D3 Michigan, N USA
Pontianak 116 C4 Borneo, C Indonesia
Pontisarae see Pontoise
Pontivy 68 A3 Morbihan, NW France
Pontoise 68 C3 anc. Briva Isarae, Cergy-Pontoise, Pontisarae. Val-d'Oise, N France

Ponziane Island – Quezaltenango

Ponziane Island *75 C5 island* C Italy
Poole *67 D7* S England, United Kingdom
Poona *see* Pune
Poopó, Lago *39 F4 var.* Lago Pampa Aullagas. *lake* W Bolivia
Popayán *36 B4* Cauca, SW Colombia
Poperinge *65 A6* West-Vlaanderen, W Belgium
Poplar Bluff *23 G5* Missouri, C USA
Popocatépetl *29 E4 volcano* S Mexico
Popper *see* Poprad
Poprad *77 D5 Ger.* Deutschendorf, *Hung.* Poprád. Prešovský Kraj, E Slovakia
Poprád *77 D5 Ger.* Popper, *Hung.* Poprád. *river* Poland/Slovakia
Porbandar *112 B4* Gujarāt, W India
Porcupine Plain *58 B3 undersea feature* E Atlantic Ocean
Pordenone *74 C2 anc.* Portenau. Friuli-Venezia Giulia, NE Italy
Poreč *78 A2 It.* Parenzo. Istra, NW Croatia
Porech'ye *see* Parechcha
Pori *63 D5 Swe.* Björneborg. Satakunta, SW Finland
Porirua *129 D5* Wellington, North Island, New Zealand
Porkhov *88 A4* Pskovskaya Oblast', W Russia
Porlamar *37 E1* Nueva Esparta, NE Venezuela
Póros *83 C6* Póros, S Greece
Póros *83 A5* Kefallinía, Iónia Nisiá, Greece, C Mediterranean Sea
Pors *see* Porsangenfijorden
Porsangenfijorden *62 D2 Lapp.* Pors. *fjord* N Norway
Porsgrunn *63 B6* Telemark, S Norway
Portachuelo *39 G4* Santa Cruz, C Bolivia
Portadown *67 B5 Ir.* Port An Dúnáin. S Northern Ireland, United Kingdom
Portalegre *70 C3 anc.* Ammaia, Amoea. Portalegre, E Portugal
Port Alexander *14 D4* Baranof Island, Alaska, USA
Port Alfred *56 D5* Eastern Cape, S South Africa
Port Amelia *see* Pemba
Port An Dúnáin *see* Portadown
Port Angeles *24 B1* Washington, NW USA
Port Antonio *32 B5* NE Jamaica
Port Arthur *27 H4* Texas, SW USA
Port Augusta *127 B6* South Australia
Port-au-Prince *32 D3 country capital* (Haiti) C Haiti
Port Blair *111 F2* Andaman and Nicobar Islands, SE India
Port Charlotte *21 E4* Florida, SE USA
Port Darwin *see* Darwin
Port d'Envalira *69 B8* E Andorra Europe
Port Douglas *126 D3* Queensland, NE Australia
Port Elizabeth *56 C5* Eastern Cape, S South Africa
Portenau *see* Pordenone
Porterville *25 C7* California, W USA
Port-Étienne *see* Nouâdhibou
Port Florence *see* Kisumu
Port-Francqui *see* Ilebo
Port-Gentil *55 A6* Ogooué-Maritime, W Gabon
Port Harcourt *53 G5* Rivers, S Nigeria
Port Hardy *14 D5* Vancouver Island, British Columbia, SW Canada
Port Harrison *see* Inukjuak
Port Hedland *124 B4* Western Australia
Port Huron *18 D3* Michigan, N USA
Portimão *70 B4 var.* Vila Nova de Portimão. Faro, S Portugal
Port Jackson *126 E1 harbour* New South Wales, E Australia
Portland *127 B7* Victoria, SE Australia
Portland *19 G2* Maine, NE USA
Portland *24 B3* Oregon, NW USA
Portland *27 G4* Texas, SW USA
Portland Bight *32 B5 bay* S Jamaica
Portlaoighise *see* Port Laoise
Port Laoise *67 B6 var.* Portlaoise, *Ir.* Portlaoighise; *prev.* Maryborough. C Ireland
Portlaoise *see* Port Laoise
Port Lavaca *27 G4* Texas, SW USA
Port Lincoln *127 A6* South Australia
Port Louis *57 H3 country capital* (Mauritius) NW Mauritius
Port-Lyautey *see* Kénitra
Port Macquarie *127 E6* New South Wales, SE Australia
Port Mahon *see* Mahón
Portmore *32 B5* C Jamaica
Port Moresby *122 B3 country capital* (Papua New Guinea) Central/National Capital District, SW Papua New Guinea
Port Natal *see* Durban
Porto *70 B2 Eng.* Oporto; *anc.* Portus Cale. Porto, NW Portugal
Porto Alegre *41 E5 var.* Pôrto Alegre. *state capital* Rio Grande do Sul, S Brazil
Porto Alegre *54 E2* São Tomé, S Sao Tome and Principe, Africa
Porto Alexandre *see* Tombua
Porto Amélia *see* Pemba
Porto Bello *see* Portobelo
Portobelo *31 G4 var.* Porto Bello, Puerto Bello. Colón, N Panama
Port O'Connor *27 G4* Texas, SW USA
Porto Edda *see* Sarandë
Portoferraio *74 B4* Toscana, C Italy
Port of Spain *33 H5 country capital* (Trinidad and Tobago) Trinidad, Trinidad and Tobago
Porto Grande *see* Mindelo
Portogruaro *74 C2* Veneto, NE Italy
Porto-Novo *53 F5 country capital* (Benin - official) S Benin
Porto Rico *see* Puerto Rico
Porto Santo *48 A2 var.* Ilha do Porto Santo. *island* Madeira, Portugal, NE Atlantic Ocean
Porto Santo, Ilha do *see* Porto Santo
Porto Torres *75 A5* Sardegna, Italy, C Mediterranean Sea
Porto Velho *40 D2 var.* Velho. *state capital* Rondônia, W Brazil
Portoviejo *38 A2 var.* Puertoviejo. Manabí, W Ecuador
Port Pirie *127 B6* South Australia
Port Rex *see* East London
Port Said *see* Būr Saʻīd
Portsmouth *67 D7* S England, United Kingdom
Portsmouth *19 G3* New Hampshire, NE USA
Portsmouth *18 D4* Ohio, N USA
Portsmouth *19 F5* Virginia, NE USA
Port Stanley *see* Stanley
Port Sudan *50 C3* Red Sea, NE Sudan
Port Swettenham *see* Klang/Pelabuhan Klang
Port Talbot *67 C7* S Wales, United Kingdom
Portugal *70 B3 off.* Portuguese Republic. *country* SW Europe
Portuguese East Africa *see* Mozambique
Portuguese Guinea *see* Guinea-Bissau
Portuguese Republic *see* Portugal
Portuguese Timor *see* East Timor
Portuguese West Africa *see* Angola
Portus Cale *see* Porto
Portus Magnus *see* Almería
Portus Magonis *see* Mahón
Port-Vila *122 D4 var.* Vila. *country capital* (Vanuatu) Éfaté, C Vanuatu
Porvenir *39 E3* Pando, NW Bolivia
Porvenir *43 B8* Magallanes, S Chile
Porvoo *63 E6 Swe.* Borgå. Uusimaa, S Finland
Porzecze *see* Parechcha
Posadas *42 D3* Misiones, NE Argentina
Poschega *see* Požega
Posen *see* Poznań
Posnania *see* Poznań
Postavy/Postawy *see* Pastavy
Posterholt *65 D5* Limburg, SE Netherlands
Postojna *73 D8 Ger.* Adelsberg, *It.* Postumia. SW Slovenia
Postumia *see* Postojna
Pöstyén *see* Piešťany
Potamós *83 C7* Antikýthira, S Greece
Potentia *see* Potenza
Potenza *75 D5 anc.* Potentia. Basilicata, S Italy
Poti *95 F2 prev.* P'ot'i. W Georgia
P'ot'i *see* Poti
Potiskum *53 G4* Yobe, NE Nigeria
Potomac River *19 E5 river* NE USA
Potosí *39 F4* Potosí, S Bolivia
Potsdam *72 D3* Brandenburg, NE Germany
Potwar Plateau *112 C2 plateau* NE Pakistan
Poŭthĭsăt *see* Pursat
Po, Valle del *see* Po Valley
Po Valley *74 C2 It.* Valle del Po. *valley* N Italy
Považská Bystrica *77 C5 Ger.* Waagbistritz, *Hung.* Vágbeszterce. Trenčiansky Kraj, W Slovakia
Poverty Bay *128 E4 inlet* North Island, New Zealand
Póvoa de Varzim *70 B2* Porto, NW Portugal
Powder River *22 D2 river* Montana/Wyoming, NW USA
Powell *22 C2* Wyoming, C USA
Powell, Lake *22 B5 lake* Utah, W USA
Požarevac *78 D4 Ger.* Passarowitz. Serbia, NE Serbia
Poza Rica *29 F4 var.* Poza Rica de Hidalgo. Veracruz-Llave, E Mexico
Poza Rica de Hidalgo *see* Poza Rica
Požega *78 D4 prev.* Slavonska Požega, *Ger.* Poschega, *Hung.* Pozsega. Požega-Slavonija, NE Croatia
Požega *78 D4* Serbia
Poznań *76 C3 Ger.* Posen, Posnania. Wielkolpolskie, C Poland
Pozoblanco *70 D4* Andalucía, S Spain
Pozsega *see* Požega
Pozsony *see* Bratislava
Pozzallo *75 C8* Sicilia, Italy, C Mediterranean Sea
Prachatice *77 A5 Ger.* Prachatitz. Jihočeský Kraj, S Czech Republic (Czechia)
Prachatitz *see* Prachatice
Prado del Ganso *see* Goose Green
Prae *see* Phrae
Prag/Praga/Prague *see* Praha
Praha *77 A5 Eng.* Prague, *Ger.* Prag, *Pol.* Praga. *country capital* (Czech Republic (Czechia)) Středočeský Kraj, NW Czech Republic (Czechia)
Praia *52 A3 country capital* (Cape Verde) Santiago, S Cape Verde
Prairie State *see* Illinois
Prathet Thai *see* Thailand
Prato *74 B3* Toscana, C Italy
Pratt *23 E5* Kansas, C USA
Prattville *20 D2* Alabama, S USA
Pravda *see* Glavinitsa
Pravia *70 C1* Asturias, N Spain
Preăh Seihânŭ *see* Sihanoukville
Preny *see* Prienai
Prenzlau *72 D3* Brandenburg, NE Germany
Prerau *see* Přerov
Přerov *77 C5 Ger.* Prerau. Olomoucký Kraj, E Czech Republic (Czechia)
Preschau *see* Prešov
Prescott *26 B2* Arizona, SW USA
Preševo *79 D5* Serbia, SE Serbia
Presidente Epitácio *41 E4* São Paulo, S Brazil
Presidente Stroessner *see* Ciudad del Este
Prešov *77 D5 var.* Preschau, *Ger.* Eperies, *Hung.* Eperjes. Prešovský Kraj, E Slovakia
Prespa, Lake *79 D6 Alb.* Liqen i Prespës, *Gk.* Límni Megáli Préspa, Límni Prespa, *Mac.* Prespansko Ezero, *Serb.* Prespansko Jezero. *lake* SE Europe
Prespa, Limni/Prespansko Ezero/Prespansko Jezero/Prespës, Liqen i *see* Prespa, Lake
Presque Isle *19 H1* Maine, NE USA
Pressburg *see* Bratislava
Preston *67 D5* NW England, United Kingdom
Prestwick *66 C4* W Scotland, United Kingdom
Pretoria *56 D4 var.* Epitoli. *country capital* (South Africa-administrative capital) Gauteng, NE South Africa
Preussisch Eylau *see* Bagrationovsk
Preußisch Holland *see* Pasłęk
Preussisch-Stargard *see* Starogard Gdański
Préveza *83 A5* Ípeiros, W Greece
Pribilof Islands *14 A3 island group* Alaska, USA
Priboj *78 C4* Serbia, W Serbia
Price *22 B4* Utah, W USA
Prichard *20 C3* Alabama, S USA
Priekulė *84 B3 Ger.* Prökuls. Klaipėda, W Lithuania
Prienai *85 B5 Pol.* Preny. Kaunas, S Lithuania
Prieska *56 C4* Northern Cape, C South Africa
Prijedor *78 B3* Republika Srpska, NW Bosnia and Herzegovina
Prijepolje *78 D4* Serbia, W Serbia
Prikaspiyskaya Nizmennost' *see* Caspian Depression
Prilep *79 D6 Turk.* Perlepe. S Macedonia
Priluki *see* Pryluky
Primorsk *84 A4 Ger.* Fischhausen. Kaliningradskaya Oblast', W Russia
Primorsko *82 E2 prev.* Keupriya. Burgas, E Bulgaria
Primorsk/Primorskoye *see* Prymors'k
Prince Albert *15 F5* Saskatchewan, S Canada
Prince Edward Island *17 F4 Fr.* Île-du Prince-Édouard. *province* SE Canada
Prince Edward Islands *47 E8 island group* S South Africa
Prince George *15 E5* British Columbia, SW Canada
Prince of Wales Island *126 B1 island* Queensland, E Australia
Prince of Wales Island *15 F2 island* Queen Elizabeth Islands, Nunavut, NW Canada
Prince of Wales Island *see* Pinang, Pulau
Prince Patrick Island *15 E2 island* Parry Islands, Northwest Territories, NW Canada
Prince Rupert *14 D4* British Columbia, SW Canada
Prince's Island *see* Príncipe
Princess Charlotte Bay *126 C2 bay* Queensland, NE Australia
Princess Elizabeth Land *132 C3 physical region* Antarctica
Príncipe *55 A5 var.* Príncipe Island, *Eng.* Prince's Island. *island* N Sao Tome and Principe
Príncipe Island *see* Príncipe
Prinzapolka *31 E3* Región Autónoma Atlántico Norte, NE Nicaragua
Pripet *85 C7 Bel.* Prypyats', *Ukr.* Pryp"yat'. *river* Belarus/Ukraine
Pripet Marshes *85 B7 wetland* Belarus/Ukraine
Prishtinë *79 D5 Eng.* Pristina, *Serb.* Priština. C Kosovo
Pristina *see* Prishtinë
Priština *see* Prishtinë
Privas *69 D5* Ardèche, E France
Privolzhskaya Vozvyshennost' *59 G3 var.* Volga Uplands. *mountain range* W Russia
Prizren *79 D5* S Kosovo
Probolinggo *116 D5* Jawa, C Indonesia
Probstberg *see* Wyszków
Progreso *29 H3* Yucatán, SE Mexico
Prokhladnyy *89 B8* Kabardino-Balkarskaya Respublika, SW Russia
Prokletije *see* North Albanian Alps
Prökuls *see* Priekulė
Prokuplje *79 D5* Serbia, SE Serbia
Prome *see* Pyay
Promyshlennyy *88 E3* Respublika Komi, NW Russia
Prościejów *see* Prostějov
Proskurov *see* Khmel 'nyts'kyy
Prossnitz *see* Prostějov
Prostějov *77 C5 Ger.* Prossnitz, *Pol.* Prościejów. Olomoucký Kraj, E Czech Republic (Czechia)
Provence *69 D6 cultural region* SE France
Providence *19 G3 state capital* Rhode Island, NE USA
Providence *see* Fort Providence
Providencia, Isla de *31 F3 island* NW Colombia, Caribbean Sea
Provideniya *133 B1* Chukotskiy Avtonomnyy Okrug, NE Russia
Provo *22 B4* Utah, W USA
Prudhoe Bay *14 D2* Alaska, USA
Prusa *see* Bursa
Pruszków *76 D3 Ger.* Kaltdorf. Mazowieckie, C Poland
Prut *86 D4 Ger.* Pruth. *river* E Europe
Pruth *see* Prut
Pružana *see* Pruzhany
Pruzhany *85 B6 Pol.* Prużana. Brestskaya Voblasts', SW Belarus
Prychornomor'ska Nyzovyna *see* Black Sea Lowland
Prydniprovs'ka Nyzovyna/Prydnyaprowskaya Nizina *see* Dnieper Lowland
Prydz Bay *132 D3 bay* Antarctica
Pryluky *87 E2 Rus.* Priluki. Chernihivs'ka Oblast', NE Ukraine
Prymors'k *87 G4 Rus.* Primorsk; *prev.* Primorskoye. Zaporiz'ka Oblast', SE Ukraine
Pryp"yat'/Prypyats' *see* Pripet
Przemyśl *77 E5 Rus.* Peremyshl. Podkarpackie, C Poland
Przheval'sk *see* Karakol
Psará *83 D5 island* E Greece
Psel *87 F2 Rus.* Psël. *river* Russia/Ukraine
Psël *see* Psel
Pskov *92 B2 Ger.* Pleskau, *Latv.* Pleskava. Pskovskaya Oblast', W Russia
Pskov, Lake *84 E3 Est.* Pihkva Järv, *Ger.* Pleskauer See, *Rus.* Pskovskoye Ozero. *lake* Estonia/Russia
Pskovskoye Ozero *see* Pskov, Lake
Ptich' *see* Ptsich
Ptsich *85 C7 Rus.* Ptich'. Homyel'skaya Voblasts', SE Belarus
Ptsich *85 C7 Rus.* Ptich'. *river* SE Belarus
Ptuj *73 E7 Ger.* Pettau; *anc.* Poetovio. NE Slovenia
Pucallpa *38 C3* Ucayali, C Peru
Puck *76 C2* Pomorskie, N Poland
Pudasjärvi *62 D4* Pohjois-Pohjanmaa, C Finland
Puebla *29 F4 var.* Puebla de Zaragoza. Puebla, S Mexico
Puebla de Zaragoza *see* Puebla
Pueblo *22 D5* Colorado, C USA
Puerto Acosta *39 E4* La Paz, W Bolivia
Puerto Aisén *43 B6* Aisén, S Chile
Puerto Ángel *29 F5* Oaxaca, SE Mexico
Puerto Argentino *see* Stanley
Puerto Ayacucho *36 D3* Amazonas, SW Venezuela
Puerto Baquerizo Moreno *38 B5 var.* Baquerizo Moreno. Galápagos Islands, Ecuador, E Pacific Ocean
Puerto Barrios *30 C2* Izabal, E Guatemala
Puerto Bello *see* Portobelo
Puerto Berrío *36 B2* Antioquia, C Colombia
Puerto Cabello *36 D1* Carabobo, N Venezuela
Puerto Cabezas *31 E2 var.* Bilwi. Región Autónoma Atlántico Norte, NE Nicaragua
Puerto Carreño *36 D3* Vichada, E Colombia
Puerto Cortés *30 C2* Cortés, NW Honduras
Puerto Cumarebo *36 C1* Falcón, N Venezuela
Puerto Deseado *43 C7* Santa Cruz, SE Argentina
Puerto Escondido *29 F5* Oaxaca, SE Mexico
Puerto Francisco de Orellana *38 B1 var.* Coca. NE Ecuador
Puerto Gallegos *see* Río Gallegos
Puerto Inírida *36 D3 var.* Obando. Guainía, E Colombia
Puerto La Cruz *37 E1* Anzoátegui, NE Venezuela
Puerto Lempira *31 E2* Gracias a Dios, E Honduras
Puerto Limón *see* Limón
Puertollano *70 D4* Castilla-La Mancha, C Spain
Puerto López *36 C1* La Guajira, N Colombia
Puerto Maldonado *39 E3* Madre de Dios, E Peru
Puerto México *see* Coatzacoalcos
Puerto Montt *43 B5* Los Lagos, C Chile
Puerto Natales *43 B7* Magallanes, S Chile
Puerto Obaldía *31 H5* Kuna Yala, NE Panama
Puerto Plata *33 E3 var.* San Felipe de Puerto Plata. N Dominican Republic
Puerto Presidente Stroessner *see* Ciudad del Este
Puerto Princesa *117 E2 off.* Puerto Princesa City. Palawan, W Philippines
Puerto Princesa City *see* Puerto Princesa
Puerto Príncipe *see* Camagüey
Puerto Rico *33 F3 off.* Commonwealth of Puerto Rico; *prev.* Porto Rico. *US commonwealth territory* C West Indies
Puerto Rico *34 B1 island* C West Indies
Puerto Rico, Commonwealth of *see* Puerto Rico
Puerto Rico Trench *34 B1 trench* NE Caribbean Sea
Puerto San José *see* San José
Puerto San Julián *43 B7 var.* San Julián. Santa Cruz, SE Argentina
Puerto Suárez *39 H4* Santa Cruz, E Bolivia
Puerto Vallarta *28 D4* Jalisco, SW Mexico
Puerto Varas *43 B5* Los Lagos, C Chile
Puerto Viejo *31 E4* Heredia, NE Costa Rica
Puertoviejo *see* Portoviejo
Puget Sound *24 B1 sound* Washington, NW USA
Puglia *75 E5 var.* Le Puglie, *Eng.* Apulia. *region* SE Italy
Pukaki, Lake *129 B6 lake* South Island, New Zealand
Pukekohe *128 D3* Auckland, North Island, New Zealand
Puket *see* Phuket
Pukhavichy *85 C6 Rus.* Pukhovichi. Minskaya Voblasts', C Belarus
Pukhovichi *see* Pukhavichy
Pula *78 A3 It.* Pola; *prev.* Pulj. Istra, NW Croatia
Pulaski *18 D5* Virginia, NE USA
Puławy *76 D4 Ger.* Neu Amerika. Lubelskie, E Poland
Pul-e Khumrī *101 E4 prev.* Pol-e Khomrī, *var.* Pul-i-Khumri. Baghlān, NE Afghanistan
Pul-i-Khumri *see* Pul-e Khumrī
Pulj *see* Pula
Pullman *24 C2* Washington, NW USA
Pułtusk *76 D3* Mazowieckie, C Poland
Puná, Isla *38 A2 island* SW Ecuador
Pune *112 C5 prev.* Poona. Mahārāshtra, W India
Punjab *112 C2 prev.* West Punjab, Western Punjab. *province* E Pakistan
Puno *39 E4* Puno, SE Peru
Punta Alta *43 C5* Buenos Aires, E Argentina
Punta Arenas *43 B8 prev.* Magallanes. Magallanes, S Chile
Punta Gorda *30 C2* Toledo, SE Belize
Punta Gorda *31 E4* Región Autónoma Atlántico Sur, SE Nicaragua
Puntarenas *30 D4* Puntarenas, W Costa Rica
Punto Fijo *36 C1* Falcón, N Venezuela
Pupuya, Nevado *39 E4 mountain* W Bolivia
Puri *113 F5 var.* Jagannath. Odisha, E India
Puriramya *see* Buriram
Purmerend *64 C3* Noord-Holland, C Netherlands
Pursat *115 D6 Khmer.* Poŭthĭsăt. Pursat, W Cambodia
Purus, Rio *40 C2 var.* Río Purús. *river* Brazil/Peru
Pusan *see* Busan
Pushkino *see* Biläsuvar
Püspökladány *77 D6* Hajdú-Bihar, E Hungary
Putorana, Gory/Putorana Mountains *see* Putorana, Plato
Putorana Mountains *93 E3 var.* Gory Putorana, *Eng.* Putorana Mountains. *mountain range* N Russia
Putrajaya *116 B3 administrative capital* (Malaysia) Kuala Lumpur, Peninsular Malaysia
Puttalam *110 C3* North Western Province, W Sri Lanka
Puttgarden *72 C2* Schleswig-Holstein, N Germany
Putumayo, Río *36 B5 var.* Içá, Rio. *river* NW South America
Puurmani *84 D2 Ger.* Talkhof. Jõgevamaa, E Estonia
Pyatigorsk *89 B7* Stavropol'skiy Kray, SW Russia
Pyatikhatki *see* P"yatykhatky
P"yatykhatky *87 F3 Rus.* Pyatikhatki. Dnipropetrovs'ka Oblast', E Ukraine
Pyay *114 B4 var.* Prome, Pye. Bago, C Myanmar (Burma)
Pye *see* Pyay
Pyetrykaw *85 C7 Rus.* Petrikov. Homyel'skaya Voblasts', SE Belarus
Pyin-Oo-Lwin *114 B3 var.* Maymyo. Mandalay, C Myanmar (Burma)
Pýlos *83 B6 var.* Pilos. Pelopónnisos, S Greece
P'yŏngyang *107 E3 var.* P'yŏngyang-si, *Eng.* Pyongyang. *country capital* (North Korea) SW North Korea
P'yŏngyang-si *see* P'yŏngyang
Pyramid Lake *25 C5 lake* Nevada, W USA
Pyrenaei Montes *see* Pyrenees
Pyrenees *80 B2 Fr.* Pyrénées, *Sp.* Pirineos; *anc.* Pyrenaei Montes. *mountain range* SW Europe
Pýrgos *83 B6 var.* Pírgos. Dytikí Elláda, S Greece
Pyritz *see* Pyrzyce
Pyryatyn *87 E2 Rus.* Piryatin. Poltavs'ka Oblast', NE Ukraine
Pyrzyce *76 B3 Ger.* Pyritz. Zachodnio-pomorskie, NW Poland
Pyu *see* Phyu
Pyuntaza *114 B4* Bago, SW Myanmar (Burma)

Q

Qāʻ al Jafr *97 C7 lake* S Jordan
Qaanaaq *60 D1 var.* Qânâq, *Dan.* Thule. Avannaarsua, N Greenland
Qabātiya *97 E6* N West Bank Asia
Qābis *see* Gabès
Qābis, Khalīj *see* Gabès, Golfe de
Qacentina *see* Constantine
Qafṣah *see* Gafsa
Qagan Us *see* Dulan
Qahremānshahr *see* Kermānshāh
Qaidam Pendi *104 C4 basin* C China
Qal'aikhum *101 F3 Rus.* Kalaikhum. S Tajikistan
Qalāt *101 E5 Per.* Kalāt. Zābol, S Afghanistan
Qal'at Bīshah *99 B5* 'Asīr, SW Saudi Arabia
Qalqiliya *see* Qalqīlya
Qalqīlya *97 D6 var.* Qalqiliya. Central, W West Bank, Asia
Qamdo *104 D5* Xizang Zizhiqu, W China
Qamishly *see* Al Qāmishlī
Qânâq *see* Qaanaaq
Qaqortoq *60 C4 Dan.* Julianehåb. Kitaa, S Greenland
Qaraghandy/Qaraghandy Oblysy *see* Karagandy
Qara Qum *see* Garagum
Qarataū *see* Karatau, Zhambyl, Kazakhstan
Qarkilik *see* Ruoqiang
Qarokŭl *101 F3 Rus.* Karakul'. E Tajikistan
Qarqannah, Juzur *see* Kerkenah, Îles de
Qars *see* Kars
Qarshi *101 E3 Rus.* Karshi; *prev.* Bek-Budi. Qashqadaryo Viloyati, S Uzbekistan
Qasigianguit *see* Qasigiannguit
Qasigiannguit *60 C3 var.* Qasigianguit, *Dan.* Christianshåb. Kitaa, C Greenland
Qaṣr al Farāfirah *50 B2 var.* Qasr Farâfra. W Egypt
Qasr Farâfra *see* Qaṣr al Farāfirah
Qaṭanā *97 B5 var.* Katana. Dimashq, S Syria
Qatar *98 C4 off.* State of Qatar, *Ar.* Dawlat Qaṭar. *country* SW Asia
Qatar, State of *see* Qatar
Qattara Depression *see* Qaṭṭārah, Munkhafaḍ al
Qaṭṭāra, Monkhafad el *see* Qaṭṭārah, Munkhafaḍ al
Qaṭṭārah, Munkhafaḍ al *50 A1 var.* Monkhafad el Qaṭṭāra, *Eng.* Qattara Depression. *desert* NW Egypt
Qausuittuq *see* Resolute
Qazaqstan/Qazaqstan Respublikasy *see* Kazakhstan
Qazimämmäd *95 H3 Rus.* Kazi Magomed. SE Azerbaijan
Qazris *see* Cáceres
Qazvīn *98 C2 var.* Kazvin. Qazvīn, N Iran
Qena *see* Qinā
Qeqertarssuaq *see* Qeqertarsuaq
Qeqertarsuaq *60 C3 var.* Qeqertarssuaq, *Dan.* Godhavn. Kitaa, S Greenland
Qeqertarsuaq *60 C3 island* W Greenland
Qeqertarsuup Tunua *60 C3 Dan.* Disko Bugt. *inlet* W Greenland
Qerveh *see* Qorveh
Qeshm *98 D4 var.* Jazīreh-ye Qeshm, Qeshm Island. *island* S Iran
Qeshm Island/Qeshm, Jazīreh-ye *see* Qeshm
Qilian Shan *104 D3 var.* Kilien Mountains. *mountain range* N China
Qimusseriarsuaq *60 C2 Dan.* Melville Bugt, *Eng.* Melville Bay. *bay* NW Greenland
Qinā *50 B2 var.* Qena; *anc.* Caene, Caenepolis. E Egypt
Qing *see* Qinghai
Qingdao *106 D4 var.* Ching-Tao, Ch'ing-tao, Tsingtao, Tsintao, *Ger.* Tsingtau. Shandong, E China
Qinghai *104 C4 var.* Chinghai, Koko Nor, Qing, Qinghai Sheng, Tsinghai. *province* C China
Qinghai Hu *104 D4 var.* Ch'ing Hai, Tsing Hai, *Mong.* Koko Nor. *lake* C China
Qinghai Sheng *see* Qinghai
Qingzang Gaoyuan *104 B4 var.* Xizang Gaoyuan, *Eng.* Plateau of Tibet. *plateau* W China
Qinhuangdao *106 D3* Hebei, E China
Qinzhou *106 B6* Guangxi Zhuangzu Zizhiqu, S China
Qiong *see* Hainan
Qiqihar *106 D2 var.* Ch'i-ch'i-ha-erh, Tsitsihar; *prev.* Lungkiang. Heilongjiang, NE China
Qira *104 B4* Xinjiang Uygur Zizhiqu, NW China
Qita Ghazzah *see* Gaza Strip
Qitai *104 C3* Xinjiang Uygur Zizhiqu, NW China
Qīzān *see* Jīzān
Qizil Orda *see* Kzylorda
Qizil Qum/Qizilqum *see* Kyzyl Kum
Qizilrabot *101 G3 Rus.* Kyzylrabot. SE Tajikistan
Qogir Feng *see* K2
Qom *98 C3 var.* Kum, Qum. Qom, N Iran
Qomolangma Feng *see* Everest, Mount
Qomul *see* Hami
Qoʻqon *101 F2 var.* Khokand, *Rus.* Kokand. Farg'ona Viloyati, E Uzbekistan
Qorveh *98 C3 var.* Qerveh, Qurveh. Kordestān, W Iran
Qostanay/Qostanay Oblysy *see* Kostanay
Qoubaïyât *96 B4 var.* Al Qubayyāt. N Lebanon
Qoussantîna *see* Constantine
Quang Ngai *115 E5 var.* Quangngai, Quang Nghia. Quang Ngai, C Vietnam
Quangngai *see* Quang Ngai
Quang Nghia *see* Quang Ngai
Quan Long *see* Ca Mau
Quanzhou *106 D6 var.* Ch'uan-chou, Tsinkiang; *prev.* Chin-chiang. Fujian, SE China
Quanzhou *106 C6* Guangxi Zhuangzu Zizhiqu, S China
Qu'Appelle *15 F5 river* Saskatchewan, S Canada
Quarles, Pegunungan *117 E4 mountain range* Sulawesi, C Indonesia
Quarnero *see* Kvarner
Quartu Sant' Elena *75 A6* Sardegna, Italy, C Mediterranean Sea
Quba *95 H2 Rus.* Kuba. N Azerbaijan
Qubba *see* Ba'qūbah
Québec *17 E4 var.* Quebec. *province capital* Québec, SE Canada
Québec *16 D3 var.* Quebec. *province* SE Canada
Queen Charlotte Islands *14 C5 Fr.* Îles de la Reine-Charlotte. *island group* British Columbia, SW Canada
Queen Charlotte Sound *14 C5 sea area* British Columbia, W Canada
Queen Elizabeth Islands *15 E1 Fr.* Îles de la Reine-Élisabeth. *island group* Nunavut, N Canada
Queensland *126 B4 state* N Australia
Queenstown *129 B7* Otago, South Island, New Zealand
Queenstown *56 D5* Eastern Cape, S South Africa
Quelimane *57 E3 var.* Kilimane, Kilmain, Quilimane. Zambézia, NE Mozambique
Quelpart *see* Jeju-do
Quepos *31 E4* Puntarenas, S Costa Rica
Que Que *see* Kwekwe
Quera *see* Chur
Querétaro *29 E4* Querétaro de Arteaga, C Mexico
Quesada *31 E4 var.* Ciudad Quesada, San Carlos. Alajuela, N Costa Rica
Quetta *112 B2* Baluchistān, SW Pakistan
Quetzalcoalco *see* Coatzacoalcos
Quezaltenango *30 A2 var.* Quetzaltenango. Quezaltenango, W Guatemala

Quibdó – Rovigo

Quibdó 36 A3 Chocó, W Colombia
Quilimane see Quelimane
Quillabamba 38 D3 Cusco, C Peru
Quilon see Kollam
Quimper 68 A3 anc. Quimper Corentin. Finistère, NW France
Quimper Corentin see Quimper
Quimperlé 68 A3 Finistère, NW France
Quincy 18 A4 Illinois, N USA
Qui Nhon/Quinhon see Quy Nhơn
Quissico 57 E4 Inhambane, S Mozambique
Quito 38 B1 country capital (Ecuador) Pichincha, N Ecuador
Qulyndy Zhazyghy see Kulunda Steppe
Qum see Qom
Qurein see Al Kuwayt
Qŭrghonteppa 101 E3 Rus. Kurgan-Tyube. SW Tajikistan
Qurlurtuuq see Kugluktuk
Qurveh see Qorveh
Quṣayr see Al Quṣayr
Quxar see Lhazê
Quy Nhơn 115 E5 var. Quinhon, Qui Nhon. Binh Đinh, C Vietnam
Qyteti Stalin see Kuçovë
Qyzylorda see Kzylorda

R

Raab 78 B1 Hung. Rába. river Austria/Hungary
Raab see Rába
Raab see Győr
Raahe 62 D4 Swe. Brahestad. Pohjois-Pohjanmaa, W Finland
Raalte 64 D3 Overijssel, E Netherlands
Raamsdonksveer 64 C4 Noord-Brabant, S Netherlands
Raasiku 84 D2 Ger. Rasik. Harjumaa, NW Estonia
Rába 77 B7 Ger. Raab. river Austria/Hungary
Rába see Raab
Rabat 48 C2 var. al Dar al Baida. country capital (Morocco) NW Morocco
Rabat 80 B5 W Malta
Rabat see Victoria
Rabbah Ammon/Rabbath Ammon see 'Ammān
Rabinal 30 B2 Baja Verapaz, C Guatemala
Rabka 77 D5 Małopolskie, S Poland
Râbniţa see Ribniţa
Rabyanah Ramlat 49 G4 var. Rebiana Sand Sea, Şahrā' Rabyānah. desert SE Libya
Rabyānah, Şahrā' see Rabyānah, Ramlat
Răcari see Durankulak
Race, Cape 17 H3 headland Newfoundland, Newfoundland and Labrador, E Canada
Rach Gia 115 D6 Kiên Giang, S Vietnam
Rach Gia, Vinh 115 D6 bay S Vietnam
Racine 18 B3 Wisconsin, N USA
Rácz-Becse see Bečej
Rădăuţi 86 C3 Ger. Radautz, Hung. Rádóc. Suceava, N Romania
Radautz see Rădăuţi
Rádeyilikóé see Fort Good Hope
Rádóc see Rădăuţi
Radom 76 D4 Mazowieckie, C Poland
Radomsko 76 D4 Rus. Novoradomsk. Łódzkie, C Poland
Radomyshl' 86 D2 Zhytomyrs'ka Oblast', N Ukraine
Radoviš 79 E6 prev. Radovište. E Macedonia
Radovište see Radoviš
Radviliškis 84 B4 Šiaulai, N Lithuania
Radzyń Podlaski 76 E4 Lubelskie, E Poland
Rae-Edzo see Edzo
Raetihi 128 D4 Manawatu-Wanganui, North Island, New Zealand
Rafa see Rafah
Rafaela 42 C3 Santa Fe, E Argentina
Rafah 97 A7 var. Rafa, Rafaḩ, Heb. Rafiaḩ, Raphiah. SW Gaza Strip
Rafḩah 98 B4 Al Ḩudūd ash Shamālīyah, N Saudi Arabia
Rafiaḩ see Rafah
Raga 51 A5 W Bahr el Ghazal, W South Sudan
Ragged Island Range 32 C2 island group S The Bahamas
Ragnit see Neman
Ragusa 75 C7 Sicilia, Italy, C Mediterranean Sea
Ragusa see Dubrovnik
Rahachow 85 D7 Rus. Rogachëv. Homyel'skaya Voblasts', SE Belarus
Rahaeng see Tak
Rahaţ, Ḩarrat 99 B5 lava flow W Saudi Arabia
Rahīmyār Khān 112 C3 Punjab, SE Pakistan
Rahovec 79 D5 Serb. Orahovac. W Kosovo
Raiatea 123 G4 island Îles Sous le Vent, W French Polynesia
Rāichūr 110 C1 Karnātaka, C India
Raidestos see Tekirdağ
Rainier, Mount 12 A4 volcano Washington, NW USA
Rainy Lake 16 A4 lake Canada/USA
Raipur 113 E4 Chhattīsgarh, C India
Rājahmundry 113 E5 Andhra Pradesh, E India
Rajang see Rajang, Batang
Rajang, Batang 116 D3 var. Rajang. river East Malaysia
Rājapālaiyam 110 C3 Tamil Nādu, SE India
Rājasthān 112 C3 cultural region NW India
Rājkot 112 C4 Gujarāt, W India
Rāj Nāndgaon 113 E4 Chhattīsgarh, C India
Rajshahi 113 G3 prev. Rampur Boalia. Rajshahi, W Bangladesh
Rakahanga 123 F3 atoll N Cook Islands
Rakaia 129 B6 river South Island, New Zealand
Rakka see Ar Raqqah
Rakke 84 E2 Lääne-Virumaa, NE Estonia
Rakvere 84 E2 Ger. Wesenberg. Lääne-Virumaa, N Estonia
Raleigh 21 F1 state capital North Carolina, SE USA
Ralik Chain 122 D1 island group Ralik Chain, W Marshall Islands
Ramadi see Ar Ramādī
Râmnicul-Sărat see Râmnicu Sărat
Râmnicu Sărat 86 C4 prev. Râmnicul-Sărat, Rîmnicu-Sărat. Buzău, E Romania
Râmnicu Vâlcea 86 B4 prev. Rîmnicu Vîlcea. Vâlcea, C Romania
Rampur Boalia see Rajshahi
Ramree Island 114 A4 island W Myanmar (Burma)
Ramtha see Ar Ramthā
Rancagua 42 B4 Libertador, C Chile
Rānchi 113 F4 Jhārkhand, N India
Randers 63 B7 Århus, C Denmark
Ráneš see Ringvassøya
Rangiora 129 C6 Canterbury, South Island, New Zealand
Rangitikei 128 D4 river North Island, New Zealand
Rangoon see Yangon
Rangpur 113 G3 Rajshahi, N Bangladesh
Rankin Inlet 15 G3 Nunavut, C Canada
Rankovićevo see Kraljevo
Ranong 115 B6 Ranong, SW Thailand
Rapa Nui see Pascua, Isla de
Raphiah see Rafah
Rapid City 22 D3 South Dakota, N USA
Rāpina 84 E3 Ger. Rappin. Põlvamaa, SE Estonia
Rapla 84 D2 Ger. Rappel. Raplamaa, NW Estonia
Rappel see Rapla
Rappin see Räpina
Rarotonga 123 G5 island S Cook Islands, C Pacific Ocean
Ras al 'Ain see Ra's al 'Ayn
Ra's al 'Ayn 96 D1 var. Ras al'Ain. Al Ḩasakah, N Syria
Ra's an Naqb 97 B7 Ma'ān, S Jordan
Raseiniai 84 B4 Kaunas, C Lithuania
Rasht 98 C2 var. Resht. Gīlān, NW Iran
Rasik see Raasiku
Râşnov 86 C4 prev. Rîşno, Rozsnyó, Hung. Barcarozsnyó. Braşov, C Romania
Rastenburg see Kętrzyn
Ratak Chain 122 D1 island group Ratak Chain, E Marshall Islands
Ratān 63 C5 Jämtland, C Sweden
Rat Buri see Ratchaburi
Ratchaburi 115 C5 var. Rat Buri. Ratchaburi, W Thailand
Ratisbon/Ratisbona/Ratisbonne see Regensburg
Rat Islands 14 A2 island group Aleutian Islands, Alaska, USA
Ratlām 112 D4 prev. Rutlam. Madhya Pradesh, C India
Ratnapura 110 D4 Sabaragamuwa Province, S Sri Lanka
Raton 26 D1 New Mexico, SW USA
Rättvik 63 C5 Dalarna, C Sweden
Raudhatain see Ar Rawḑatayn
Raufarhöfn 61 E4 Nordhurland Eystra, NE Iceland
Raukawa see Cook Strait
Raukumara Range 128 E3 mountain range North Island, New Zealand
Rāulakela see Rāurkela
Rauma 63 D5 Swe. Raumo. Satakunta, SW Finland
Raumo see Rauma
Rāurkela 113 F4 var. Rāulakela, Rourkela. Odisha, E India
Ravenna 74 C3 Emilia-Romagna, N Italy
Ravi 112 C2 river India/Pakistan
Rāwalpindi 112 C1 Punjab, NE Pakistan
Rawa Mazowiecka 76 D4 Łódzkie, C Poland
Rawicz 76 C4 Ger. Rawitsch. Wielkopolskie, C Poland
Rawitsch see Rawicz
Rawlins 22 C3 Wyoming, C USA
Rawson 43 C6 Chubut, SE Argentina
Rayak 96 B4 var. Rayaq, Riyāq. E Lebanon
Rayaq see Rayak
Rayong 115 C5 Rayong, S Thailand
Razazah, Buhayrat ar 98 B3 var. Baḩr al Milḩ. lake C Iraq
Razdolnoye see Rozdol'ne
Razelm, Lacul see Razim, Lacul
Razgrad 82 D2 Razgrad, N Bulgaria
Razim, Lacul 86 D5 prev. Lacul Razelm. lagoon NW Black Sea
Reading 67 D7 S England, United Kingdom
Reading 19 F4 Pennsylvania, NE USA
Realicó 42 C4 La Pampa, C Argentina
Reăng Kesei 115 D5 Battambang, W Cambodia
Greater Antarctica see East Antarctica
Rebecca, Lake 125 C6 lake Western Australia
Rebiana Sand Sea see Rabyānah, Ramlat
Rebun-to 108 C2 island NE Japan
Rechitsa see Rechytsa
Rechytsa 85 D7 Rus. Rechitsa. Brestskaya Voblasts', SW Belarus
Recife 41 G2 prev. Pernambuco. state capital Pernambuco, E Brazil
Recklinghausen 72 A4 Nordrhein-Westfalen, W Germany
Recogne 65 C7 Luxembourg, SE Belgium
Reconquista 42 D3 Santa Fe, C Argentina
Red Deer 15 E5 Alberta, SW Canada
Redding 25 B5 California, W USA
Redon 68 B4 Ille-et-Vilaine, NW France
Red River 23 E1 river Canada/USA
Red River 114 C2 var. Yuan, Chin. Yuan Jiang, Vtn. Sông Hông Hà. river China/Vietnam
Red River 13 C6 river S USA
Red River 20 B3 river Louisiana, S USA
Red Sea 50 C3 var. Sinus Arabicus. sea Africa/Asia
Red Wing 23 G2 Minnesota, N USA
Reefton 129 C5 West Coast, South Island, New Zealand
Reese River 25 C5 river Nevada, W USA
Refahiye 95 E3 Erzincan, C Turkey
Regensburg 73 C6 Eng. Ratisbon, Fr. Ratisbonne, hist. Ratisbona; anc. Castra Regina, Reginum. Bayern, SE Germany
Regenstauf 73 C6 Bayern, SE Germany
Rēgestān 100 D5 prev. Rīgestān, var. Registan. desert region S Afghanistan
Reggane 48 D3 C Algeria
Reggio see Reggio nell'Emilia
Reggio Calabria see Reggio di Calabria
Reggio di Calabria 75 D7 var. Reggio Calabria, Gk. Rhegion; anc. Regium, Rhegium. Calabria, SW Italy
Reggio Emilia see Reggio nell'Emilia
Reggio nell'Emilia 74 B2 var. Reggio Emilia, abbrev. Reggio; anc. Regium Lepidum. Emilia-Romagna, N Italy
Reghin 86 C4 Ger. Sächsisch-Reen, Hung. Szászrégen; prev. Reghinul Săsesc, Ger. Sächsisch-Regen. Mureş, C Romania
Reghinul Săsesc see Reghin
Regina 15 F5 province capital Saskatchewan, S Canada
Reginum see Regensburg
Registan see Rēgestān
Regium see Reggio di Calabria
Regium Lepidum see Reggio nell'Emilia
Rehoboth 56 B3 Hardap, C Namibia
Rehovot 97 A6 ; prev. Reḩovot. Central, C Israel
Reḩovot see Rehovot
Reichenau see Bogatynia, Poland
Reichenberg see Liberec
Reid 125 D6 Western Australia
Reikjavik see Reykjavík
Ré, Île de 68 A4 island W France
Reims 68 D3 Eng. Rheims; anc. Durocortorum, Remi. Marne, N France
Reindeer Lake 15 F4 lake Manitoba/Saskatchewan, C Canada
Reine-Charlotte, Îles de la see Queen Charlotte Islands
Reine-Élisabeth, Îles de la see Queen Elizabeth Islands
Reinga, Cape 128 C1 headland North Island, New Zealand
Reinosa 70 D1 Cantabria, N Spain
Reka see Rijeka
Rekhovot see Rehovot
Reliance 15 F4 Northwest Territories, C Canada
Remi see Reims
Rendina see Rentína
Rendsburg 72 B2 Schleswig-Holstein, N Germany
Rengat 116 B4 Sumatera, W Indonesia
Reni 86 D4 Odes'ka Oblast', SW Ukraine
Rennell 122 C4 var. Mu Nggava. island S Solomon Islands
Rennes 68 B3 Bret. Roazon; anc. Condate. Ille-et-Vilaine, NW France
Reno 25 C5 Nevada, W USA
Renqiu 106 C4 Hebei, E China
Rentína 83 B5 var. Rendina. Thessalía, C Greece
Reps see Rupea
Repulse Bay 15 G3 Northwest Territories, N Canada
Reschitza see Reşiţa
Resht see Rasht
Resicabánya see Reşiţa
Resistencia 42 D3 Chaco, NE Argentina
Reşiţa 86 A4 Ger. Reschitza, Hung. Resicabánya. Caraş-Severin, W Romania
Resolute 15 F2 Inuit Qausuittuq. Cornwallis Island, Nunavut, N Canada
Resolution Island 17 E1 island Nunavut, NE Canada
Resolution Island 129 A7 island SW New Zealand
Réunion 57 H4 off. La Réunion. French overseas department W Indian Ocean
Réunion 119 B5 island W Indian Ocean
Reus 71 F2 Cataluña, E Spain
Reutlingen 73 B6 Baden-Württemberg, S Germany
Reuver 65 D5 Limburg, SE Netherlands
Reval/Revel see Tallinn
Revillagigedo Island 28 B5 island Alexander Archipelago, Alaska, USA
Rexburg 24 E3 Idaho, NW USA
Reyes 39 F3 Beni, NW Bolivia
Rey, Isla del 31 G5 island Archipiélago de las Perlas, SE Panama
Reykjanes Basin 60 C5 var. Irminger Basin. undersea basin N Atlantic Ocean
Reykjanes Ridge 58 A1 undersea ridge N Atlantic Ocean
Reykjavík 61 E5 var. Reikjavik. country capital (Iceland) Höfudhborgarsvaedhi, W Iceland
Reynosa 29 E2 Tamaulipas, C Mexico
Reẕā'īyeh, Daryācheh-ye see Orūmīyeh, Daryācheh-ye
Rezé 68 A4 Loire-Atlantique, NW France
Rēzekne 84 D4 Ger. Rositten; prev. Rus. Rezhitsa. Rēzekne, SE Latvia
Rezhitsa see Rēzekne
Rezovo 82 E3 Turk. Rezve. Burgas, E Bulgaria
Rezve see Rezovo
Rhaedestus see Tekirdağ
Rhegion/Rhegium see Reggio di Calabria
Rheims see Reims
Rhein see Rhine
Rheine 72 A3 var. Rheine in Westfalen. Nordrhein-Westfalen, NW Germany
Rheine in Westfalen see Rheine
Rheinisches Schiefergebirge 73 A5 var. Rhine State Uplands, Eng. Rhenish Slate Mountains. mountain range W Germany
Rhenish Slate Mountains see Rheinisches Schiefergebirge
Rhin see Rhine
Rhine 58 D4 Dut. Rijn, Fr. Rhin, Ger. Rhein. river W Europe
Rhinelander 18 B2 Wisconsin, N USA
Rhine State Uplands see Rheinisches Schiefergebirge
Rho 74 B2 Lombardia, N Italy
Rhode Island 19 G3 off. State of Rhode Island and Providence Plantations, also known as Little Rhody, Ocean State. state NE USA
Rhodes 83 E7 var. Ródhos, Eng. Rhodes, It. Rodi; anc. Rhodos. island Dodekánisa, Greece, Aegean Sea
Rhodes see Ródos
Rhodesia see Zimbabwe
Rhodope Mountains 82 C3 var. Rodhópi Óri, Bul. Rhodope Planina, Rodopi, Gk. Orosirá Rodhópis, Turk. Dospad Dagh. mountain range Bulgaria/Greece
Rhodope Planina see Rhodope Mountains
Rhône 58 C4 river France/Switzerland
Rhum 66 B3 var. Rum. island W Scotland, United Kingdom
Ribble 67 D5 river NW England, United Kingdom
Ribeira see Santa Uxía de Ribeira
Ribeirão Preto 41 F4 São Paulo, S Brazil
Riberalta 39 F2 Beni, N Bolivia
Ribniţa 86 D3 var. Râbniţa, Rus. Rybnitsa. NE Moldova
Rice Lake 18 A2 Wisconsin, N USA
Richard Toll 52 B3 N Senegal
Richfield 22 B4 Utah, W USA
Richland 24 C2 Washington, NW USA
Richmond 129 C5 Tasman, South Island, New Zealand
Richmond 18 C5 Kentucky, S USA
Richmond 19 E5 state capital Virginia, NE USA
Richmond Range 129 C5 mountain range South Island, New Zealand
Ricobayo, Embalse de 70 C2 reservoir NW Spain
Ricomagus see Riom
Ridder 92 D4 Kaz. Lenīngor, prev. Leninogorsk. Vostochnyy Kazakhstan, E Kazakhstan
Ridgecrest 25 C7 California, W USA
Ried see Ried im Innkreis
Ried im Innkreis 73 D6 var. Ried. Oberösterreich, NW Austria
Riemst 65 D6 Limburg, NE Belgium
Riesa 72 D4 Sachsen, E Germany
Rift Valley see Great Rift Valley
Rīga 84 C3 Eng. Riga. country capital (Latvia) Rīga, C Latvia
Rigaer Bucht see Riga, Gulf of
Riga, Gulf of 84 C3 Est. Liivi Laht, Ger. Rigaer Bucht, Latv. Rīgas Jūras Līcis, Rus. Rizhskiy Zaliv; prev. Est. Riia Laht. gulf Estonia/Latvia
Rīgas Jūras Līcis see Riga, Gulf of
Rīgestān see Rēgestān
Riia Laht see Riga, Gulf of
Riihimäki 63 D5 Kanta-Häme, S Finland
Rijeka 78 A2 Ger. Sankt Veit am Flaum, It. Fiume, Slvn. Reka; anc. Tarsatica. Primorje-Gorski Kotar, NW Croatia
Rijn see Rhine
Rijssel see Lille
Rijssen 64 E3 Overijssel, E Netherlands
Rimah, Wadi ar 98 B4 var. Wādī ar Rummah. dry watercourse C Saudi Arabia
Rimini 74 C3 anc. Ariminum. Emilia-Romagna, N Italy
Rîmnicu-Sărat see Râmnicu Sărat
Rîmnicu Vîlcea see Râmnicu Vâlcea
Rimouski 17 E4 Québec, SE Canada
Ringebu 63 B5 Oppland, S Norway
Ringen see Rõngu
Ringkøbing Fjord 63 A7 fjord W Denmark
Ringvassøya 62 C2 Lapp. Ráneš. island N Norway
Rio see Rio de Janeiro
Riobamba 38 B1 Chimborazo, C Ecuador
Rio Branco 34 B3 state capital Acre, W Brazil
Rio Branco, Território de see Roraima
Río Bravo 29 E2 Tamaulipas, C Mexico
Río Cuarto 42 C4 Córdoba, C Argentina
Rio de Janeiro 41 F4 var. Rio. state capital Rio de Janeiro, SE Brazil
Río Gallegos 43 B7 var. Gallegos, Puerto Gallegos. Santa Cruz, S Argentina
Rio Grande 41 E5 var. São Pedro do Rio Grande do Sul. Rio Grande do Sul, S Brazil
Río Grande 28 D3 Zacatecas, C Mexico
Rio Grande do Norte 41 G2 off. Estado do Rio Grande do Norte. state/region E Brazil
Rio Grande do Norte, Estado do see Rio Grande do Norte
Rio Grande do Sul 41 E5 off. Estado do Rio Grande do Sul. state/region S Brazil
Rio Grande do Sul, Estado do see Rio Grande do Sul
Rio Grande Plateau see Rio Grande Rise
Rio Grande Rise 35 E6 var. Rio Grande Plateau. undersea plateau SW Atlantic Ocean
Ríohacha 36 B1 La Guajira, N Colombia
Río Lagartos 29 H3 Yucatán, SE Mexico
Riom 69 C5 anc. Ricomagus. Puy-de-Dôme, C France
Río Verde 29 E4 var. Rioverde. San Luis Potosí, C Mexico
Rioverde see Río Verde
Ripoll 71 G2 Cataluña, NE Spain
Rishiri-to 108 C2 var. Risiri Tô. island NE Japan
Risiri Tô see Rishiri-tō
Rîşno see Râşnov
Risti 84 D2 Ger. Kreuz. Läänemaa, W Estonia
Rivas 30 D4 Rivas, SW Nicaragua
Rivera 42 D3 Rivera, NE Uruguay
River Falls 18 A2 Wisconsin, N USA
Riverside 25 C7 California, W USA
Riverton 129 A7 Southland, South Island, New Zealand
Riverton 22 C3 Wyoming, C USA
Rivière-du-Loup 17 E4 Québec, SE Canada
Rivne 86 C2 Pol. Równe, Rus. Rovno. Rivnens'ka Oblast', NW Ukraine
Rivoli 74 A2 Piemonte, NW Italy
Riyadh/Riyāḑ, Minţaqat ar see Ar Riyāḑ
Riyāq see Rayak
Rize 95 E2 Rize, NE Turkey
Rizhao 106 D4 Shandong, E China
Rizhskiy Zaliv see Riga, Gulf of
Rkîz 52 C3 Trarza, W Mauritania
Road Town 33 F3 dependent territory capital (British Virgin Islands) Tortola, C British Virgin Islands
Roanne 69 C5 anc. Rodunma. Loire, E France
Roanoke 19 E5 Virginia, NE USA
Roanoke River 21 F1 river North Carolina/ Virginia, SE USA
Roatán 30 C2 var. Coxen Hole, Coxin Hole. Islas de la Bahía, N Honduras
Roat Kampuchea see Cambodia
Roazon see Rennes
Robbie Ridge 121 E3 undersea ridge W Pacific Ocean
Robert Williams see Caála
Robinson Range 125 B5 mountain range Western Australia
Robson, Mount 15 E5 mountain British Columbia, SW Canada
Robstown 27 G4 Texas, SW USA
Roca Partida, Isla 28 B5 island W Mexico
Rocas, Atol das 41 G2 island E Brazil
Rochefort 65 C7 Namur, SE Belgium
Rochefort 68 B4 var. Rochefort sur Mer. Charente-Maritime, W France
Rochefort sur Mer see Rochefort
Rochester 23 G3 Minnesota, N USA
Rochester 19 G2 New Hampshire, NE USA
Rocheuses, Montagnes/Rockies see Rocky Mountains
Rockall Bank 58 B2 undersea bank N Atlantic Ocean
Rockall Trough 58 B2 trough N Atlantic Ocean
Rockdale 126 E2 Texas, SW USA
Rockford 18 B3 Illinois, N USA
Rockhampton 126 D4 Queensland, E Australia
Rock Hill 21 E1 South Carolina, SE USA
Rockingham 125 A6 Western Australia
Rock Island 18 B3 Illinois, N USA
Rock Sound 32 C1 Eleuthera Island, C The Bahamas
Rock Springs 22 C3 Wyoming, C USA
Rockstone 37 F3 C Guyana
Rocky Mount 21 F1 North Carolina, SE USA
Rocky Mountains 12 B4 var. Rockies, Fr. Montagnes Rocheuses. mountain range Canada/USA
Roden 64 E2 Drenthe, NE Netherlands
Rodez 69 C5 anc. Segodunum. Aveyron, S France
Rodhópi Óri see Rhodope Mountains
Ródhos/Rodi see Ródos
Rodopi see Rhodope Mountains
Rodosto see Tekirdağ
Rodunma see Roanne
Roermond 65 D5 Limburg, SE Netherlands
Roeselare 65 A6 Fr. Roulers; prev. Rousselaere. West-Vlaanderen, W Belgium
Rogachëv see Rahachow
Rogatica 78 C4 Republika Srpska, SE Bosnia and Herzegovina
Rogers 20 A1 Arkansas, C USA
Roger Simpson Island see Abemama
Roi Ed see Roi Et
Roi Et 115 D5 var. Muang Roi Et, Roi Ed. Roi Et, E Thailand
Roja 84 C2 Talsi, NW Latvia
Rokiškis 84 C4 Panevėžys, NE Lithuania
Rokycany 77 A5 Ger. Rokytzan. Plzeňský Kraj, W Czech Republic (Czechia)
Rokytzan see Rokycany
Rôlas, Ilha das 54 E2 island S Sao Tome and Principe, Africa, E Atlantic Ocean
Rolla 23 G5 Missouri, C USA
Röm see Rømø
Roma 74 C4 Eng. Rome. country capital (Italy) Lazio, C Italy
Roma 127 D5 Queensland, E Australia
Roman 82 C2 Vratsa, NW Bulgaria
Roman 86 C4 Hung. Románvásár. Neamţ, NE Romania
Romania 86 B4 Bul. Rumŭniya, Ger. Rumänien, Hung. Románia, Rom. România, SCr. Rumunjska, Ukr. Rumuniya, prev. Republica Socialistă România, Roumania, Rumania, Socialist Republic of Romania, prev.Rom. Romînia. country SE Europe
România, Republica Socialistă see Romania
Romania, Socialist Republic of see Romania
Románvásár see Roman
Rome 20 D2 Georgia, SE USA
Rome see Roma
Romînia see Romania
Romny 87 F2 Sums'ka Oblast', NE Ukraine
Rømø 63 A7 Ger. Röm. island SW Denmark
Roncador, Serra do 34 D4 mountain range C Brazil
Ronda 70 D5 Andalucía, S Spain
Rondônia 40 D3 off. Estado de Rondônia; prev. Território de Rondônia. state/region W Brazil
Rondônia, Estado de see Rondônia
Rondônia, Território de see Rondônia
Rondonópolis 41 E3 Mato Grosso, W Brazil
Rongelap Atoll 122 D1 var. Rönlap. atoll Ralik Chain, NW Marshall Islands
Rõngu 84 D3 Ger. Ringen. Tartumaa, SE Estonia
Rönlap see Rongelap Atoll
Rønne 63 B8 Bornholm, E Denmark
Ronne Ice Shelf 132 A3 ice shelf Antarctica
Roosendaal 65 C5 Noord-Brabant, S Netherlands
Roosevelt Island 132 B4 island Antarctica
Roraima 40 D1 off. Estado de Roraima; prev. Território de Rio Branco, Território de Roraima. state/region N Brazil
Roraima, Estado de see Roraima
Roraima, Mount 37 E3 mountain N South America
Roraima, Território de see Roraima
Røros 63 B5 Sør-Trøndelag, S Norway
Ross 129 B6 West Coast, South Island, New Zealand
Rosa, Lake 32 D2 lake Great Inagua, S The Bahamas
Rosario 42 D4 Santa Fe, C Argentina
Rosario 42 D2 San Pedro, C Paraguay
Rosario see Rosarito
Rosarito 28 A1 var. Rosario. Baja California Norte, NW Mexico
Roscianum see Rossano
Roscommon 18 C2 Michigan, N USA
Roseau 33 G4 prev. Charlotte Town. country capital (Dominica) SW Dominica
Roseburg 24 B4 Oregon, NW USA
Rosenau see Rožňava
Rosenberg 27 G4 Texas, SW USA
Rosenberg see Ružomberok, Slovakia
Rosengarten 72 B3 Niedersachsen, N Germany
Rosenheim 73 C6 Bayern, S Germany
Rosia 71 H5 W Gibraltar Europe
Rosia Bay 71 H5 bay SW Gibraltar Europe W Mediterranean Sea Atlantic Ocean
Roşiori de Vede 86 B5 Teleorman, S Romania
Rositten see Rēzekne
Roslavl' 89 A5 Smolenskaya Oblast', W Russia
Rosmalen 64 C4 Noord-Brabant, S Netherlands
Rossano 75 E6 anc. Roscianum. Calabria, SW Italy
Ross Ice Shelf 132 B4 ice shelf Antarctica
Rossiyskaya Federatsiya see Russia
Rosso 52 B3 Trarza, SW Mauritania
Rossosh' 89 B6 Voronezhskaya Oblast', W Russia
Ross Sea 132 B4 sea Antarctica
Rostak see Ar Rustāq
Rostock 72 C2 Mecklenburg-Vorpommern, NE Germany
Rostov see Rostov-na-Donu
Rostov-na-Donu 89 B7 var. Rostov, Eng. Rostov-on-Don. Rostovskaya Oblast', SW Russia
Rostov-on-Don see Rostov-na-Donu
Roswell 26 D2 New Mexico, SW USA
Rota 122 B1 island S Northern Mariana Islands
Rotcher Island see Tamana
Rothera 132 A2 UK research station Antarctica
Rotomagus see Rouen
Rotorua 128 D3 Bay of Plenty, North Island, New Zealand
Rotorua, Lake 128 D3 lake North Island, New Zealand
Rotterdam 64 C4 Zuid-Holland, SW Netherlands
Rottweil 73 B6 Baden-Württemberg, S Germany
Rotuma 123 E4 island NW Fiji Oceania, S Pacific Ocean
Roubaix 68 C2 Nord, N France
Rouen 68 C3 anc. Rotomagus. Seine-Maritime, N France
Roulers see Roeselare
Roumania see Romania
Round Rock 27 G3 Texas, SW USA
Rourkela see Rāurkela
Rousselaere see Roeselare
Roussillon 69 C6 cultural region S France
Rouyn-Noranda 16 D4 Québec, SE Canada
Rovaniemi 62 D3 Lappi, N Finland
Rovigno see Rovinj
Rovigo 74 C2 Veneto, NE Italy

Rovinj – San Juan Mountains

Rovinj 78 A3 It. Rovigno. Istra, NW Croatia
Rovno see Rivne
Rovuma, Rio 57 F2 var. Ruvuma. river Mozambique/Tanzania
Rovuma, Rio see Ruvuma
Równe see Rivne
Roxas City 117 E2 Panay Island, C Philippines
Royale, Isle 18 B1 island Michigan, N USA
Royan 69 B5 Charente-Maritime, W France
Rozdol'ne 87 F4 Rus. Razdolnoye. Respublika Krym, S Ukraine
Rožňava 77 D6 Ger. Rosenau, Hung. Rozsnyó. Košický Kraj, E Slovakia
Rózsahegy see Ružomberok
Rozsnyó see Râşnov, Romania
Rozsnyó see Rožňava, Slovakia
Ruanda see Rwanda
Ruapehu, Mount 128 D4 volcano North Island, New Zealand
Ruapuke Island 129 B8 island SW New Zealand
Ruatoria 128 E3 Gisborne, North Island, New Zealand
Ruawai 128 D2 Northland, North Island, New Zealand
Rubezhnoye see Rubizhne
Rubizhne 87 H3 Rus. Rubezhnoye. Luhans'ka Oblast', E Ukraine
Ruby Mountains 25 D5 mountain range Nevada, W USA
Rucava 84 B3 Liepāja, SW Latvia
Rudensk see Rudzyensk
Rūdiškės 85 B5 Vilnius, S Lithuania
Rudnik see Dolni Chiflik
Rudny see Rudnyy
Rudnyy 92 C4 var. Rudny. Kostanay, N Kazakhstan
Rudolf, Lake see Turkana, Lake
Rudolfswert see Novo mesto
Rudzyensk 85 C6 Rus. Rudensk. Minskaya Voblasts', C Belarus
Rufiji 51 C7 river E Tanzania
Rufino 42 C4 Santa Fe, C Argentina
Rugāji 84 D4 Balvi, E Latvia
Rügen 72 D2 headland NE Germany
Ruggell 72 E1 N Liechtenstein Europe
Ruhja see Rūjiena
Ruhnu 84 C2 var. Ruhnu Saar, Swe. Runö. island SW Estonia
Ruhnu Saar see Ruhnu
Rujen see Rūjiena
Rūjiena 84 D3 Est. Ruhja, Ger. Rujen. Valmiera, N Latvia
Rukwa, Lake 51 B7 lake SE Tanzania
Rum see Rhum
Ruma 78 D3 Vojvodina, N Serbia
Rumadiya see Ar Ramādī
Rumania/Rumänien see Romania
Rumbek 51 B5 El Buhayrat, C South Sudan
Rum Cay 32 D2 island C The Bahamas
Rumia 76 C2 Pomorskie, N Poland
Rummah, Wādī ar see Rimah, Wādī ar
Rummelsburg in Pommern see Miastko
Rumuniya/Rumŭniya/Rumunjska see Romania
Runanga 129 B5 West Coast, South Island, New Zealand
Rundu 56 C3 var. Runtu. Okavango, NE Namibia
Runö see Ruhnu
Runtu see Rundu
Ruoqiang 104 C3 var. Jo-ch'iang, Uigh. Charkhlik, Charkhliq, Qarkilik. Xinjiang Uygur Zizhiqu, NW China
Rupea 86 C4 Ger. Reps, Hung. Kőhalom; prev. Cohalm. Braşov, C Romania
Rupel 65 B5 river N Belgium
Rupella see la Rochelle
Rupert, Rivière de 16 D3 river Québec, C Canada
Rusaddir see Melilla
Ruschuk/Rusçuk see Ruse
Ruse 82 D1 var. Ruschuk, Rustchuk, Turk. Rusçuk. Ruse, N Bulgaria
Russadir see Melilla
Russellville 20 A1 Arkansas, C USA
Russia 90 D2 off. Russian Federation, Latv. Krievija, Rus. Rossiyskaya Federatsiya. country Asia/Europe
Russian America see Alaska
Russian Federation see Russia
Rustaq see Ar Rustāq
Rustavi 95 G2 prev. Rust'avi. SE Georgia
Rust'avi see Rustavi
Rustchuk see Ruse
Ruston 20 B2 Louisiana, S USA
Rutanzige, Lake see Edward, Lake
Rutba see Ar Ruṭbah
Rutlam see Ratlām
Rutland 19 F2 Vermont, NE USA
Rutog 104 A4 var. Rutög, Rutok. Xizang Zizhiqu, W China
Rutok see Rutog
Ruvuma see Rovuma, Rio
Ruwenzori 55 E5 mountain range Dem. Rep. Congo/Uganda
Ruzhany 85 B6 Brestskaya Voblasts', SW Belarus
Ružomberok 77 C5 Ger. Rosenberg, Hung. Rózsahegy. Žilinský Kraj, N Slovakia
Rwanda 51 B6 off. Rwandese Republic; prev. Ruanda. country C Africa
Rwandese Republic see Rwanda
Ryazan' 89 B5 Ryazanskaya Oblast', W Russia
Rybach'ye see Balykchy
Rybinsk 88 B4 prev. Andropov. Yaroslavskaya Oblast', W Russia
Rybnik 77 C5 Śląskie, S Poland
Rybnitsa see Rîbnița
Ryde 126 E1 United Kingdom
Ryki 76 D4 Lubelskie, E Poland
Rykovo see Yenakiyeve
Rypin 76 C3 Kujawsko-pomorskie, C Poland
Ryssel see Lille
Rysy 77 C5 mountain S Poland
Ryukyu Islands see Nansei-shotō
Ryukyu Trench 103 F3 var. Nansei Syotō Trench. trench S East China Sea
Rzeszów 77 E5 Podkarpackie, SE Poland
Rzhev 88 B4 Tverskaya Oblast', W Russia

S

Saale 72 C4 river C Germany
Saalfeld 73 C5 var. Saalfeld an der Saale. Thüringen, C Germany
Saalfeld an der Saale see Saalfeld
Saarbrücken 73 A6 Fr. Sarrebruck. Saarland, SW Germany
Sääre 84 C2 var. Sjar. Saaremaa, W Estonia
Saare see Saaremaa
Saaremaa 84 C2 Ger. Oesel, Ösel; prev. Saare. island W Estonia
Saariselkä 62 D2 Lapp. Suoločielgi. Lappi, N Finland
Sab' Ābār 96 C4 var. Sab'a Biyar, Sa'b Bi'ār. Ḥimş, C Syria
Sab'a Biyar see Sab' Ābār
Šabac 78 D3 Serbia, W Serbia
Sabadell 71 G2 Cataluña, E Spain
Sabah 116 D3 prev. British North Borneo, North Borneo. state East Malaysia
Sabanalarga 36 B1 Atlántico, N Colombia
Sabaneta 36 C1 Falcón, N Venezuela
Sabaria see Szombathely
Sab'atayn, Ramlat as 99 C6 desert C Yemen
Sabaya 39 F4 Oruro, S Bolivia
Sa'b Bi'ār see Sab' Ābār
Saberi, Hamun-e 100 C5 var. Daryācheh-ye Hāmun, Daryācheh-ye Sīstān. lake Afghanistan/Iran
Sabhā 49 F3 C Libya
Sabi see Save
Sabinas 29 E2 Coahuila, NE Mexico
Sabinas Hidalgo 29 E2 Nuevo León, NE Mexico
Sabine River 27 H3 river Louisiana/Texas, SW USA
Sabkha see As Sabkhah
Sable, Cape 21 E5 headland Florida, SE USA
Sable Island 17 G4 island Nova Scotia, SE Canada
Şabyā 99 B6 Jīzān, SW Saudi Arabia
Sabzawar see Sabzevār
Sabzevār 98 D2 var. Sabzawar. Khorāsān-Razavī, NE Iran
Sachsen 72 D4 Eng. Saxony, Fr. Saxe. state E Germany
Sachs Harbour 15 E2 var. Ikaahuk. Banks Island, Northwest Territories, N Canada
Sächsisch-Reen/Sächsisch-Regen see Reghin
Sacramento 25 B5 state capital California, W USA
Sacramento Mountains 26 D2 mountain range New Mexico, SW USA
Sacramento River 25 B5 river California, W USA
Sacramento Valley 25 B5 valley California, W USA
Sá da Bandeira see Lubango
Şa'dah 99 B6 NW Yemen
Sado 109 C5 var. Sadoga-shima. island C Japan
Sadoga-shima see Sado
Saena Julia see Siena
Safad see Tsefat
Şafāqis see Sfax
Şafāshahr 98 D3 var. Deh Bīd. Fārs, C Iran
Safed see Tsefat
Säffle 63 B6 Värmland, C Sweden
Safford 26 C3 Arizona, SW USA
Safi 48 B2 W Morocco
Selseleh-ye Safīd Kuh 100 D4 Eng. Paropamisus Range. mountain range W Afghanistan
Sagaing 114 B3 Sagaing, C Myanmar (Burma)
Sagami-nada 109 D6 inlet SW Japan
Sagan see Żagań
Sāgar 112 D4 prev. Saugor. Madhya Pradesh, C India
Sagarmāthā see Everest, Mount
Sagebrush State see Nevada
Saghez see Saqqez
Saginaw 18 C3 Michigan, N USA
Saginaw Bay 18 D2 lake bay Michigan, N USA
Sagua la Grande 32 B2 Villa Clara, C Cuba
Sagunto 71 F3 Cat. Sagunt, Ar. Murviedro; anc. Saguntum. País Valenciano, E Spain
Sagunt/Saguntum see Sagunto
Sahara 46 B3 desert Libya/Algeria
Sahara el Gharbîya see Şaḥrā' al Gharbīyah
Saharan Atlas see Atlas Saharien
Sahel 52 D3 physical region C Africa
Sāḥilīyah, Jibāl as 96 B3 mountain range NW Syria
Sāhīwāl 112 C2 prev. Montgomery. Punjab, E Pakistan
Şaḥrā' al Gharbīyah 50 B2 var. Sahara el Gharbîya, Eng. Western Desert. desert C Egypt
Şaḥrā' ash Sharqīyah 81 H5 Eng. Arabian Desert, Eastern Desert. desert E Egypt
Saïda 97 A5 var. Şaydā, Sayida; anc. Sidon. W Lebanon
Sa'īdābād see Sīrjān
Saidpur 113 G3 var. Syedpur. Rajshahi, NW Bangladesh
Saidu Sharif 112 C1 var. Mingora, Mongora. North-West Frontier Province, N Pakistan
Saigon see Hô Chi Minh
Saimaa 63 E5 lake SE Finland
St Albans 67 E6 anc. Verulamium. E England, United Kingdom
Saint Albans 18 D5 West Virginia, NE USA
St Andrews 66 C4 E Scotland, United Kingdom
Saint Anna Trough see Svyataya Anna Trough
St. Ann's Bay 32 B4 C Jamaica
St. Anthony 17 G3 Newfoundland and Labrador, SE Canada
Saint Augustine 21 E3 Florida, SE USA
St Austell 67 C7 SW England, United Kingdom
St.Botolph's Town see Boston
St-Brieuc 68 A3 Côtes d'Armor, NW France
St. Catharines 16 D5 Ontario, S Canada
St-Chamond 69 D5 Loire, E France
Saint Christopher and Nevis, Federation of see Saint Kitts and Nevis
Saint Christopher-Nevis see Saint Kitts and Nevis
Saint Clair, Lake 18 D3 var. Lac à L'Eau Claire. lake Canada/USA
St-Claude 69 D5 anc. Condate. Jura, E France
Saint Cloud 23 F2 Minnesota, N USA
Saint Croix 33 F3 island S Virgin Islands (US)
Saint Croix River 18 A2 river Minnesota/ Wisconsin, N USA
St David's Island 20 B5 island E Bermuda
St-Denis 57 G4 dependent territory capital (Réunion) NW Réunion
St-Dié 68 E4 Vosges, NE France
St-Egrève 69 D5 Isère, E France
Sainte Marie, Cap see Vohimena, Tanjona
Saintes 69 B5 anc. Mediolanum. Charente-Maritime, W France
St-Étienne 69 D5 Loire, E France
St-Flour 69 C5 Cantal, C France
St-Gall/Saint Gall/St. Gallen see Sankt Gallen
St-Gaudens 69 B6 Haute-Garonne, S France
Saint George 127 D5 Queensland, E Australia
St George 20 B4 N Bermuda
Saint George 22 A5 Utah, W USA
St. George's 33 G5 country capital (Grenada) SW Grenada
St-Georges 17 E4 Québec, SE Canada
St-Georges 37 H3 E French Guiana
Saint George's Channel 67 B6 channel Ireland/ Wales, United Kingdom
St George's Island 20 B4 island E Bermuda
St Helena see St Helena, Ascension and Tristan da Cunha
St Helena, Ascension and Tristan da Cunha 47 A6 UK overseas territory C Atlantic Ocean
St Helier 67 D8 dependent territory capital (Jersey) S Jersey, Channel Islands
St.Iago de la Vega see Spanish Town
Saint Ignace 18 C2 Michigan, N USA
St-Jean, Lac 17 E4 lake Québec, SE Canada
Saint Joe River 24 D2 river Idaho, NW USA
St. John 17 F4 New Brunswick, SE Canada
Saint John see Saint John
Saint John 19 H1 Fr. Saint-John. river Canada/USA
St John's 33 G3 country capital (Antigua and Barbuda) Antigua, Antigua and Barbuda
St. John's 17 H3 province capital Newfoundland and Labrador, E Canada
Saint Joseph 23 F4 Missouri, C USA
St Julian's see San Giljan
St Kilda 66 A3 island NW Scotland, United Kingdom
Saint Kitts and Nevis 33 F3 off. Federation of Saint Christopher and Nevis, var. Saint Christopher-Nevis. country E West Indies
St-Laurent see St-Laurent-du-Maroni
St-Laurent-du-Maroni 37 H3 var. St-Laurent. NW French Guiana
St-Laurent, Fleuve see St. Lawrence
St. Lawrence 17 E4 Fr. Fleuve St-Laurent. river Canada/USA
St. Lawrence, Gulf of 17 F3 gulf NW Atlantic Ocean
Saint Lawrence Island 14 B2 island Alaska, USA
St-Lô 68 B3 anc. Briovera, Laudus. Manche, N France
St-Louis 68 E4 Haut-Rhin, NE France
Saint Louis 52 B3 NW Senegal
Saint Louis 23 G4 Missouri, C USA
Saint Lucia 33 E1 country SE West Indies
Saint Lucia Channel 33 H4 channel Martinique/ Saint Lucia
St-Malo 68 B3 Ille-et-Vilaine, NW France
St-Malo, Golfe de 68 A3 gulf NW France
Saint Martin see Sint Maarten
St.Matthew's Island see Zadetkyi Kyun
St.Matthias Group 122 B3 island group NE Papua New Guinea
St. Moritz 73 B7 Ger. Sankt Moritz, Rmsch. San Murezzan. Graubünden, SE Switzerland
St-Nazaire 68 A4 Loire-Atlantique, NW France
Saint Nicholas see São Nicolau
Saint-Nicolas see Sint-Niklaas
St-Omer 68 C2 Pas-de-Calais, N France
Saint Paul 23 F2 state capital Minnesota, N USA
St-Paul, Île 119 C6 var. St.Paul Island. island Île St-Paul, NE French Southern and Antarctic Lands Antarctica Indian Ocean
St.Paul Island see St-Paul, Île
St Peter Port 67 D8 dependent territory capital (Guernsey) C Guernsey, Channel Islands
Saint Petersburg 21 E4 Florida, SE USA
Saint Petersburg see Sankt-Peterburg
St-Pierre and Miquelon 17 G4 Fr. Îles St-Pierre et Miquelon. French territorial collectivity NE North America
St-Quentin 68 C3 Aisne, N France
Saint Thomas see São Tomé, Sao Tome and Principe
Saint Thomas see Charlotte Amalie, Virgin Islands (US)
Saint Ubes see Setúbal
Saint Vincent 33 G4 island N Saint Vincent and the Grenadines
Saint Vincent see São Vicente
Saint Vincent and the Grenadines 33 H4 country SE West Indies
Saint Vincent, Cape see São Vicente, Cabo de
Saint Vincent Passage 33 H4 passage Saint Lucia/ Saint Vincent and the Grenadines
Saint Yves see Setúbal
Saipan 120 B1 island/country capital (Northern Mariana Islands) S Northern Mariana Islands
Saishū see Jeju-do
Sajama, Nevado 39 F4 mountain W Bolivia
Sajószentpéter 77 D6 Borsod-Abaúj-Zemplén, NE Hungary
Sakākah 98 B4 Al Jawf, NW Saudi Arabia
Sakakawea, Lake 22 D1 reservoir North Dakota, N USA
Sak'art'velo see Georgia
Sakata 108 D4 Yamagata, Honshū, C Japan
Sakhalin 93 G4 var. Sakhalin. island SE Russia
Sakhalin see Sakhalin, Ostrov
Sakhon Nakhon see Sakon Nakhon
Şäki 95 G2 Rus. Sheki; prev. Nukha. NW Azerbaijan
Saki see Saky
Sakishima-shoto 108 A3 var. Sakisima Syotō. island group SW Japan
Sakisima Syotō see Sakishima-shotō
Sakiz see Saqqez
Sakiz-Adasi see Chíos
Sakon Nakhon 114 D4 var. Muang Sakon Nakhon, Sakhon Nakhon. Sakon Nakhon, E Thailand
Saky 87 F5 Rus. Saki. Respublika Krym, S Ukraine
Sal 52 A3 island Ilhas de Barlavento, NE Cape Verde
Sala 63 C6 Västmanland, C Sweden
Salacgrīva 84 C3 Est. Salatsi. Limbaži, N Latvia
Sala Consilina 75 D5 Campania, S Italy
Salado, Río 40 D5 river E Argentina
Salado, Río 42 C3 river C Argentina
Şalālah 99 D6 SW Oman
Salamá 30 B2 Baja Verapaz, C Guatemala
Salamanca 42 B4 Coquimbo, C Chile
Salamanca 70 D2 anc. Helmantica, Salmantica. Castilla-León, NW Spain
Salamīyah 96 B3 var. As Salamīyah. Ḥamāh, W Syria
Salang see Phuket
Salantai 84 B3 Klaipėda, NW Lithuania
Salatsi see Salacgrīva
Salavan 115 D5 var. Saravan, Saravane. Salavan, S Laos
Salavat 89 D6 Respublika Bashkortostan, W Russia
Sala y Gomez 131 F4 island Chile, E Pacific Ocean
Sala y Gomez Fracture Zone see Sala y Gomez Ridge
Sala y Gomez Ridge 131 G4 var. Sala y Gomez Fracture Zone. fracture zone SE Pacific Ocean
Salazar see N'Dalatando
Šalčininkai 85 C5 Vilnius, SE Lithuania
Salduba see Zaragoza
Saldus 84 B3 Ger. Frauenburg. Saldus, W Latvia
Sale 127 C7 Victoria, SE Australia
Salé 48 C2 NW Morocco
Salekhard 92 D3 prev. Obdorsk. Yamalo-Nenetskiy Avtonomnyy Okrug, N Russia
Salem 110 C2 Tamil Nādu, SE India
Salem 24 B3 state capital Oregon, NW USA
Salerno 75 D5 anc. Salernum. Campania, S Italy
Salerno, Gulf of 75 C5 Eng. Gulf of Salerno. gulf S Italy
Salerno, Gulf of see Salerno, Golfo di
Salernum see Salerno
Salihorsk 85 C7 Rus. Soligorsk. Minskaya Voblasts', S Belarus
Salima 57 E2 Central, C Malawi
Salina 23 E5 Kansas, C USA
Salina Cruz 29 F5 Oaxaca, SE Mexico
Salinas 38 A2 Guayas, W Ecuador
Salinas 25 B6 California, W USA
Salisbury 67 D7 var. New Sarum. S England, United Kingdom
Salisbury see Harare
Sällan see Sørøya
Salliq see Coral Harbour
Sallyana see Şalyān
Salmantica see Salamanca
Salmon River 24 D3 river Idaho, NW USA
Salmon River Mountains 24 D3 mountain range Idaho, NW USA
Salo 63 D6 Länsi-Suomi, SW Finland
Salon-de-Provence 69 D6 Bouches-du-Rhône, SE France
Salonica/Salonika see Thessaloníki
Salonta 86 A3 Hung. Nagyszalonta. Bihor, W Romania
Sal'sk 89 B7 Rostovskaya Oblast', SW Russia
Salt see As Salṭ
Salta 42 C2 Salta, NW Argentina
Saltash 67 C7 SW England, United Kingdom
Saltillo 29 E3 Coahuila, NE Mexico
Salt Lake City 22 B4 state capital Utah, W USA
Salto 42 D4 Salto, N Uruguay
Salton Sea 25 D8 lake California, W USA
Salvador 41 G3 prev. São Salvador. state capital Bahia, E Brazil
Salween 102 C2 Bur. Thanlwin, Chin. Nu Chiang, Nu Jiang. river SE Asia
Şalyān 113 E3 var. Sallyana. Mid Western, W Nepal
Salzburg 73 D6 anc. Juvavum. Salzburg, N Austria
Salzgitter 72 C4 prev. Watenstedt-Salzgitter. Niedersachsen, C Germany
Salzwedel 72 C3 Sachsen-Anhalt, N Germany
Šamac see Bosanski Šamac
Samakhixai see Attapu
Samalayuca 28 C1 Chihuahua, N Mexico
Samar 117 F2 island C Philippines
Samara 92 B3 prev. Kuybyshev. Samarskaya Oblast', W Russia
Samarang see Semarang
Samarinda 116 D4 Borneo, C Indonesia
Samarkand see Samarqand
Samarkandski/Samarkandskoye see Temirtau
Samarobriva see Amiens
Samarqand 101 E2 Rus. Samarkand. Samarqand Viloyati, C Uzbekistan
Samawa see As Samāwah
Şamaxı 95 H2 E Azerbaijan
Sambalpur 113 F4 Odisha, E India
Sambava 57 G2 Antsiranana, NE Madagascar
Sambir 86 B2 Rus. Sambor. L'vivs'ka Oblast', NW Ukraine
Sambor see Sambir
Sambre 68 D2 river Belgium/France
Samfya 56 D2 Luapula, N Zambia
Saminatal 72 E2 valley Austria/Liechtenstein, Europe
Samnān see Semnān
Sam Neua see Xam Nua
Samoa 123 E4 off. Independent State of Western Samoa, var. Sāmoa; prev. Western Samoa. country W Polynesia
Sāmoa see Samoa
Samoa Basin 121 E3 undersea basin W Pacific Ocean
Samobor 78 A2 Zagreb, N Croatia
Sámos 83 E6 prev. Limín Vathéos. Sámos, Dodekánisa, Greece, Aegean Sea
Sámos 83 D6 island Dodekánisa, Greece, Aegean Sea
Samothrace see Samothráki
Samothráki 82 D4 Samothráki, NE Greece
Samothraki 82 C4 anc. Samothrace. island NE Greece
Sampit 116 C4 Borneo, C Indonesia
Sâmraông 115 D5 prev. Phumĭ Sâmraông, Phum Samrong. Siĕmréab, NW Cambodia
Samsun 94 D2 anc. Amisus. Samsun, N Turkey
Samt'redia 95 F2 W Georgia
Samui, Ko 115 C6 island SW Thailand
Samut Prakan 115 C5 var. Muang Samut Prakan, Paknam. Samut Prakan, C Thailand
San 52 D3 Ségou, C Mali
San 77 E5 river SE Poland
Şan'ā' 99 B6 Eng. Sanaa. country capital (Yemen) W Yemen
Sana 78 B3 river NW Bosnia and Herzegovina
Sanaa see Şan'ā'
Sanae 132 B2 South African research station Antarctica
Sanaga 55 B5 river C Cameroon
San Ambrosio, Isla 35 A5 Eng. San Ambrosio Island. island W Chile
San Ambrosio Island see San Ambrosio, Isla
Sanandaj 98 C3 prev. Sinneh. Kordestān, W Iran
San Andrés, Isla de 31 F3 island NW Colombia, Caribbean Sea
San Andrés Tuxtla 29 F4 var. Tuxtla. Veracruz-Llave, E Mexico
San Angelo 27 F3 Texas, SW USA
San Antonio 30 B2 Toledo, S Belize
San Antonio 42 B4 Valparaíso, C Chile
San Antonio 27 F4 Texas, SW USA
San Antonio Oeste 43 C5 Río Negro, E Argentina
San Antonio River 27 G4 river Texas, SW USA
Sanāw 99 C6 var. Sanaw. NE Yemen
San Benedicto, Isla 28 B4 island W Mexico
San Benito 30 B1 Petén, N Guatemala
San Benito 27 G5 Texas, SW USA
San Bernardino 25 C7 California, W USA
San Blas 28 C3 Sinaloa, C Mexico
San Blas, Cape 20 D3 headland Florida, SE USA
San Blas, Cordillera de 31 G4 mountain range NE Panama
San Carlos 30 D4 Río San Juan, S Nicaragua
San Carlos 26 B2 Arizona, SW USA
San Carlos see Quesada, Costa Rica
San Carlos de Ancud see Ancud
San Carlos de Bariloche 43 B5 Río Negro, SW Argentina
San Carlos del Zulia 36 C2 Zulia, W Venezuela
San Clemente Island 25 B8 island Channel Islands, California, W USA
San Cristóbal 36 C2 Táchira, W Venezuela
San Cristobal 122 C4 var. Makira. island SE Solomon Islands
San Cristóbal see San Cristóbal de Las Casas
San Cristóbal de Las Casas 29 G5 var. San Cristóbal. Chiapas, SE Mexico
San Cristóbal, Isla 38 B5 var. Chatham Island. island Galápagos Islands, Ecuador, E Pacific Ocean
Sancti Spíritus 32 B2 Sancti Spíritus, C Cuba
Sandakan 116 D3 Sabah, East Malaysia
Sandalwood Island see Sumba, Pulau
Sandanski 82 C3 prev. Sveti Vrach. Blagoevgrad, SW Bulgaria
Sanday 66 D2 island NE Scotland, United Kingdom
Sanders 26 C2 Arizona, SW USA
Sand Hills 22 D3 mountain range Nebraska, C USA
San Diego 25 C8 California, W USA
Sandnes 63 A6 Rogaland, S Norway
Sandomierz 76 D4 Rus. Sandomir. Świętokrzyskie, C Poland
Sandomir see Sandomierz
Sandoway see Thandwe
Sandpoint 24 C1 Idaho, NW USA
Sand Springs 27 G1 Oklahoma, C USA
Sandusky 18 D3 Ohio, N USA
Sandvika 63 A6 Akershus, S Norway
Sandviken 63 C6 Gävleborg, C Sweden
Sandwich Island see Efate
Sandwich Islands see Hawaiian Islands
Sandy Bay 71 H5 Saskatchewan, C Canada
Sandy City 22 B4 Utah, W USA
Sandy Lake 16 B3 lake Ontario, C Canada
San Esteban 30 D2 Olancho, C Honduras
San Eugenio/San Eugenio del Cuareim see Artigas
San Felipe 36 D1 Yaracuy, NW Venezuela
San Felipe de Puerto Plata see Puerto Plata
San Félix, Isla 35 A5 Eng. San Felix Island. island W Chile
San Felix Island see San Félix, Isla
San Fernando 70 C5 prev. Isla de León. Andalucía, S Spain
San Fernando 33 H5 Trinidad, Trinidad and Tobago
San Fernando 24 D1 California, W USA
San Fernando 36 D2 var. San Fernando de Apure. Apure, C Venezuela
San Fernando de Apure see San Fernando
San Fernando del Valle de Catamarca 42 C3 var. Catamarca. Catamarca, NW Argentina
San Fernando de Monte Cristi see Monte Cristi
San Francisco 25 B6 California, W USA
San Francisco del Oro 28 C2 Chihuahua, N Mexico
San Francisco de Macorís 33 E3 C Dominican Republic
San Fructuoso see Tacuarembó
San Gabriel 38 B1 Carchi, N Ecuador
San Gabriel Mountains 24 E1 mountain range California, USA
Sangihe, Kepulauan see Sangir, Kepulauan
San Giljan 80 B5 var. St Julian's. N Malta
Sangir, Kepulauan 117 F3 var. Kepulauan Sangihe. island group N Indonesia
Sāngli 110 B1 Mahārāshtra, W India
Sangmélima 55 B5 Sud, S Cameroon
Sangre de Cristo Mountains 26 D1 mountain range Colorado/New Mexico, C USA
San Ignacio 30 B1 prev. Cayo, El Cayo. Cayo, W Belize
San Ignacio 39 F3 Beni, N Bolivia
San Ignacio 28 B2 Baja California Sur, NW Mexico
San Joaquin Valley 25 B7 valley California, W USA
San Jorge, Golfo 43 C6 var. Gulf of San Jorge. gulf S Argentina
San Jorge, Gulf of see San Jorge, Golfo
San José 31 E4 country capital (Costa Rica) San José, C Costa Rica
San José 39 G3 var. San José de Chiquitos. Santa Cruz, E Bolivia
San José 30 B3 var. Puerto San José. Escuintla, S Guatemala
San Jose 25 B6 California, W USA
San José see San José del Guaviare, Colombia
San José de Chiquitos see San José
San José de Cúcuta see Cúcuta
San José del Guaviare 36 C4 var. San José. Guaviare, S Colombia
San Juan 42 B4 San Juan, W Argentina
San Juan 33 F3 dependent territory capital (Puerto Rico) NE Puerto Rico
San Juan see San Juan de los Morros
San Juan Bautista 42 D3 Misiones, S Paraguay
San Juan Bautista see Villahermosa
San Juan Bautista Tuxtepec see Tuxtepec
San Juan de Alicante see Sant Joan d'Alacant
San Juan del Norte 31 E4 var. Greytown. Río San Juan, SE Nicaragua
San Juan de los Morros 36 D2 var. San Juan. Guárico, N Venezuela
San Juanito, Isla 28 C4 island C Mexico
San Juan Mountains 26 D1 mountain range Colorado, C USA

Severnaya Dvina – Sognefjorden

Severnaya Dvina *88 C4 var.* Northern Dvina. *river* NW Russia
Severnaya Zemlya *93 E2 var.* Nicholas II Land. *island group* N Russia
Severnyy *88 E3* Respublika Komi, NW Russia
Severodonetsk *see* Syeverodonets'k
Severodvinsk *88 C3 prev.* Molotov, Sudostroy. Arkhangel'skaya Oblast', NW Russia
Severomorsk *88 C2* Murmanskaya Oblast', NW Russia
Seversk *92 D4* Tomskaya Oblast', C Russia
Sevier Lake *22 A4 lake* Utah, W USA
Sevilla *70 C4 Eng.* Seville; *anc.* Hispalis. Andalucía, SW Spain
Seville *see* Sevilla
Sevlievo *82 D2* Gabrovo, N Bulgaria
Sevluš/Sevlyush *see* Vynohradiv
Seward's Folly *see* Alaska
Seychelles *57 G1 off.* Republic of Seychelles. *country* W Indian Ocean
Seychelles, Republic of *see* Seychelles
Seyðisfjörður *61 E5* Austurland, E Iceland
Seýdi *100 D2 Rus.* Seydi; *prev.* Neftezavodsk. Lebap Welaýaty, E Turkmenistan
Seyhan *see* Adana
Sfákia *see* Chóra Sfakíon
Sfântu Gheorghe *86 C4 Ger.* Sankt-Georgen, *Hung.* Sepsiszentgyörgy; *prev.* Şepşi-Sângeorz, Sfîntu Gheorghe. Covasna, C Romania
Sfax *49 F2 Ar.* Şafāqis. E Tunisia
Sfîntu Gheorghe *see* Sfântu Gheorghe
's-Gravenhage *64 B4 var.* Den Haag, *Eng.* The Hague, *Fr.* La Haye. *country capital* (Netherlands-seat of government) Zuid-Holland, W Netherlands
's-Gravenzande *64 B4* Zuid-Holland, W Netherlands
Shaan/Shaanxi Sheng *see* Shaanxi
Shaanxi *106 B5 var.* Shaan, Shaanxi Sheng, Shan-hsi, Shenshi, Shensi. *province* C China
Shabani *see* Zvishavane
Shabeelle, Webi *see* Shebeli
Shache *104 A3 var.* Yarkant. Xinjiang Uygur Zizhiqu, NW China
Shacheng *see* Huailai
Shackleton Ice Shelf *132 D3 ice shelf* Antarctica
Shaddādī *see* Ash Shadādah
Shāhābād *see* Eslāmābād
Sha Hi *see* Orūmīyeh, Daryācheh-ye
Shahjahanabad *see* Delhi
Shahr-e Kord *98 C3 var.* Shahr Kord. Chahār MaḤall va Bakhtiārī, C Iran
Shahr Kord *see* Shahr-e Kord
Shāhrūd *98 D2 prev.* Emāmrūd, Emāmshahr. Semnān, N Iran
Shalkar *92 B4 var.* Chelkar. Aktyubinsk, W Kazakhstan
Shām, Bādiyat ash *see* Syrian Desert
Shana *see* Kuril'sk
Shandi *see* Shendi
Shandong *106 D4 var.* Lu, Shandong Sheng, Shantung. *province* E China
Shandong Sheng *see* Shandong
Shanghai *106 D5 var.* Shang-hai. Shanghai Shi, E China
Shangrao *106 D5* Jiangxi, S China
Shan-hsi *see* Shaanxi, China
Shan-hsi *see* Shanxi, China
Shannon *67 A6 Ir.* An tSionainn. *river* W Ireland
Shan Plateau *114 B3 plateau* E Myanmar (Burma)
Shansi *see* Shanxi
Shantar Islands *see* Shantarskiye Ostrova
Shantarskiye Ostrova *93 G3 Eng.* Shantar Islands. *island group* E Russia
Shantou *106 D6 var.* Shan-t'ou, Swatow. Guangdong, S China
Shan-t'ou *see* Shantou
Shantung *see* Shandong
Shanxi *106 C4 var.* Jin, Shan-hsi, Shansi, Shanxi Sheng. *province* C China
Shan Xian *see* Sanmenxia
Shanxi Sheng *see* Shanxi
Shaoguan *106 C6 var.* Shao-kuan, *Cant.* Kukong; *prev.* Ch'u-chiang. Guangdong, S China
Shao-kuan *see* Shaoguan
Shaqrā' *98 B4* Ar Riyāḑ, C Saudi Arabia
Shaqrā *see* Shuqrah
Shar *92 D5 var.* Charsk. Vostochnyy Kazakhstan, E Kazakhstan
Shari *108 D2* Hokkaidō, NE Japan
Shari *see* Chari
Sharjah *see* Ash Shāriqah
Shark Bay *125 A5 bay* Western Australia
Sharqī, Al Jabal ash/Sharqi, Jebel esh *see* Anti-Lebanon
Shashe *56 D3 var.* Shashi. *river* Botswana/Zimbabwe
Shashi *see* Shashe
Shatskiy Rise *103 G1 undersea rise* N Pacific Ocean
Shawnee *27 G1* Oklahoma, C USA
Shaykh, Jabal ash *see* Hermon, Mount
Shchadryn *85 D7 Rus.* Shchedrin. Homyel'skaya Voblasts', SE Belarus
Shchedrin *see* Shchadryn
Shcheglovsk *see* Kemerovo
Shchëkino *89 B5* Tul'skaya Oblast', W Russia
Shchors *see* Snovs'k
Shchuchin *see* Shchuchyn
Shchuchinsk *92 C4 prev.* Shchuchye. Akmola, N Kazakhstan
Shchuchye *see* Shchuchinsk
Shchuchyn *85 B5 Pol.* Szczuczyn Nowogródzki, *Rus.* Shchuchin. Hrodzyenskaya Voblasts', W Belarus
Shebekino *89 A6* Belgorodskaya Oblast', W Russia
Shebelē Wenz, Wabē *see* Shebeli
Shebeli *51 D5 Amh.* Wabē Shebelē Wenz, *It.* Scebeli, *Som.* Webi Shabeelle. *river* Ethiopia/Somalia
Sheberghān *see* Shibirghān
Sheboygan *18 B2* Wisconsin, N USA
Shebshi Mountains *54 A4 var.* Schebschi Mountains. *mountain range* E Nigeria
Shechem *see* Nablus
Shedadi *see* Ash Shadādah
Sheffield *67 D5* N England, United Kingdom
Shekhem *see* Nablus
Sheki *see* Şäki
Shelby *22 B1* Montana, NW USA
Sheldon *23 F3* Iowa, C USA
Shelekhov Gulf *see* Shelikhova, Zaliv
Shelikhova, Zaliv *93 G2 Eng.* Shelekhov Gulf. *gulf* E Russia
Shendi *50 C4 var.* Shandī. River Nile, NE Sudan
Shengking *see* Liaoning
Shenking *see* Liaoning
Shenshi/Shensi *see* Shaanxi
Shenyang *106 D3 Chin.* Shen-yang, *Eng.* Moukden, Mukden; *prev.* Fengtien. *province capital* Liaoning, NE China
Shen-yang *see* Shenyang
Shepetivka *86 D2 Rus.* Shepetovka. Khmel'nyts'ka Oblast', NW Ukraine
Shepetovka *see* Shepetivka
Shepparton *127 C7* Victoria, SE Australia
Sherbrooke *17 E4* Québec, SE Canada
Shereik *50 C3* River Nile, N Sudan
Sheridan *22 C2* Wyoming, C USA
Sherman *27 G2* Texas, SW USA
's-Hertogenbosch *64 C4 Fr.* Bois-le-Duc, *Ger.* Herzogenbusch. Noord-Brabant, S Netherlands
Shetland Islands *66 D1 island group* NE Scotland, United Kingdom
Shevchenko *see* Aktau
Shibergān/Shiberghan *see* Shibirghān
Shibirghān *101 E3 var.* Sheberghān, Shiberghan, Shiberghān. Jowzjān, N Afghanistan
Shibetsu *108 D2 var.* Sibetu. Hokkaidō, NE Japan
Shibh Jazīrat Sīnā' *see* Sinai
Shibushi-wan *109 B8 bay* SW Japan
Shigatse *see* Xigazê
Shih-chia-chuang/Shihmen *see* Shijiazhuang
Shihezi *104 C2* Xinjiang Uygur Zizhiqu, NW China
Shiichi *see* Shyichy
Shijiazhuang *106 C4 var.* Shih-chia-chuang; *prev.* Shihmen. *province capital* Hebei, E China
Shikārpur *112 B3* Sind, S Pakistan
Shikoku *109 C7 var.* Sikoku. *island* SW Japan
Shikoku Basin *103 F2 var.* Sikoku Basin. *undersea basin* N Philippine Sea
Shikotan, Ostrov *108 E2 Jap.* Shikotan-tō. *island* NE Russia
Shikotan-tō *see* Shikotan, Ostrov
Shilabo *51 D5* Sumalē, E Ethiopia
Shiliguri *113 F3 prev.* Siliguri. West Bengal, NE India
Shilka *93 F4 river* S Russia
Shillong *113 G3 state capital* Meghālaya, NE India
Shimanto *see* Nakamura
Shimbir Berris *see* Shimbiris
Shimbiris *50 E4 var.* Shimbir Berris. *mountain* N Somalia
Shimoga *see* Shivamogga
Shimonoseki *109 A7 var.* Simonoseki, *hist.* Akamagaseki, Bakan. Yamaguchi, Honshū, SW Japan
Shinano-gawa *109 C5 var.* Sinano Gawa. *river* Honshū, C Japan
Shīndand *100 D4 prev.* Shīndand. Herāt, W Afghanistan
Shīndand *see* Shīndand
Shingū *109 C6 var.* Singû. Wakayama, Honshū, SW Japan
Shinjō *108 D4 var.* Sinzyô. Yamagata, Honshū, C Japan
Shinyanga *51 C7* Shinyanga, NW Tanzania
Shiprock *26 C1* New Mexico, SW USA
Shirāz *98 D4 var.* Shīrāz. Fārs, S Iran
Shishchitsy *see* Shyshchytsy
Shivamogga *110 C2 prev.* Shimoga. Karnātaka, W India
Shivpuri *112 D3* Madhya Pradesh, C India
Shizugawa *108 D4* Miyagi, Honshū, NE Japan
Shizuoka *109 D6 var.* Sizuoka. Shizuoka, Honshū, S Japan
Shklov *see* Shklow
Shklow *85 D6 Rus.* Shklov. Mahilyowskaya Voblasts', E Belarus
Shkodër *79 C5 var.* Shkodra, *It.* Scutari, *SCr.* Skadar. Shkodër, NW Albania
Shkodra *see* Shkodër
Shkodrës, Liqeni i *see* Scutari, Lake
Shkumbinit, Lumi i *79 C6 var.* Shkumbî, Shkumbin. *river* C Albania
Shkumbî/Shkumbin *see* Shkumbinit, Lumi i
Sholāpur *see* Solāpur
Shostka *87 F1* Sums'ka Oblast', NE Ukraine
Show Low *26 B2* Arizona, SW USA
Show Me State *see* Missouri
Shpola *87 E3* Cherkas'ka Oblast', N Ukraine
Shqipëria/Shqipërisë, Republika e *see* Albania
Shreveport *20 A2* Louisiana, S USA
Shrewsbury *67 D6 hist.* Scrobesbyrig'. W England, United Kingdom
Shu *92 C5 Kaz.* Shū. Zhambyl, SE Kazakhstan
Shuang-liao *see* Liaoyuan
Shū, Kazakhstan *see* Shu
Shumagin Islands *14 B3 island group* Alaska, USA
Shumen *82 D2* Shumen, NE Bulgaria
Shumilina *85 E5 Rus.* Shumilino. Vitsyebskaya Voblasts', NE Belarus
Shumilino *see* Shumilina
Shunsen *see* Chuncheon
Shuqrah *99 B7 var.* Shaqrā. SW Yemen
Shwebo *114 B3* Sagaing, C Myanmar (Burma)
Shyichy *85 C7 Rus.* Shiichi. Homyel'skaya Voblasts', SE Belarus
Shymkent *92 B5 prev.* Chimkent. Yuzhnyy Kazakhstan, S Kazakhstan
Shyshchytsy *85 C6 Rus.* Shishchitsy. Minskaya Voblasts', C Belarus
Siam *see* Thailand
Siam, Gulf of *see* Thailand, Gulf of
Sian *see* Xi'an
Siang *see* Brahmaputra
Siangtan *see* Xiangtan
Šiauliai *84 B4 Ger.* Schaulen. Šiauliai, N Lithuania
Siazan' *see* Siyäzän
Sibay *89 D6* Respublika Bashkortostan, W Russia
Šibenik *78 B4 It.* Sebenico. Šibenik-Knin, S Croatia
Siberia *see* Sibir'
Siberoet *see* Siberut, Pulau
Siberut, Pulau *116 A4 prev.* Siberoet. *island* Kepulauan Mentawai, W Indonesia
Sibi *112 B2* Baluchistān, SW Pakistan
Sibir' *93 E3 var.* Siberia. *physical region* NE Russia
Sibiti *55 B6* Lékoumou, S Congo
Sibiu *86 B4 Ger.* Hermannstadt, *Hung.* Nagyszeben. Sibiu, C Romania
Sibolga *116 B3* Sumatera, W Indonesia
Sibu *116 D3* Sarawak, East Malaysia
Sibut *54 C4 prev.* Fort-Sibut. Kémo, S Central African Republic
Sibuyan Sea *117 E2 sea* W Pacific Ocean
Sichon *115 C6 var.* Ban Sichon, Si Chon. Nakhon Si Thammarat, SW Thailand
Si Chon *see* Sichon
Sichuan *106 B5 var.* Chuan, Sichuan Sheng, Ssu-ch'uan, Szechuan, Szechwan. *province* C China
Sichuan Pendi *106 B5 basin* C China
Sichuan Sheng *see* Sichuan
Sicilian Channel *see* Sicily, Strait of
Sicily *75 C7 Eng.* Sicily; *anc.* Trinacria. *island* Italy, C Mediterranean Sea
Sicily, Strait of *75 B7 var.* Sicilian Channel. *strait* C Mediterranean Sea
Sicuani *39 E4* Cusco, S Peru
Sidári *82 A4* Kérkyra, Iónia Nisiá, Greece, C Mediterranean Sea
Sidas *116 C4* Borneo, C Indonesia
Siderno *75 D7* Calabria, SW Italy
Sidhirókastro *see* Sidirókastro
Sîdi Barrâni *50 A1* NW Egypt
Sidi Bel Abbès *48 D2 var.* Sidi bel Abbès, Sidi-Bel-Abbès. NW Algeria
Sidirókastro *82 C3 prev.* Sidhirókastron. Kentrikí Makedonía, NE Greece
Sidley, Mount *132 B4 mountain* Antarctica
Sidney *22 D1* Montana, NW USA
Sidney *22 D4* Nebraska, C USA
Sidney *18 C4* Ohio, N USA
Sidon *see* Saïda
Sidra *see* Surt
Sidra/Sidra, Gulf of *see* Surt, Khalīj, N Libya
Siebenbürgen *see* Transylvania
Siedlce *76 E3 Ger.* Sedlez, *Rus.* Sesdlets. Mazowieckie, C Poland
Siegen *72 B4* Nordrhein-Westfalen, W Germany
Siemiatycze *76 E3* Podlaskie, NE Poland
Siena *74 B3 Fr.* Sienne; *anc.* Saena Julia. Toscana, C Italy
Sienne *see* Siena
Sieradz *76 C4* Sieradz, C Poland
Sierpc *76 D3* Mazowieckie, C Poland
Sierra Leone *52 C4 off.* Republic of Sierra Leone. *country* W Africa
Sierra Leone Basin *44 C4 undersea basin* E Atlantic Ocean
Sierra Leone, Republic of *see* Sierra Leone
Sierra Leone Ridge *see* Sierra Leone Rise
Sierra Leone Rise *44 C4 var.* Sierra Leone Ridge, Sierra Leone Schwelle. *undersea rise* E Atlantic Ocean
Sierra Leone Schwelle *see* Sierra Leone Rise
Sierra Vista *26 B3* Arizona, SW USA
Sífnos *83 C6 anc.* Siphnos. *island* Kykládes, Greece, Aegean Sea
Sigli *116 A3* Sumatera, W Indonesia
Siglufjörður *61 E4* Nordhurland Vestra, N Iceland
Signal Peak *26 A2 mountain* Arizona, SW USA
Signan *see* Xi'an
Signy *132 A2 UK research station* South Orkney Islands, Antarctica
Siguatepeque *30 C2* Comayagua, W Honduras
Siguiri *52 D4* NE Guinea
Sihanoukville *115 D6 Khmer.* Preăh Seihânŭ; *prev.* Kompong Som. Sihanoukville, SW Cambodia
Siilinjärvi *62 E4* Pohjois-Savo, C Finland
Siirt *95 F4 var.* Sert; *anc.* Tigranocerta. Siirt, SE Turkey
Sikandarabad *see* Secunderābād
Sikasso *52 D4* Sikasso, S Mali
Sikeston *23 H5* Missouri, C USA
Sikhote-Alin', Khrebet *93 G4 mountain range* SE Russia
Siking *see* Xi'an
Siklós *77 C7* Baranya, SW Hungary
Sikoku *see* Shikoku
Sikoku Basin *see* Shikoku Basin
Šilalė *84 B4* Tauragė, W Lithuania
Silchar *113 G3* Assam, NE India
Silesia *76 B4 physical region* SW Poland
Silifke *94 C4 anc.* Seleucia. İçel, S Turkey
Siliguri *see* Shiliguri
Siling Co *104 C5 lake* W China
Silinhot *see* Xilinhot
Silistra *82 E1 var.* Silistria; *anc.* Durostorum. Silistra, NE Bulgaria
Silistria *see* Silistra
Sillamäe *84 E2 Ger.* Sillamäggi. Ida-Virumaa, NE Estonia
Sillamäggi *see* Sillamäe
Sillein *see* Žilina
Šilutė *84 B4 Ger.* Heydekrug. Klaipėda, W Lithuania
Silvan *95 E4* Dıyarbakır, SE Turkey
Silva Porto *see* Kuito
Silver State *see* Colorado
Silver State *see* Nevada
Simanichy *85 C7 Rus.* Simonichi. Homyel'skaya Voblasts', SE Belarus
Simav *94 B3* Kütahya, W Turkey
Simav Çayı *94 A3 river* NW Turkey
Simbirsk *see* Ul'yanovsk
Simeto *75 C7 river* Sicilia, Italy, C Mediterranean Sea
Simeulue, Pulau *116 A3 island* NW Indonesia
Simferopol' *87 F5* Respublika Krym, S Ukraine
Simitla *82 C3* Blagoevgrad, SW Bulgaria
Şimlăul Silvaniei/Şimleul Silvaniei *see* Şimleu Silvaniei
Şimleu Silvaniei *86 B3 Hung.* Szilágysomlyó; *prev.* Şimlăul Silvaniei, Şimleul Silvaniei. Sălaj, NW Romania
Simonichi *see* Simanichy
Simonoseki *see* Shimonoseki
Simpelveld *65 D6* Limburg, SE Netherlands
Simplon Pass *73 B8 pass* S Switzerland
Simpson *see* Fort Simpson
Simpson Desert *126 B4 desert* Northern Territory/South Australia
Sinai *50 C2 var.* Sinai Peninsula, *Ar.* Shibh Jazīrat Sīnā', Sīnā'. *physical region* NE Egypt
Sinaia *86 C4* Prahova, SE Romania
Sinano Gawa *see* Shinano-gawa
Sīnā/Sinai Peninsula *see* Sinai
Sincelejo *36 B2* Sucre, NW Colombia
Sind *112 B3 var.* Sindh. *province* SE Pakistan
Sindelfingen *73 B6* Baden-Württemberg, SW Germany
Sindh *see* Sind
Sindi *84 D2 Ger.* Zintenhof. Pärnumaa, SW Estonia
Sines *70 B4* Setúbal, S Portugal
Singan *see* Xi'an
Singapore *116 B3 country capital* (Singapore) S Singapore
Singapore *116 A1 off.* Republic of Singapore. *country* SE Asia
Singapore, Republic of *see* Singapore
Singen *73 B6* Baden-Württemberg, S Germany
Singida *51 C7* Singida, C Tanzania
Singkang *117 E4* Sulawesi, C Indonesia
Singkawang *116 C3* Borneo, C Indonesia
Singora *see* Songkhla
Singû *see* Shingū
Sining *see* Xining
Siniscola *75 A5* Sardegna, Italy, C Mediterranean Sea
Sinj *78 B4* Split-Dalmacija, SE Croatia
Sinkiang/Sinkiang Uighur Autonomous Region *see* Xinjiang Uygur Zizhiqu
Sinnamarie *see* Sinnamary
Sinnamary *37 H3 var.* Sinnamarie. N French Guiana
Sinneh *see* Sanandaj
Sînnicolau Mare *see* Sânnicolau Mare
Sinoe, Lacul *see* Sinoie, Lacul
Sinoie, Lacul *86 D5 prev.* Lacul Sinoe. *lagoon* SE Romania
Sinop *94 D2 anc.* Sinope. Sinop, N Turkey
Sinope *see* Sinop
Sinsheim *73 B6* Baden-Württemberg, SW Germany
Sint Maarten *33 G3 Eng.* Saint Martin. *island* Lesser Antilles
Sint-Michielsgestel *64 C4* Noord-Brabant, S Netherlands
Sîn-Miclăuş *see* Gheorgheni
Sint-Niklaas *65 B5 Fr.* Saint-Nicolas. Oost-Vlaanderen, N Belgium
Sint-Pieters-Leeuw *65 B6* Vlaams Brabant, C Belgium
Sintra *70 B3 prev.* Cintra. Lisboa, W Portugal
Sinujiif *51 E5* Nugaal, NE Somalia
Sinus Aelaniticus *see* Aqaba, Gulf of
Sinus Gallicus *see* Lion, Golfe du
Sinyang *see* Xinyang
Sinzyô *see* Shinjō
Sion *73 A7 Ger.* Sitten; *anc.* Sedunum. Valais, SW Switzerland
Sioux City *23 F3* Iowa, C USA
Sioux Falls *23 F3* South Dakota, N USA
Sioux State *see* North Dakota
Siphnos *see* Sífnos
Siping *106 D3 var.* Ssu-p'ing, Szeping; *prev.* Ssu-p'ing-chieh. Jilin, NE China
Siple, Mount *132 A4 mountain* Siple Island, Antarctica
Siquirres *31 E4* Limón, E Costa Rica
Siracusa *75 D7 Eng.* Syracuse. Sicilia, Italy, C Mediterranean Sea
Sir Edward Pellew Group *126 B2 island group* Northern Territory, NE Australia
Siret *86 C3 var.* Siretul, *Ger.* Sereth, *Rus.* Seret. *river* Romania/Ukraine
Siretul *see* Siret
Siria *see* Syria
Sirikit Reservoir *114 C4 lake* N Thailand
Sīrjān *98 D4 prev.* Sa'īdābād. Kermān, S Iran
Sirna *see* Sýrna
Şırnak *95 F4* Şırnak, SE Turkey
Síros *see* Sýros
Sirte *see* Surt
Sirti, Gulf of *see* Surt, Khalīj
Sisak *78 B3 var.* Siscia, *Ger.* Sissek, *Hung.* Sziszek; *anc.* Segestica. Sisak-Moslavina, C Croatia
Siscia *see* Sisak
Sisimiut *60 C3 var.* Holsteinborg, Holsteinsborg, Holstenborg, Holstensborg. Kitaa, S Greenland
Sissek *see* Sisak
Sīstān, Daryācheh-ye *see* Şāberī, Hāmūn-e
Sitaş Cristuru *see* Cristuru Secuiesc
Siteía *83 D8 var.* Sitía. Kríti, Greece, E Mediterranean Sea
Sitges *71 G2* Cataluña, NE Spain
Sitía *see* Siteía
Sittang *see* Sittoung
Sittard *65 D5* Limburg, SE Netherlands
Sitten *see* Sion
Sittoung *114 B4 var.* Sittang. *river* S Myanmar (Burma)
Sittwe *114 A3 var.* Akyab. Rakhine State, W Myanmar (Burma)
Siuna *30 D3* Región Autónoma Atlántico Norte, NE Nicaragua
Siut *see* Asyūţ
Sivas *94 D3 anc.* Sebastia, Sebaste. Sivas, C Turkey
Siverek *95 E4* Şanlıurfa, S Turkey
Siwa *see* Sīwah
Sīwah *50 A2 var.* Siwa. NW Egypt
Six Counties, The *see* Northern Ireland
Six-Fours-les-Plages *69 D6* Var, SE France
Siyäzän *95 H2 Rus.* Siazan'. NE Azerbaijan
Sjar *see* Sääre
Sjenica *79 D5 Turk.* Seniça. Serbia, SW Serbia
Skadar *see* Shkodër
Skadarsko Jezero *see* Scutari, Lake
Skagerak *see* Skagerrak
Skagerrak *63 A6 var.* Skagerak. *channel* N Europe
Skagit River *24 B1 river* Washington, NW USA
Skalka *62 C3 lake* N Sweden
Skarżysko-Kamienna *76 D4* Świętokrzyskie, C Poland
Skaudvilė *84 B4* Tauragė, SW Lithuania
Skegness *67 E6* E England, United Kingdom
Skellefteå *62 D4* Västerbotten, N Sweden
Skellefteälven *62 C4 river* N Sweden
Ski *63 B6* Akershus, S Norway
Skíathos *83 C5* Skíathos, Vóreies Sporádes, Greece, Aegean Sea
Skidal' *85 B5 Rus.* Skidel'. Hrodzyenskaya Voblasts', W Belarus
Skidel' *see* Skidal'
Skiermûntseach *see* Schiermonnikoog
Skierniewice *76 D3* Łódzkie, C Poland
Skiftet *84 C1 strait* Finland Atlantic Ocean Baltic Sea Gulf of Bothnia/Gulf of Finland
Skíros *see* Skýros
Skópelos *83 C5* Skópelos, Vóreies Sporádes, Greece, Aegean Sea
Skopje *79 D6 var.* Üsküb, *Turk.* Üsküp, *prev.* Skoplje; *anc.* Scupi. *country capital* (FYR Macedonia) N FYR Macedonia
Skoplje *see* Skopje
Skovorodino *93 F4* Amurskaya Oblast', SE Russia
Skudnesfjorden *63 A6 fjord* S Norway
Skuodas *84 B3 Ger.* Schoden, *Pol.* Szkudy. Klaipėda, NW Lithuania
Skye, Isle of *66 B3 island* NW Scotland, United Kingdom
Skylge *see* Terschelling
Skýros *83 C5 var.* Skíros. Skńyros, Vóreies Sporádes, Greece, Aegean Sea
Skýros *83 C5 var.* Skíros; *anc.* Scyros. *island* Vóreies Sporádes, Greece, Aegean Sea
Slagelse *63 B7* Vestsjælland, E Denmark
Slatina *78 C3 Hung.* Szlatina; *prev.* Podravska Slatina. Virovitica-Podravina, NE Croatia
Slatina *86 B5* Olt, S Romania
Slavgorod *see* Slawharad
Slavonski Brod *78 C3 Ger.* Brod, *Hung.* Bród; *prev.* Brod, Brod na Savi. Brod-Posavina, NE Croatia
Slavuta *86 C2* Khmel'nyts'ka Oblast', NW Ukraine
Slavyansk *see* Slov"yans'k
Slawharad *85 E7 Rus.* Slavgorod. Mahilyowskaya Voblasts', E Belarus
Sławno *76 C2* Zachodnio-pomorskie, NW Poland
Slēmānī *see* As Sulaymānīyah
Sliema *80 B5* N Malta
Sligo *67 A5 Ir.* Sligeach. Sligo, NW Ireland
Sliven *82 D2 var.* Slivno. Sliven, C Bulgaria
Slivnitsa *82 B2* Sofiya, W Bulgaria
Slivno *see* Sliven
Slobozia *86 C5* Ialomiţa, SE Romania
Slonim *85 B6 Pol.* Słonim. Hrodzyenskaya Voblasts', W Belarus
Słonim *see* Slonim
Slovakia *77 C6 off.* Slovenská Republika, *Ger.* Slowakei, *Hung.* Szlovákia, *Slvk.* Slovensko. *country* C Europe
Slovak Ore Mountains *see* Slovenské rudohorie
Slovenia *73 D8 off.* Republic of Slovenia, *Ger.* Slowenien, *Slvn.* Slovenija. *country* SE Europe
Slovenia, Republic of *see* Slovenia
Slovenija *see* Slovenia
Slovenská Republika *see* Slovakia
Slovenské rudohorie *77 D6 Eng.* Slovak Ore Mountains, *Ger.* Slowakisches Erzgebirge, Ungarisches Erzgebirge. *mountain range* C Slovakia
Slovensko *see* Slovakia
Slov"yans'k *87 G3 Rus.* Slavyansk. Donets'ka Oblast', E Ukraine
Slowakei *see* Slovakia
Slowakisches Erzgebirge *see* Slovenské rudohorie
Slowenien *see* Slovenia
Słubice *76 B3 Ger.* Frankfurt. Lubuskie, W Poland
Sluch *86 D1 river* NW Ukraine
Słupsk *76 C2 Ger.* Stolp. Pomorskie, N Poland
Slutsk *85 C6* Minskaya Voblasts', S Belarus
Smallwood Reservoir *17 F2 lake* Newfoundland and Labrador, S Canada
Smara *48 B3 var.* Es Semara. N Western Sahara
Smarhon' *85 C5 Pol.* Smorgonie, *Rus.* Smorgon'. Hrodzyenskaya Voblasts', W Belarus
Smederevo *78 D4 Ger.* Semendria. Serbia, N Serbia
Smederevska Palanka *78 D4* Serbia, C Serbia
Smela *see* Smila
Smila *87 E2 Rus.* Smela. Cherkas'ka Oblast', C Ukraine
Smilten *see* Smiltene
Smiltene *84 D3 Ger.* Smilten. Valka, N Latvia
Smøla *62 A4 island* W Norway
Smolensk *89 A5* Smolenskaya Oblast', W Russia
Smorgon'/Smorgonie *see* Smarhon'
Smyrna *see* İzmir
Snake *12 B4 river* Yukon, NW Canada
Snake River *24 C3 river* NW USA
Snake River Plain *24 D4 plain* Idaho, NW USA
Sneek *64 D2* Friesland, N Netherlands
Sneeuw-gebergte *see* Maoke, Pegunungan
Sněžka *76 B4 Ger.* Schneekoppe, *Pol.* Śnieżka. *mountain* N Czech Republic (Czechia)/Poland
Śniardwy, Jezioro *76 D2 Ger.* Spirdingsee. *lake* NE Poland
Sniečkus *see* Visaginas
Śnieżka *see* Sněžka
Snina *77 E5 Hung.* Szinna. Prešovský Kraj, E Slovakia
Snovs'k *87 E1 Rus.* Shchors. Chernihivs'ka Oblast', N Ukraine
Snowdonia *67 C6 mountain range* NW Wales, United Kingdom
Snowdrift *see* Łutselk'e
Snow Mountains *see* Maoke, Pegunungan
Snyder *27 F3* Texas, SW USA
Sobradinho, Barragem de *see* Sobradinho, Represa de
Sobradinho, Represa de *41 F2 var.* Barragem de Sobradinho. *reservoir* E Brazil
Sochi *89 A7* Krasnodarskiy Kray, SW Russia
Société, Îles de la/Society Islands *see* Société, Archipel de la
Society Islands *123 G4 var.* Archipel de Tahiti, Îles de la Société, *Eng.* Society Islands. *island group* W French Polynesia
Soconusco, Sierra de *see* Madre, Sierra
Socorro *26 D2* New Mexico, SW USA
Socorro, Isla *28 B5 island* W Mexico
Socotra *see* Suquṭrā
Soc Trăng *115 D6 var.* Khanh Hung. Soc Trăng, S Vietnam
Socuéllamos *71 E3* Castilla-La Mancha, C Spain
Sodankylä *62 D3* Lappi, N Finland
Sodari *see* Sodiri
Söderhamn *63 C5* Gävleborg, C Sweden
Södertälje *63 C6* Stockholm, C Sweden
Sodiri *50 B4 var.* Sawdirī, Sodari. Northern Kordofan, C Sudan
Soekaboemi *see* Sukabumi
Soemba *see* Sumba, Pulau
Soengaipenoeh *see* Sungaipenuh
Soerabaja *see* Surabaya
Soerakarta *see* Surakarta
Sofia *see* Sofiya
Sofiya *82 C2 var.* Sophia, *Eng.* Sofia, *Lat.* Serdica. *country capital* (Bulgaria) (Bulgaria) Sofiya-Grad, W Bulgaria
Sogamoso *36 B3* Boyacá, C Colombia
Sognefjorden *63 A5 fjord* NE North Sea

Sohâg – Suzhou

Sohâg *see* Sawhāj
Sohar *see* Şuḩār
Sohm Plain *44 B3 abyssal plain* NW Atlantic Ocean
Sohrau *see* Żory
Sokal' *86 C2 Rus.* Sokal. L'vivs'ka Oblast', NW Ukraine
Söke *94 A4* Aydın, SW Turkey
Sokhumi *95 E1 Rus.* Sukhumi. NW Georgia
Sokodé *53 F4* C Togo
Sokol *88 C4* Vologodskaya Oblast', NW Russia
Sokółka *76 E3* Podlaskie, NE Poland
Sokolov *77 A5 Ger.* Falkenau an der Eger; *prev.* Falknov nad Ohří. Karlovarský Kraj, W Czech Republic (Czechia)
Sokone *52 B3* W Senegal
Sokoto *53 F3* Sokoto, NW Nigeria
Sokoto *53 F4 river* NW Nigeria
Sokotra *see* Suquţrā
Solāpur *102 B3 var.* Sholāpur. Mahārāshtra, W India
Solca *86 C3 Ger.* Solka. Suceava, N Romania
Sol, Costa del *70 D5 coastal region* S Spain
Soldeu *69 B7* NE Andorra Europe
Solec Kujawski *76 C3* Kujawsko-pomorskie, C Poland
Soledad *36 B1* Atlántico, N Colombia
Isla Soledad *see* East Falkland
Soligorsk *see* Salihorsk
Solikamsk *92 C3* Permskaya Oblast', NW Russia
Sol'-Iletsk *89 D6* Orenburgskaya Oblast', W Russia
Solingen *72 A4* Nordrhein-Westfalen, W Germany
Solka *see* Solca
Sollentuna *63 C6* Stockholm, C Sweden
Solo *see* Surakarta
Solok *116 B4* Sumatera, W Indonesia
Solomon Islands *122 C3 prev.* British Solomon Islands Protectorate. *country* W Solomon Islands N Melanesia W Pacific Ocean
Solomon Islands *122 C3 island group* Papua New Guinea/Solomon Islands
Solomon Sea *122 B3 sea* W Pacific Ocean
Soltau *72 B3* Niedersachsen, NW Germany
Sol'tsy *88 A4* Novgorodskaya Oblast', W Russia
Solun *see* Thessaloníki
Solwezi *56 D2* North Western, NW Zambia
Sōma *108 D4* Fukushima, Honshū, C Japan
Somalia *51 D5 off.* Somali Democratic Republic, *Som.* Jamuuriyada Demuqraadiga Soomaaliyeed, Soomaaliya; *prev.* Italian Somaliland, Somaliland Protectorate. *country* E Africa
Somali Basin *47 E5 undersea basin* W Indian Ocean
Somali Democratic Republic *see* Somalia
Somaliland *51 D5 disputed territory* N Somalia
Somaliland Protectorate *see* Somalia
Sombor *78 C3 Hung.* Zombor. Vojvodina, NW Serbia
Someren *65 D5* Noord-Brabant, SE Netherlands
Somerset *20 A5 var.* Somerset Village. W Bermuda
Somerset *18 C5* Kentucky, S USA
Somerset Island *20 A5 island* W Bermuda
Somerset Island *15 F2 island* Queen Elizabeth Islands, Nunavut, NW Canada
Somerset Village *see* Somerset
Somers Islands *see* Bermuda
Somerton *26 A2* Arizona, SW USA
Someş *86 B3 river* Hungary/Romania Europe
Somme *68 C2 river* N France
Sommerfeld *see* Lubsko
Somotillo *30 C3* Chinandega, NW Nicaragua
Somoto *30 D3* Madriz, NW Nicaragua
Songea *51 C8* Ruvuma, S Tanzania
Sŏngjin *see* Kimch'aek
Songkhla *115 C7 var.* Songkla, *Mal.* Singora. Songkhla, SW Thailand
Songkla *see* Songkhla
Sonoran Desert *26 A3 var.* Desierto de Altar. *desert* Mexico/USA
Sonsonate *30 B3* Sonsonate, W El Salvador
Soochow *see* Suzhou
Soomaaliya/Soomaaliyeed, Jamuuriyada Demuqraadiga *see* Somalia
Soome Laht *see* Finland, Gulf of
Sop Hao *114 D3* Houaphan, N Laos
Sophia *see* Sofiya
Sopianae *see* Pécs
Sopot *76 C2 Ger.* Zoppot. Pomorskie, N Poland
Sopron *77 B6 Ger.* Ödenburg. Győr-Moson-Sopron, NW Hungary
Sorau/Sorau in der Niederlausitz *see* Żary
Sorgues *69 D6* Vaucluse, SE France
Sorgun *94 D3* Yozgat, C Turkey
Soria *71 E2* Castilla-León, N Spain
Soroca *86 D3 Rus.* Soroki. N Moldova
Sorochino *see* Sarochyna
Soroki *see* Soroca
Sorokyne *87 H3 Rus.* Krasnodon. Luhans'ka Oblast', E Ukraine
Sorong *117 F4* Papua, E Indonesia
Sørøy *see* Sørøya
Sørøya *62 C2 var.* Sørøy, *Lapp.* Sállan. *island* N Norway
Sortavala *88 B3 prev.* Serdobol'. Respublika Kareliya, NW Russia
Sotavento, Ilhas de *52 A3 var.* Leeward Islands. *island group* S Cape Verde
Sotkamo *62 E4* Kainuu, C Finland
Souanké *55 B5* Sangha, NW Congo
Soueida *see* As Suwaydā'
Souflí *82 D3 prev.* Souflíon. Anatolikí Makedonía kai Thráki, NE Greece
Souflíon *see* Souflí
Soufrière *33 F2* W Saint Lucia
Soukhné *see* As Sukhnah
Sŏul *see* Seoul
Soûr *97 A5 var.* Şūr; *anc.* Tyre. SW Lebanon
Souris River *23 E1 var.* Mouse River. *river* Canada/USA
Soúrpi *83 B5* Thessalía, C Greece
Sousse *49 F2 var.* Süsah. NE Tunisia
South Africa *56 C4 off.* Republic of South Africa, *Afr.* Suid-Afrika. *country* S Africa
South Africa, Republic of *see* South Africa
South America *34 continent*
Southampton *67 D7 hist.* Hamwih, *Lat.* Clausentum. S England, United Kingdom
Southampton Island *15 G3 island* Nunavut, NE Canada
South Andaman *111 F2 island* Andaman Islands, India, NE Indian Ocean
South Australia *127 A5 state* S Australia
South Australian Basin *120 B5 undersea basin* SW Indian Ocean
South Bend *18 C3* Indiana, N USA
South Beveland *see* Zuid-Beveland
South Bruny Island *127 C8 island* Tasmania, SE Australia
South Carolina *21 E2 off.* State of South Carolina, *also known as* The Palmetto State. *state* SE USA
South Carpathians *see* Carpaţii Meridionalii
South China Basin *103 E4 undersea basin* SE South China Sea
South China Sea *103 E4 Chin.* Nan Hai, *Ind.* Laut Cina Selatan, *Vtn.* Biển Đông. *sea* SE Asia
South Dakota *22 D2 off.* State of South Dakota, *also known as* The Coyote State, Sunshine State. *state* N USA
Southeast Indian Ridge *119 D7 undersea ridge* Indian Ocean/Pacific Ocean
Southeast Pacific Basin *131 E5 var.* Belling Hausen Mulde. *undersea basin* SE Pacific Ocean
South East Point *127 C7 headland* Victoria, S Australia
Southend-on-Sea *67 E6* E England, United Kingdom
Southern Alps *129 B6 mountain range* South Island, New Zealand
Southern Cook Islands *123 F4 island group* S Cook Islands
Southern Cross *125 B6* Western Australia
Southern Indian Lake *15 F4 lake* Manitoba, C Canada
Southern Ocean *45 B7 ocean* Atlantic Ocean/Indian Ocean/Pacific Ocean
Southern Uplands *66 C4 mountain range* S Scotland, United Kingdom
South Fiji Basin *120 D4 undersea basin* S Pacific Ocean
South Geomagnetic Pole *132 B3 pole* Antarctica
South Georgia *35 D8 island* South Georgia and the South Sandwich Islands, SW Atlantic Ocean
South Goulburn Island *124 E2 island* Northern Territory, N Australia
South Huvadhu Atoll *110 A5 atoll* S Maldives
South Indian Basin *119 D7 undersea basin* Indian Ocean/Pacific Ocean
South Island *129 C6 island* S New Zealand
South Korea *107 E4 off.* Republic of Korea, *Kor.* Taehan Min'guk. *country* E Asia
South Lake Tahoe *25 C5* California, W USA
South Orkney Islands *132 A2 island group* Antarctica
South Ossetia *95 F2 former autonomous region* SW Georgia
South Pacific Basin *see* Southwest Pacific Basin
South Platte River *22 D4 river* Colorado/Nebraska, C USA
South Pole *132 B3 pole* Antarctica
South Sandwich Islands *35 D8 island group* SW Atlantic Ocean
South Sandwich Trench *35 E8 trench* SW Atlantic Ocean
South Shetland Islands *132 A2 island group* Antarctica
South Shields *66 D4* NE England, United Kingdom
South Sioux City *23 F3* Nebraska, C USA
South Sudan *50 B5 off.* Republic of South Sudan, *country* N Africa
South Taranaki Bight *128 C4 bight* SE Tasman Sea
South Tasmania Plateau *see* Tasman Plateau
South Uist *66 B3 island* NW Scotland, United Kingdom
South-West Africa/South West Africa *see* Namibia
South West Cape *129 A8 headland* Stewart Island, New Zealand
Southwest Indian Ocean Ridge *see* Southwest Indian Ridge
Southwest Indian Ridge *119 B6 var.* Southwest Indian Ocean Ridge. *undersea ridge* SW Indian Ocean
Southwest Pacific Basin *121 E4 var.* South Pacific Basin. *undersea basin* SE Pacific Ocean
Sovereign Base Area *80 C5 uk military installation* S Cyprus
Soweto *56 D4* Gauteng, NE South Africa
Sōya-kaikyō *see* La Pérouse Strait
Spain *70 D3 off.* Kingdom of Spain, *Sp.* España; *anc.* Hispania, Iberia, *Lat.* Hispana. *country* SW Europe
Spain, Kingdom of *see* Spain
Spalato *see* Split
Spanish Town *32 B5 hist.* St.Iago de la Vega. C Jamaica
Sparks *25 C5* Nevada, W USA
Sparta *see* Spárti
Spartanburg *21 E1* South Carolina, SE USA
Spárti *83 B6 Eng.* Sparta. Pelopónnisos, S Greece
Spearfish *22 D2* South Dakota, N USA
Speightstown *33 G1* NW Barbados
Spencer *23 F3* Iowa, C USA
Spencer Gulf *127 B6 gulf* South Australia
Spey *66 C3 river* NE Scotland, United Kingdom
Spice Islands *see* Maluku
Spiess Seamount *45 C7 seamount* S Atlantic Ocean
Spijkenisse *64 B4* Zuid-Holland, SW Netherlands
Spíli *83 C8* Kríti, Greece, E Mediterranean Sea
Spin Būldak *101 E5* Kandahār, S Afghanistan
Spirdingsee *see* Śniardwy, Jezioro
Spitsbergen *61 F2 island* NW Svalbard
Split *78 B4 It.* Spalato. Split-Dalmacija, S Croatia
Špoģi *84 D4* Daugvapils, SE Latvia
Spokane *24 C2* Washington, NW USA
Spratly Islands *116 B2 Chin.* Nansha Qundao. *disputed territory* SE Asia
Spree *72 D4 river* E Germany
Springbok *56 B5* NE South Africa
Springfield *18 B4 state capital* Illinois, N USA
Springfield *19 G3* Massachusetts, NE USA
Springfield *23 G5* Missouri, C USA
Springfield *18 C4* Ohio, N USA
Springfield *24 B3* Oregon, NW USA
Spring Garden *37 F2* NE Guyana
Spring Hill *21 E4* Florida, SE USA
Springs Junction *129 C5* West Coast, South Island, New Zealand
Springsure *126 D4* Queensland, E Australia
Sprottau *see* Szprotawa
Spruce Knob *19 E4 mountain* West Virginia, NE USA
Srbinje *see* Foča
Srbobran *78 D3 var.* Bácsszenttamás, *Hung.* Szenttamás. Vojvodina, N Serbia
Srebrenica *78 C4* Republika Srpska, E Bosnia and Herzegovina
Sredets *82 D2 prev.* Syulemeshlii. Stara Zagora, C Bulgaria
Sredets *82 E2 prev.* Grudovo. Burgas, E Bulgaria
Srednerusskaya Vozvyshennost' *87 G1 Eng.* Central Russian Upland. *mountain range* W Russia
Sremska Mitrovica *78 C3 prev.* Mitrovica, *Ger.* Mitrowitz. Vojvodina, NW Serbia
Srepok, Sông *see* Srêpôk, Tônle
Srêpôk, Tônle *115 E5 var.* Sông Srepok. *river* Cambodia/Vietnam
Sri Aman *116 C3* Sarawak, East Malaysia
Sri Jayawardanapura *see* Sri Jayewardenapura Kotte
Sri Jayewardenapura Kotte *110 D3 administrative capital* (Sri Lanka) *var.* Sri Jayawardanapura. Western Province, W Sri Lanka
Srīkākulam *113 F5* Andhra Pradesh, E India
Sri Lanka *110 D3 off.* Democratic Socialist Republic of Sri Lanka; *prev.* Ceylon. *country* S Asia
Sri Lanka, Democratic Socialist Republic of *see* Sri Lanka
Srinagarind Reservoir *115 C5 lake* W Thailand
Srpska, Republika *78 B3 republic* Bosnia and Herzegovina
Ssu-ch'uan *see* Sichuan
Ssu-p'ing/Ssu-p'ing-chieh *see* Siping
Stabroek *65 B5* Antwerpen, N Belgium
Stade *72 B3* Niedersachsen, NW Germany
Stadskanaal *64 E2* Groningen, NE Netherlands
Stafford *67 D6* C England, United Kingdom
Staicele *84 D3* Limbaži, N Latvia
Ştaierdorf-Anina *see* Anina
Stájerlakanina *see* Anina
Stakhanov *see* Kadiyivka
Stalin *see* Varna
Stalinabad *see* Dushanbe
Stalingrad *see* Volgograd
Stalino *see* Donets'k
Stalinobod *see* Dushanbe
Stalinsk *see* Novokuznetsk
Stalinski Zaliv *see* Varnenski Zaliv
Stalin, Yazovir *see* Iskŭr, Yazovir
Stalowa Wola *76 E4* Podkarpackie, SE Poland
Stamford *19 F3* Connecticut, NE USA
Stampalia *see* Astypálaia
Stanislau *see* Ivano-Frankivs'k
Stanislav *see* Ivano-Frankivs'k
Stanisławów *see* Ivano-Frankivs'k
Stanke Dimitrov *see* Dupnitsa
Stanley *43 D7 var.* Port Stanley, Puerto Argentino. *dependent territory capital* (Falkland Islands) East Falkland, Falkland Islands
Stanleyville *see* Kisangani
Stann Creek *see* Dangriga
Stanovoy Khrebet *91 E3 mountain range* SE Russia
Stanthorpe *127 D5* Queensland, E Australia
Staphorst *64 D2* Overijssel, E Netherlands
Starachowice *76 D4* Świętokrzyskie, C Poland
Stara Kanjiža *see* Kanjiža
Stara Pazova *78 D3 Ger.* Altpasua, *Hung.* Ópazova. Vojvodina, N Serbia
Stara Planina *see* Balkan Mountains
Stara Zagora *82 D2 Lat.* Augusta Trajana. Stara Zagora, C Bulgaria
Starbuck Island *123 G3 prev.* Volunteer Island. *island* E Kiribati
Stargard in Pommern *see* Stargard Szczeciński
Stargard Szczeciński *76 B3 Ger.* Stargard in Pommern. Zachodnio-pomorskie, NW Poland
Stari Bečej *see* Bečej
Starobel'sk *see* Starobil's'k
Starobil's'k *87 H2 Rus.* Starobel'sk. Luhans'ka Oblast', E Ukraine
Starobin *85 C7 var.* Starobyn. Minskaya Voblasts', S Belarus
Starobyn *see* Starobin
Starogard Gdański *76 C2 Ger.* Preussisch-Stargard. Pomorskie, N Poland
Starokonstantinov *see* Starokostyantyniv
Starokostyantyniv *86 D2 Rus.* Starokonstantinov. Khmel'nyts'ka Oblast', NW Ukraine
Starominskaya *89 A7* Krasnodarskiy Kray, SW Russia
Staryya Darohi *85 C6 Rus.* Staryye Dorogi. Minskaya Voblasts', S Belarus
Staryye Dorogi *see* Staryya Darohi
Staryy Oskol *89 B6* Belgorodskaya Oblast', W Russia
State College *19 E4* Pennsylvania, NE USA
Staten Island *see* Estados, Isla de los
Statesboro *21 E2* Georgia, SE USA
States, The *see* United States of America
Staunton *19 E5* Virginia, NE USA
Stavanger *63 A6* Rogaland, S Norway
Stavers Island *see* Vostok Island
Stavropol' *89 B7 prev.* Voroshilovsk. Stavropol'skiy Kray, SW Russia
Stavropol' *see* Tol'yatti
Steamboat Springs *22 C4* Colorado, C USA
Steenwijk *64 D2* Overijssel, N Netherlands
Steier *see* Steyr
Steierdorf/Steierdorf-Anina *see* Anina
Steinamanger *see* Szombathely
Steinkjer *62 B4* Nord-Trøndelag, C Norway
Stejarul *see* Karapelit
Stendal *72 C3* Sachsen-Anhalt, C Germany
Stepanakert *see* Xankändi
Stephenville *27 F3* Texas, SW USA
Sterling *22 D4* Colorado, C USA
Sterling *18 B3* Illinois, N USA
Sterlitamak *92 B3* Respublika Bashkortostan, W Russia
Stettin *see* Szczecin
Stettiner Haff *see* Szczeciński, Zalew
Stevenage *67 E6* E England, United Kingdom
Stevens Point *18 B2* Wisconsin, N USA
Stewart Island *129 A8 island* S New Zealand
Steyerlak-Anina *see* Anina
Steyr *73 D6 var.* Steier. Oberösterreich, N Austria
St. Helena Bay *56 B5 bay* SW South Africa
Stif *see* Sétif
Stillwater *27 G1* Oklahoma, C USA
Štip *79 E6* E FYR Macedonia
Stirling *66 C4* C Scotland, United Kingdom
Stjørdalshalsen *62 B4* Nord-Trøndelag, C Norway
St-Maur-des-Fossés *68 E2* Val-de-Marne, Île-de-France, N France Europe
Stockach *73 B6* Baden-Württemberg, S Germany
Stockholm *63 C6 country capital* (Sweden) Stockholm, C Sweden
Stockmannshof *see* Pļaviņas
Stockton *25 B6* California, W USA
Stockton Plateau *27 E4 plain* Texas, SW USA
Stœng Trĕng *see* Stung Treng
Stoke *see* Stoke-on-Trent
Stoke-on-Trent *67 D6 var.* Stoke. C England, United Kingdom
Stolbce *see* Stowbtsy
Stolbtsy *see* Stowbtsy
Stolp *see* Słupsk
Stolpmünde *see* Ustka
Stómio *82 B4* Thessalía, C Greece
Store Bælt *see* Storebælt
Storebelt *see* Storebælt
Støren *63 B5* Sør-Trøndelag, S Norway
Storfjorden *61 G2 fjord* S Norway
Storhammer *see* Hamar
Stornoway *66 B2* NW Scotland, United Kingdom
Storsjön *63 B5 lake* C Sweden
Storuman *62 C4* Västerbotten, N Sweden
Storuman *62 C4 lake* N Sweden
Stowbtsy *85 C6 Pol.* Stolbce, *Rus.* Stolbtsy. Minskaya Voblasts', C Belarus
Strabane *67 B5 Ir.* An Srath Bán. W Northern Ireland, United Kingdom
Strakonice *77 A5 Ger.* Strakonitz. Jihočeský Kraj, S Czech Republic (Czechia)
Strakonitz *see* Strakonice
Stralsund *72 D2* Mecklenburg-Vorpommern, NE Germany
Stranraer *67 C5* S Scotland, United Kingdom
Strasbourg *68 E3 Ger.* Strassburg; *anc.* Argentoratum. Bas-Rhin, NE France
Strășeni *86 D3 var.* Strasheny. C Moldova
Strasheny *see* Strășeni
Strassburg *see* Strasbourg, France
Strassburg *see* Aiud, Romania
Stratford *128 D4* Taranaki, North Island, New Zealand
Strathfield *126 E2* New South Wales, E Australia
Straubing *73 C6* Bayern, SE Germany
Strehaia *86 B5* Mehedinţi, SW Romania
Strelka *92 D4* Krasnoyarskiy Kray, C Russia
Strigonium *see* Esztergom
Strofilia *see* Strofyliá
Strofyliá *83 C5 var.* Strofilia. Évvoia, C Greece
Stromboli *75 D6 island* Isole Eolie, S Italy
Stromeferry *66 C3* N Scotland, United Kingdom
Strömstad *63 B6* Västra Götaland, S Sweden
Strömsund *62 C4* Jämtland, C Sweden
Struga *79 D6* SW FYR Macedonia
Struma *see* Strymónas
Strumica *79 E6* E FYR Macedonia
Strumyani *82 C3* Blagoevgrad, SW Bulgaria
Strymónas *82 C3 Bul.* Struma. *river* Bulgaria/Greece
Stryy *86 B2* L'vivs'ka Oblast', NW Ukraine
Studholme *129 B6* Canterbury, South Island, New Zealand
Stuhlweissenberg *see* Székesfehérvár
Stung Treng *115 D5 Khmer.* Stœng Trĕng. Stung Treng, NE Cambodia
Sturgis *22 D3* South Dakota, N USA
Stuttgart *73 B6* Baden-Württemberg, SW Germany
Stykkishólmur *61 E4* Vesturland, W Iceland
Styr *86 C1 Rus.* Styr'. *river* Belarus/Ukraine
Su *see* Jiangsu
Suakin *50 C3 var.* Sawakin. Red Sea, NE Sudan
Subačius *84 C4* Panevėžys, NE Lithuania
Subaykhān *96 E3* Dayr az Zawr, E Syria
Subotica *78 D2 Ger.* Maria-Theresiopel, *Hung.* Szabadka. Vojvodina, N Serbia
Suceava *86 C3 Ger.* Suczawa, *Hung.* Szucsava. Suceava, NE Romania
Su-chou *see* Suzhou
Suchow *see* Suzhou, Jiangsu, China
Suchow *see* Xuzhou, Jiangsu, China
Sucker State *see* Illinois
Sucre *39 F4 hist.* Chuquisaca, La Plata. *country capital* (Bolivia-legal capital) Chuquisaca, S Bolivia
Suczawa *see* Suceava
Sudan *50 A4 off.* Republic of Sudan, *Ar.* Jumhuriyat as-Sudan; *prev.* Anglo-Egyptian Sudan. *country* N Africa
Sudanese Republic *see* Mali
Sudan, Jumhuriyat as- *see* Sudan
Sudan, Republic of *see* Sudan
Sudbury *16 C4* Ontario, S Canada
Sudd *51 B5 swamp region* N South Sudan
Sudeten *76 B4 var.* Sudetes, Sudetic Mountains, *Cz./Pol.* Sudety. *mountain range* Czech Republic (Czechia)/Poland
Sudetes/Sudetic Mountains/Sudety *see* Sudeten
Südkarpaten *see* Carpaţii Meridionalii
Südliche Morava *see* Južna Morava
Sudong, Pulau *116 A2 island* SW Singapore Asia
Sudostroy *see* Severodvinsk
Sue *51 B5 river* S South Sudan
Sueca *71 F3* País Valenciano, E Spain
Sue Wood Bay *20 B5 bay* W Bermuda North America W Atlantic Ocean
Suez *50 B1 Ar.* As Suways, El Suweis. NE Egypt
Suez Canal *50 B1 Ar.* Qanāt as Suways. *canal* NE Egypt
Suez, Gulf of *see* Khalīj as Suways
Suğla Gölü *94 C4 lake* SW Turkey
Şuḩār *99 D5 var.* Sohar. N Oman
Sühbaatar *105 E1* Selenge, N Mongolia
Suhl *73 C5* Thüringen, C Germany
Suicheng *see* Suixi
Suid-Afrika *see* South Africa
Suidwes-Afrika *see* Namibia
Suixi *106 C6 var.* Suicheng. Guangdong, S China
Sujāwal *112 B3* Sind, SE Pakistan
Sukabumi *116 C5 prev.* Soekaboemi. Jawa, C Indonesia
Sukagawa *109 D5* Fukushima, Honshū, C Japan
Sukarnapura *see* Jayapura
Sukarno, Puntjak *see* Jaya, Puncak
Sukhne *see* As Sukhnah
Sukhona *88 C4 var.* Tot'ma. *river* NW Russia
Sukhumi *see* Sokhumi
Sukkertoppen *see* Maniitsoq
Sukkur *112 B3* Sind, SE Pakistan
Sukumo *109 B7* Kōchi, Shikoku, SW Japan
Sulaimaniya *see* As Sulaymānīyah
Sulaiman Range *112 C2 mountain range* C Pakistan
Sula, Kepulauan *117 E4 island group* C Indonesia
Sulawesi *117 E4 Eng.* Celebes. *island* C Indonesia
Sulawesi, Laut *see* Celebes Sea
Sulechów *76 B3 Ger.* Züllichau. Lubuskie, W Poland
Suliag *see* Sawhāj
Sullana *38 B2* Piura, NW Peru
Sullivan Island *see* Lanbi Kyun
Sulphur Springs *27 G2* Texas, SW USA
Sultānābād *see* Arāk
Sulu Archipelago *117 E3 island group* SW Philippines
Sülüktü *see* Sulyukta
Sulu, Laut *see* Sulu Sea
Sulu Sea *117 E2 var.* Laut Sulu. *sea* SW Philippines
Sulyukta *101 E2 Kir.* Sülüktü. Batkenskaya Oblast', SW Kyrgyzstan
Sumatera *115 B8 Eng.* Sumatra. *island* W Indonesia
Sumatra *see* Sumatera
Šumava *see* Bohemian Forest
Sumba, Pulau *117 E5 Eng.* Sandalwood Island; *prev.* Soemba. *island* Nusa Tenggara, C Indonesia
Sumba, Selat *117 E5 strait* Nusa Tenggara, S Indonesia
Sumbawanga *51 B7* Rukwa, W Tanzania
Sumbe *56 B2 var.* N'Gunza, *Port.* Novo Redondo. Cuanza Sul, W Angola
Sumeih *51 B5* Southern Darfur, S Sudan
Sumgait *see* Sumqayıt, Azerbaijan
Summer Lake *24 B4 lake* Oregon, NW USA
Summit *71 H5* Alaska, USA
Sumqayıt *95 H2 Rus.* Sumgait. E Azerbaijan
Sumy *87 F2* Sums'ka Oblast', NE Ukraine
Sunbury *127 C7* Victoria, SE Australia
Sunda Islands *see* Greater Sunda Islands
Sunda, Selat *116 B5 strait* Jawa/Sumatera, SW Indonesia
Sunda Trench *see* Java Trench
Sunderland *66 D4 var.* Wearmouth. NE England, United Kingdom
Sundsvall *63 C5* Västernorrland, C Sweden
Sunflower State *see* Kansas
Sungaipenuh *116 B4 prev.* Soengaipenoeh. Sumatera, W Indonesia
Sunnyvale *25 A6* California, W USA
Sunset State *see* Oregon
Sunshine State *see* Florida
Sunshine State *see* New Mexico
Sunshine State *see* South Dakota
Suntar *93 F3* Respublika Sakha (Yakutiya), NE Russia
Sunyani *53 E5* W Ghana
Suoločielgi *see* Saariselkä
Suomenlahti *see* Finland, Gulf of
Suomen Tasavalta/Suomi *see* Finland
Suomussalmi *62 E4* Kainuu, E Finland
Suŏng *115 D6* Kampong Cham, C Cambodia
Suoyarvi *88 B3* Respublika Kareliya, NW Russia
Supe *38 C3* Lima, W Peru
Supérieur, Lac *see* Superior, Lake
Superior *18 A1* Wisconsin, N USA
Superior, Lake *18 B1 Fr.* Lac Supérieur. *lake* Canada/USA
Suqrah *see* Şawqirah
Suquţrā *99 C7 var.* Sokotra, *Eng.* Socotra. *island* SE Yemen
Şūr *99 E5* NE Oman
Şūr *see* Soûr
Surabaja *see* Surabaya
Surabaya *116 D5 prev.* Surabaja, Soerabaja. Jawa, C Indonesia
Surakarta *116 C5 Eng.* Solo; *prev.* Soerakarta. Jawa, S Indonesia
Šurany *77 C6 Hung.* Nagysurány. Nitriansky Kraj, SW Slovakia
Sūrat *112 C4* Gujarāt, W India
Suratdhani *see* Surat Thani
Surat Thani *115 C6 var.* Suratdhani. Surat Thani, SW Thailand
Surazh *85 E5* Vitsyebskaya Voblasts', NE Belarus
Surdulica *79 E5* Serbia, SE Serbia
Sûre *65 D7 var.* Sauer. *river* W Europe
Surendranagar *112 C4* Gujarāt, W India
Surfers Paradise *127 E5* Queensland, E Australia
Surgut *92 D3* Khanty-Mansiyskiy Avtonomnyy Okrug-Yugra, C Russia
Surin *115 D5* Surin, E Thailand
Surinam *see* Suriname
Suriname *37 G3 off.* Republic of Suriname, *var.* Surinam; *prev.* Dutch Guiana, Netherlands Guiana. *country* N South America
Suriname, Republic of *see* Suriname
Sūrīya/Sūrīyah, Al-Jumhūrīyah al-'Arabīyah as- *see* Syria
Surkhab, Darya-i- *see* Kahmard, Daryā-ye
Surkhob *101 F3 river* C Tajikistan
Surt *49 G2 var.* Sirte. N Libya
Surt, Khalīj *49 F2 Eng.* Gulf of Sidra, Gulf of Sirti, Sidra. *gulf* N Libya
Surtsey *61 E5 island* S Iceland
Suruga-wan *109 D6 bay* SE Japan
Susa *74 A2* Piemonte, NE Italy
Sūsah *see* Sousse
Susanville *25 B5* California, W USA
Susitna *14 C3* Alaska, USA
Susteren *65 D5* Limburg, SE Netherlands
Susuman *93 G3* Magadanskaya Oblast', E Russia
Sutlej *112 C2 river* India/Pakistan
Suur Munamägi *84 D3 var.* Munamägi, *Ger.* Eier-Berg. *mountain* SE Estonia
Suur Väin *84 C2 Ger.* Grosser Sund. *strait* W Estonia
Suva *123 E4 country capital* (Fiji) Viti Levu, W Fiji
Suvalkai/Suvalki *see* Suwałki
Suvorovo *82 E2 prev.* Vetrino. Varna, E Bulgaria
Suwałki *76 E2 Lith.* Suvalkai, *Rus.* Suvalki. Podlaskie, NE Poland
Şuwār *see* Aş Şuwār
Suways, Qanāt as *see* Suez Canal
Suweida *see* As Suwaydā'
Suzhou *106 D5 var.* Soochow, Su-chou, Suchow; *prev.* Wuhsien. Jiangsu, E China

Svalbard – Teseney

Svalbard 61 E1 constituent part of Norway. island group Arctic Ocean
Svartisen 62 C3 glacier C Norway
Svay Rieng 115 D6 Khmer. Svay Riĕng. Svay Rieng, S Cambodia
Svay Riĕng see Svay Rieng
Sveg 63 B5 Jämtland, C Sweden
Svietlahorsk 85 D7 Rus. Svetlogorsk. Homyel'skaya Voblasts', SE Belarus
Svenstavik 63 C5 Jämtland, C Sweden
Sverdlovsk see Yekaterinburg
Sverige see Sweden
Sveti Vrach see Sandanski
Svetlogorsk see Svietlahorsk
Svetlograd 89 B7 Stavropol'skiy Kray, SW Russia
Svetlovodsk see Svitlovods'k
Svetozarevo see Jagodina
Svilengrad 82 D3 prev. Mustafa-Pasha. Khaskovo, S Bulgaria
Svitlovods'k 87 F3 Rus. Svetlovodsk. Kirovohrads'ka Oblast', C Ukraine
Svizzera see Switzerland
Svobodnyy 93 G4 Amurskaya Oblast', SE Russia
Svyataya Anna Trough 133 C4 var. Saint Anna Trough. trough N Kara Sea
Swabian Jura see Schwäbische Alb
Swakopmund 56 B3 Erongo, W Namibia
Swan Islands 31 E1 island group NE Honduras North America
Swansea 67 C7 Wel. Abertawe. S Wales, United Kingdom
Swarzędz 76 C3 Poznań, W Poland
Swatow see Shantou
Swaziland 56 D4 off. Kingdom of Swaziland. country S Africa
Swaziland, Kingdom of see Swaziland
Sweden 62 B4 off. Kingdom of Sweden, Swe. Sverige. country N Europe
Sweden, Kingdom of see Sweden
Sweetwater 27 F3 Texas, SW USA
Świdnica 76 B4 Ger. Schweidnitz. Wałbrzych, SW Poland
Świdwin 76 B2 Ger. Schivelbein. Zachodnio-pomorskie, NW Poland
Świebodzice 76 B4 Ger. Freiburg in Schlesien, Swiebodzice. Walbrzych, SW Poland
Świebodzin 76 B3 Ger. Schwiebus. Lubuskie, W Poland
Świecie 76 C3 Ger. Schwertberg. Kujawsko-pomorskie, C Poland
Swindon 67 D7 S England, United Kingdom
Swinemünde see Świnoujście
Świnoujście 76 B2 Ger. Swinemünde. Zachodnio-pomorskie, NW Poland
Swiss Confederation see Switzerland
Switzerland 73 A7 off. Swiss Confederation, Fr. La Suisse, Ger. Schweiz, It. Svizzera; anc. Helvetia. country C Europe
Sycaminum see Hefa
Sydenham Island see Nonouti
Sydney 126 D1 state capital New South Wales, SE Australia
Sydney 17 G4 Cape Breton Island, Nova Scotia, SE Canada
Sydney Island see Manra
Syedpur see Saidpur
Syemyezhava 85 C6 Rus. Semechevo. Minskaya Voblasts', C Belarus
Syene see Aswân
Syeverodonets'k 87 H3 Rus. Severodonetsk. Luhans'ka Oblast', E Ukraine
Syktyvkar 88 D4 prev. Ust'-Sysol'sk. Respublika Komi, NW Russia
Sylhet 113 G3 Sylhet, NE Bangladesh
Synel'nykove 87 G3 Dnipropetrovs'ka Oblast', E Ukraine
Syowa 132 C2 Japanese research station Antarctica
Syracuse 19 E3 New York, NE USA
Syracuse see Siracusa
Syrdar'ya 92 B4 Sirdaryo Viloyati, E Uzbekistan
Syria 96 B3 off. Syrian Arab Republic, var. Siria, Syrie, Ar. Al-Jumhūrīyah al-'Arabīyah as-Sūrīyah, Sūrīya. country SW Asia
Syrian Arab Republic see Syria
Syrian Desert 97 D5 Ar. Al Hamad, Bādiyat ash Shām. desert SW Asia
Syrie see Syria
Sýrna 83 E7 var. Sirna. island Kykládes, Greece, Aegean Sea
Sýros 83 C6 var. Síros. island Kykládes, Greece, Aegean Sea
Syulemeshlii see Sredets
Syvash, Zaliv see Syvash, Zatoka
Syvash, Zatoka 87 F4 Rus. Zaliv Syvash. inlet S Ukraine
Syzran' 89 C6 Samarskaya Oblast', W Russia
Szabadka see Subotica
Szamotuły 76 B3 Poznań, W Poland
Szászrégen see Reghin
Szatmárnémeti see Satu Mare
Száva see Sava
Szczecin 76 B3 Eng./Ger. Stettin. Zachodnio-pomorskie, NW Poland
Szczecinek 76 B2 Ger. Neustettin. Zachodnio-pomorskie, NW Poland
Szczeciński, Zalew 76 A2 var. Stettiner Haff, Ger. Oderhaff. bay Germany/Poland
Szczuczyn Nowogródzki see Shchuchyn
Szczytno 76 D3 Ger. Ortelsburg. Warmińsko-Mazurskie, NE Poland
Szechuan/Szechwan see Sichuan
Szeged 77 D7 Ger. Szegedin, Rom. Seghedin. Csongrád, SE Hungary
Szegedin see Szeged
Székelykeresztúr see Cristuru Secuiesc
Székesfehérvár 77 C6 Ger. Stuhlweissenberg; anc. Alba Regia. Fejér, W Hungary
Szeklerburg see Miercurea-Ciuc
Szekler Neumarkt see Târgu Secuiesc
Szekszárd 77 C7 Tolna, S Hungary
Szempcz/Szenc see Senec
Szenice see Senica
Szenttamás see Srbobran
Szeping see Siping
Szilágysomlyó see Şimleu Silvaniei
Szinna see Snina
Sziszek see Sisak
Szitás-Keresztúr see Cristuru Secuiesc
Szkudy see Skuodas
Szlatina see Slatina
Szlovákia see Slovakia
Szolnok 77 D6 Jász-Nagykun-Szolnok, C Hungary
Szombathely 77 B6 Ger. Steinamanger; anc. Sabaria, Savaria. Vas, W Hungary
Szprotawa 76 B4 Ger. Sprottau. Lubuskie, W Poland
Sztálinváros see Dunaújváros
Szucsava see Suceava

T

Tabariya, Bahrat see Tiberias, Lake
Table Rock Lake 27 G1 reservoir Arkansas/Missouri, C USA
Tábor 77 B5 Jihočeský Kraj, S Czech Republic (Czechia)
Tabora 51 B7 Tabora, W Tanzania
Tabrīz 98 C2 var. Tebriz; anc. Tauris. Āžarbāyjān-e Sharqī, NW Iran
Tabuaeran 123 G2 prev. Fanning Island. atoll Line Islands, E Kiribati
Tabūk 98 A4 Tabūk, NW Saudi Arabia
Täby 63 C6 Stockholm, C Sweden
Tachau see Tachov
Tachov 77 A5 Ger. Tachau. Plveňský Kraj, W Czech Republic (Czechia)
Tacloban 117 F2 off. Tacloban City. Leyte, C Philippines
Tacloban City see Tacloban
Tacna 39 E4 Tacna, SE Peru
Tacoma 24 B2 Washington, NW USA
Tacuarembó 42 D4 prev. San Fructuoso. Tacuarembó, C Uruguay
Tademaït, Plateau du 48 D3 plateau C Algeria
Tadmor/Tadmur see Tudmur
Tādpatri 110 C2 Andhra Pradesh, E India
Tadzhikistan see Tajikistan
Taegu see Daegu
Taehan-haehyŏp see Korea Strait
Taehan Min'guk see South Korea
Taejŏn see Daejeon
Tafassâsset, Ténéré du 53 G2 desert N Niger
Tafila/Ţafilah, Muḩāfaẓat aṭ see Aṭ Ţafilah
Taganrog 89 A7 Rostovskaya Oblast', SW Russia
Taganrog, Gulf of 87 G4 Rus. Taganrogskiy Zaliv, Ukr. Tahanroz'ka Zatoka. gulf Russia/Ukraine
Taganrogskiy Zaliv see Taganrog, Gulf of
Taguatinga 41 F3 Tocantins, C Brazil
Tagus 70 C3 Port. Rio Tejo, Sp. Río Tajo. river Portugal/Spain
Tagus Plain 58 A4 abyssal plain E Atlantic Ocean
Tahanroz'ka Zatoka see Taganrog, Gulf of
Tahat 49 E4 mountain SE Algeria
Tahiti 123 H4 island Îles du Vent, W French Polynesia
Tahiti, Archipel de see Société, Archipel de la
Tahlequah 27 G1 Oklahoma, C USA
Tahoe, Lake 25 B5 lake California/Nevada, W USA
Tahoua 53 F3 Tahoua, W Niger
Taibei see Taipei
Taichū see Taichung
Taichung 106 D6 Jap. Taichū; var. Taizhong, Taiwan. C Taiwan
Taiden see Daejeon
Taieri 129 B7 river South Island, New Zealand
Taihape 128 D4 Manawatu-Wanganui, North Island, New Zealand
Taihoku see Taibei
Taikyū see Daegu
Tailem Bend 127 B7 South Australia
T'ainan see Tainan
Tainan 106 D6 prev. T'ainan, Dainan. S Taiwan
Taipei 106 D6 var. Taibei, T'aipei; Jap. Taihoku; prev. Daihoku. country capital (Taiwan) N Taiwan
Taiping 116 B3 Perak, Peninsular Malaysia
Taiwan 106 D6 off. Republic of China, var. Formosa, Formo'sa. country E Asia
Taiwan see Taichung
T'aiwan Haihsia/Taiwan Haixia see Taiwan Strait
Taiwan Strait 106 D6 var. Formosa Strait, Chin. T'aiwan Haihsia, Taiwan Haixia. strait China/Taiwan
Taiyuan 106 C4 var. T'ai-yuan, T'ai-yüan; prev. Yangku. province capital Shanxi, C China
T'ai-yuan/T'ai-yüan see Taiyuan
Taizhong see Tiachung
Ta'izz 99 B7 SW Yemen
Tajikistan 101 E3 off. Republic of Tajikistan, Rus. Tadzhikistan, Taj. Jumhurii Tojikiston; prev. Tajik S.S.R. country C Asia
Tajikistan, Republic of see Tajikistan
Tajik S.S.R see Tajikistan
Tajo, Río see Tagus
Tak 114 C4 var. Rahaeng. Tak, W Thailand
Takao see Kaohsiung
Takaoka 109 C5 Toyama, Honshū, SW Japan
Takapuna 128 D2 Auckland, North Island, New Zealand
Takeshima see Liancourt Rocks
Takhiatash see Taxiatosh
Takhtakupyr see Taxtako'pir
Takikawa 108 D2 Hokkaidō, NE Japan
Takla Makan Desert see Taklimakan Shamo
Taklimakan Shamo 104 B3 Eng. Takla Makan Desert. desert NW China
Takow see Kaohsiung
Takutea 123 G4 island S Cook Islands
Talabriga see Aveiro, Portugal
Talabriga see Talavera de la Reina, Spain
Talachyn 85 D6 Rus. Tolochin. Vitsyebskaya Voblasts', NE Belarus
Talamanca, Cordillera de 31 E5 mountain range S Costa Rica
Talara 38 B2 Piura, NW Peru
Talas 101 F2 Talasskaya Oblast', NW Kyrgyzstan
Talaud, Kepulauan 117 F3 island group E Indonesia
Talavera de la Reina 70 D3 anc. Caesarobriga, Talabriga. Castilla-La Mancha, C Spain
Talca 42 B4 Maule, C Chile
Talcahuano 43 B5 Bío Bío, C Chile
Taldykorgan 92 C5 Kaz. Taldyqorghan; prev. Taldy-Kurgan. Taldykorgan, SE Kazakhstan
Taldy-Kurgan/Taldyqorghan see Taldykorgan
Ta-lien see Dalian
Taliq-an see Tāloqān
Tal'ka 85 C6 Minskaya Voblasts', C Belarus
Talkhof see Puurmani
Tallahassee 20 D3 prev. Muskogean. state capital Florida, SE USA
Tall al Abyaḑ see At Tall al Abyaḑ
Tallin see Tallinn
Tallinn 84 D2 Ger. Reval, Rus. Tallin; prev. Revel. country capital (Estonia) Harjumaa, NW Estonia
Tall Kalakh 96 B4 var. Tell Kalakh. Ḩimş, C Syria
Tallulah 20 B2 Louisiana, S USA
Talnakh 92 D3 Taymyrskiy (Dolgano-Nenetskiy) Avtonomnyy Okrug, N Russia
Tal'ne 87 E3 Rus. Tal'noye. Cherkas'ka Oblast', C Ukraine
Tal'noye see Tal'ne
Taloga 27 F1 Oklahoma, C USA
Tāloqān 101 E3 var. Taliq-an. Takhār, NE Afghanistan
Talsen see Talsi
Talsi 84 C3 Ger. Talsen. Talsi, NW Latvia
Taltal 42 B2 Antofagasta, N Chile
Talvik 62 D2 Finnmark, N Norway
Tamabo, Banjaran 116 D3 mountain range East Malaysia
Tamale 53 E4 C Ghana
Tamana 123 E3 prev. Rotcher Island. atoll Tungaru, W Kiribati
Tamanrasset 49 E4 var. Tamenghest. S Algeria
Tamar 67 C7 river SW England, United Kingdom
Tamar see Tudmur
Tamatave see Toamasina
Tamazunchale 29 E4 San Luis Potosí, C Mexico
Tambacounda 52 C3 SE Senegal
Tambov 89 B6 Tambovskaya Oblast', W Russia
Tambura 51 B5 W Equatoria, SW South Sudan
Tamchaket see Tâmchekkeṭ
Tâmchekkeṭ 52 C3 var. Tamchaket. Hodh el Gharbi, S Mauritania
Tamenghest see Tamanrasset
Tamil Nādu 110 C3 prev. Madras. cultural region SE India
Tam Ky 115 E5 Quang Nam-Đa Năng, C Vietnam
Tammerfors see Tampere
Tampa 21 E4 Florida, SE USA
Tampa Bay 21 E4 bay Florida, SE USA
Tampere 63 D5 Swe. Tammerfors. Pirkanmaa, W Finland
Tampico 29 E3 Tamaulipas, C Mexico
Tamworth 127 D6 New South Wales, SE Australia
Tanabe 109 C7 Wakayama, Honshū, SW Japan
Tana Bru 62 D2 Finnmark, N Norway
T'ana Hāyk' 50 C4 var. Lake Tana. lake NW Ethiopia
Tanais see Don
Tana, Lake see T'ana Hāyk'
Tanami Desert 124 D3 desert Northern Territory, N Australia
Tananarive see Antananarivo
Ţăndărei 86 D5 Ialomiţa, SE Romania
Tandil 43 D5 Buenos Aires, E Argentina
Tandjoengkarang see Bandar Lampung
Tanega-shima 109 B8 island Nansei-shotō, SW Japan
Tanen Taunggyi see Tane Range
Tane Range 114 B4 Bur. Tanen Taunggyi. mountain range W Thailand
Tanezrouft 48 D4 desert Algeria/Mali
Ţanf, Jabal aṭ 96 D4 mountain SE Syria
Tanga 51 C7 Tanga, E Tanzania
Tanganyika and Zanzibar see Tanzania
Tanganyika, Lake 51 B7 lake E Africa
Tanger 48 C2 var. Tangiers, Tangier, Fr./Ger. Tangerk, Sp. Tánger; anc. Tingis. NW Morocco
Tangerk see Tanger
Tanggula Shan 104 C4 mountain W China
Tangier see Tanger
Tangiers see Tanger
Tangra Yumco 104 B5 var. Tangro Tso. lake W China
Tangro Tso see Tangra Yumco
Tangshan 106 D3 var. T'ang-shan. Hebei, E China
T'ang-shan see Tangshan
Tanimbar, Kepulauan 117 F5 island group Maluku, E Indonesia
Tanintharyi 115 B6 prev. Tenasserim. S Myanmar (Burma)
Tanjungkarang/Tanjungkarang-Telukbetung see Bandar Lampung
Tanna 122 D4 island S Vanuatu
Tannenhof see Krynica
Tan-Tan 48 B3 SW Morocco
Tan-tung see Dandong
Tanzania 51 C7 off. United Republic of Tanzania, Swa. Jamhuri ya Muungano wa Tanzania; prev. German East Africa, Tanganyika and Zanzibar. country E Africa
Tanzania, Jamhuri ya Muungano wa see Tanzania
Tanzania, United Republic of see Tanzania
Taoudenit see Taoudenni
Taoudenni see Tombouctou
Tapa 84 E2 Ger. Taps. Lääne-Virumaa, NE Estonia
Tapachula 29 G5 Chiapas, SE Mexico
Tapaiu see Gvardeysk
Tapajós, Rio 41 E2 var. Tapajóz. river NW Brazil
Tapajóz see Tapajós, Rio
Taps see Tapa
Ţarābulus 49 F2 var. Ţarābulus al Gharb, Eng. Tripoli. country capital (Libya) NW Libya
Ţarābulus al Gharb see Ţarābulus
Tarābulus/Ţarābulus ash Shām see Tripoli
Taraclia 86 D4 Rus. Tarakilya. S Moldova
Tarakilya see Taraclia
Taranaki, Mount 128 C4 var. Egmont. volcano North Island, New Zealand
Tarancón 71 E3 Castilla-La Mancha, C Spain
Taranto 75 E5 var. Tarentum. Puglia, SE Italy
Taranto, Gulf of 75 E6 Eng. Gulf of Taranto. gulf S Italy
Taranto, Gulf of see Taranto, Golfo di
Tarapoto 38 C2 San Martín, N Peru
Tarare 69 D5 Rhône, E France
Tarascon 69 D6 Bouches-du-Rhône, SE France
Tarawa 122 D2 atoll and capital (Kiribati) Tungaru, W Kiribati
Taraz 92 C5 prev. Aulie Ata, Auliye-Ata, Dzhambul, Zhambyl. Zhambyl, S Kazakhstan
Tarazona 71 E2 Aragón, NE Spain
Tarbes 69 B6 anc. Bigorra. Hautes-Pyrénées, S France
Tarcoola 127 A6 South Australia
Taree 127 D6 New South Wales, SE Australia
Tarentum see Taranto
Târgovişte 86 C5 prev. Tîrgovişte. Dâmboviţa, S Romania
Targu Jiu 86 B4 prev. Tîrgu Jiu. Gorj, W Romania
Târgul-Neamţ see Târgu-Neamţ
Târgul-Săcuiesc see Târgu Secuiesc
Târgu Mureş 86 B4 prev. Oşorhei, Tîrgu Mures, Ger. Neumarkt, Hung. Marosvásárhely. Mureş, C Romania
Târgu-Neamţ 86 C3 var. Târgul-Neamţ; prev. Tîrgu-Neamţ. Neamţ, NE Romania
Târgu Ocna 86 C4 Hung. Aknavásár; prev. Tîrgu Ocna. Bacău, E Romania
Târgu Secuiesc 86 C4 Ger. Neumarkt, Szekler Neumarkt, Hung. Kezdivásárhely; prev. Chezdi-Oşorheiu, Târgul-Săcuiesc, Tîrgu Secuiesc. Covasna, E Romania
Tar Heel State see North Carolina
Tarija 39 G5 Tarija, S Bolivia
Tarim 99 C6 C Yemen
Tarim Basin see Tarim Pendi
Tarim Pendi 102 C2 Eng. Tarim Basin. basin NW China
Tarim He 104 B3 river NW China
Tarma 38 C3 Junín, C Peru
Tarn 69 C6 cultural region S France
Tarn 69 C6 river S France
Tarnobrzeg 76 D4 Podkarpackie, SE Poland
Tarnopol see Ternopil'
Tarnów 77 D5 Małopolskie, S Poland
Tarraco see Tarragona
Tarragona 71 G2 anc. Tarraco. Cataluña, E Spain
Tarrasa see Terrassa
Tàrrega 71 F2 var. Tarrega. Cataluña, NE Spain
Tarsatica see Rijeka
Tarsus 94 C4 İçel, S Turkey
Tartous/Tartouss see Ţarţūs
Tartu 84 D3 Ger. Dorpat; prev. Rus. Yurev, Yury'ev. Tartumaa, SE Estonia
Ţarţūs 96 A3 off. Muḩāfaẓat Ţarţūs, var. Tartous, Tartus. governorate W Syria
Ţarţūs, Muḩāfaẓat see Ţarţūs
Ta Ru Tao, Ko 115 B7 island S Thailand Asia
Tarvisio 74 D2 Friuli-Venezia Giulia, NE Italy
Tarvisium see Treviso
Tashauz see Daşoguz
Tashi Chho Dzong see Thimphu
Tashkent see Toshkent
Tash-Kömür see Tash-Kumyr
Tash-Kumyr 101 F2 Kir. Tash-Kömür. Dzhalal-Abadskaya Oblast', W Kyrgyzstan
Tashqurghan see Khulm
Tasikmalaja see Tasikmalaya
Tasikmalaya 116 C5 prev. Tasikmalaja. Jawa, C Indonesia
Tasman Basin 120 C5 var. East Australian Basin. undersea basin S Tasman Sea
Tasman Bay 129 C5 inlet South Island, New Zealand
Tasmania 127 B8 prev. Van Diemen's Land. state SE Australia
Tasmania 130 B4 island SE Australia
Tasman Plateau 120 C5 var. South Tasmania Plateau. undersea plateau SW Tasman Sea
Tasman Sea 120 C5 sea SW Pacific Ocean
Tassili-n-Ajjer 49 E4 plateau E Algeria
Tatabánya 77 C6 Komárom-Esztergom, NW Hungary
Tatar Pazardzhik see Pazardzhik
Tathlith 99 B5 'Asīr, S Saudi Arabia
Tatra Mountains 77 D5 Ger. Tatra, Hung. Tátra, Pol./Slvk. Tatry. mountain range Poland/Slovakia
Tatra/Tátra see Tatra Mountains
Tatry see Tatra Mountains
Ta-t'ung/Tatung see Datong
Tatvan 95 F3 Bitlis, SE Turkey
Ta'ū 123 F4 var. Tau. island Manua Islands, E American Samoa
Taukum, Peski 101 G1 desert SE Kazakhstan
Taumarunui 128 D4 Manawatu-Wanganui, North Island, New Zealand
Taungdwingyi 114 B3 Magway, C Myanmar (Burma)
Taunggyi 114 B3 Shan State, C Myanmar (Burma)
Taungoo 114 B4 Bago, C Myanmar (Burma)
Taunton 67 C7 SW England, United Kingdom
Taupo 128 D3 Waikato, North Island, New Zealand
Taupo, Lake 128 D3 lake North Island, New Zealand
Tauragė 84 B4 Ger. Tauroggen. Tauragė, SW Lithuania
Tauranga 128 D3 Bay of Plenty, North Island, New Zealand
Tauris see Tabrīz
Tauroggen see Tauragė
Taurus Mountains see Toros Dağları
Tavas 94 B4 Denizli, SW Turkey
Tavastehus see Hämeenlinna
Tavira 70 C5 Faro, S Portugal
Tavoy see Dawei
Tavoy Island see Mali Kyun
Ta Waewae Bay 129 A7 bay South Island, New Zealand
Tawakoni, Lake 27 G2 reservoir Texas, SW USA
Tawau 116 D3 Sabah, East Malaysia
Ţawkar see Tokar
Tawzar see Tozeur
Taxco 29 E4 var. Taxco de Alarcón. Guerrero, S Mexico
Taxco de Alarcón see Taxco
Taxiatosh 100 C2 Rus. Takhiatash. Qoraqalpog'iston Respublikasi, W Uzbekistan
Taxtako'pir 100 D1 Rus. Takhtakupyr. Qoraqalpog'iston Respublikasi, NW Uzbekistan
Tay 66 C3 river C Scotland, United Kingdom
Taylor 27 G3 Texas, SW USA
Taymā' 98 A4 Tabūk, NW Saudi Arabia
Taymyr, Ozero 93 E2 lake N Russia
Taymyr, Poluostrov 93 E2 peninsula N Russia
Taz 92 D3 river N Russia
Tbilisi 95 G2 var. T'bilisi, Eng. Tiflis. country capital (Georgia) SE Georgia
T'bilisi see **Tbilisi**
Tchad see Chad
Tchad, Lac see Chad, Lake
Tchien see Zwedru
Tchongking see Chongqing
Tczew 76 C2 Ger. Dirschau. Pomorskie, N Poland
Te Anau 129 A7 Southland, South Island, New Zealand
Te Anau, Lake 129 A7 lake South Island, New Zealand
Teapa 29 G4 Tabasco, SE Mexico
Teate see Chieti
Tebingtinggi 116 B3 Sumatera, N Indonesia
Tebriz see Tabrīz
Techirghiol 86 D5 Constanţa, SE Romania
Tecomán 28 D4 Colima, SW Mexico
Tecpan 29 E5 var. Tecpan de Galeana. Guerrero, S Mexico
Tecpan de Galeana see Tecpan
Tecuci 86 C4 Galaţi, E Romania
Tedzhen see Harīrūd/Tejen
Tedzhen see Tejen
Tees 67 D5 river N England, United Kingdom
Tefé 40 D2 Amazonas, N Brazil
Tegal 116 C4 Jawa, C Indonesia
Tegelen 65 D5 Limburg, SE Netherlands
Tegucigalpa 30 C3 country capital (Honduras) Francisco Morazán, SW Honduras
Teheran see Tehrān
Tehrān 98 C3 var. Teheran. country capital (Iran) Tehrān, N Iran
Tehuacán 29 F4 Puebla, S Mexico
Tehuantepec 29 F5 var. Santo Domingo Tehuantepec. Oaxaca, SE Mexico
Tehuantepec, Golfo de 29 F5 var. Gulf of Tehuantepec. gulf S Mexico
Tehuantepec, Gulf of see Tehuantepec, Golfo de
Tehuantepec, Isthmus of see Tehuantepec, Istmo de
Tehuantepec, Istmo de 29 F5 var. Isthmus of Tehuantepec. isthmus SE Mexico
Tejen 100 C3 Rus. Tedzhen. Ahal Welaýaty, S Turkmenistan
Tejen see Harīrūd
Tejo, Rio see Tagus
Te Kao 128 C1 Northland, North Island, New Zealand
Tekax 29 H4 var. Tekax de Álvaro Obregón. Yucatán, SE Mexico
Tekax de Álvaro Obregón see Tekax
Tekeli 92 C5 Almaty, SE Kazakhstan
Tekirdağ 94 A2 It. Rodosto; anc. Bisanthe, Raidestos, Rhaedestus. Tekirdağ, NW Turkey
Te Kuiti 128 D3 Waikato, North Island, New Zealand
Tela 30 C2 Atlántida, NW Honduras
Telanaipura see Jambi
Telangana 112 D5 cultural region SE India
Tel Aviv-Jaffa see Tel Aviv-Yafo
Tel Aviv-Yafo 97 A6 var. Tel Aviv-Jaffa. Tel Aviv, C Israel
Teles Pirés see São Manuel, Rio
Telish 82 C2 prev. Azizie. Pleven, N Bulgaria
Tell Abiad/Tell Abyad see At Tall al Abyaḑ
Tell Kalakh see Tall Kalakh
Tell Shedadi see Ash Shadādah
Tel'man/Tel'mansk see Gubadag
Teloekbetoeng see Bandar Lampung
Telo Martius see Toulon
Telschen see Telšiai
Telšiai 84 B3 Ger. Telschen. Telšiai, NW Lithuania
Telukbetung see Bandar Lampung
Temerin 78 D3 Vojvodina, N Serbia
Temeschburg/Temeschwar see Timişoara
Temesvár/Temeswar see Timişoara
Temirtau 92 C4 prev. Samarkandski, Samarkandskoye. Karagandy, C Kazakhstan
Tempio Pausania 75 A5 Sardegna, Italy, C Mediterranean Sea
Temple 27 G3 Texas, SW USA
Temuco 43 B5 Araucanía, C Chile
Temuka 129 B6 Canterbury, South Island, New Zealand
Tenasserim see Tanintharyi
Ténenkou 52 D3 Mopti, C Mali
Ténéré 53 G3 physical region C Niger
Tenerife 48 A3 island Islas Canarias, Spain, NE Atlantic Ocean
Tengger Shamo 105 E3 desert N China
Tengréla 52 D4 var. Tingréla. N Ivory Coast
Tenkodogo 53 E4 S Burkina
Tennant Creek 126 A3 Northern Territory, C Australia
Tennessee 20 C1 off. State of Tennessee, also known as The Volunteer State. state SE USA
Tennessee River 20 C1 river S USA
Tenos see Tínos
Tepelena see Tepelenë
Tepelenë 79 C7 var. Tepelena, It. Tepeleni. Gjirokastër, S Albania
Tepeleni see Tepelenë
Tepic 28 D4 Nayarit, C Mexico
Teplice 76 A4 Ger. Teplitz; prev. Teplice-Šanov, Teplitz-Schönau. Ústecký Kraj, NW Czech Republic (Czechia)
Teplice-Šanov/Teplitz/Teplitz-Schönau see Teplice
Tequila 28 D4 Jalisco, SW Mexico
Teraina 123 G2 prev. Washington Island. atoll Line Islands, E Kiribati
Teramo 74 C4 anc. Interamna. Abruzzi, C Italy
Tercan 95 E3 Erzincan, NE Turkey
Terceira 70 A5 var. Ilha Terceira. island Azores, Portugal, NE Atlantic Ocean
Terceira, Ilha see Terceira
Terekhovka see Tsyerakhowka
Teresina 41 F2 var. Therezina. state capital Piauí, NE Brazil
Termez see Termiz
Termia see Kýthnos
Términos, Laguna de 29 G4 lagoon SE Mexico
Termiz 101 E3 Rus. Termez. Surkhondaryo Viloyati, S Uzbekistan
Termoli 74 D4 Molise, C Italy
Terneuzen 65 B5 var. Neuzen. Zeeland, SW Netherlands
Terni 74 C4 anc. Interamna Nahars. Umbria, C Italy
Ternopil' 86 C2 Pol. Tarnopol, Rus. Ternopol'. Ternopil's'ka Oblast', W Ukraine
Ternopol' see Ternopil'
Terracina 75 C5 Lazio, C Italy
Terranova di Sicilia see Gela
Terranova Pausania see Olbia
Terrassa 71 G2 Cast. Tarrasa. Cataluña, E Spain
Terre Adélie 132 C4 physical region Antarctica
Terre Haute 18 B4 Indiana, N USA
Terre Neuve see Newfoundland and Labrador
Terschelling 64 C1 Fris. Skylge. island Waddeneilanden, N Netherlands
Teruel 71 F3 anc. Turba. Aragón, E Spain
Tervel 82 E1 prev. Kurtbunar, Rom. Curtbunar. Dobrich, NE Bulgaria
Tervueren see Tervuren
Tervuren 65 C6 var. Tervueren. Vlaams Brabant, C Belgium
Teseney 50 C4 var. Tessenei. W Eritrea

Van Wert – Wanki, Río

Van Wert *18 C4* Ohio, N USA
Vapincum *see* Gap
Varakļāni *84 D4* Madona, C Latvia
Vārānasi *113 E3* *prev.* Banaras, Benares, *hist.* Kasi. Uttar Pradesh, N India
Varangerfjorden *62 E2* *Lapp.* Várjjatvuotna. *fjord* N Norway
Varangerhalvøya *62 D2* *Lapp.* Várnjárga. *peninsula* N Norway
Varannó *see* Vranov nad Topl'ou
Varasd *see* Varaždin
Varaždin *78 B2* *Ger.* Warasdin, *Hung.* Varasd. Varaždin, N Croatia
Varberg *63 B7* Halland, S Sweden
Vardar *79 E6* *Gk.* Axiós. *river* FYR Macedonia/Greece
Varde *63 A7* Ribe, W Denmark
Vareia *see* Logroño
Varėna *85 B5* *Pol.* Orany. Alytus, S Lithuania
Varese *74 B2* Lombardia, N Italy
Vârful Moldoveanu *86 B4* *var.* Moldoveanul; *prev.* Vîrful Moldoveanu. *mountain* C Romania
Várjjatvuotna *see* Varangerfjorden
Varkaus *63 E5* Pohjois-Savo, C Finland
Varna *82 E2* *prev.* Stalin; *anc.* Odessus. Varna, E Bulgaria
Varnenski Zaliv *82 E2* *prev.* Stalinski Zaliv. *bay* E Bulgaria
Várnjárga *see* Varangerhalvøya
Varshava *see* Warszawa
Vasa *see* Vaasa
Vasilikí *83 A5* Lefkáda, Iónia Nisiá, Greece, C Mediterranean Sea
Vasilishki *85 B5* *Pol.* Wasiliszki. Hrodzyenskaya Voblasts', W Belarus
Vasil'kov *see* Vasyl'kiv
Vaslui *86 D4* Vaslui, C Romania
Västerås *63 C6* Västmanland, C Sweden
Vasyl'kiv *87 E2* *var.* Vasil'kov. Kyyivs'ka Oblast', N Ukraine
Vaté *see* Efate
Vatican City *75 A7* *off.* Vatican City. *country* S Europe
Vatnajökull *61 E5* *glacier* SE Iceland
Vatter, Lake *see* Vättern
Vättern *63 B6* *Eng.* Lake Vatter; *prev.* Lake Vetter. *lake* S Sweden
Vaughn *26 D2* New Mexico, SW USA
Vaupés, Río *36 C4* *var.* Rio Uaupés. *river* Brazil/Colombia
Vava'u Group *123 E4* *island group* N Tonga
Vavuniya *110 D3* Northern Province, N Sri Lanka
Vawkavysk *85 B6* *Pol.* Wołkowysk, *Rus.* Volkoysk. Hrodzyenskaya Voblasts', W Belarus
Växjö *63 C7* *var.* Vexiö. Kronoberg, S Sweden
Vaygach, Ostrov *88 E2* *island* NW Russia
Veendam *64 E2* Groningen, NE Netherlands
Veenendaal *64 D4* Utrecht, C Netherlands
Vega *62 B4* *island* C Norway
Veglia *see* Krk
Veisiejai *85 B5* Alytus, S Lithuania
Vejer de la Frontera *70 C5* Andalucía, S Spain
Veldhoven *65 D5* Noord-Brabant, S Netherlands
Velebit *78 A3* *mountain range* C Croatia
Velenje *73 E7* *Ger.* Wöllan. N Slovenia
Veles *79 E6* *Turk.* Köprülü. C FYR Macedonia
Velho *see* Porto Velho
Velika Kikinda *see* Kikinda
Velika Morava *78 D4* *var.* Glavn'a Morava, Morava, *Ger.* Grosse Morava. *river* C Serbia
Velikaya *91 G2* *river* NE Russia
Veliki Bečkerek *see* Zrenjanin
Velikiye Luki *88 A4* Pskovskaya Oblast', W Russia
Velikiy Novgorod *88 B4* *prev.* Novgorod. Novgorodskaya Oblast', W Russian Federation
Veliko Tŭrnovo *82 D2* *prev.* Tirnovo, Trnovo, Tŭrnovo. Veliko Tŭrnovo, N Bulgaria
Velingrad *82 C3* Pazardzhik, C Bulgaria
Vel'ký Krtíš *77 D6* Banskobystrický Kraj, C Slovakia
Vellore *110 C2* Tamil Nādu, SE India
Velobriga *see* Viana do Castelo
Velsen *see* Velsen-Noord
Velsen-Noord *64 C3* *var.* Velsen. Noord-Holland, W Netherlands
Vel'sk *88 C4* *var.* Velsk. Arkhangel'skaya Oblast', NW Russia
Velvendós *see* Velvéntos
Velvéntos *82 B4* *var.* Velvendós. C Greece
Velykyy Tokmak *see* Tokmak
Vendôme *68 C4* Loir-et-Cher, C France
Venedig *see* Venezia
Vener, Lake *see* Vänern
Venetia *see* Venezia
Venezia *74 C2* *Eng.* Venice, *Fr.* Venise, *Ger.* Venedig; *anc.* Venetia. Veneto, NE Italy
Venezia, Golfo di *see* Venice, Gulf of
Venezuela *36 D2* *off.* Republic of Venezuela; *prev.* Estados Unidos de Venezuela, United States of Venezuela. *country* N South America
Venezuela, Estados Unidos de *see* Venezuela
Venezuela, Gulf of *36 C1* *Eng.* Gulf of Maracaibo, Gulf of Venezuela. *gulf* NW Venezuela
Venezuela, Gulf of *see* Venezuela, Golfo de
Venezuelan Basin *34 B1* *undersea basin* E Caribbean Sea
Venezuela, Republic of *see* Venezuela
Venezuela, United States of *see* Venezuela
Venice *20 C4* Louisiana, S USA
Venice *see* Venezia
Venice, Gulf of *74 C2* *It.* Golfo di Venezia, *Slvn.* Beneški Zaliv. *gulf* N Adriatic Sea
Venise *see* Venezia
Venlo *65 D5* *prev.* Venloo. Limburg, SE Netherlands
Venloo *see* Venlo
Venta *84 B3* *Ger.* Windau. *river* Latvia/Lithuania
Venta Belgarum *see* Winchester
Ventia *see* Valence
Ventimiglia *74 A3* Liguria, NW Italy
Ventspils *84 B2* *Ger.* Windau. Ventspils, NW Latvia
Vera *42 D3* Santa Fe, C Argentina
Veracruz *29 F4* *var.* Veracruz Llave. Veracruz-Llave, E Mexico
Veracruz Llave *see* Veracruz
Vercellae *see* Vercelli
Vercelli *74 A2* *anc.* Vercellae. Piemonte, NW Italy
Verdal *see* Verdalsøra
Verdalsøra *62 B4* *var.* Verdal. Nord-Trøndelag, C Norway
Verde, Cabo *see* Cape Verde
Verde, Costa *70 D1* *coastal region* N Spain
Verden *72 B3* Niedersachsen, NW Germany
Veria *see* Véroia
Verkhnedvinsk *see* Vyerkhnyadzvinsk
Verkhneudinsk *see* Ulan-Ude
Verkhoyanskiy Khrebet *93 F3* *mountain range* NE Russia
Vermillion *23 F3* South Dakota, N USA
Vermont *19 F2* *off.* State of Vermont, *also known as* Green Mountain State. *state* NE USA
Vernadsky *132 A2* *Ukrainian research station* Antarctica
Vernal *22 B4* Utah, W USA
Vernon *27 F2* Texas, SW USA
Verőcze *see* Virovitica
Véroia *82 B4* *var.* Veria, Vérroia, *Turk.* Karaferiye. Kentrikí Makedonía, N Greece
Verona *74 C2* Veneto, NE Italy
Vérroia *see* Véroia
Versailles *68 D1* Yvelines, N France
Versecz *see* Vršac
Verulamium *see* St Albans
Verviers *65 D6* Liège, E Belgium
Vesdre *65 D6* *river* E Belgium
Veselinovo *82 D2* Shumen, NE Bulgaria
Vesontio *see* Besançon
Vesoul *68 D4* *anc.* Vesulium, Vesulum. Haute-Saône, E France
Vesterålen *62 B2* *island group* N Norway
Vestfjorden *62 C3* *fjord* C Norway
Vestmannaeyjar *61 E5* Sudhurland, S Iceland
Vesulium/Vesulum *see* Vesoul
Vesuna *see* Périgueux
Vesuvio *75 D5* *Eng.* Vesuvius. *volcano* S Italy
Vesuvius *see* Vesuvio
Veszprém *77 C7* *Ger.* Veszprim. Veszprém, W Hungary
Veszprim *see* Veszprém
Vetrino *see* Suvorovo
Vetrino *see* Vyetryna
Vetter, Lake *see* Vättern
Veurne *65 A5* *var.* Furnes. West-Vlaanderen, W Belgium
Vexiö *see* Växjö
Viacha *39 F4* La Paz, W Bolivia
Viana de Castelo *see* Viana do Castelo
Viana do Castelo *70 B2* *var.* Viana de Castelo; *anc.* Velobriga. Viana do Castelo, NW Portugal
Vianen *64 C4* Utrecht, C Netherlands
Viangchan *114 C4* *Eng./Fr.* Vientiane. *country capital* (Laos) C Laos
Viangphoukha *114 C3* *var.* Vieng Pou Kha. Louang Namtha, N Laos
Viareggio *74 B3* Toscana, C Italy
Viborg *63 A7* Viborg, NW Denmark
Vic *71 G2* *var.* Vich; *anc.* Ausa, Vicus Ausonensis. Cataluña, NE Spain
Vicentia *see* Vicenza
Vicenza *74 C2* *anc.* Vicentia. Veneto, NE Italy
Vich *see* Vic
Vichy *69 C5* Allier, C France
Vicksburg *20 B2* Mississippi, S USA
Victoria *57 H1* *country capital* (Seychelles) Mahé, SW Seychelles
Victoria *14 D5* *province capital* Vancouver Island, British Columbia, SW Canada
Victoria *80 A5* *var.* Rabat. Gozo, NW Malta
Victoria *27 G4* Texas, SW USA
Victoria *127 C7* *state* SE Australia
Victoria *see* Masvingo, Zimbabwe
Victoria Bank *see* Vitória Seamount
Victoria de Durango *see* Durango
Victoria de las Tunas *see* Las Tunas
Victoria Falls *56 C3* Matabeleland North, W Zimbabwe
Victoria Falls *56 C2* *waterfall* Zambia/Zimbabwe
Victoria Falls *see* Iguaçu, Saltos do
Victoria Island *15 F3* *island* Northwest Territories/Nunavut, NW Canada
Victoria, Lake *51 B6* *var.* Victoria Nyanza. *lake* E Africa
Victoria Land *132 C4* *physical region* Antarctica
Victoria Nyanza *see* Victoria, Lake
Victoria River *124 D3* *river* Northern Territory, N Australia
Victorville *25 C7* California, W USA
Vicus Ausonensis *see* Vic
Vicus Elbii *see* Viterbo
Vidalia *21 E2* Georgia, SE USA
Videm-Krško *see* Krško
Vídeň *see* Wien
Vidin *82 B1* *anc.* Bononia. Vidin, NW Bulgaria
Vidzy *85 C5* Vitsyebskaya Voblasts', NW Belarus
Viedma *43 C5* Río Negro, E Argentina
Vieja, Sierra *26 D3* *mountain range* Texas, SW USA
Vieng Pou Kha *see* Viangphoukha
Vienna *see* Wien, Austria
Vienna *see* Vienne, France
Vienne *69 D5* *anc.* Vienna. Isère, E France
Vienne *68 B4* *river* W France
Vientiane *see* Viangchan
Vientos, Paso de los *see* Windward Passage
Vierzon *68 C4* Cher, C France
Viesīte *84 C4* *Ger.* Eckengraf. Jēkabpils, S Latvia
Vietnam *114 D4* *off.* Socialist Republic of Vietnam, *Vtn.* Công Hoa Xa Hôi Chu Nghia Viêt Nam. *country* SE Asia
Vietnam, Socialist Republic of *see* Vietnam
Vietri *see* Viêt Tri
Viêt Tri *114 D3* *var.* Vietri. Vinh Phu, N Vietnam
Vieux Fort *33 F2* S Saint Lucia
Vigo *70 B2* Galicia, NW Spain
Viipuri *see* Vyborg
Vijayawāda *110 D1* *prev.* Bezwada. Andhra Pradesh, SE India
Vila *see* Port-Vila
Vila Artur de Paiva *see* Cubango
Vila da Ponte *see* Cubango
Vila de João Belo *see* Xai-Xai
Vila de Mocímboa da Praia *see* Mocímboa da Praia
Vila do Conde *70 B2* Porto, NW Portugal
Vila do Zumbo *56 D2* *prev.* Vila do Zumbu, Zumbo. Tete, NW Mozambique
Vila do Zumbu *see* Vila do Zumbo
Vilafranca del Penedès *71 G2* *var.* Villafranca del Panadés. Cataluña, NE Spain
Vila General Machado *see* Camacupa
Vila Henrique de Carvalho *see* Saurimo
Viļaka *84 D4* *Ger.* Marienhausen. Balvi, NE Latvia
Vilalba *70 C1* Galicia, NW Spain
Vila Marechal Carmona *see* Uíge
Vila Nova de Gaia *70 B2* Porto, NW Portugal
Vila Nova de Portimão *see* Portimão
Vila Pereira de Eça *see* N'Giva
Vila Real *70 C2* *var.* Vila Rial. Vila Real, N Portugal
Vila Rial *see* Vila Real
Vila Robert Williams *see* Caála
Vila Salazar *see* N'Dalatando
Vila Serpa Pinto *see* Menongue
Vileyka *see* Vilyeyka
Vilhelmina *62 C4* Västerbotten, N Sweden
Vilhena *40 D3* Rondônia, W Brazil
Vília *83 C5* Attikí, C Greece
Viliya *85 C5* *Lith.* Neris. *river* W Belarus
Viliya *see* Neris
Viljandi *84 D2* *Ger.* Fellin. Viljandimaa, S Estonia
Vilkaviškis *84 B4* *Pol.* Wyłkowyszki. Marijampolė, SW Lithuania
Villa Acuña *28 D2* *var.* Ciudad Acuña. Coahuila, NE Mexico
Villa Bella *39 F2* Beni, N Bolivia
Villacarrillo *71 E4* Andalucía, S Spain
Villa Cecilia *see* Ciudad Madero
Villach *73 D7* *Slvn.* Beljak. Kärnten, S Austria
Villacidro *75 A5* Sardegna, Italy, C Mediterranean Sea
Villa Concepción *see* Concepción
Villa del Pilar *see* Pilar
Villafranca de los Barros *70 C4* Extremadura, W Spain
Villafranca del Panadés *see* Vilafranca del Penedès
Villahermosa *29 G4* *prev.* San Juan Bautista. Tabasco, SE Mexico
Villajoyosa *71 F4* *Cat.* La Vila Joíosa. País Valenciano, E Spain
Villa María *42 C4* Córdoba, C Argentina
Villa Martín *39 F5* Potosí, SW Bolivia
Villa Mercedes *42 C4* San Juan, Argentina
Villanueva *28 D3* Zacatecas, C Mexico
Villanueva de la Serena *70 C3* Extremadura, W Spain
Villanueva de los Infantes *71 E4* Castilla-La Mancha, C Spain
Villarrica *42 D2* Guairá, SE Paraguay
Villavicencio *36 B3* Meta, C Colombia
Villaviciosa *70 D1* Asturias, N Spain
Villazón *39 G5* Potosí, S Bolivia
Villena *71 F4* País Valenciano, E Spain
Villeurbanne *69 D5* Rhône, E France
Villingen-Schwenningen *73 B6* Baden-Württemberg, S Germany
Villmanstrand *see* Lappeenranta
Vilna *see* Vilnius
Vilnius *85 C5* *Pol.* Wilno, *Ger.* Wilna; *prev. Rus.* Vilna. *country capital* (Lithuania) Vilnius, SE Lithuania
Vilvoorde *65 C6* *Fr.* Vilvorde. Vlaams Brabant, C Belgium
Vilvorde *see* Vilvoorde
Vilyeyka *85 C5* *Pol.* Wilejka, *Rus.* Vileyka. Minskaya Voblasts', NW Belarus
Vilyuy *93 F3* *river* NE Russia
Viña del Mar *42 B4* Valparaíso, C Chile
Vinaròs *71 F3* País Valenciano, E Spain
Vincennes *18 B4* Indiana, N USA
Vindhya Mountains *see* Vindhya Range
Vindhya Range *112 D4* *var.* Vindhya Mountains. *mountain range* N India
Vindobona *see* Wien
Vineland *19 F4* New Jersey, NE USA
Vinh *114 D4* Nghê An, N Vietnam
Vinh Loi *see* Bac Liêu
Vinishte *82 C2* Montana, NW Bulgaria
Vinita *27 G1* Oklahoma, C USA
Vinkovci *78 C3* *Ger.* Winkowitz, *Hung.* Vinkovcze. Vukovar-Srijem, E Croatia
Vinkovcze *see* Vinkovci
Vinnitsa *see* Vinnytsya
Vinnytsya *86 D2* *Rus.* Vinnitsa. Vinnyts'ka Oblast', C Ukraine
Vinogradov *see* Vynohradiv
Vinson Massif *132 A3* *mountain* Antarctica
Viranşehir *95 E4* Şanlıurfa, SE Turkey
Vîrful Moldoveanu *see* Vârful Moldoveanu
Virginia *23 G1* Minnesota, N USA
Virginia *19 E5* *off.* Commonwealth of Virginia, *also known as* Mother of Presidents, Mother of States, Old Dominion. *state* NE USA
Virginia Beach *19 F5* Virginia, NE USA
Virgin Islands *see* British Virgin Islands
Virgin Islands (US) *33 F3* *var.* Virgin Islands of the United States; *prev.* Danish West Indies. *US unincorporated territory* E West Indies
Virgin Islands of the United States *see* Virgin Islands (US)
Virôchey *115 E5* Ratanakiri, NE Cambodia
Virovitica *78 C2* *Ger.* Virovititz, *Hung.* Verőcze; *prev. Ger.* Werowitz. Virovitica-Podravina, NE Croatia
Virovititz *see* Virovitica
Virton *65 D8* Luxembourg, SE Belgium
Virtsu *84 D2* *Ger.* Werder. Läänemaa, W Estonia
Vis *78 B4* *It.* Lissa; *anc.* Issa. *island* S Croatia
Vis *see* Fish
Visaginas *84 C4* *prev.* Sniečkus. Utena, E Lithuania
Visākhapatnam *113 E5* *var.* Vishakhapatnam. Andhra Pradesh, SE India
Visalia *25 C6* California, W USA
Visby *63 C7* *Ger.* Wisby. Gotland, SE Sweden
Viscount Melville Sound *15 F2* *prev.* Melville Sound. *sound* Northwest Territories, N Canada
Visé *65 D6* Liège, E Belgium
Viseu *70 C2* *prev.* Vizeu. Viseu, N Portugal
Vishakhapatnam *see* Visākhapatnam
Vislinskiy Zaliv *see* Vistula Lagoon
Visoko *78 C4* Federacija Bosna I Hercegovina, C Bosnia and Herzegovina
Visttasjohka *62 D3* *river* N Sweden
Vistula *76 C2* *Eng.* Vistula, *Ger.* Weichsel. *river* C Poland
Vistula *see* Wisła
Vistula Lagoon *76 C2* *Ger.* Frisches Haff, *Pol.* Zalew Wiślany, *Rus.* Vislinskiy Zaliv. *lagoon* Poland/Russia
Vitebsk *see* Vitsyebsk
Viterbo *74 C4* *anc.* Vicus Elbii. Lazio, C Italy
Viti *see* Fiji
Viti Levu *123 E4* *island* W Fiji
Vitim *93 F4* *river* C Russia
Vitória *41 F4* *state capital* Espírito Santo, SE Brazil
Vitoria *see* Vitoria-Gasteiz
Vitoria Bank *see* Vitória Seamount
Vitória da Conquista *41 F3* Bahia, E Brazil
Vitoria-Gasteiz *71 E1* *var.* Vitoria, *Eng.* Vittoria. País Vasco, N Spain
Vitória Seamount *45 B5* *var.* Victoria Bank, Vitoria Bank. *seamount* C Atlantic Ocean
Vitré *68 B3* Ille-et-Vilaine, NW France
Vitsyebsk *85 E5* *Rus.* Vitebsk. Vitsyebskaya Voblasts', NE Belarus
Vittoria *75 C7* Sicilia, Italy, C Mediterranean Sea
Vittoria *see* Vitoria-Gasteiz
Vizcaya, Golfo de *see* Biscay, Bay of
Vizianagaram *113 E5* *var.* Vizianagram. Andhra Pradesh, E India
Vizianagram *see* Vizianagaram
Vjosës, Lumi i *79 C7* *var.* Vijosa, Vijosë, *Gk.* Aóos. *river* Albania/Greece
Vlaanderen *see* Flanders
Vlaardingen *64 B4* Zuid-Holland, SW Netherlands
Vladikavkaz *89 B8* *prev.* Dzaudzhikau, Ordzhonikidze. Respublika Severnaya Osetiya, SW Russia
Vladimir *89 B5* Vladimirskaya Oblast', W Russia
Vladimirovka *see* Yuzhno-Sakhalinsk
Vladimir-Volynskiy *see* Volodymyr-Volyns'kyy
Vladivostok *93 G5* Primorskiy Kray, SE Russia
Vlagtwedde *64 E2* Groningen, NE Netherlands
Vlasotince *79 E5* Serbia, SE Serbia
Vlieland *64 C1* *Fris.* Flylân. *island* Waddeneilanden, N Netherlands
Vlijmen *64 C4* Noord-Brabant, S Netherlands
Vlissingen *65 B5* *Eng.* Flushing, *Fr.* Flessingue. Zeeland, SW Netherlands
Vlodava *see* Włodawa
Vlonë/Vlora *see* Vlorë
Vlorë *79 C7* *prev.* Vlonë, *It.* Valona, Vlora. Vlorë, SW Albania
Vlotslavsk *see* Włocławek
Vöcklabruck *73 D6* Oberösterreich, NW Austria
Vogelkop *see* Doberai, Jazirah
Vohimena, Tanjona *57 F4* *Fr.* Cap Sainte Marie. *headland* S Madagascar
Voiron *69 D5* Isère, E France
Vojvodina *78 D3* *Ger.* Wojwodina. Vojvodina, N Serbia
Volga *89 B7* *river* NW Russia
Volga Uplands *see* Privolzhskaya Vozvyshennost'
Volgodonsk *89 B7* Rostovskaya Oblast', SW Russia
Volgograd *89 B7* *prev.* Stalingrad, Tsaritsyn. Volgogradskaya Oblast', SW Russia
Volkhov *88 B4* Leningradskaya Oblast', NW Russia
Volkoysk *see* Vawkavysk
Volmari *see* Valmiera
Volnovakha *87 G3* Donets'ka Oblast', SE Ukraine
Volodymyr-Volyns'kyy *86 C1* *Pol.* Włodzimierz, *Rus.* Vladimir-Volynskiy. Volyns'ka Oblast', NW Ukraine
Vologda *88 B4* Vologodskaya Oblast', W Russia
Vólos *83 B5* Thessalía, C Greece
Volozhin *see* Valozhyn
Vol'sk *89 C6* Saratovskaya Oblast', W Russia
Volta *53 E5* *river* SE Ghana
Volta Blanche *see* White Volta
Volta, Lake *53 E5* *reservoir* SE Ghana
Volta Noire *see* Black Volta
Volturno *75 D5* *river* S Italy
Volunteer Island *see* Starbuck Island
Volzhskiy *89 B6* Volgogradskaya Oblast', SW Russia
Võnnu *84 E3* *Ger.* Wendau. Tartumaa, SE Estonia
Voorst *64 D3* Gelderland, E Netherlands
Voranava *85 C5* *Pol.* Werenów, *Rus.* Voronovo. Hrodzyenskaya Voblasts', W Belarus
Vorderrhein *73 B7* *river* SE Switzerland
Vóreies Sporádes *83 C5* *var.* Vóreioi Sporádes, Vórioi Sporádhes, *Eng.* Northern Sporades. *island group* E Greece
Vóreioi Sporádes *see* Vóreies Sporádes
Vórioi Sporádhes *see* Vóreies Sporádes
Vorkuta *92 C2* Respublika Komi, NW Russia
Vormsi *84 C2* *var.* Vormsi Saar, *Ger.* Worms, *Swed.* Ormsö. *island* W Estonia
Vormsi Saar *see* Vormsi
Voronezh *89 B6* Voronezhskaya Oblast', W Russia
Voronovo *see* Voranava
Voroshilov *see* Ussuriysk
Voroshilovgrad *see* Luhans'k, Ukraine
Voroshilovsk *see* Stavropol', Russia
Võru *84 D3* *Ger.* Werro. Võrumaa, SE Estonia
Vosges *68 E4* *mountain range* NE France
Vostochno-Sibirskoye More *93 F1* *Eng.* East Siberian Sea. *sea* Arctic Ocean
Vostochnyy Sayan *93 E4* *Mong.* Dzüün Soyonï Nuruu, *Eng.* Eastern Sayans. *mountain range* Mongolia/Russia
Vostock Island *see* Vostok Island
Vostok *132 C3* *Russian research station* Antarctica
Vostok Island *123 G3* *var.* Vostock Island; *prev.* Stavers Island. *island* Line Islands, SE Kiribati
Voznesens'k *87 E3* *Rus.* Voznesensk. Mykolayivs'ka Oblast', S Ukraine
Vranje *79 E5* Serbia, SE Serbia
Vranov *see* Vranov nad Topl'ou
Vranov nad Topl'ou *77 D5* *var.* Vranov, *Hung.* Varannó. Prešovský Kraj, E Slovakia
Vratsa *82 C2* Vratsa, NW Bulgaria
Vrbas *78 C3* Vojvodina, NW Serbia
Vrbas *78 C3* *river* N Bosnia and Herzegovina
Vršac *78 E3* *Ger.* Werschetz, *Hung.* Versecz. Vojvodina, NE Serbia
Vsetín *77 C5* *Ger.* Wsetin. Zlínský Kraj, E Czech Republic (Czechia)
Vučitrn *see* Vushtrri
Vukovar *78 C3* *Hung.* Vukovár. Vukovar-Srijem, E Croatia
Vulcano, Isola *75 C7* *island* Isole Eolie, S Italy
Vung Tau *115 E6* *prev. Fr.* Cape Saint Jacques, Cap Saint-Jacques. Ba Ria-Vung Tau, S Vietnam
Vushtrri *79 D5* *Serb.* Vučitrn. N Kosovo
Vyatka *89 C5* *river* NW Russia
Vyatka *see* Kirov
Vyborg *88 B3* *Fin.* Viipuri. Leningradskaya Oblast', NW Russia
Vyerkhnyadzvinsk *85 D5* *Rus.* Verkhnedvinsk. Vitsyebskaya Voblasts', N Belarus
Vyetryna *85 D5* *Rus.* Vetrino. Vitsyebskaya Voblasts', N Belarus
Vynohradiv *86 B3* *Cz.* Sevluš, *Hung.* Nagyszöllös, *Rus.* Vinogradov; *prev.* Sevlyush. Zakarpats'ka Oblast', W Ukraine

W

Wa *53 E4* NW Ghana
Waag *see* Váh
Waagbistritz *see* Považská Bystrica
Waal *64 C4* *river* S Netherlands
Wabash *18 C4* Indiana, N USA
Wabash River *18 B5* *river* N USA
Waco *27 G3* Texas, SW USA
Wad Al-Hajarah *see* Guadalajara
Waddān *49 F3* NW Libya
Waddeneilanden *64 C1* *Eng.* West Frisian Islands. *island group* N Netherlands
Waddenzee *64 C1* *var.* Wadden Zee. *sea* SE North Sea
Wadden Zee *see* Waddenzee
Waddington, Mount *14 D5* *mountain* British Columbia, SW Canada
Wādī as Sīr *97 B6* *var.* Wadi es Sir. 'Ammān, NW Jordan
Wadi es Sir *see* Wādī as Sīr
Wadi Halfa *50 B3* *var.* Wādī Ḥalfā'. Northern, N Sudan
Wādī Mūsā *97 B7* *var.* Petra. Ma'ān, S Jordan
Wad Madanī *see* Wad Medani
Wad Medani *50 C4* *var.* Wad Madanī. Gezira, C Sudan
Waflia *117 F4* Pulau Buru, E Indonesia
Wagadugu *see* Ouagadougou
Wagga Wagga *127 C7* New South Wales, SE Australia
Wagin *125 B7* Western Australia
Wāh *112 C1* Punjab, NE Pakistan
Wahai *117 F4* Pulau Seram, E Indonesia
Wahaybah, Ramlat Al *see* Wahībah, Ramlat Āl
Wahiawā *25 A8* *var.* Wahiawa. O'ahu, Hawaii, USA, C Pacific Ocean
Wahībah, Ramlat Ahl *see* Wahībah, Ramlat Āl
Wahibah Sands *99 E5* *var.* Ramlat Ahl Wahībah, Ramlat Al Wahaybah, *Eng.* Wahībah Sands. *desert* N Oman
Wahībah Sands *see* Wahībah, Ramlat Āl
Wahpeton *23 F2* North Dakota, N USA
Wahran *see* Oran
Waiau *129 A7* *river* South Island, New Zealand
Waigeo, Pulau *117 G4* *island* Maluku, E Indonesia
Waikaremoana, Lake *128 E4* *lake* North Island, New Zealand
Wailuku *25 B8* Maui, Hawaii, USA, C Pacific Ocean
Waimate *129 B6* Canterbury, South Island, New Zealand
Waiouru *128 D4* Manawatu-Wanganui, North Island, New Zealand
Waipara *129 C6* Canterbury, South Island, New Zealand
Waipawa *128 E4* Hawke's Bay, North Island, New Zealand
Waipukurau *128 D4* Hawke's Bay, North Island, New Zealand
Wairau *129 C5* *river* South Island, New Zealand
Wairoa *128 E4* Hawke's Bay, North Island, New Zealand
Wairoa *128 D2* *river* North Island, New Zealand
Waitaki *129 B6* *river* South Island, New Zealand
Waitara *128 D4* Taranaki, North Island, New Zealand
Waitzen *see* Vác
Waiuku *128 D3* Auckland, North Island, New Zealand
Wakasa-wan *109 C6* *bay* C Japan
Wakatipu, Lake *129 A7* *lake* South Island, New Zealand
Wakayama *109 C6* Wakayama, Honshū, SW Japan
Wake Island *130 C2* *US unincorporated territory* NW Pacific Ocean
Wake Island *120 D1* *atoll* NW Pacific Ocean
Wakkanai *108 C1* Hokkaidō, NE Japan
Walachei/Walachia *see* Wallachia
Wałbrzych *76 B4* *Ger.* Waldenburg, Waldenburg in Schlesien. Dolnośląskie, SW Poland
Walcourt *65 C7* Namur, S Belgium
Wałcz *76 B3* *Ger.* Deutsch Krone. Zachodnio-pomorskie, NW Poland
Waldenburg/Waldenburg in Schlesien *see* Wałbrzych
Waldia *see* Weldiya
Wales *14 C2* Alaska, USA
Wales *67 C6* *Wel.* Cymru. *cultural region* Wales, United Kingdom
Walgett *127 D5* New South Wales, SE Australia
Walk *see* Valga, Estonia
Walk *see* Valka, Latvia
Walker Lake *25 C5* *lake* Nevada, W USA
Wallachia *86 B5* *var.* Walachia, *Ger.* Walachei, *Rom.* Valachia. *cultural region* S Romania
Walla Walla *24 C2* Washington, NW USA
Wallenthal *see* Haţeg
Wallis and Futuna *123 E4* *Fr.* Territoire de Wallis et Futuna. *French overseas territory* C Pacific Ocean
Wallis et Futuna, Territoire de *see* Wallis and Futuna
Walnut Ridge *20 B1* Arkansas, C USA
Waltenberg *see* Zalău
Walthamstow *67 B7* Waltham Forest, SE England, United Kingdom
Walvisbaai *see* Walvis Bay
Walvis Bay *56 A4* *Afr.* Walvisbaai. Erongo, NW Namibia
Walvish Ridge *see* Walvis Ridge
Walvis Ridge *47 B7* *var.* Walvish Ridge. *undersea ridge* E Atlantic Ocean
Wan *see* Anhui
Wanaka *129 B6* Otago, South Island, New Zealand
Wanaka, Lake *129 A6* *lake* South Island, New Zealand
Wanchuan *see* Zhangjiakou
Wandel Sea *61 E1* *sea* Arctic Ocean
Wandsworth *67 A8* Wandsworth, SE England, United Kingdom
Wanganui *128 D4* Manawatu-Wanganui, North Island, New Zealand
Wangaratta *127 C7* Victoria, SE Australia
Wankie *see* Hwange
Wanki, Río *see* Coco, Río

Wanlaweyn – Yerevan

Wanlaweyn 51 D6 var. Wanle Weyn, It. Uanle Uen. Shabeellaha Hoose, SW Somalia
Wanle Weyn see Wanlaweyn
Wanxian see Wanzhou
Wanzhou 106 B5 var. Wanxian. Chongqing, C China
Warangal 113 E5 Telangana, C India
Warasdin see Varaždin
Warburg 72 B4 Nordrhein-Westfalen, W Germany
Ware 15 E4 British Columbia, W Canada
Waremme 65 C6 Liège, E Belgium
Waren 72 C3 Mecklenburg-Vorpommern, NE Germany
Wargla see Ouargla
Warkworth 128 D2 Auckland, North Island, New Zealand
Warnemünde 72 C2 Mecklenburg-Vorpommern, NE Germany
Warner 27 G1 Oklahoma, C USA
Warnes 39 G4 Santa Cruz, C Bolivia
Warrego River 127 C5 seasonal river New South Wales/Queensland, E Australia
Warren 18 D3 Michigan, N USA
Warren 18 D3 Ohio, N USA
Warren 19 E3 Pennsylvania, NE USA
Warri 53 F5 Delta, S Nigeria
Warrnambool 127 B7 Victoria, SE Australia
Warsaw/Warschau see Warszawa
Warszawa 76 D3 Eng. Warsaw, Ger. Warschau, Rus. Varshava. country capital (Poland) Mazowieckie, C Poland
Warta 76 B3 Ger. Warthe. river W Poland
Wartberg see Senec
Warthe see Warta
Warwick 127 E5 Queensland, E Australia
Wasa 132 B2 Swedish research station Antarctica
Washington 22 A2 NE England, United Kingdom
Washington D.C. 19 E4 country capital (USA) District of Columbia, NE USA
Washington Island see Teraina
Washington, Mount 19 G2 mountain New Hampshire, NE USA
Wash, The 67 E6 inlet E England, United Kingdom
Wasiliszki see Vasilishki
Waspam 31 E2 var. Waspán. Región Autónoma Atlántico Norte, NE Nicaragua
Waspán see Waspam
Watampone 117 E4 var. Bone. Sulawesi, C Indonesia
Watenstedt-Salzgitter see Salzgitter
Waterbury 19 F3 Connecticut, NE USA
Waterford 67 B6 Ir. Port Láirge. Waterford, S Ireland
Waterloo 23 G3 Iowa, C USA
Watertown 19 F2 New York, NE USA
Watertown 23 F2 South Dakota, N USA
Waterville 19 G2 Maine, NE USA
Watford 67 A7 E England, United Kingdom
Watlings Island see San Salvador
Watsa 55 E5 Orientale, NE Dem. Rep. Congo
Watts Bar Lake 20 D1 reservoir Tennessee, S USA
Wau 51 B5 var. Wāw. Western Bahr el Ghazal, C South Sudan
Waukegan 18 B3 Illinois, N USA
Waukesha 18 B3 Wisconsin, N USA
Wausau 18 B2 Wisconsin, N USA
Waverly 23 G3 Iowa, C USA
Wavre 65 C6 Walloon Brabant, C Belgium
Wāw see Wau
Wawa 16 C4 Ontario, S Canada
Waycross 21 E3 Georgia, SE USA
Wearmouth see Sunderland
Webfoot State see Oregon
Webster City 23 F3 Iowa, C USA
Weddell Plain 132 A2 abyssal plain SW Atlantic Ocean
Weddell Sea 132 A2 sea SW Atlantic Ocean
Weener 72 A3 Niedersachsen, NW Germany
Weert 65 D5 Limburg, SE Netherlands
Weesp 64 C3 Noord-Holland, C Netherlands
Węgorzewo 76 D2 Ger. Angerburg. Warmińsko-Mazurskie, NE Poland
Weichsel see Wisła
Weimar 72 C4 Thüringen, C Germany
Weissenburg see Alba Iulia, Romania
Weissenburg in Bayern 73 C6 Bayern, SE Germany
Weissenstein see Paide
Weisskirchen see Bela Crkva
Weiswampach 65 D7 Diekirch, N Luxembourg
Wejherowo 76 C2 Pomorskie, NW Poland
Welchman Hall 33 G1 C Barbados
Weldiya 50 C4 var. Waldia, It. Valdia. Āmara, N Ethiopia
Welkom 56 D4 Free State, C South Africa
Welle see Uele
Wellesley Islands 126 B2 island group Queensland, N Australia
Wellington 129 D5 country capital (New Zealand) Wellington, North Island, New Zealand
Wellington 23 F5 Kansas, C USA
Wellington see Wellington, Isla
Wellington, Isla 43 A7 var. Wellington. island S Chile
Wells 24 D4 Nevada, W USA
Wellsford 128 D2 Auckland, North Island, New Zealand
Wells, Lake 125 C5 lake Western Australia
Wels 73 D6 anc. Ovilava. Oberösterreich, N Austria
Wembley 67 A8 Alberta, W Canada
Wemmel 65 B6 Vlaams Brabant, C Belgium
Wenatchee 24 B2 Washington, NW USA
Wenchi 53 E4 W Ghana
Wen-chou/Wenchow see Wenzhou
Wendau see Võnnu
Wenden see Cēsis
Wenzhou 106 D5 var. Wen-chou, Wenchow. Zhejiang, SE China
Werda 56 C4 Kgalagadi, S Botswana
Werder see Virtsu
Werenów see Voranava
Werkendam 64 C4 Noord-Brabant, S Netherlands
Werowitz see Virovitica
Werro see Võru
Werschetz see Vršac
Wesenberg see Rakvere
Weser 72 B3 river NW Germany
Wessel Islands 126 B1 island group Northern Territory, N Australia
West Antarctica 132 A3 var. Lesser Antarctica. physical region Antarctica
West Australian Basin see Wharton Basin
West Bank 97 A6 disputed region SW Asia
West Bend 18 B3 Wisconsin, N USA
West Bengal 113 F4 cultural region NE India
West Cape 129 A7 headland South Island, New Zealand
West Des Moines 23 F3 Iowa, C USA
Westerland 72 B2 Schleswig-Holstein, N Germany
Western Australia 124 B4 state W Australia
Western Bug see Bug
Western Carpathians 77 E7 mountain range W Romania Europe
Western Desert see Şaḩrā' al Gharbīyah
Western Dvina 63 E7 Bel. Dzvina, Ger. Düna, Latv. Daugava, Rus. Zapadnaya Dvina. river W Europe
Western Ghats 112 C5 mountain range SW India
Western Isles see Outer Hebrides
Western Punjab see Punjab
Western Sahara 48 B3 disputed territory administered by Morocco N Africa
Western Samoa see Samoa
Western Samoa, Independent State of see Samoa
Western Sayans see Zapadnyy Sayan
Western Scheldt see Westerschelde
Western Sierra Madre see Madre Occidental, Sierra
Westerschelde 65 B5 Eng. Western Scheldt; prev. Honte. inlet S North Sea
West Falkland 43 C7 var. Gran Malvina, Isla Gran Malvina. island W Falkland Islands
West Fargo 23 F2 North Dakota, N USA
West Frisian Islands see Waddeneilanden
West Irian see Papua
Westliche Morava see Zapadna Morava
West Mariana Basin 120 B1 var. Perece Vela Basin. undersea feature W Pacific Ocean
West Memphis 20 B1 Arkansas, C USA
West New Guinea see Papua
Weston-super-Mare 67 D7 SW England, United Kingdom
West Palm Beach 21 F4 Florida, SE USA
West Papua see Papua
Westport 129 C5 West Coast, South Island, New Zealand
West Punjab see Punjab
West River see Xi Jiang
West Siberian Plain see Zapadno-Sibirskaya Ravnina
West Virginia 18 D4 off. State of West Virginia, also known as Mountain State. state NE USA
Wetar, Pulau 117 F5 island Kepulauan Damar, E Indonesia
Wetzlar 73 B5 Hessen, W Germany
Wevok 14 C2 var. Wewuk. Alaska, USA
Wewuk see Wevok
Wexford 67 B6 Ir. Loch Garman. SE Ireland
Weyburn 15 F5 Saskatchewan, S Canada
Weymouth 67 D7 S England, United Kingdom
Wezep 64 D3 Gelderland, E Netherlands
Whakatane 128 E3 Bay of Plenty, North Island, New Zealand
Whale Cove 15 G3 var. Tikirjuaq. Nunavut, C Canada
Whangarei 128 D2 Northland, North Island, New Zealand
Wharton Basin 119 D5 var. West Australian Basin. undersea feature E Indian Ocean
Whataroa 129 B6 West Coast, South Island, New Zealand
Wheatland 22 D3 Wyoming, C USA
Wheeler Peak 26 D1 mountain New Mexico, SW USA
Wheeling 18 D4 West Virginia, NE USA
Whitby 67 D5 N England, United Kingdom
Whitefish 22 B1 Montana, NW USA
Whitehaven 67 C5 NW England, United Kingdom
Whitehorse 14 D4 territory capital Yukon, W Canada
White Nile 50 B4 Ar. Al Baḩr al Abyaḑ, An Nil al Abyaḑ, Bahr el Jebel. river C South Sudan
White River 22 D3 river South Dakota, N USA
White Sea see Beloye More
White Volta 53 E4 var. Nakambé, Fr. Volta Blanche. river Burkina/Ghana
Whitianga 128 D2 Waikato, North Island, New Zealand
Whitney, Mount 25 C6 mountain California, W USA
Whitsunday Group 126 D3 island group Queensland, E Australia
Whyalla 127 B6 South Australia
Wichita 23 F5 Kansas, C USA
Wichita Falls 27 F2 Texas, SW USA
Wichita River 27 F2 river Texas, SW USA
Wickenburg 26 B2 Arizona, SW USA
Wicklow 67 B6 Ir. Cill Mhantáin. county E Ireland
Wicklow Mountains 67 B6 Ir. Sléibhte Chill Mhantáin. mountain range E Ireland
Wieliczka 77 D5 Małopolskie, S Poland
Wieluń 76 C4 Sieradz, C Poland
Wien 73 E6 Eng. Vienna, Hung. Bécs, Slvk. Vídeň, Slvn. Dunaj; anc. Vindobona. country capital (Austria) Wien, NE Austria
Wiener Neustadt 73 E6 Niederösterreich, E Austria
Wierden 64 E3 Overijssel, E Netherlands
Wiesbaden 73 B5 Hessen, W Germany
Wieselburg and Ungarisch-Altenburg/Wieselburg-Ungarisch-Altenburg see Mosonmagyaróvár
Wiesenhof see Ostrołęka
Wight, Isle of 67 D7 island United Kingdom
Wigorna Ceaster see Worcester
Wijchen 64 D4 Gelderland, SE Netherlands
Wijk bij Duurstede 64 D4 Utrecht, C Netherlands
Wilcannia 127 C6 New South Wales, SE Australia
Wilejka see Vilyeyka
Wilhelm, Mount 122 B3 mountain C Papua New Guinea
Wilhelm-Pieck-Stadt see Guben
Wilhelmshaven 72 B3 Niedersachsen, NW Germany
Wilia/Wilja see Neris
Wilkes Barre 19 F3 Pennsylvania, NE USA
Wilkes Land 132 C4 physical region Antarctica
Wilkomierz see Ukmergė
Willard 26 D2 New Mexico, SW USA
Willcox 26 C3 Arizona, SW USA
Willebroek 65 B5 Antwerpen, C Belgium
Willemstad 33 E5 dependent territory capital (Curaçao) Lesser Antilles, S Caribbean Sea
Williston 22 D1 North Dakota, N USA
Wilmington 19 F4 Delaware, NE USA
Wilmington 21 F2 North Carolina, SE USA
Wilmington 18 C4 Ohio, N USA
Wilna/Wilno see Vilnius
Wilrijk 65 C5 Antwerpen, N Belgium
Winchester 67 D7 hist. Wintanceaster, Lat. Venta Belgarum. S England, United Kingdom
Winchester 19 E4 Virginia, NE USA
Windau see Ventspils, Latvia
Windau see Venta, Latvia/Lithuania
Windhoek 56 B3 Ger. Windhuk. country capital (Namibia) Khomas, C Namibia
Windhuk see Windhoek
Windorah 126 C4 Queensland, C Australia
Windsor 16 C5 Ontario, S Canada
Windsor 67 D7 S England, United Kingdom
Windsor 19 G3 Connecticut, NE USA
Windward Islands 33 H4 island group E West Indies
Windward Islands see Barlavento, Ilhas de, Cape Verde
Windward Passage 32 D3 Sp. Paso de los Vientos. channel Cuba/Haiti
Winisk 16 C2 river Ontario, C Canada
Winkowitz see Vinkovci
Winnebago, Lake 18 B2 lake Wisconsin, N USA
Winnemucca 25 C5 Nevada, W USA
Winnipeg 15 G5 province capital Manitoba, S Canada
Winnipeg, Lake 15 G5 lake Manitoba, C Canada
Winnipegosis, Lake 16 A3 lake Manitoba, C Canada
Winona 23 G3 Minnesota, N USA
Winschoten 64 E2 Groningen, NE Netherlands
Winsen 72 B3 Niedersachsen, N Germany
Winston Salem 21 E1 North Carolina, SE USA
Winsum 64 D1 Groningen, NE Netherlands
Wintanceaster see Winchester
Winterswijk 64 E4 Gelderland, E Netherlands
Winterthur 73 B7 Zürich, NE Switzerland
Winton 126 C4 Queensland, E Australia
Winton 129 A7 Southland, South Island, New Zealand
Wisby see Visby
Wisconsin 18 A2 off. State of Wisconsin, also known as Badger State. state N USA
Wisconsin Rapids 18 B2 Wisconsin, N USA
Wisconsin River 18 B3 river Wisconsin, N USA
Wisła see Vistula
Wislany, Zalew see Vistula Lagoon
Wismar 72 C2 Mecklenburg-Vorpommern, N Germany
Wittenberge 72 C3 Brandenburg, N Germany
Wittlich 73 A5 Rheinland-Pfalz, SW Germany
Wittstock 72 C3 Brandenburg, NE Germany
W. J. van Blommesteinmeer 37 G3 reservoir E Suriname
Władysławowo 76 C2 Pomorskie, N Poland
Włocławek 76 C3 Ger./Rus. Vlotslavsk. Kujawsko-pomorskie, C Poland
Włodawa 76 E4 Rus. Vlodava. Lubelskie, SE Poland
Włodzimierz see Volodymyr-Volyns'kyy
Wlotzkasbaken 56 B3 Erongo, W Namibia
Wodonga 127 C7 Victoria, SE Australia
Wodzisław Śląski 77 C5 Ger. Loslau. Śląskie, S Poland
Wojerecy see Hoyerswerda
Wōjjā see Wotje Atoll
Wojwodina see Vojvodina
Woking 67 D7 SE England, United Kingdom
Wolf, Isla 38 A4 island Galápagos Islands, W Ecuador South America
Wolfsberg 73 D7 Kärnten, SE Austria
Wolfsburg 72 C3 Niedersachsen, N Germany
Wolgast 72 D2 Mecklenburg-Vorpommern, NE Germany
Wołkowysk see Vawkavysk
Wöllan see Velenje
Wollaston Lake 15 F4 Saskatchewan, C Canada
Wollongong 127 D6 New South Wales, SE Australia
Wolmar see Valmiera
Wołożyn see Valozhyn
Wolvega 64 D2 Fris. Wolvegea. Friesland, N Netherlands
Wolvegea see Wolvega
Wolverhampton 67 D6 C England, United Kingdom
Wolverine State see Michigan
Wŏnsan 107 E3 SE North Korea
Woodburn 24 B3 Oregon, NW USA
Woodland 25 B5 California, W USA
Woodruff 18 B2 Wisconsin, N USA
Woods, Lake of the 16 A3 Fr. Lac des Bois. lake Canada/USA
Woodville 128 D4 Manawatu-Wanganui, North Island, New Zealand
Woodward 27 F1 Oklahoma, C USA
Worcester 56 C5 Western Cape, SW South Africa
Worcester 67 D6 hist. Wigorna Ceaster. W England, United Kingdom
Worcester 19 G3 Massachusetts, NE USA
Workington 67 C5 NW England, United Kingdom
Worland 22 C3 Wyoming, C USA
Wormatia see Worms
Worms 73 B5 anc. Augusta Vangionum, Borbetomagus, Wormatia. Rheinland-Pfalz, SW Germany
Worms see Vormsi
Worthington 23 F3 Minnesota, N USA
Wotje Atoll 122 D1 var. Wōjjā. atoll Ratak Chain, E Marshall Islands
Woudrichem 64 C4 Noord-Brabant, S Netherlands
Wrangel Island 93 F1 Eng. Wrangel Island. island NE Russia
Wrangel Island see Vrangelya, Ostrov
Wrangel Plain 133 B2 undersea feature Arctic Ocean
Wrocław 76 C4 Eng./Ger. Breslau. Dolnośląskie, SW Poland
Września 76 C3 Wielkopolskie, C Poland
Wsetin see Vsetín
Wuchang see Wuhan
Wuday'ah 99 C6 spring/well S Saudi Arabia
Wuhai 105 E3 var. Haibowan. Nei Mongol Zizhiqu, N China
Wuhan 106 C5 var. Han-kou, Han-k'ou, Hanyang, Wuchang, Wu-han; prev. Hankow. province capital Hubei, C China
Wu-han see Wuhan
Wuhsien see Suzhou
Wuhsi/Wu-his see Wuxi
Wuhu 106 D5 var. Wu-na-mu. Anhui, E China
Wujlān see Ujelang Atoll
Wukari 53 G4 Taraba, E Nigeria
Wuliang Shan 106 A6 mountain range SW China
Wu-lu-k'o-mu-shi/Wu-lu-mu-ch'i see Ürümqi
Wu-na-mu see Wuhu
Wuppertal 72 A4 prev. Barmen-Elberfeld. Nordrhein-Westfalen, W Germany
Würzburg 73 B5 Bayern, SW Germany
Wusih see Wuxi
Wuxi 106 D5 var. Wuhsi, Wu-hsi, Wusih. Jiangsu, E China
Wuyi Shan 103 E3 mountain range SE China
Wye 67 C6 Wel. Gwy. river England/Wales, United Kingdom
Wyłkowyszki see Vilkaviškis
Wyndham 124 D3 Western Australia
Wyoming 18 C3 Michigan, N USA
Wyoming 22 B3 off. State of Wyoming, also known as Equality State. state C USA
Wyszków 76 D3 Ger. Probstberg. Mazowieckie, NE Poland

X

Xaafuun, Raas 50 E4 var. Ras Hafun. cape NE Somalia
Xaçmaz 95 H2 Rus. Khachmas. N Azerbaijan
Xaignabouli 114 C4 prev. Muang Xaignabouri, Fr. Sayaboury. Xaignabouli, N Laos
Xai-Xai 57 E4 prev. João Belo, Vila de João Belo. Gaza, S Mozambique
Xalapa 29 F4 Veracruz-Llave, Mexico
Xam Nua 114 D3 var. Sam Neua. Houaphan, N Laos
Xankändi 95 G3 Rus. Khankendi; prev. Stepanakert. SW Azerbaijan
Xánthi 82 C3 Anatolikí Makedonía kai Thráki, NE Greece
Xàtiva 71 F3 Cas. Xàtiva; anc. Setabis, var. Jativa. País Valenciano, E Spain
Xauen see Chefchaouen
Xäzär Dänizi see Caspian Sea
Xeres see Jeréz de la Frontera
Xiaguan see Dali
Xiamen 106 D6 var. Hsia-men; prev. Amoy. Fujian, SE China
Xi'an 106 C4 var. Changan, Sian, Signan, Siking, Singan, Xian. province capital Shaanxi, C China
Xiang see Hunan
Xiangkhoang see Phônsaven
Xiangtan 106 C5 var. Hsiang-t'an, Siangtan. Hunan, S China
Xiao Hinggan Ling 106 D2 Eng. Lesser Khingan Range. mountain range NE China
Xichang 106 B5 Sichuan, C China
Xieng Khouang see Phônsaven
Xieng Ngeun see Muong Xiang Ngeun
Xigazê 104 C5 var. Jih-k'a-tse, Shigatse, Xigaze. Xizang Zizhiqu, W China
Xi Jiang 102 D3 var. Hsi Chiang, Eng. West River. river S China
Xilinhot 105 F2 var. Silinhot. Nei Mongol Zizhiqu, N China
Xilokastro see Xylókastro
Xin see Xinjiang Uygur Zizhiqu
Xingkai Hu see Khanka, Lake
Xingu, Rio 41 E2 river C Brazil
Xingxingxia 104 D3 Xinjiang Uygur Zizhiqu, NW China
Xining 105 E4 var. Hsining, Hsi-ning, Sining. province capital Qinghai, C China
Xinjiang see Xinjiang Uygur Zizhiqu
Xinjiang Uygur Zizhiqu 104 B3 var. Sinkiang, Sinkiang Uighur Autonomous Region, Xin, Xinjiang. autonomous region NW China
Xinpu see Lianyungang
Xinxiang 106 C4 Henan, C China
Xinyang 106 C5 var. Hsin-yang, Sinyang. Henan, C China
Xinzo de Limia 70 C2 Galicia, NW Spain
Xiqing Shan 102 D2 mountain range C China
Xiva 100 D2 Rus. Khiva, Khiwa. Xorazm Viloyati, W Uzbekistan
Xixón see Gijón
Xizang see Xizang Zizhiqu
Xizang Gaoyuan see Qingzang Gaoyuan
Xizang Zizhiqu 104 B4 var. Thibet, Tibetan Autonomous Region, Xizang, Eng. Tibet. autonomous region W China
Xolotlán see Managua, Lago de
Xucheng see Xuwen
Xuddur 51 D5 var. Hudur, It. Oddur. Bakool, SW Somalia
Xuwen 106 C7 var. Xucheng. Guangdong, S China
Xuzhou 106 D4 var. Hsu-chou, Suchow, Tongshan; prev. T'ung-shan. Jiangsu, E China
Xylókastro 83 B5 var. Xilokastro. Pelopónnisos, S Greece

Y

Ya'an 106 B5 var. Yaan. Sichuan, C China
Yabēlo 51 C5 Oromīya, C Ethiopia
Yablis 31 E2 Región Autónoma Atlántico Norte, NE Nicaragua
Yablonovyy Khrebet 93 F4 mountain range S Russia
Yabrai Shan 105 E3 mountain range NE China
Yafran 49 F2 NW Libya
Yaghan Basin 45 B7 undersea feature S E Pacific Ocean
Yagotin see Yahotyn
Yahotyn 87 E2 Rus. Yagotin. Kyyivs'ka Oblast', N Ukraine
Yahualica 28 D4 Jalisco, SW Mexico
Yakima 24 B2 Washington, NW USA
Yakima River 24 B2 river Washington, NW USA
Yakoruda 82 C3 Blagoevgrad, SW Bulgaria
Yaku-shima 109 B8 island Nansei-shotō, SW Japan
Yakutat 14 D4 Alaska, USA
Yakutsk 93 F3 Respublika Sakha (Yakutiya), NE Russia
Yala 115 C7 Yala, SW Thailand
Yalizava 85 D6 Rus. Yelizovo. Mahilyowskaya Voblasts', E Belarus
Yalong Jiang 106 A5 river C China
Yalova 94 B3 Yalova, NW Turkey
Yalpug, Ozero see Yalpuh, Ozero
Yalpuh, Ozero 86 D4 Rus. Ozero Yalpug. lake SW Ukraine
Yalta 87 F5 Respublika Krym, S Ukraine
Yalu 103 E2 Chin. Yalu Jiang, Jap. Oryokko, Kor. Amnok-kang. river China/North Korea
Yalu Jiang see Yalu
Yamaguchi 109 B7 var. Yamaguti. Yamaguchi, Honshū, SW Japan
Yamal, Poluostrov 92 D2 peninsula N Russia
Yamanīyah, Al Jumhūrīyah al see Yemen
Yambio 51 B5 var. Yambiyo. Western Equatoria, S South Sudan
Yambiyo see Yambio
Yambol 82 D2 Turk. Yanboli. Yambol, E Bulgaria
Yamdena, Pulau 117 G5 prev. Jamdena. island Kepulauan Tanimbar, E Indonesia
Yamoussoukro 52 D5 country capital (Ivory Coast) C Ivory Coast
Yamuna 112 D3 prev. Jumna. river N India
Yana 93 F2 river NE Russia
Yanboli see Yambol
Yanbu 'al Baḩr 99 A5 Al Madīnah, W Saudi Arabia
Yangambi 55 D5 Orientale, N Dem. Rep. Congo
Yangchow see Yangzhou
Yangiyoʻl 101 E2 Rus. Yangiyul'. Toshkent Viloyati, E Uzbekistan
Yangiyul' see Yangiyoʻl
Yangku see Taiyuan
Yangon 114 B4 Eng. Rangoon. Yangon, S Myanmar (Burma)
Yangtze 106 B5 var. Yangtze Kiang, Eng. Yangtze. river C China
Yangtze see Chang Jiang
Yangtze Kiang see Chang Jiang
Yangzhou 106 D5 var. Yangchow. Jiangsu, E China
Yankton 23 E3 South Dakota, N USA
Yany Kapu 87 F4 Rus. Krasnoperekops'k. Respublika Krym, S Ukraine
Yannina see Ioánnina
Yanskiy Zaliv 91 F2 bay N Russia
Yantai 106 D4 var. Yan-t'ai; prev. Chefoo, Chih-fu. Shandong, E China
Yaoundé 55 B5 var. Yaunde. country capital (Cameroon) Centre, S Cameroon
Yap 122 A1 island Caroline Islands, W Micronesia
Yapanskoye More East Sea/Japan, Sea of
Yapen, Pulau 117 G4 prev. Japen. island E Indonesia
Yap Trench 120 B2 var. Yap Trough. undersea feature SE Philippine Sea
Yap Trough see Yap Trench
Yapurá see Caquetá, Río, Brazil/Colombia
Yapurá see Japurá, Rio, Brazil/Colombia
Yaqui, Río 28 C2 river NW Mexico
Yaransk 89 C5 Kirovskaya Oblast', NW Russia
Yarega 88 D4 Respublika Komi, NW Russia
Yaren 122 D2 de facto country capital (Nauru) Nauru, SW Pacific
Yarkant see Shache
Yarlung Zangbo Jiang see Brahmaputra
Yarmouth 17 F5 Nova Scotia, SE Canada
Yarmouth see Great Yarmouth
Yaroslav see Jarosław
Yaroslavl' 88 B4 Yaroslavskaya Oblast', W Russia
Yarumal 36 B2 Antioquia, NW Colombia
Yasyel'da 85 B7 river Brestskaya Voblasts', SW Belarus Europe
Yatsushiro 109 A7 var. Yatusiro. Kumamoto, Kyūshū, SW Japan
Yatusiro see Yatsushiro
Yaunde see Yaoundé
Yavarí see Javari, Rio
Rio Yavari 40 C2 var. Yavarí. river Brazil/Peru
Yaviza 31 H5 Darién, SE Panama
Yavoriv 86 B2 Pol. Jaworów, Rus. Yavorov. L'vivs'ka Oblast', NW Ukraine
Yavorov see Yavoriv
Yazd 98 D3 var. Yezd. Yazd, C Iran
Yazoo City 20 B2 Mississippi, S USA
Yding Skovhøj 63 A7 hill C Denmark
Ýdra 83 C6 var. Ídhra. island Ýdra, S Greece
Ye 115 B5 Mon State, S Myanmar (Burma)
Yecheng 104 A3 var. Kargilik. Xinjiang Uygur Zizhiqu, NW China
Yefremov 89 B5 Tul'skaya Oblast', W Russia
Yekaterinburg 92 C3 prev. Sverdlovsk. Sverdlovskaya Oblast', C Russia
Yekaterinodar see Krasnodar
Yekaterinoslav see Dnipro
Yelets 89 B5 Lipetskaya Oblast', W Russia
Yelisavetpol see Gäncä
Yelizavetgrad see Kropyvnyts'kyy
Yelizovo see Yalizava
Yell 66 D1 island NE Scotland, United Kingdom
Yellowhammer State see Alabama
Yellowknife 15 E4 territory capital Northwest Territories, W Canada
Yellow River see Huang He
Yellow Sea 106 D4 Chin. Huang Hai, Kor. Hwang-Hae. sea E Asia
Yellowstone River 22 C2 river Montana/Wyoming, NW USA
Yel'sk 85 C7 Homyel'skaya Voblasts', SE Belarus
Yelwa 53 F4 Kebbi, W Nigeria
Yemen 99 C7 off. Republic of Yemen, Ar. Al Jumhūrīyah al Yamanīyah, Al Yaman. country SW Asia
Yemen, Republic of see Yemen
Yemva 88 D4 prev. Zheleznodorozhnyy. Respublika Komi, NW Russia
Yenakiyeve 87 G3 Rus. Yenakiyevo; prev. Ordzhonikidze, Rykovo. Donets'ka Oblast', E Ukraine
Yenakiyevo see Yenakiyeve
Yenangyaung 114 A3 Magway, W Myanmar (Burma)
Yendi 53 E4 NE Ghana
Yengisar 104 A3 Xinjiang Uygur Zizhiqu, NW China
Yenierenköy 80 D4 var. Yialousa, Gk. Agialoúsa. NE Cyprus
Yenipazar see Novi Pazar
Yenisey 92 D3 river Mongolia/Russia
Yenping see Nanping
Yeovil 67 D7 SW England, United Kingdom
Yeppoon 126 D4 Queensland, E Australia
Yerevan 95 F3 Eng. Erivan. country capital (Armenia) C Armenia